Environmental Applications of Carbon Nanomaterials-Based Devices

Environmental Applications of Carbon Nanomaterials-Based Devices

Edited by

Shadpour Mallakpour
Chaudhery Mustansar Hussain

WILEY-VCH

Editors

Shadpour Mallakpour
Department of Chemistry
Isfahan University of Technology
8415683111 Khomeini Shahr
Isfahan
Iran

Chaudhery Mustansar Hussain
Department of Chemistry and
Environmental Science
New Jersey Institute of Technology
Newark, New Jersey
United States

Cover Design: Wiley
Cover Image: © Rost-9D/Getty Images

Library of Congress Card No.:
applied for

British Library Cataloguing-in-Publication Data
A catalogue record for this book is available from the British Library.

Bibliographic information published by the Deutsche Nationalbibliothek
The Deutsche Nationalbibliothek lists this publication in the Deutsche Nationalbibliografie; detailed bibliographic data are available on the Internet at <http://dnb.d-nb.de>.

Print ISBN: 978-3-527-34865-7
ePDF ISBN: 978-3-527-83096-1
ePub ISBN: 978-3-527-83098-5
oBook ISBN: 978-3-527-83097-8

Typesetting Straive, Chennai, India
Printing and Binding CPI Group (UK) Ltd, Croydon, CR0 4YY

Printed on acid-free paper

C110340_060921

Contents

1

Graphene-Based Nanomembranes for Sustainable Water Purification Applications

Uluvangada T. Uthappa[1], *Dusan Losic*[2,3], *and Mahaveer D. Kurkuri*[1]

[1] *JAIN University, Centre for Nano and Material Sciences, JAIN Global Campus, Bengaluru, Karnataka, 562112, India*
[2] *The University of Adelaide, School of Chemical Engineering, Adelaide, SA, 5005, Australia*
[3] *The University of Adelaide, ARC Research Hub for Graphene Enabled Industry Transformation, Adelaide, SA, 5005, Australia*

1.1 Introduction

The global crisis on water scarcity ranks top among the world's major problems. Provided with recent statistics, two-thirds of the world population will face major difficulties by 2025 due to the lack of water facilities. Due to the rapid population expansion, industrial development, and water pollution, it incessantly depleted freshwater resources. The extreme consumption of freshwater resources, particularly for developing countries, due to the growth of agriculture and industrial activities completely dependent on water resources and continuous utilization leads to severe ecological disasters. Thus, the shortage of water resources is an alarming issue that needs urgent resolution. To address these problems, it is crucial to conserve water resources [1].

On the other hand, membrane technology is a separation process that permeates certain species to pass via membrane while controlling the others, which is completely dependent on the properties of the membrane material. Based on their properties, membranes can be categorized as dense and porous membranes according to the driving force and ionic and molecular movement transported through membranes. Usually, it is well recognized that diverse species permeate based on the specific diffusion rates via the membranes, which are generally influenced based on solubility, diffusivity, and the membranes' porosity, as represented in Figure 1.1.

It is believed that the smaller molecules might pass through the membrane and whereas others could be blocked based on the pore sizes of the membranes. The ceramic and polymers are the most widely used membranes for water purifications plants. Typically, ceramic membranes retain better thermal and mechanical properties. However, it is difficult to control the pore size accurately with low-cost approaches. In contrast, polymeric membranes are economically and extensively utilized in microfiltration (MF), ultrafiltration (UF), nanofiltration (NF), and reverse osmosis (RO) process due to the rapid permeation and high

Environmental Applications of Carbon Nanomaterials-Based Devices, First Edition.
Edited by Shadpour Mallakpour and Chaudhery Mustansar Hussain.

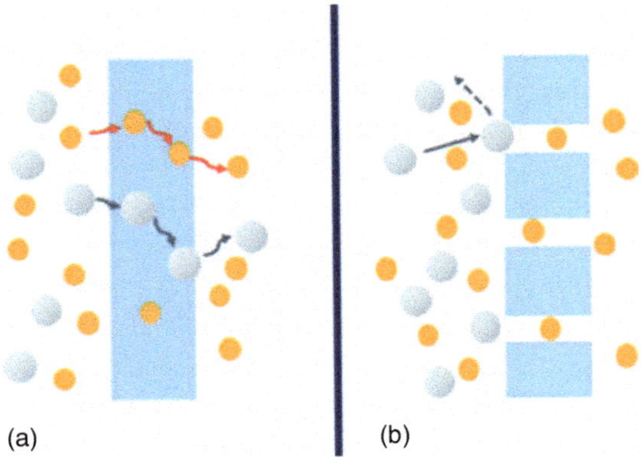

Figure 1.1 Transport mechanism for dense (a) and porous membrane (b). Source: Lyu et al. [2]. © 2018 Royal Society of Chemistry.

selectivity. However, polymers are hydrophobic that can easily adsorb target pollutants that lead to membrane fouling. Compared to ceramic and polymeric membranes, the new astonishing classic example used in membrane technology is recently discovered graphene oxide (GO). Also, reduced graphene oxide (rGO) is an important alternative class that belongs to the graphene derivative that has been explored for the development of a new generation of membrane separation technology [2]. Since from past few years' research expansion in graphene known as two-dimensional (2D) carbon allotrope and their oxidized derivatives (GO) has rapidly developed by the drastic increase in publications and patents. It is important to mention that graphene and GO have distinctive favorable material advantages compared to carbon nanotubes (CNTs) concerning separation and purification technologies. GO can be easily fabricated and engineered with favorable size, surface chemistry/modification, and structural properties. Moreover, its industrial process consumes a lesser amount of energy, making their production cost economic (for example: 500–1000 MJ/kg by solvent exfoliation of graphite or chemical reduction of GO, when compared to CNTs 100 000 MJ/kg) [3].

This chapter summarizes the very recent advances on graphene-based nanomembranes for water purifications. The properties of GO-based membranes and fabrication strategies of various graphene-based nanomembranes and their applications in desalination and dyes and some highlights on heavy metals treatment are elucidated and discussed. Finally, a conclusion and future perspectives have been provided. This current book chapter could help the researchers and experts in membrane technology working on GO-based nanomembranes for water treatment applications.

1.2 Graphene and GO-Based Membrane Characteristics and Properties

For the first time, graphene was witnessed in electron microscopes in 1962 supported on the metal surfaces. Further, it was again rediscovered in 2004 by Novoselov and Geim and awarded Nobel Prize in Physics 2010 due to their groundbreaking experiments on 2D material graphene. Since 2010, graphene has fascinated major attractions in several applications, including water purification. Graphene contains carbon atoms bonded to each other in hexagonal patterns. The monolayered and double-layered graphene is very thin, and so it can be regarded as 2D material. Graphene's flat based honeycomb arrangements are responsible for favorable characteristic features such as strongest, lightest, most conductive, transparent materials [4], high specific surface area (2600 m^2/g), flexibility [5, 6], and large scale production from cheap raw material (graphite) [7]. GO is an oxidized form of graphene, made up of carbon atoms bonded in hexagonal honeycomb lattice structures. Based on strong oxidations surrounding environment and based on the synthetic protocol such as Hummers or Staudenmaier method, enormous groups of oxygen epoxide, hydroxyl, and carboxyl acid functional groups present in GO are represented in Figure 1.2.

Such diverse functional groups are responsible for better hydrophilicity and permit better dispersion of GO flakes in aqueous media. These substantial features significantly enable GO deposition from solution using water as an environmentally benign solvent. From the past few years, GO showed unlimited attention as a novel 2D membrane in water treatment applications due to its exceptional mechanical [8] and antifouling properties [9], atomically thin thickness, and outstanding dispersion in water [8]. Various composite membranes are obtained by mixing GO with different polymers and other surface enriched materials. Such studies have been carried to enhance surface properties such as conductivity, tensile strength, and or elasticity of materials [4]. These surface properties impart GO-based functionalized nanomembranes for advanced water treatment technologies. The selective permeation path of graphene-based membranes that allow separation via the nanopore structure originated within the basal plane of hexagonal crystalline structures.

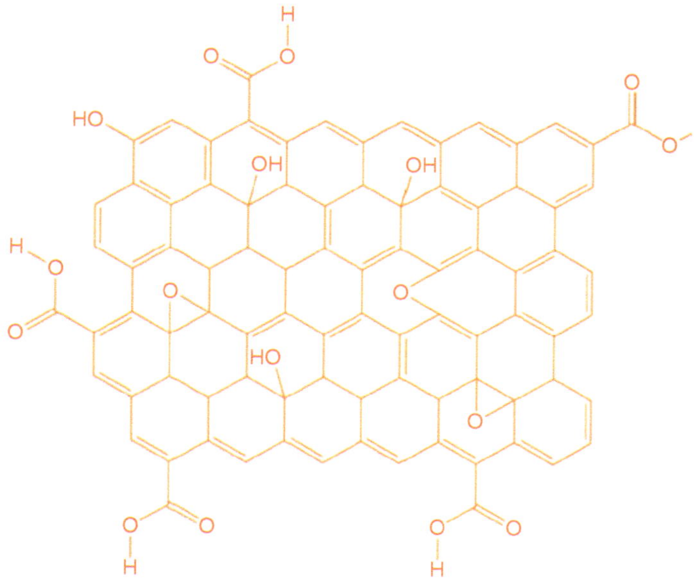

Figure 1.2 Typical chemical structure of GO showing graphene sheet derivatized by phenyl epoxide, hydroxyl groups on the basal plane, and carboxylic group on the edge.

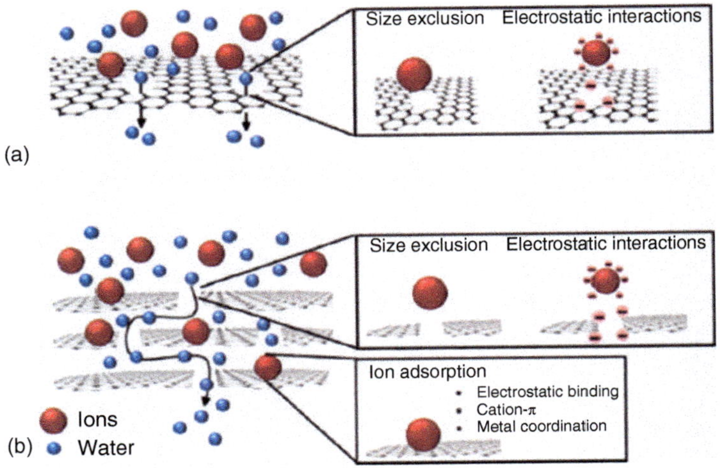

Figure 1.3 Selective permeation routes in nanoporous graphene and stacked sheets of GO. (a) Nanoporous graphene membrane and (b) stacked graphene oxide membrane. Source: Dervin et al. [9]. © 2016 Royal Society of Chemistry.

On the other hand, ions or molecules are selectively transported or passed via interlayer spacing of 2D graphene-based materials. The stacked nanosheets of GO form a multilayer laminate that shows abundant mechanical strength for application in pressure-assisted water filtration due to strong hydrogen bonds among distinct sheets. The oxygen comprising functional groups deposited haphazardly beside the edges of GO sheets, which maintain both significant interlayer spacing between and empty spaces among non-oxidized areas, constituting a network of nanocapillaries within the film. These nanochannels allow penetration of water molecules and successive transport laterally on hydrophobic non-oxidized regions of membranes that could support the rapid flow of water molecules. Thus liquid, vapors, and gases are opposed, as shown in Figure 1.3. In addition, water-soluble oxygen functional groups present beside the GO sheets, which adsorb water molecules and further diffused between the non-polar hydrocarbon present on the backbone of GO. Such permeation of water escalates the interlayer spacing in the middle of stacked GO sheets and helps in water flux over nanochannels at a higher flow rate. It has also been recognized that smaller ions enter GO membranes considerably superior to simple diffusion. Conversely, enlargement of nanochannels during the hydrated state only accepts ions of analogous sizes [9].

1.3 Fabrication of Graphene-Based Nanomembranes for Water Treatment Applications

1.3.1 Desalination

The membrane-based desalination method is classified based on membrane pore size and rejection mechanisms [10]. Desalination is a process of removing salts from seawater or brackish water, and it is known to be the core technology for improving freshwater scarcity [11]. To improve the desalination process and enhance water treatments, innovative technological classes of membrane systems are advanced. However, to overcome the costs and energy demands accompanied by membrane treatment, advanced, innovative, and economical membranes are highly desirable [9].

1.3.2 Treatment for Dyes

Many dyes are used to color the products from various industries such as food, paints, leather, plastics, and textiles. The significant amount of effluents from those industries creates serious ecological complications to the environment for most of the underground and surface water bodies. Moreover, dye waste turbidity results in blocking sunlight and reducing diffusion into the water that could obstruct the biological process. As far as concerned, organic dyes are highly stable, toxic, carcinogenic, and mutagenic. In some cases, these properties lead to organ damage and abnormalities in living creatures. As a result, it is crucial to treat or remove such dyes before discharging them into environments [12].

1.3.3 Graphene Nanomembranes for Salt and Dye Rejection

In this report, charged nanochannels produced in polyelectrolyte intercalated amine reduced graphene oxide membranes (PE-ArGO membrane) were reported. The overall synthetic process is shown in Figure 1.4. The PE-ArGO membrane with a rejection sheet of around 160 nm in thickness with laminate structural topographies with a smooth top surface of low roughness of around 17.2 nm. The PE-ArGO membrane was *in situ* modified

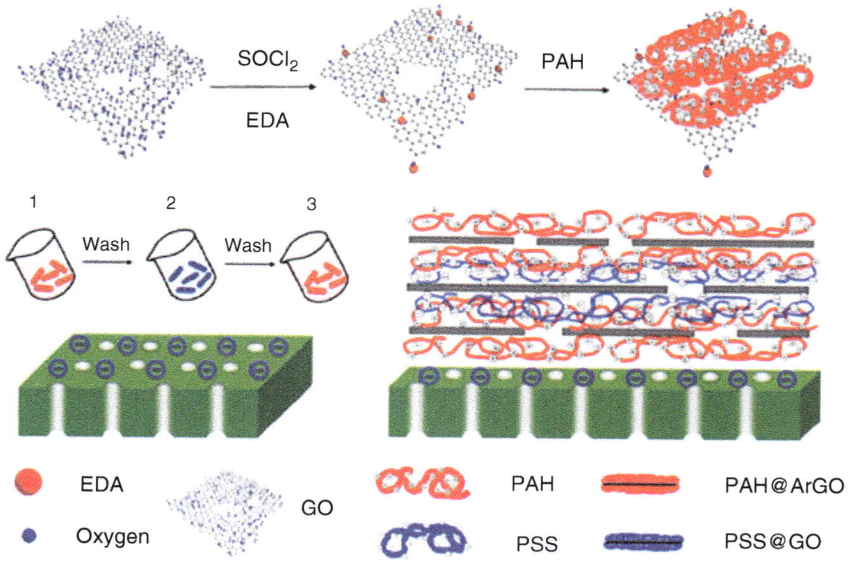

Figure 1.4 Schematic showing synthetic for PE-ArGO membrane. Source: Song et al. [13]. © 2017 American Chemical Society.

(m-PE-ArGO) using free chlorine scavenging to adjust the charge, though keeping modification to the layered structure in a nominal way. The PE-ArGO and m-PE-ArGO membrane surface charge were +4.37 and −4.28 mC/m^2. Using the pressure-assisted technique, pure water permeability for PE@ArGO and m-PE@ArGO membrane was 2.8 and 10.8 LM^2/h/bar. The salt rejection is extremely dependent on the charge density of membranes, valence of co-ions, and size of hydrated ions. The rejection of slats for positively charged PE-ArGO membrane follows in the order of: R ($MgCl_2$), 93% > R (NaCl), 88% = R ($MgSO_4$), 88% > R (Na_2SO_4), 65%, whereas R (Na_2SO_4), 90% > R (NaCl), 85% > R ($MgSO_4$), 68% > R ($MgCl_2$), 42% followed trend for negatively charged m-PE-ArGO membranes. Altogether, this is the first report on high density positive/negative charge gated ion transport behavior studies reported in GO-based membranes [13].

Stationary repulsion force and π–π contact among GO nanosheets have been tuned using interlayer functionalization and external treatment. Negatively charged congo red (CR) dye with π orbital adsorbed on the nonoxide domain of GO nanosheets, interlayers cross-linked by Ca^{2+} ions via chelating reaction. The GO–CR nanosheets π–π stacking interactions altered by thermal reduction of GO/Ca/CR membrane using the hot pressing method. The interlayer spacing of GO/Ca/CR membranes tuned from 3.20 to 5.7 Å under dry conditions is observed. In wet conditions, 7.7–11 Å. In the meantime, modified GO membranes' swelling behavior decreased compared to the bare GO bare membrane. Notably, the prepared membrane showed a permeation flux of 17 LM^2/h/bar, and the rejection for methylene blue (MB) was 99.5% [14].

Wang et al. demonstrated exfoliated hydrotalcite/graphene oxide (EHT/GO) nanosheet combined through flocculation deposition and homogeneously dispersed in polyethyleneimine (PEI) solution used as an aqueous phase for interfacial polymerization. An overview of the synthetic process is shown in Figure 1.5. The prepared hybrid membranes EHT/GO displayed improved positively charged property after integrating EHT/GO into polyamide (PA) membrane compared with pristine NF and GO incorporated NF membrane. EHT/GO exhibited a more promising transport channel of water molecules

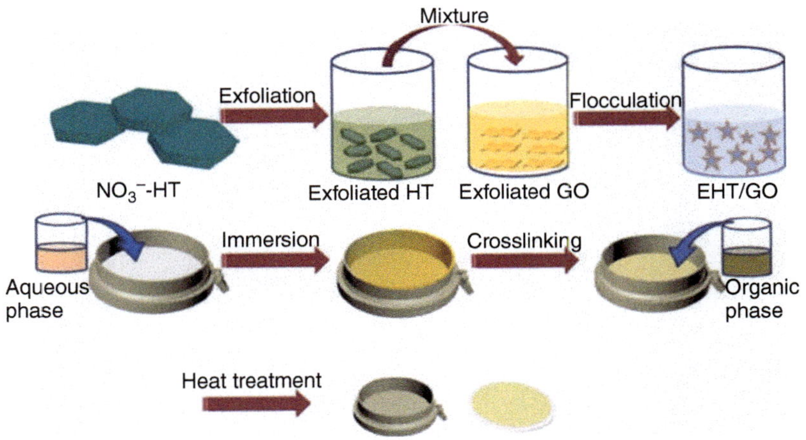

Figure 1.5 Overview of the synthetic process of EHT/GO hybrid nanosheets and nanofiltration membranes. Source: Wang et al. [15]. © 2019 Elsevier.

resulting in improved water flux of 26 LM^2/h/bar at 0.8 MPa. To this, the membrane rejection performances were 97% and 32% for 1000 ppm of Mg_2Cl_2 and Na_2SO_4 feed solutions, presenting prospective use in hard water softening applications [15].

The high rejection of organic dyes graphene oxide/attapulgite (GO/APT) composite nanofiltration membrane was developed using the vacuum-assisted filtration method. The composite membrane was fabricated by a meticulous assembly of APT nanorods incorporated into the interlayers of GO nanosheets. The APT nanorods are easily intercalated into the stacked GO lamellar layers to form stable C—O—Si bonds undergoing 3D network laminate assembly through the surface grafting functionalization. It showed a significant role in enhanced performance in water permeability. Compared with the bare GO membrane, the GO/APT membrane showed fourfold water flux performance while maintaining a high rejection rate of almost 100% for rhodamine b (RhB) dye molecules. The reason could be due to d-spacing, hydrophilicity, and 3D network lamellar structural properties. The separation mechanism was also evaluated for the GO/APT membrane with different positively/negatively charged and neutral dyes. Based on the size exclusion and electrostatic interactions, selective permeation could be done using GO/APT membranes [16]. An overview of the synthetic process is shown in Figure 1.6.

Thin-film nanocomposite (TFN) acts as an ideal platform to enhance water permeability of commercially available PA membranes in RO desalination. Thus, nitrogen-doped graphene oxide quantum dots (N-GOQD) fabricated PA/N-GOQD TFN membranes by interfacial polymerization for RO desalination applications. It is noteworthy to mention after the addition of 0.02 wt/v% N-GOQD into the PA membrane, there was a sudden enhancement of water permeability by almost three times, retaining identical salt rejection of pristine PA membrane. Improved water permeability leads to enhanced hydrophilicity of the

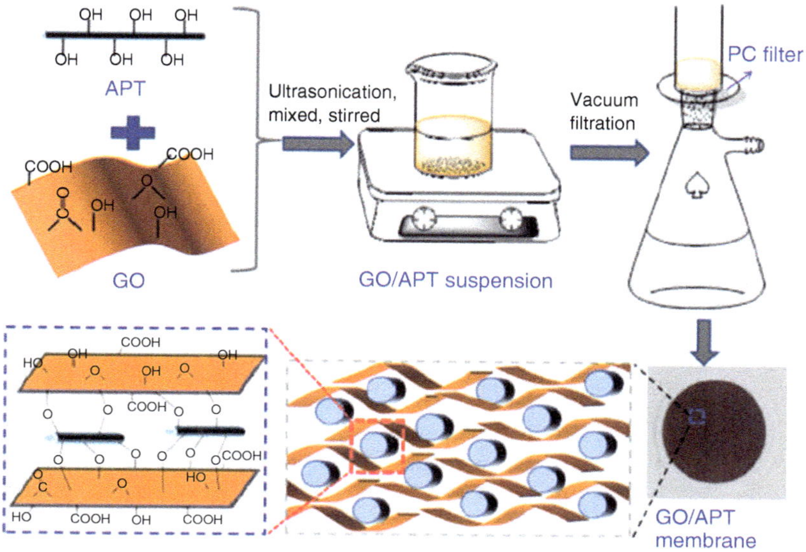

Figure 1.6 Synthetic process for GO/APT nanocomposite membrane. Source: Wang et al. [16]. © 2019 Elsevier.

membrane surface, enhanced actual membrane surface area, and a large cavity at the interface between N-GOQD and PA matrix [17].

Cheng et al. fabricated reduced graphene oxide-gold nanoparticle (rGO-AuNP) membrane by vacuum filtration for wastewater dye treatment applications. Using the TEM technique, the size of AuNPs on the rGO surface measured about 8–10 nm in size. The layer by layer (LBL) stacking of rGO-AuNP showed a high flux of 204 LM^2/h/bar and retention of 99% for RhB dye molecules [18].

GO membranes are recognized as potential candidates in separation technology. Nonetheless, their instability in aqueous environments is the drawback in real-time applications. Ma et al. designed facile eco-friendly intercalation of montmorillonite (MMT) nano patches on GO to form stable membranes, addressing such complications. The developed stable membranes showed a high water flux of 139.5 LM^2/h/bar under 0.01 MPa. Further, for various organic dye molecules such as congo red (CR), RhB, crystal violet (CV), methylene blue (MB) showed 98.75%, 99.44%, 99.90%, and 99.94% respectively [19].

In this study, carboxylation of GO nanosheets was carried out through a series of multiple oxidation methods and nucleophilic substitution reactions. Graphene acidic (GA) membrane fabricated using carboxylation of GO nanosheets through the multiple oxidation process. The developed GA membranes showed high water permeability and enhanced salt rejections when associated with bare GO and modified GO membranes with surface negative charge and hydrophilicity. GA membrane showed pure water flux of 73.5 LM^2/h/bar and peak rejection of 64% and 95% for 2000 mg/l NaCl and Na_2SO_4 feed solutions at 1.5 MPa [20].

In an aqueous solution in the absence of organic solvents GO, rGO, and graphene oxide/polyacrylamide (GO/PAM), nanofiltration membranes are prepared using vacuum filtration. The hydrophilicity of such developed graphene-based membranes was analyzed using contact angle measurements by using X-ray diffraction interlayer spacing of various membranes such as GO (0.85 nm), GO/PAM (0.68 nm), and rGO (0.36 nm). The performance studies of all these graphene-based membranes were carried out using dead-end filtration mode. The water flux and retention for RhB dye showed 399 LM^2/h/bar, 85% for GO and 188.89 LM^2/h/bar, 95.43% for GO/PAM and 85.85 LM^2/h/bar, 97% for rGO membranes. The GO/PAM membranes showed the best separation performance due to the actual fine-tuning of interlayer spacing. These could be potential membranes for high performance in water purification applications [21].

To address global water issues, forward osmosis (FO) based water treatment showed an immense impact. However, it has some limitations in practical application, like the lack of FO membranes with high permeability and selectivity. The novel reduced GO/CNT hollow fiber through electrophoretic deposition coupling with chemical reduction process with high FO performances was evaluated. Two main factors were noted. One is due to the ultralow friction, and distinct interlayer spacing of rGO showed higher water permeability and selectivity. The second factor is high porosity, and improved wettability warranted the CNT hollow fiber with low internal concentration polarization ensured and enhanced water flux. Using deionized water feed at a 0.5 M NaCl draw solution (DS), reduced GO/CNT showed exceptional water flux of 22.6 LMH, indicating 3.3 times greater than the commercial membrane. In the meantime, its reverse salt flux displayed only 1.6 gMH compared to 2.2 gMH for commercial membrane. The obtained results suggest that the rGO/CNT membrane could be a substituent for clean water supply in the FO process [22].

Cellulose acetate (CA)/GO nanocomposites were fabricated for seawater desalination using the phase inversion method. GO nanosheets with grain size 2.7 nm were added to the CA matrix and casting CA/GO onto polyester fabric to enact asymmetric CA (RO) membranes to enhance thermal and mechanical stability. Further, the addition of GO into the CA matrix altered the porosity structure. Meanwhile, synthetic seawater rejection of CA membrane modified with 1 wt% GO showed almost 90% (80% more salt rejection of bare CA membrane) for 25 000 ppm NaCl at 25 bar, which is 1.8 more when compared to bare CA membrane. Also, the flux of the CA membrane showed minor reduction (30% decrease) in the presence of 1 wt% GO showing prospective in seawater desalination [23].

Kang et al. developed graphene oxide-oxidized carbon nanotubes-layer by layer assembly (GO-OCNTs-LBL) with different bilayers on polyethersulfone (PES) membranes. The insertion of OCNTs on the GO interlayer for GO-OCNTs-LBL membrane controls the enlargement of GO. Few other studies were inspected using various salts such as NaCl, $MgCl_2$, Na_2SO_4, sucrose as DSs, and deionized water as feed. GO-OCNTs-LBL membranes showed high water flux than GO-LBL (control) for NaCl, $MgCl_2$, Na_2SO_4, and sucrose as DS. The solute flux of GO-OCNTs-LBL membrane for NaCl decreased by 70% than that of GO-LBL membrane. Consequently, the GO-OCNTs-LBL membrane showed superior rejection performances for NaCl than the GO-LBL membrane alone [24].

A polydopamine (PDA) functionalized reduced graphene oxide-metal organic frameworks (rGO-MOFs-HKUST-1) nanocomposites were developed with expeditious rejection capacity for organic dye. In this work, MOFs-HKUST-1 based on rGO was functionalized with PDA. These PDA-rGO-MOFs-HKUST-1 nanocomposites formed on the surface of commercially available CA membrane. The addition of HKUST-1 and cross-linking functionalization of PDA showed excellent rejection, flux, hydrophilic, and antifouling performances. The prepared nanocomposite membrane (PDA-rGO-MOFs-HKUST-1) could successfully reject methylene blue (99.8%) and congo red (CR) 89% efficiently. Further, it is important to note that the flux of the PDA-rGO-MOFs-HKUST-1 membrane showed a drastic increase of 31-fold due to HKUST-1 and PDA [25]. The schematic showing mechanisms involved in PDA-rGO-MOFs-HKUST nanocomposites shown in Figure 1.7.

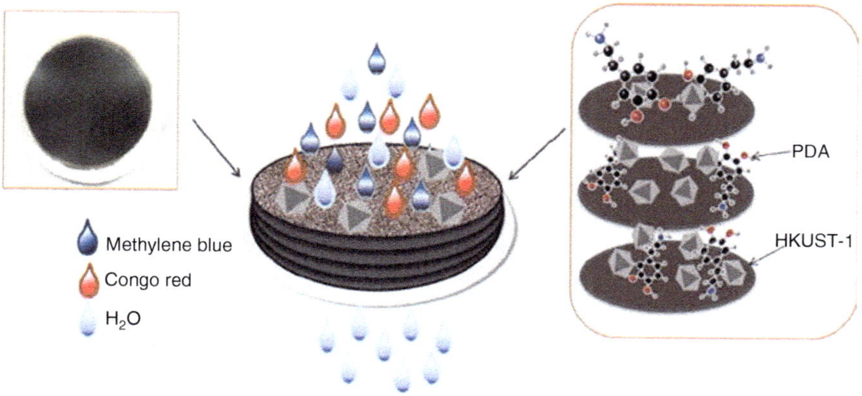

Figure 1.7 Schematic showing mechanisms involved in PDA-rGO-MOFs-HKUST nanocomposites. Source: Liu et al. [25]. © 2019 Elsevier.

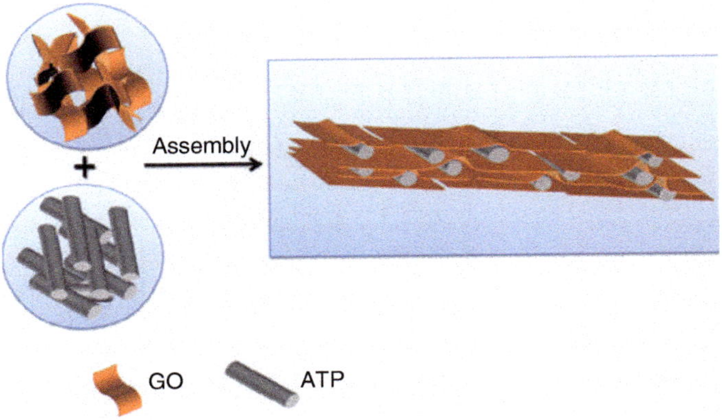

Figure 1.8 Schematic self-assembly of GO/ATP membranes. Source: Luo et al. [26]. © 2019 Elsevier.

Lee et al. effectively surface-functionalized EDTA-rGO membranes for nanofiltration. The EDTA-rGO membranes' surface was eventually treated with oxygen plasma to boost its water permeability and denoted (P-EDTA-rGO). The P-EDTA-rGO showed 80% of rejection for NaCl salt and water permeance of 150 LM^2/h/bar, showing 15 times greater than before plasma surface treatment. The developed P-EDTA-rGO nanofiltration membrane caused high rejection rates. Permeabilities of the small interlayer spacing (5.7 Å) is due to EDTA functionalization and improved hydrophilicity from enhanced plasma treatment [25].

In this work, the attapulgite (ATP) nanofibers/GO composite membrane is self-assembled by mixed solutions containing both ATP and GO shown in Figure 1.8. ATP nanofibers intercalated on GO nanosheets interlayers, acting as laminate structures for effectual trails for water molecules GO. Besides, ATP nanofibers able to widen mass transfer channels and able to promote hydration capacity. The permeate flux from 28.64 LM^2/h/bar for GO membranes (control) to 221 LM^2/h/bar for ATP/GO membranes was observed and showed extraordinary rejection rates for organic dye molecules such as CBB R250 99.88% and 99.93% for TMPyP [26].

In membrane studies, particularly in water treatment, stability is one of the major problems. In this work, Fang et al. constructed a highly water stable GO-based metal-organic framework (MOF) (UiO66) composite membrane was designed. The construction of substrate involves two main steps: (i) doping GO nanosheets into the membrane casting solution comprising polyacrylonitrile (PAN) that forms 2D–3D inter-connecting pores by phase inversion mode designated as GO-PAN, and (ii) immersing GO-PAN into dopamine solution, which undergoes self-polymerization undergoes to form better compatible and flexible substrate denoted as PGP. The UiO66 layer was assembled by vacuum-assisted filtration technique on porous PGP support illustrated in Figure 1.9. The stable GO-based UiO66/PAN, known as thin-film composite membrane, showed higher water flux 31.33 LM^2/h/bar and maintaining an effective rejection rate of >94% for various dyes such as congo red, methyl orange (MO), methylene blue, and RhB [27].

In other work, the graphene oxide framework (GOF) was designed via "from" and "to" cross-linking approaches. GOF with compact nanochannels were fabricated by LBL using branched polyethyleneimine (BPEI), GO, and thiourea particles as the well-defined

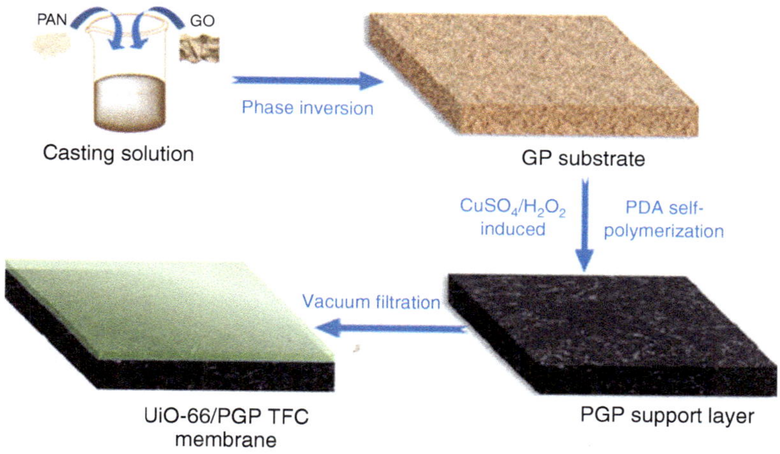

Figure 1.9 Overview process involved in the UiO66/PGP-TFC NF membrane process. Source: Fang et al. [27]. © 2020 Elsevier.

building blocks. The surface morphology of BPEI/GO/thiourea membranes revealed with compact GOF arrangement with very narrow interlamellar spacing (0.49 nm) when matched to BPEI/GO membranes (0.52 nm). Structural interpretation revealed thiourea partially reduced GO molecules mediated as "to" cross-linker linking adjacent GO nanosheets. While the GO deposition BPEI accomplished as "from" cross-linker by binding thiourea linked GO laminates to form highly stable membranes. The GOF membrane performances showed rejection of 99.5% for both anionic and cationic dyes like methyl orange and RhB. The BPEI/GO/thiourea membrane synthesized with 12 bilayers using 0.25 mg/ml of GO concentrations showed pure water flux of 24 LM^2/h/bar and Na_2SO_4 rejection of 94%, showing permeability 2.5 times when compared with commercially available nanofiltration membrane. More importantly, BPEI/GO/thiourea showed excellent stability in both acid and basic aqueous solutions [28].

A 3D graphene oxide hydrogel membranes (GOHMs) designed by gelation of GO with ferrous ions through vacuum aided filtration and their application shown in Figure 1.10. In this work, ferrous ions assisted as cross-linkers, which enhanced bonding strength between

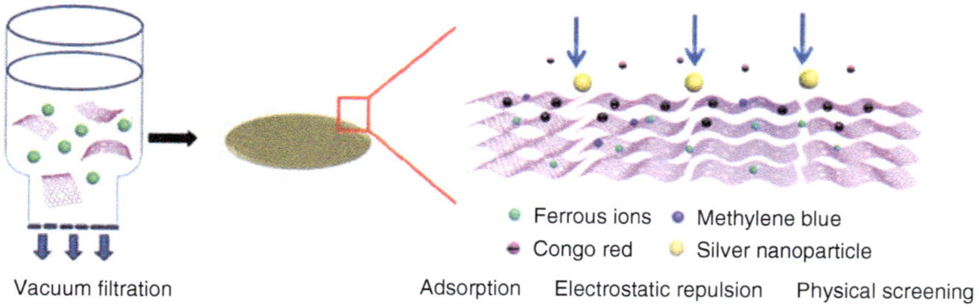

Figure 1.10 Schematic of GOHMs membrane preparation and retention mechanisms of target pollutants. Source: Chen et al. [29]. © 2019 Royal Society of Chemistry.

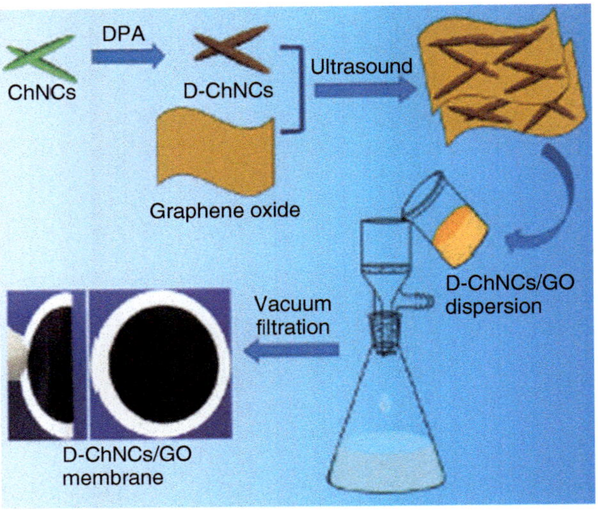

Figure 1.11 Schematic illustration of D-ChNCs and D-ChNCs/GO composite membranes. Source: Ou et al. [30]. © 2019 American Chemical Society.

GO nanosheets to prompt microstructure formation of GO via cation π interactions to form 3D lamellar porous structures. In contrast, with pristine GO membranes, the GOHMs exhibited superior water stability and enhanced water permeability of 111 LM^2/h/bar and retention performances of >99% for methylene blue and congo red dyes [29].

In this work, free-standing chitin nanocrystals (ChNCs) was modified by dopamine (D) to form GO composites and labeled as D-ChNCs-GO, and their schematic is shown in Figure 1.11. The hydrophilicity is purely based on the contents of D-ChNCs that displays a very sharp increase in water flux from 8.8 for bare GO to 135.6 LM^2/h/bar for D-ChNCs-GO composite membranes. Additionally, the rejection rates for methylene blue and congo red dye showed 99.3% and 98.3% of removal performances [30].

Januario et al. modified the PES microfiltration membrane with sulfuric acid, titanium dioxide (TiO_2), and GO solutions using the LBL self-assembly method through an electrostatic interactions pressure-driven filtration system. The separation performances of the designed membrane for various charged dye molecules were twilight yellow (TY) (69.98%), red bordeaux (BR) (93%), and safranin orange (SO) (100%) based on several factors such as size exclusion, electrostatic repulsion, and adsorptive processes [31].

The graphene oxide/MIL-88A(Fe) (GO/M88A) membrane was fabricated using the vacuum filtration method. The intercalation of M88A on graphene nanosheets showed enhanced separation performances. The designed membrane was utilized for the photo-Fenton degradation process showing 99.81% separation efficiency for methylene blue (MB) dye with enhanced 26.3 LM^2/h/bar of flux rate [32].

In this report, a nominal strategy was adapted to design graphene oxide-methylene blue (GO/MB) membrane composites. GO was surface modified by MB through electrostatic interactions and π–π conjugation systems. The rejection study for GO/MB membrane for various dyes (methyl orange, disperse black9, and RhB showed 93%, 99.85%, and 82.56% of rejections, and pure water flux was 7.67 LM^2/h/bar [33].

The GO with the 2D lamellar arrangement and single-atom thickness showed several benefits in water treatment applications assembled with the continuous membrane. Nevertheless, a suitable process to advance the improved separation property of GO membrane is highly imperative. Also, damage to the membrane during the large-scale application route, which results in a significant loss, are the main drawbacks that to be addressed. In this work, a hierarchically self-assembled GO hybrid membrane was designed to achieve superior efficiency in the water purification process and self-healing property, which leads to the synergetic effect of MOF and hydrophilic coating of chitosan biopolymer.

Intercalated MOF effectually extended the available channel space of GO and capable of molecule sieving properties. In the meantime, the chitosan-coated layer could particularly adsorb water molecules and attain self-healing via cross-linking reaction. Developed GO hybrid membrane showed enhanced water flux 14.62 LM^2/h/bar, showing an effective 344% improvement when compared with water flux of GO alone. Meanwhile, the GO hybrid membrane was able to show a rejection rate of >99% for methylene blue (MB), methyl orange (MO), acid orange (AO), and direct red 80 (DR) dyes. The damaged GO hybrid membranes can also recover their original water purification 95% for water flux and 99% of rejection ratios [34]. The schematic process of MOF–GO–CS composite membranes used for the water purification process is shown in Figure 1.12.

Due to the extraordinary permeance of GO, it has an inherent application in membrane separation. Fang et al. designed rGO by green hydrothermal reduction method. Further, the rGO membrane was synthesized on polyvinylidene fluoride (PVDF) supporting membranes by vacuum-assisted filtration. At this point, the interlayer structure of GO and the permeability of rGO membranes can be fine-tuned by altering the reduction degree of GO via altering the hydrothermal reaction time. Due to the defects of rGO nanosheets, less oxygen-bearing functional groups and improved hydrophobicity enable transport of water molecular in relatively smaller interlayer space due to hydrothermal reduction. In contrast, with the bare GO membrane, the hydrothermal reduced rGO membrane revealed excellent permeability with a rejection ratio of 99% for methyl blue, congo red, and crystal violet dyes [35].

Li et al. have designed polysulfone grafted graphene oxide nanosheets (GO-g-PSf) through nucleophilic substitution reaction between hydroxyl groups present in GO and chloromethyl groups of chloromethylated polysulfone (CMPSf). The GO-g-PSf dispersion in *N*-methyl pyrrolidone (NMP) is used to produce via vacuum filtration using PA

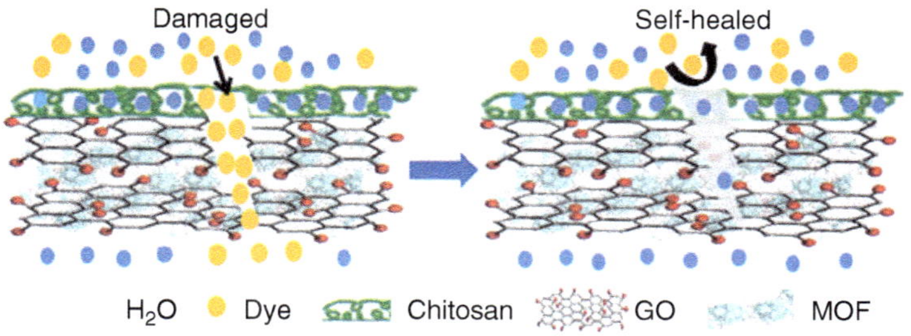

Figure 1.12 Schematic process of MOF-GO-CS composite membranes used for water purification process. Source: Chang et al. [34]. © 2019 American Chemical Society.

composite membrane as supports. GO-g-PSf hybrid membrane with a deposition rate of 31.8 μg/cm^2 showed superior water flux of 88 LM^2/h than pristine GO composite membrane 12.8 LM^2/h at 2 bar the enlarged interlayer spacing among the developed GO-g-PSf nanosheets. The rejection rate of GO-g-PSf membranes displayed 90% for acid black 1, 99.5% for congo red, and 99.8% for methylene blue dyes under cross-flow process at 2 bar. Notably, the water permeance of GO-g-PSf membrane endured at 25.9 LM^2/h after 60 hours of operation process for congo red separation. It is also worth noting that GO-g-PSf membrane showed enhanced structural stability after two hours of ultrasonic irradiation due to the physical enlargement of PSf chains [36]. The preparation strategies involved for GO-g-PSf and their chemical reaction showed in Figure 1.13.

A covalent organic framework-1 (COF-1) with systematically arranged structure and controlled pore size was covalently functionalized on the GO surface through *in situ* mode and schematically shown in Figure 1.14. The fabricated GO/COF-1 nanocomposites enhanced the dispersivity and stability in water over COF-1, permitting GO/COF-1 nanocomposites in water purification applications. The nanocomposites assembled on the PAN support to form ultrafiltration nanomembranes through dead-end filtration mode. The enhanced water permeation of 310 LM^2/h and high rejection rates for various dyes such as congo red, direct red 23, and methylene blue showing >99% of rejection rate and superior permeability for ion salts, were observed [37].

A nano graphene oxide particles (NGO), by the aid of multi-scale arrangement, were designed. In this work, divalent metal cations (Ca^{2+}, Cu^{2+}, and Mg^{2+}) engaged as cross-linking agents in order to stack nano NGO particles on both surface and pore channels of

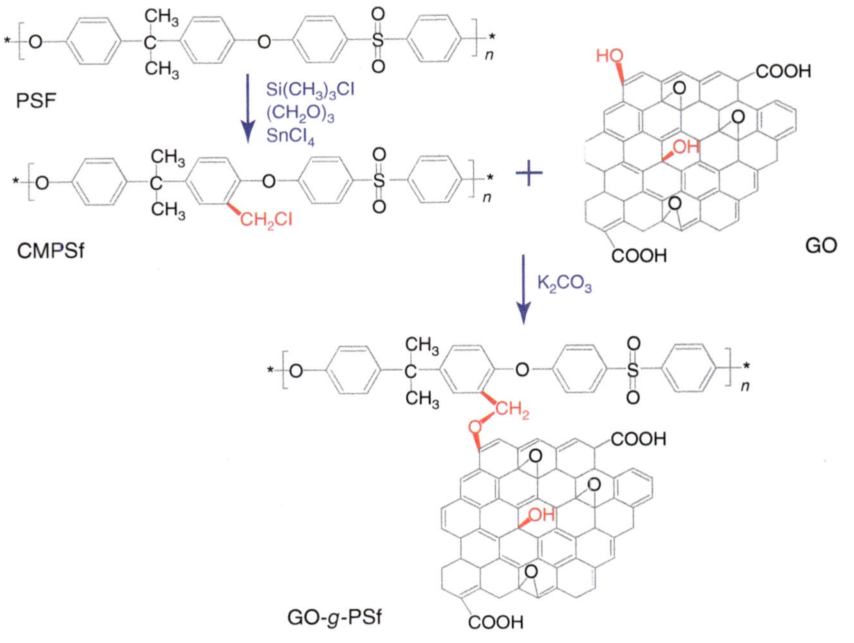

Figure 1.13 Preparation strategies involved for GO-*g*-PSf. Source: Li et al. [36]. © 2020 Elsevier.

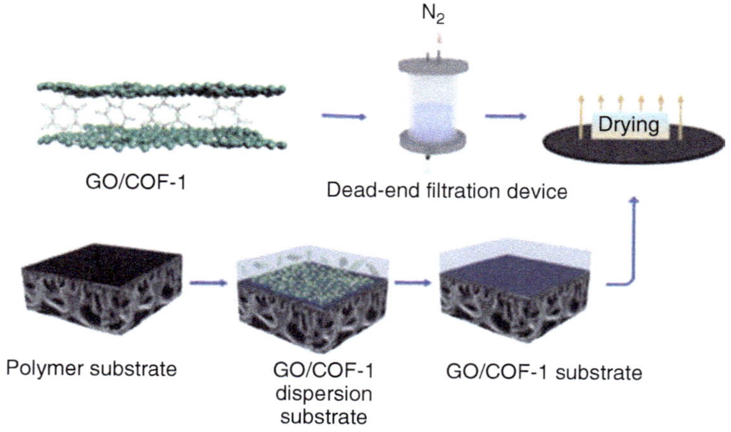

Figure 1.14 Schematic overview of GO/COF-1 membrane fabrication. Source: Zhang et al. [37]. © 2019 Elsevier.

poly(*p*-phenylene sulfide) (PPS) formed over electrostatic and coordination bonds. Metal NGO frameworks presented identical MOFs structure with nanoscale openings, and Metal-NGO/PSS membranes displayed outstanding water permeability and superior dye rejections. Compared with the transversal interlayer molecular transport chamber present in existing GO membranes, Metal-NGO/PSS membranes exhibited condensed permeable channel, maximum pore opening attaining designed membrane with rapid water molecular transport rate and superior pure water flux of 3109 LMH/MPa. Also, the Metal-NGO/PSS membranes showed a high rejection rate >98% for various dyes [38].

In another work, hybrid nanosheets were obtained by grafting GO and covalent organic frameworks (COFs) and organized building blocks. The covalent triazine framework (CTF) COFs exfoliated into nanosheets and reacted with GO to form GO-CTF hybrid nanosheets. The addition of CTF helps to change the interlayer distance between GO-CTF nanomembranes exhibiting high rejection for organic dyes >90%. Additionally, GO-CTF membranes showed a water flux of 226 LM^2/h, which is more than 12-fold higher than pristine GO membranes. Besides, chemical bonds between GO and COFs showed enhanced stability [39].

Multifunctional thin-film nanocomposites (MTFNs) nanofiltration membrane is designed using the combination of rGO@TiO_2@Ag into the PA layer was developed. The definite features of the graphene-based nanocomposites were obtained by microwave irradiation process, which recommended water channelization and provided with super hydrophilicity features. The MTFN-rGO@TiO_2@Ag preparation process is shown in Figure 1.15. Moreover, high permeability of 52 LM^2/h, desalination performances of 96% for 1 g/l Na_2SO_4 feed solution, and dye rejection of 98% for rose bengal dye feed solution of 0.5 g/l were observed [40].

In other work, unique, low-pressure nanofiltration membranes (LMPs) consisting of MoS_2 nano-supporting spacer among the GO layers are synthesized by pressure-driven assembly and heat treatment. The GO/MoS_2 showed pure water permeability of 10 LM^2/h at low pressure (2 bar), having greater than 13.6 times more when compared to pristine GO

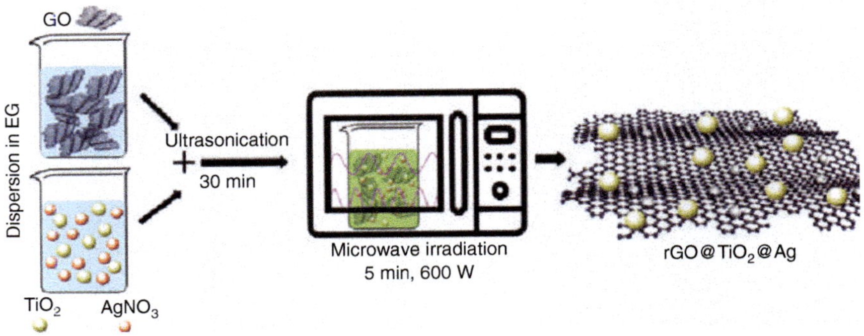

Figure 1.15 Schematic showing fabrication of rGO-TiO_2-Ag nanocomposites through microwave irradiation. Source: Abadikhah et al. [40]. © 2019 American Chemical Society.

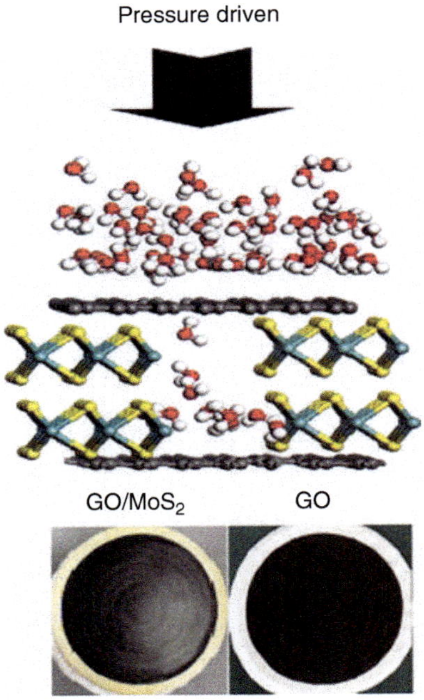

Figure 1.16 Schematic showing fabrication of GO and GO/MoS_2. Source: Zhang et al. [41]. © 2019 Elsevier.

membrane 0.75 LM^2/h. Also, high rejection for different charged dyes >95% was observed. Also, moderate salt rejections of 43%, 65%, 26.5%, and 23% for NaCl, Na_2SO_4, $MgCl_2$, and $MgSO_4$ respectively [41]. The formation of GO and GO/MoS_2 are shown in Figure 1.16.

It is well-known that the structural instability and flux deterioration factors hinder the real-time application of GO-based membranes. Thus, an innovative GO/DTiO_2-PDA-PEI dopamine modified TiO_2 hybrid nanofiltration membrane was obtained through multi couple strategy, self-assembly, co-polymerization, and surface grafting method. Fine-tuning the nanoscale interspace among the GO nanosheets and fabricating the membrane with a positively charged surface and negatively charged subsurface with a pore size of 0.87 nm was developed. The designed membranes showed water permeance of 41.675 LM^2/h. The rejection rate for the high molecular weight of positively charged crystal violet dye 99.9%, positively charged safranine 96%, negatively charged EBT 99.9%, and low molecular weight of negative dye Alizarin yellow GG 89.4% of less performance was observed [42].

Yang et al. designed a novel nanofiltration membrane via rapid co-deposition of metal phenolic networks (MPNs) and piperazine (PIP) on a porous substrate of trimesoyl chloride using a cross-linking approach. Motivated by the catechol chemistry, tannic acid surface modified to tether PIP monomers are formed by co-deposition through coordination bonds with Fe^{3+} ions on the substrate to form Fe^{3+}/TA-PIP complex. The designed membrane showed improved water permeability of 13.73 LM^2/h and 89.52% of $MgSO_4$ rejection. Furthermore, to enhance the membrane performance, modified GO nanosheets were embedded on Fe^{3+}/TA-PIP complex, exhibiting improved water permeability of 21.66 LM^2/h and $MgSO_4$ rejection of 91% [43].

Another work revealed polyethersulfone micro-filtration membranes (mPES) functionalized with polyethilenimine and GO by LBL assembly approach through electrostatic interactions pressure-driven filtration system. The highly positively charged polyethilenimine allows easy to assemble on PES substrate to accept negative layers of GO. Finally, the designed combinations of GO/PEI showed the best performance for Blue Corazol (BC) with dye rejection of 97.8% and pure water permeability of 99.4 LM^2/h. Also, the membrane showed a flux recovery of >80% after hydraulically cleaning. Also, rejection of BC dye in real dye bath wastewater showed outstanding performance with >96% of rejection rate was observed [44].

In other reported work, ultra-thin GO films with better salt rejection made-up through contra-diffusion-induced cross-linking constructed on cation-π and electrostatic interactions between cations and GO flakes. Moreover, it is a rapid process. Only five minutes is required to design a defect-free GO film. Several cations such as Na^+, K^+, Mg^{2+}, Ca^{2+}, and Fe^{3+} are used as "cross-linkers" to design GO membranes with high separation performances. The designed membranes showed a 90% rejection for Na_2SO_4 salts with a permeance of 4.3 $LM^2/h/bar$ [45].

In this report, UV reduced graphene oxide (UrGO) membranes supported on PVDF was designed for wastewater treatment applications. The weak reduction GO membrane showed optimal performances for removing pollutants. Due to the weak reduction of GO membrane, pristine graphitic sp^2 domains increased with a minor decrease in d-spacing. Therefore, UrGO membranes revealed a high water flux of 38.27 $LM^2/h/bar$ by enhanced flux more than 270% compared to GO membrane alone. Also, the rejection rate of 99% was observed for RhB dye, and for other dyes such as EBT, methylene blue, crystal violet showed >96% of rejection performances [46].

To sieve small inorganic salt ions number of reports are projected by reducing the interlayer spacing down to several angstroms. On the other hand, the major challenge for such compacted graphene oxide membranes (GOMs) is a very low water flux of <0.1 $LM^2/h/bar$ that hinders real-world applications. Planar heterogeneous graphene oxide membranes (PHGOM) with definite salt rejection and high water flux is reported. A horizontal ion transport via oppositely charged GO multi-layer heterojunction showed bi-unipolar transport performance by blocking the conduction of both cations and anions. Furthermore, supported by the forward electric field, salt concentrations are depleted in the neutral transition area of PHGOM. In this condition, deionized water is extracted from the depletion zone. Further, undergoing all these mechanisms high rejection rate for NaCl and water flux of 1529 LM^2/h/bar was observed [47].

In other work, starch modified GO was added to the polypiperazinamide network to give superior flux hyper-branched starch functionalized graphene oxide composites (HGOST) nanofiltration membranes. The modified GO starch composites are embedded in the PA top layer by esterification. The membranes are negatively charged with the top layer and showed high rejection with 96% for Na_2SO_4 salts and pure water flux of 10 LM^2/h/bar [48].

Wang et al. proposed new multiple tailored assembly ethylenediamine cross-linked GO nanocomposites membrane on a macroporous tubular ceramic substrate. In this work, the GO nanosheets are well orderly stacked, showing the construction of efficient nanochannels for mass transfer. The subsequent membrane showed permeance of 375.75 LM^2/h/bar, which is 5.62 times superior to GO membranes, and rejection of 98.6% for EBT dye. The flux for Na_2SO_4 salt was 232.72 LM^2/h/bar, and a 78% rejection rate was observed [49].

Nonetheless, effectual tuning of rGO laminar arrangements for improved membrane performance remains a challenging task. In this work, novel 1-D graphitic carbon nitride nanotubes (g-C3N4-NTs) intercalated rGO nanofiltration (NF) membrane with the better photo-induced self-cleaning property. The g-C_3N_4-NT enlarged the rGO interlayer spacing for enhanced water permeability and has visible light photocatalytic properties to remove dye pollutants. The g-C_3N_4-NT/rGO membranes exhibited water permeability of 4.87 LM^2/h/bar and RhB removal rate of >98% for long term operation [50].

It is known that rGO membranes with narrow channels display salt rejection compared to conventional nanofiltration (NF) membranes. However, their water permeance is less due to the high tortuosity of water permeation. This report includes a simple approach to creating in-plane nanopores on the GO sheets before reduction, decreasing the tortuosity, increasing water permeance, and maintaining effective salt rejection. Holey graphene oxide (HGO) nanosheets synthesized by chemical etching using hydrogen peroxide by depressing porous support by vacuum filtration and reduction through contact with hydriodic acid solutions to create r-HGO membranes. Formation of nanopores improves water permeance from 0.4 to 6.6 LM^2/h/bar and with a Na_2SO_4 rejection rate of 98.5%. More importantly, water permeance and Na_2SO_4 rejection reached the first time to the level of commercial PA-based NF membranes [51].

In another work, the enhanced nanofiltration ability of graphene-based membranes was designed by adding graphene oxide nanoribbons (GONRs) to the graphene nanochannels. The designed membranes could efficiently filter small molecules with a high rejection rate of >95% for various dyes (methyl red, methyl orange, methylene blue, and rose bengal) with high water permeance 312.8 LM^2/h/bar. The improved nanofiltration was observed due to the electrically charged and expanded nanochannels [52].

The GO membranes in the aqueous solution limit their wastewater treatment in membrane separation applications. The nanocomposite membrane made up of sarium-MOFs and GO nanosheets were synthesized through the filtration of Sm-MOFs dispersions via *in situ* growth. In this work, GO sheets disallowed the adjacent GO layers expanding in aqueous solutions, undergoes a stable skeleton membrane structure. Sm-MOFs' addition helps in the enlargement of composite membrane space, which would favorable for higher permeance. On the other hand, the designed composite membranes showed high permeance of 26 LM^2/h/bar indicating three times superior to pristine GO membrane; >91% high rejection rate for RhB and methylene blue dyes proved with long term stability performances [53].

Generally, poor chlorine resistance reduces the service life of the PA membrane in RO. In this work, N-GOQDs were grafted and integrated on the PA membrane surface. Usually, grafting investigations reduces water flux. However, in this work, the flux of the grafted membrane showed enhanced performance compared with commercial bare membranes. The grafted membrane with concentrations of 0.002 mg/ml showed better desalination of water flux of 40 LM^2/h/bar and NaCl rejection of 96.2% compared to bare membrane 36 LM^2/h/bar with 94% rejection rate [54].

Yang et al. developed a GO/AgNPs nanocomposites membrane prepared to determine the side effects of AgNPs for nanofiltration performances. Different AgNPs sizes such as 8, 20, and 33 nm were coated on GO membranes, and the filtration process was evaluated on dead-end filtration mode. Among different sized AgNPs, only 20 nm AgNPs coated GO membranes showed the highest water flux of 106 LM^2/h/bar and 97.73% rejection rate for RhB dye molecules. The mechanism involved in nanofiltration is interlaying spacing and defect size of designed GO/AgNPs membranes [55].

The swelling of GO membranes is one of the disadvantages of liquid separation applications. To avoid this, cross-linking is commonly used to stabilize GO membranes and helps in the enlargement of d-spacing. The hybrid GO/boron nitride membrane is formed through physical and chemical bond formations. Changing the GO nanosheets and amino-modified boron nitride facilitates plane-plane confirmation, edge–edge covalent bonding, which helps develop compactly packed laminar structure, preventing membrane swelling. Further, the designed stable membrane was able to show 99.98% of rejection rate and 4.15 LM^2/h/bar of flux performances for methylene blue target pollutant [56].

It is important to note that most GO laminar membranes have poor water flux and high rejections for both dyes and salts, which is usually not recommended for dye/salt separations. In this work, NH_2-Fe_3O_4 was introduced on GO nanosheets through vacuum filtration method. The advantages of NH_2-Fe_3O_4 act as a rigid spherical nanospacer to fine-tune GO interlayer spacing and cross-linkers to enhance the stability of GO membranes. The 8 wt% loading of NH_2-Fe_3O_4 exhibited improved water flux of 78 LM^2/h/bar, which is 4.8 superior to pristine GO membrane. Also, it showed 94% of congo red rejection rate and low NaCl rejection performances suggesting effective dye/salt mixture separation [57].

Wu et al. fabricated glycine mediated GO/g-C_3N_4 lamellar membranes for nanofiltration. In this work, glycine acts as a linker for the improved interactions among GO and g-C_3N_4. Also, g-C_3N_4 and glycine help to increase the interlayer spacing of GO membranes. When compared to bare GO membranes, composite membranes showed rapid water transportations. Furthermore, surface modifications by introducing hyperbranched polyethyleneimine (HPEI) coating showed excellent separation performances. The developed membranes showed rejection rates for RhB 77.7, Alcian blue (AB) 89.5, Eosin Y (EY) 91, Evas blue (EB) 99.4%, respectively, with permeance of 15.7 and 42.9 LM^2/h/bar [58].

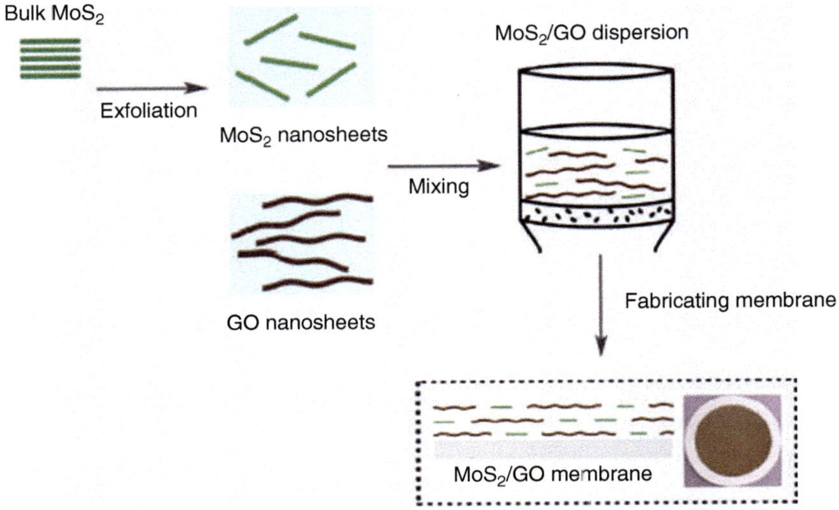

Figure 1.17 Fabrication of MoS_2/GO nanocomposite membrane. Source: Ma et al. [59]. © 2020 Elsevier.

A facile method was established to decrease the repulsive hydration among GO sheets by intercalating MoS_2 on GO nanosheets, and their fabrication process is shown in Figure 1.17. The strong van der Waals forces among MoS_2 and GO, weak interactions between MoS_2 and water molecules and swelling of GO membranes were prominently reduced. The smooth surface of MoS_2 provides with high water flux of 48 LM^2/h/bar. Additionally, separation performance with rejection rates for various dyes 99.99%, 99.90%, 99.76%, 99.66% for DR80, CR, RhB, and MB dyes. Also, rejection rates for various salts 96.85% $K_3Fe(CN)_6$, 66.76% Na_2SO_4, 58.31% $MgSO_4$, 56.00% NaCl, and 44.85% for $MgCl_2$ was observed [59].

In another work, solvothermal induced 2D rGO-TiO_2 with laminar structure was constructed as building blocks for the enhanced separation permeance. The specific properties of designed rGO-TiO_2 membranes such as high surface area, interfacial effects, and narrow layer spacing for size exclusion showed >97% rejection rates for various dye molecules RhB, methyl orange, methylene blue, and congo red. Also, it attained good permeance of 9.36 LM^2/h/bar under the cross-flow filtration process [60].

The graphene oxide nanomembranes (GONMs) with different reduction degrees from diverse graphitic resources proved for water purification performances. The main differences in various functionalities present on the GO surfaces such as (carboxyl (–COOH) or C-GO, hydroxyl (–OH) or H-GO, and epoxy (–COC) or E-GO which is not clearly understood the role of oxygen-bearing groups in the purification process. Thus, in this report, a systematic study was performed on C-GO, H-GO, and E-GO among three oxygen-bearing functional groups. To know the surface properties (Zeta analysis) and d-spacing (XRD) was performed. The obtained results showed oxygen functional groups with diverse hydrophilicity and electricity changed the d-spacing of GONMs and C-GONMs with more negative

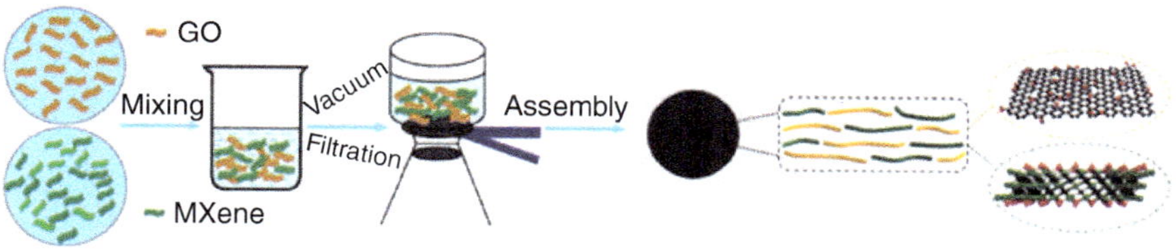

Figure 1.18 Fabrication and assembly of GO/MXene via vacuum filtration process. Source: Liu [62]. © 2020 Elsevier.

charges showed best performances >99% of rejection rate for Trypan blue (TB) dye due to electrostatic effects [61].

In this work, a novel composite membrane based on GO and MXene is demonstrated, and fabrication strategies are represented in Figure 1.18. The important heterogeneous structure of GO/MXene membranes with a different synergistic effect in substrate rejection and permeability was observed. The designed GO/MXene based membranes showed a higher water flux of 71.9 LM^2/h/bar than the pristine GO membrane showed only 6.5 LM^2/h/bar. The rejection of various organic dye molecules such as neutral red (NR), methylene blue (MB), crystal violet (CV), and brilliant blue (BB) exhibited >99.5% efficiencies indicating effective membranes in water treatment [62].

A hierarchical carbon membrane (HCM) was fabricated through a successive vacuum filtration of carbon fiber (CF) substrate, a CNT intermediate supportive layer, and a GO layered membrane. Fabricated HCM membranes were evaluated by capillary flow porometry, thermogravimetric analysis, and membrane performance evaluation. HCM membranes showed an 85% rejection rate for sulfate ions and 1.3 LM^2/h/bar of permeability [63].

In this report, hydrophobic nanofibrous composites for vacuum membrane distillation (VMD) comprising of hydrophobic PVDF nanofibrous layer and hydrophobic polypropylene (PP) nonwoven fabric (NWF) substrate were prepared. The PVDF nanofibrous layer was directly fabricated on the surface of the PP NWF substrate through the electrospinning method. 1*H*, 1*H*, 2*H*, 2*H*-perfluorooctyltriethoxysilane (FTES) modified GO nanosheets were integrated with the PVDF nanofiber layer electrospinning process to improve the hydrophobicity and water flux in VMD. For the designed membranes, the water flux was 36.4 LM^2/h/bar, and NaCl salt rejection was 99.9% [64]. The preparation process and hydrophobic modification of GO nanosheets with FTES are shown in Figure 1.19.

Halakoo and Feng reported surface functionalization of thin-film composite PA membranes for desalination of high salinity water through pervaporation. The surface modification includes treating the PA membrane with chlorine via LBL assembly of positively charged PEI and negatively charged GO. The chlorination treatment improved water permeability, and combining PEI and GO to the chlorine-treated PA membrane helps in enhanced salt rejections. Designed PEI/GO/LBL membranes for the various aqueous salts such as NaCl, Na_2SO_4, $MgSO_4$, and $MgCl_2$ showed water flux of 8 kg/m^2 h with superior slat rejections of >99.9% [65].

In this work, the significance of nanowrinkles in rGOMs for non-selective water and salt transport by fine-tuning 2D nanochannels and nanowrinkles with three reduction methods

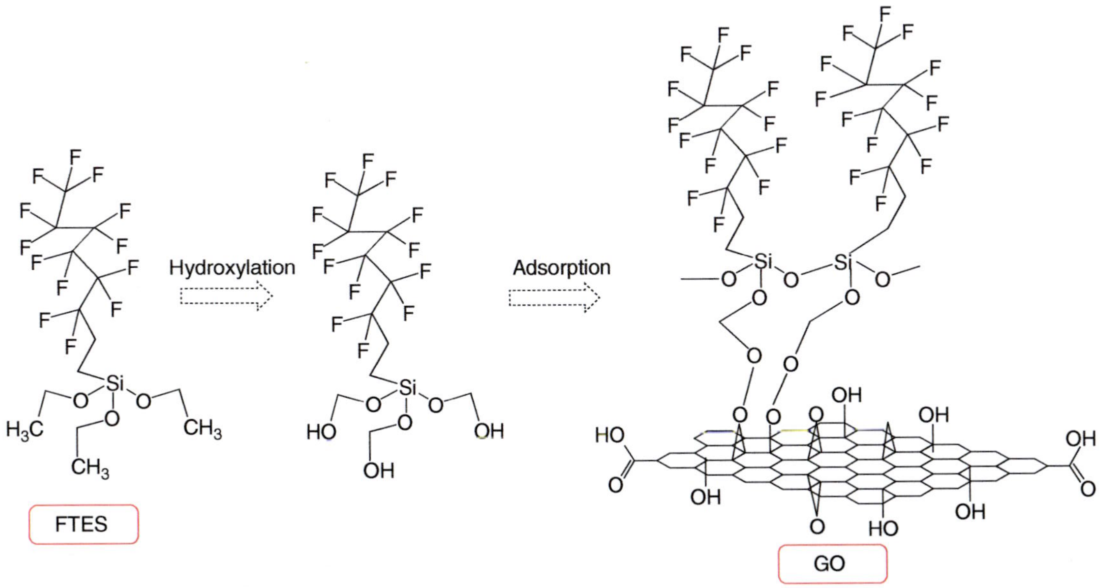

Figure 1.19 Schematic showing a hydrophobic modification of GO nanosheets with FTES. Source: Li et al. [64].

were demonstrated. In the oxidizing atmosphere of air, non-selective nanowrinkles are reduced mild and moderate air thermal reduction process, which is very important for specific water and salt separations. Thus, designed Air-rGO membranes exhibited a high 83% NaCl rejection with enhanced water flux through a hydraulic pressure-driven separation system [66].

In this current work, improving the stability of GO membranes surface was modified by intercalating ionic polymers rich in benzene and cationic imidazolium (or anionic sulfonic) functionalities helps in cross-linking the adjacent nanosheets through non-covalent interactions. The 2D C-GO (imidazolium interactions/cationic) and A-GO (sulfonic interactions/anionic) in the solvated state could preserve their constancy for a longer period. In contrast, solvated GO membranes would instantly decompose when immersed in water. In nanofiltration applications, both 2D C-GO and A-GO showed high permeances of 1111, and 600 LM^2/h/bar under 5 bar compared to pristine GO membrane showed only 65 LM^2/h/bar. Also, in desalination, the 2D C-GO membranes with high water flux of 2.49 LM^2/h/bar and NaCl salt rejection of 95% compared with GO membrane 1.66 LM^2/h/bar and rejection of 76% was observed [67].

GO and rGO significantly used as membranes in water treatment applications. Nevertheless, the main drawbacks are swollen nature and unstable after contact with aqueous solutions. In this work, membranes produced from GO were synthesized using liquid-phase exfoliation, having less oxygen content, unlike the GO/rGO systems used. Its application in ion sieving is reported. Less oxygen content graphene membranes formed from flakes changing size are used to know the effects of flake morphology on ion transport. Especially, reducing flake length and thickness helps improve the number of tortuosity of nanochannels among the layers, resulting in a substantial ion transport reduction. The smaller flakes showed better surface charge due to defects that inhibit chloride mobility permitting for physical sieving and charge repulsions. As designed, graphene membranes showed outstanding Na^+ rejection 97% and water permeance ~10 times higher than those reported GO-based membranes with higher stability with no witnessed swelling [68].

1.3.4 Translation of Graphene Nanomembranes for Real Applications

In this chapter, many modification concepts and applications are reported, proof of concept stage in researcher labs. There are no many examples of their translation to industrial scale, and the main reason is their scalability and competitiveness compared with established and low-cost nanofiltration membranes (https://www.graphene.manchester.ac.uk/learn/applications/membranes). In recent years, there are several successful examples of scalable production of GO membranes fabricated by coating commercial filtering material. One of the most successful and advanced examples of this development comes from Monash University (Professor M. Majumder) group who collaborated with local graphene company Ionic Industries and developed the first scalable GO membrane production in Figure 1.20. The demonstration and production were able to process and manufacture several hundreds of long membranes using this process. This company is currently in the process of applying this technology to clean water production.

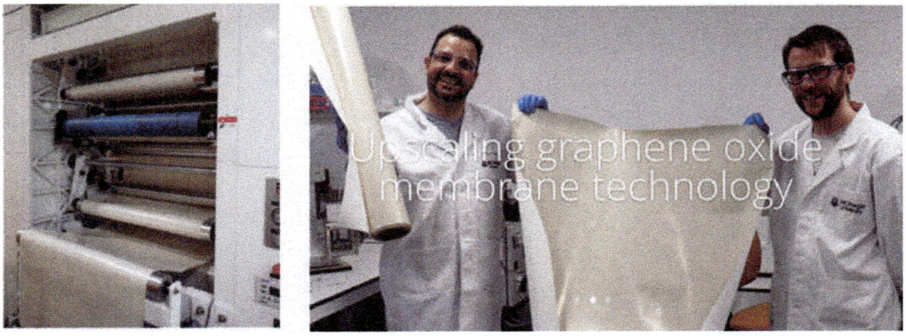

Figure 1.20 Photos showing industrial-scale production of GO membranes in Melbourne, Australia, developed by Monash University and Ionic Industries. Source: https://www.monash.edu/engineering/about-us/news-events/latest-news/articles/2018/monash-pours-enery-into-clean-water-future. Reproduced with permission.

1.4 Graphene Nanomembranes for Heavy Metals Treatment

1.4.1 Heavy Metals

Heavy metallic ions present in the water system are very troublesome to biodegrade. It can easily enter the human body via food chain, instigating a series of irreversible physiological diseases. For example, mercury ions can severely damage the central nervous system, which leads to headaches and gastroenteritis. Lead ions can damage brain tissue, and the excessive amount of lead ions in the body leads to loss of appetite, developmental delay, and hearing impairment. Similarly, other heavy metals have their disadvantages, so it is very important to treat such toxic pollutants safely [69].

An investigation based on the influence of GO nanosheet lateral size on the microstructure and liquid separation performance of ethylene diamine (EDA) cross-linked membranes was studied. Three groups of GO nanosheets with different lateral sizes separated by centrifugation used for membrane preparation in the presence of cross-linker EDA. The designed membrane performed long term stability and exhibited with higher flux rate and >99% of rejection for Cr(VI), showing membrane advantages with constricted d-spacing and enlarged ion size in acidic conditions [70].

The GO and GOMs display excellent rejection for ions, particularly in water treatment. Though, processing and scale-up of GOMs with stability issues persists as one of the major limitations. In this report, fabrication of GOMs with blade casting technique with various cations (K^+, Na^+, Li^+, Mg^{2+}, Ca^{2+}, and Ba^{2+}) intercalation with variable thickness and interlaying spacing. The average water flux obtained for GO-K^+, GO-Ba^{2+} GO-Ca^{2+}, GO-Mg^{2+}, and was 106.35, 46.59, 39.7, and 36.64 LMH. The rejection of Na^+ was 82.57% (GOM-K^+) > 78.6% (GOM-Ba^{2+}) > 75.4% (GOM-Ca^{2+}) > 72.3% (GOM-Mg^{2+}) whereas rejection for Mg^{2+} 98.84% (GOM-K^+) > 97.7% (GOM-Ba^{2+}) > 96.6% (GOM-Ca^{2+}) > 95.57% (GOM-Mg^{2+}). In addition, GOM-Ca^{2+}-R showed highest flux rates and rejection rates for Mg^{2+} (99.5%) followed by Cu^{2+} (98.4%), Ni^{2+} (97%), Pb^{2+} (96%), Na^+ (88.6%) due to the interlayer spacing when associated to other GOMs [71].

Chen et al. reported GO and PES membrane fabricated for separation performance of Hg^{2+} ions, where the GO membrane assembled on PES microfiltration membrane through

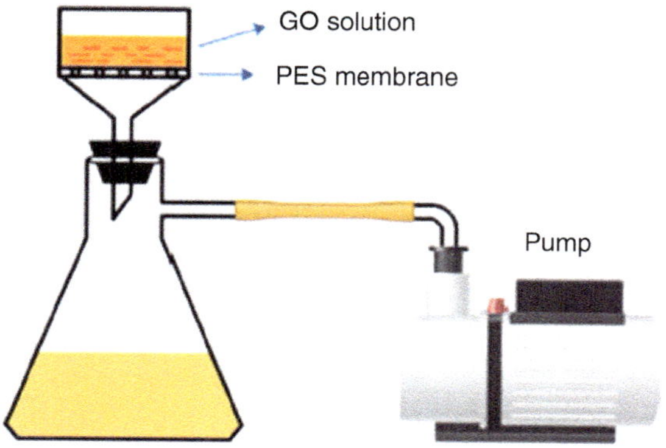

Figure 1.21 GO-PES vacuum-assisted process. Source: Chen et al. [72]. © 2019 American Chemical Society.

vacuum-assisted technique shown in Figure 1.21. The designed GO/PES membranes showed Hg^{2+} retention of 80% and flux of 26.2 LM^2/h bar. Further, the membranes with three circulations maintained the retention rate of 60.63% [72].

In other work, a dual functionalized P_mG_n@PVDF membrane was fabricated by doping GO and followed by PDA coating. The developed membranes showed pure water flux of 188 LMH/h, and Pb^{2+} could remove successfully in the water volume of 2106 LM^2/h/bar. The doping of GO helped maintain Pb^{2+} rejection, which can be ascribed due to the chelation of amino groups present on the PDA layer. Also, actual Pb^{2+} removal was attained to 5029 LM^{-2} due to the addition of citric acid [73].

In this work, TFN FO as a support layer and fabricated in the presence of different concentrations (0–8 wt%) of polyethylene glycol 400, polysulfone, and 1-methyl-2-pyrrolidone through phase inversion process. An active layer is obtained by integrating different weight ratios of (0–0.012 wt%) GO, 1,3-Phenylenediamine, and 1,3,5-benzene trichloride into the PA layer via interfacial polymerization process. The water flux for TFN membrane was 34.3 LMH, enhanced by 174% and 129% compared to TFC membranes 12.5 LMH and commercial FO membrane 15 LMH. Also, the rejection ratios for various metal ions followed the trend lead (99.9%), cadmium (99.7%), and chromium (98.3%) [74].

Harun et al. demonstrated the use of jatropha oil in the fabrication of membrane and their comparison with GO into the polymer matrix. The jatropha oil-based polyol (JOL) was blended with hexamethylene diisocyanate (HDI) to yield jatropha polyurethane membranes (JPU). Further, GO with different loading was prepared to obtain a JPU/GO composite membrane. The water flux enhanced from 223 to 523 LM^2/h. Later, filtration for copper ions showed rejections from 33% to 71.60% [75].

1.5 Conclusion and Future Perspectives

Membrane separation technology authenticated a significant role in wastewater treatments. Many research studies have been focused on separation performances using GO and GO modified/functionalized membranes in water treatment applications in recent years. It has been demonstrated that GO alone and GO modified membranes showed enhanced membrane properties in terms of mechanical strength, selectivity, flux, rejection, and thermal properties. Various fabrication strategies adapted and reported for GO/GO surface-modified nanomembranes in the desalination process (removal of salts), various dye and heavy metals separation or treatment showed inspiring results. In addition, some of the most promising results paved the information that GO-based nanomembranes could replace some of the existing commercial membranes is demonstrated. Even though several important challenges need to be addressed, such as long-term stability issues and industrial productions, graphene-based nanomembranes are potential low-cost candidates in water treatments. However, the primary key success includes a proper balanced direction toward production costs, and facile manufacturing operations are highly required. Accordingly, it is imperative to advance the expansion of research and development on GO-based nanomembranes. Finally, successful development in GO-based nanomembranes will provide one of the most promising platforms for overcoming the expected global water crisis near future.

Acknowledgments

MDK acknowledges DST, India (DST/BDTD/EAG/2019) and DST Nanomission, India (SR/NM/NS-20/2014) for financial support. DL acknowledges support from ARC Research Hub for Graphene Enabled industry transformation (IH150100003). The authors also thank Jain University, India, for providing facilities.

Important Websites

https://www.google.com/search?q=Graphene+based+nanomembranes+for+sustainable+water+purification+applications&rlz=1C1CHBF_enUS858US858&oq=Graphene+based+nanomembranes+for+sustainable+water+purification+applications&aqs=chrome..69i57.1483j0j7&sourceid=chrome&ie=UTF-8.

https://www.google.com/search?q=Graphene+based+nanomembranes+for+sustainable+water+purification+applications&newwindow=1&rlz=1C1CHBF_enUS858US858&sxsrf=ALeKk03uoBKVyKKhJQH-BtDrCM_2hYmkpA:1603719599866&source=lnms&tbm=isch&sa=X&ved=2ahUKEwjF0MaIsdLsAhWQilwKHcI_AAcQ_AUoAnoECAUQBA&biw=1368&bih=802&dpr=2.

References

1 Du, Y.C., Huang, L.J., Wang, Y.X. et al. (2019). Recent developments in graphene-based polymer composite membranes: preparation, mass transfer mechanism, and applications. *J. Appl. Polym. Sci.* 136: 47761.

2 Lyu, J., Wen, X., Kumar, U. et al. (2018). Separation and purification using GO and r-GO membranes. *RSC Adv.* 8: 23130–23151.

3 Jiang, Y., Biswas, P., and Fortner, J.D. (2016). A review of recent developments in graphene-enabled membranes for water treatment. *Environ. Sci. Water Res. Technol.* 2: 915–922.

4 Boretti, A., Al-Zubaidy, S., Vaclavikova, M. et al. (2018). Outlook for graphene-based desalination membranes. *NPJ Clean Water* 1: 1–11.

5 Kabiri, S., Tran, D.N., Cole, M.A., and Losic, D. (2016). Functionalized three-dimensional (3D) graphene composite for high efficiency removal of mercury. *Environ. Sci. Water Res. Technol.* 2: 390–402.

6 Uthappa, U., Kurkuri, M.D., and Kigga, M. (2019). Nanotechnology advances for the development of various drug carriers. In: *Nanobiotechnology in Bioformulations*, 187–224. Springer.

7 Yap, P.L., Kabiri, S., Tran, D.N., and Losic, D. (2018). Multifunctional binding chemistry on modified graphene composite for selective and highly efficient adsorption of mercury. *ACS Appl. Mater. Interfaces* 11: 6350–6362.

8 Fathizadeh, M., Xu, W.L., Zhou, F. et al. (2017). Graphene oxide: a novel 2-dimensional material in membrane separation for water purification. *Adv. Mater. Interfaces* 4: 1600918.

9 Dervin, S., Dionysiou, D.D., and Pillai, S.C. (2016). 2D nanostructures for water purification: graphene and beyond. *Nanoscale* 8: 15115–15131.

10 Hegab, H.M. and Zou, L. (2015). Graphene oxide-assisted membranes: fabrication and potential applications in desalination and water purification. *J. Membr. Sci.* 484: 95–106.

11 Babu, P., Nambiar, A., He, T. et al. (2018). A review of clathrate hydrate based desalination to strengthen energy–water nexus. *ACS Sustainable Chem. Eng.* 6: 8093–8107.

12 Sriram, G., Uthappa, U., Kigga, M. et al. (2019). Xerogel activated diatoms as an effective hybrid adsorbent for the efficient removal of malachite green. *New J. Chem.* 43: 3810–3820.

13 Song, X., Zambare, R.S., Qi, S. et al. (2017). Charge-gated ion transport through polyelectrolyte intercalated amine reduced graphene oxide membranes. *ACS Appl. Mater. Interfaces* 9: 41482–41495.

14 Zhao, S., Zhu, H., Wang, H. et al. (2019). Free-standing graphene oxide membrane with tunable channels for efficient water pollution control. *J. Hazard. Mater.* 366: 659–668.

15 Wang, X., Wang, H., Wang, Y. et al. (2019). Hydrotalcite/graphene oxide hybrid nanosheets functionalized nanofiltration membrane for desalination. *Desalination* 451: 209–218.

16 Wang, C.-Y., Zeng, W.-J., Jiang, T.-T. et al. (2019). Incorporating attapulgite nanorods into graphene oxide nanofiltration membranes for efficient dyes wastewater treatment. *Sep. Purif. Technol.* 214: 21–30.

17 Fathizadeh, M., Tien, H.N., Khivantsev, K. et al. (2019). Polyamide/nitrogen-doped graphene oxide quantum dots (N-GOQD) thin film nanocomposite reverse osmosis membranes for high flux desalination. *Desalination* 451: 125–132.

18 Cheng, M.-m., L-j, H., Wang, Y.-x. et al. (2019). Reduced graphene oxide–gold nanoparticle membrane for water purification. *Sep. Sci. Technol.* 54: 1079–1085.

19 Ma, J., He, Y., Shi, H. et al. (2019). Stable graphene oxide-based composite membranes intercalated with montmorillonite nanoplatelets for water purification. *J. Mater. Sci.* 54: 2241–2255.

20 Khorramdel, H., Dabiri, E., Tabrizi, F.F., and Galehdari, M. (2019). Synthesis and characterization of graphene acid membrane with ultrafast and selective water transport channels. *Sep. Purif. Technol.* 212: 497–504.

21 Cheng, M.-m., L-j, H., Wang, Y.-x. et al. (2019). Synthesis of graphene oxide/polyacrylamide composite membranes for organic dyes/water separation in water purification. *J. Mater. Sci.* 54: 252–264.

22 Fan, X., Liu, Y., and Quan, X. (2019). A novel reduced graphene oxide/carbon nanotube hollow fiber membrane with high forward osmosis performance. *Desalination* 451: 117–124.

23 Ghaseminezhad, S.M., Barikani, M., and Salehirad, M. (2019). Development of graphene oxide-cellulose acetate nanocomposite reverse osmosis membrane for seawater desalination. *Compos. B. Eng.* 161: 320–327.

24 Kang, H., Wang, W., Shi, J. et al. (2019). Interlamination restrictive effect of carbon nanotubes for graphene oxide forward osmosis membrane via layer by layer assembly. *Appl. Surf. Sci.* 465: 1103–1106.

25 Lee, B., Suh, D.W., Hong, S.P., and Yoon, J. (2019). A surface-modified EDTA-reduced graphene oxide membrane for nanofiltration and anti-biofouling prepared by plasma post-treatment. *Environ. Sci. Nano.* 6: 2292–2298.

26 Luo, Z., Fang, Q., Xu, X. et al. (2019). Attapulgite nanofibers and graphene oxide composite membrane for high-performance molecular separation. *J. Colloid Interface Sci.* 545: 276–281.

27 Fang, S.-Y., Zhang, P., Gong, J.-L. et al. (2020). Construction of highly water-stable metal-organic framework UiO-66 thin-film composite membrane for dyes and antibiotics separation. *Chem. Eng.* 385: 123400.

28 Rajesh, S. and Bose, A.B. (2019). Development of Graphene Oxide Framework Membranes via the "from" and "to" Cross-Linking Approach for Ion-Selective Separations. *ACS Appl. Mater. Interfaces* 11: 27706–27716.

29 Chen, Z., Wang, J., Duan, X. et al. (2019). Facile fabrication of 3D ferrous ion crosslinked graphene oxide hydrogel membranes for excellent water purification. *Environ. Sci. Nano.* 6: 3060–3071.

30 Ou, X., Yang, X., Zheng, J., and Liu, M. (2019). Free-standing graphene oxide–chitin nanocrystal composite membrane for dye adsorption and oil/water separation. *ACS Sustain. Chem. Eng.* 7: 13379–13390.

31 Januário, E.F.D., Beluci, N.C.L., Vidovix, T.B. et al. (2020). Functionalization of membrane surface by layer-by-layer self-assembly method for dyes removal. *Process Saf. Environ. Prot.* 134: 140–148.

32 Xie, A., Cui, J., Yang, J. et al. (2020). Graphene oxide/Fe (III)-based metal-organic framework membrane for enhanced water purification based on synergistic separation and photo-Fenton processes. *Appl. Catal. B.* 264: 118548.

33 Hou, J., Chen, Y., Shi, W. et al. (2020). Graphene oxide/methylene blue composite membrane for dyes separation: Formation mechanism and separation performance. *Appl. Surf. Sci.* 505: 144145.

34 Chang, R., Ma, S., Guo, X. et al. (2019). Hierarchically assembled graphene oxide composite membrane with self-healing and high-efficiency water purification performance. *ACS Appl. Mater. Interfaces* 11: 46251–46260.

35 Fan, X., Cai, C., Gao, J. et al. (2020). Hydrothermal reduced graphene oxide membranes for dyes removing. *Sep. Purif. Technol.* 241: 116730.

36 Li, J., Hu, M., Pei, H. et al. (2020). Improved water permeability and structural stability in a polysulfone-grafted graphene oxide composite membrane used for dye separation. *J. Membr. Sci.* 595: 117547.

37 Zhang, X., Li, H., Wang, J. et al. (2019). In-situ grown covalent organic framework nanosheets on graphene for membrane-based dye/salt separation. *J. Membr. Sci.* 581: 321–330.

38 Gao, Y., Su, K., Wang, X., and Li, Z. (2019). A metal-nano GO frameworks/PPS membrane with super water flux and high dyes interception. *J. Membr. Sci.* 574: 55–64.

39 Khan, N.A., Yuan, J., Wu, H. et al. (2019). Mixed nanosheet membranes assembled from chemically grafted graphene oxide and covalent organic frameworks for ultra-high water flux. *ACS Appl. Mater. Interfaces* 11: 28978–28986.

40 Abadikhah, H., Naderi Kalali, E., Khodi, S. et al. (2019). Multifunctional thin-film nanofiltration membrane incorporated with reduced graphene oxide@TiO_2@Ag nanocomposites for high desalination performance, dye retention, and antibacterial properties. *ACS Appl. Mater. Interfaces* 11: 23535–23545.

41 Zhang, P., Gong, J.-L., Zeng, G.-M. et al. (2019). Novel "loose" GO/MoS_2 composites membranes with enhanced permeability for effective salts and dyes rejection at low pressure. *J. Membr. Sci.* 574: 112–123.

42 Xu, Y., Peng, G., Liao, J. et al. (2020). Preparation of molecular selective GO/$DTiO_2$-PDA-PEI composite nanofiltration membrane for highly pure dye separation. *J. Membr. Sci.* 601: 117727.

43 Yang, Y., Li, Y., Li, Q. et al. (2019). Rapid co-deposition of graphene oxide incorporated metal-phenolic network/piperazine followed by crosslinking for high flux nanofiltration membranes. *J. Membr. Sci.* 588: 117203.

44 Homem, N.C., Beluci, N.d.C.L., Amorim, S. et al. (2019). Surface modification of a polyethersulfone microfiltration membrane with graphene oxide for reactive dyes removal. *Appl. Surf. Sci.* 486: 499–507.

45 Du, Y., Zhang, X., Yang, J. et al. (2020). Ultra-thin graphene oxide films via contra-diffusion method: fast fabrication for ion rejection. *J. Membr. Sci.* 595: 117586.

46 Yu, H., He, Y., Xiao, G. et al. (2020). Weak-reduction graphene oxide membrane for improving water purification performance. *J. Mater. Sci. Technol.* 39: 106–112.

47 Wen, Q., Jia, P., Cao, L. et al. (2020). Electric-field-induced ionic sieving at planar graphene oxide heterojunctions for miniaturized water desalination. *Adv. Mater.*: 1903954.

48 Ambre, J.P., Dhopte, K.B., Nemade, P.R., and Dalvi, V.H. (2019). High flux hyperbranched starch-graphene oxide piperazinamide composite nanofiltration membrane. *J. Environ. Chem. Eng.* 7: 103300.

49 Wang, Q., Zhao, G., Li, C., and Meng, H. (2019). Orderly stacked ultrathin graphene oxide membranes on a macroporous tubular ceramic substrate. *J. Membr. Sci.* 586: 177–184.

50 Wei, Y., Zhu, Y., and Jiang, Y. (2019). Photocatalytic self-cleaning carbon nitride nanotube intercalated reduced graphene oxide membranes for enhanced water purification. *Chem. Eng. J.* 356: 915–925.

51 Chen, X., Feng, Z., Gohil, J. et al. (2019). Reduced holey graphene oxide membranes for desalination with improved water permeance. *ACS Appl. Mater. Interfaces* 12: 1387–1394.

52 Cho, K.M., Lee, H.-J., Nam, Y.T. et al. (2019). Ultrafast-selective nanofiltration of an hybrid membrane comprising laminated reduced graphene oxide/graphene oxide nanoribbons. *ACS Appl. Mater. Interfaces* 11: 27004–27010.

53 Yang, G., Zhang, D., Zhu, G. et al. (2020). A Sm-MOF/GO nanocomposite membrane for efficient organic dye removal from wastewater. *RSC Adv.* 10: 8540–8547.

54 Yi, Z., Shao, F., Yu, L. et al. (2020). Chemical grafting N-GOQD of polyamide reverse osmosis membrane with improved chlorine resistance, water flux and NaCl rejection. *Desalination* 479: 114341.

55 Yang, K., Huang, L.-j., Wang, Y.-x. et al. (2020). Graphene oxide nanofiltration membranes containing silver nanoparticles: tuning separation efficiency via nanoparticle size. *Nanomaterials* 10: 454.

56 Lin, H., Mehra, N., Li, Y., and Zhu, J. (2020). Graphite oxide/boron nitride hybrid membranes: the role of cross-plane laminar bonding for a durable membrane with large water flux and high rejection rate. *J. Membr. Sci.* 593: 117401.

57 Dong, L., Li, M., Zhang, S. et al. (2020). NH_2-Fe_3O_4-regulated graphene oxide membranes with well-defined laminar nanochannels for desalination of dye solutions. *Desalination* 476: 114227.

58 Wu, Z., Gao, L., Wang, J. et al. (2020). Preparation of glycine mediated graphene oxide/g-C_3N_4 lamellar membranes for nanofiltration. *J. Membr. Sci.* 601: 117948.

59 Ma, J., Tang, X., He, Y. et al. (2020). Robust stable MoS_2/GO filtration membrane for effective removal of dyes and salts from water with enhanced permeability. *Desalination* 480: 114328.

60 Yu, J., Zhang, Y., Chen, J. et al. (2020). Solvothermal-induced assembly of 2D-2D rGO-TiO_2 nanocomposite for the construction of nanochannel membrane. *J. Membr. Sci.* 600: 117870.

61 Yu, H., He, Y., Xiao, G. et al. (2020). The roles of oxygen-containing functional groups in modulating water purification performance of graphene oxide-based membrane. *Chem. Eng. J.* 389: 124375.

62 Liu, T., Liu, X., Graham, N. et al. (2020). Two-dimensional MXene incorporated graphene oxide composite membrane with enhanced water purification performance. *J. Membr. Sci.* 593: 117431.

63 Amadei, C.A., Arribas, P., Cruzado, L., and Vecitis, C. (2020). Graphene oxide membranes on a hierarchical elemental carbon-based support. *Environ. Sci.: Nano.*

64 Li, H., Shi, W., Zeng, X. et al. (2020). Improved desalination properties of hydrophobic GO-incorporated PVDF electrospun nanofibrous composites for vacuum membrane distillation. *Sep. Purif. Technol.* 230: 115889.

65 Halakoo, E. and Feng, X. (2020). Layer-by-layer assembly of polyethyleneimine/graphene oxide membranes for desalination of high-salinity water via pervaporation. *Sep. Purif. Technol.* 234: 116077.

66 Yuan, S., Li, Y., Xia, Y. et al. (2020). Minimizing non-selective nanowrinkles of reduced graphene oxide laminar membranes for enhanced NaCl rejection. *Environ. Sci. Technol. Lett.* 7: 273–279.

67 Ran, J., Chu, C., Pan, T. et al. (2019). Non-covalent cross-linking to boost the stability and permeability of graphene-oxide-based membranes. *J. Mater. Chem. A* 7: 8085–8091.

68 Hirunpinyopas, W., Iamprasertkun, P., Bissett, M.A., and Dryfe, R.A. (2020). Tunable charge/size selective ion sieving with ultrahigh water permeance through laminar graphene membranes. *Carbon* 156: 119–129.

69 Wang, Z., Wu, A., Colombi Ciacchi, L. et al. (2018). Recent advances in nanoporous membranes for water purification. *Nanomaterials* 8: 65.

70 Lin, H., Li, Y., and Zhu, J. (2020). Cross-linked GO membranes assembled with GO nanosheets of differently sized lateral dimensions for organic dye and chromium separation. *J. Membr. Sci.* 598: 117789.

71 Ghaffar, A., Zhang, L., Zhu, X., and Chen, B. (2019). Scalable graphene oxide membranes with tunable water channels and stability for ion rejection. *Environ. Sci.: Nano* 6: 904–915.

72 Chen, H., Wu, H., Wang, Q. et al. (2019). Separation performance of Hg^{2+} in desulfurization wastewater by the graphene oxide polyethersulfone membrane. *Energy Fuels* 33: 9241–9248.

73 Zhang, Y., Feng, Y., Xiang, Q. et al. (2020). A high-flux and anti-interference dual-functional membrane for effective removal of Pb(II) from natural water. *J. Hazard. Mater.* 384: 121492.

74 Saeedi-Jurkuyeh, A., Jafari, A.J., Kalantary, R.R., and Esrafili, A. (2020). A novel synthetic thin-film nanocomposite forward osmosis membrane modified by graphene oxide and polyethylene glycol for heavy metals removal from aqueous solutions. *React. Funct. Polym.* 146: 104397.

75 Harun, N.H., Abidin, Z.Z., Abdullah, A.H., and Othaman, R. (2020). Sustainable jatropha oil-based membrane with graphene oxide for potential application in Cu(II) ion removal from aqueous solution. *Processes* 8: 230.

2

Magnetic Graphene Oxide and Its Composite Nanomaterials: Application in Environmental Decontamination

Karan Chaudhary and Dhanraj T. Masram

University of Delhi, Department of Chemistry, Delhi, 110 007, India

2.1 Introduction

In the modernized world, the increasing number of agricultural and industrial activities by humans has been the reason behind the increased concern toward environmental issues such as contamination of soil, air, and water along with climate change [1]. The contamination of water by different toxic pollutants as well as the concentration level of these contaminants is one of the important environmental issues for the world. For the survival of all living beings on the planet, the most vital natural resource is water. Human activities causing water pollution are a serious threat for the survival of living beings [2]. Industrial activities, mining operations, and the use of various organic and metal-based compounds are the main sources for water pollution [3, 4]. From consuming contaminated water, humans have suffered from various waterborne diseases such as typhoid, kidney and liver damage, giardiasis, hookworm, ascariasis, Alzheimer's disease, multiple sclerosis, and hormonal problems along with other problems like heart disease, cancer, and Parkinson's disease [5]. Due to the effects of the polluted water on human health, it is important to develop technologies in the field of water purification. Many technological approaches have been developed for the decontamination of water, such as nanofiltration, electrochemical degradation, membrane filtration, adsorption, oxidation, and photocatalytic degradation [6]. For a technique to be sustainable, cost-effectiveness, and water pollutant withdrawal ability without generating harmful by-products are important factors to be considered [7]. The most promising and efficient technique among these is adsorption, as it fulfills the criteria of efficient removal of water pollutants without producing harmful by-products and is easily operable and less cost-effective [8].

The advancement in the field of nanotechnology has proved to be beneficial for sustainable water treatment and other environmental applications and has engrossed great attention worldwide [9]. Treatment of water pollutants is a big challenge, and nanomaterials are important for their removal as well as the environmental management of these nanomaterials [10]. Various nanomaterial and functionalized nanomaterials have been used for

Environmental Applications of Carbon Nanomaterials-Based Devices, First Edition.
Edited by Shadpour Mallakpour and Chaudhery Mustansar Hussain.

several industries that include environmental industries having one of the applications as wastewater treatment [11–13]. As adsorption technique is one of the major techniques for the removal of toxic contaminants, various nanomaterials such as carbon-based nanomaterials, silica-based nanomaterials, clays, metal oxide nanoparticles (NPs), metallic nanomaterials, polymer-based nanomaterials, layered double hydroxides, and metal–organic frameworks have been extensively employed as adsorbents for the removal of toxic contaminants from water [14–19]. In the adsorption technique, the removal of toxic contaminants from water–carbon-based nanomaterials has engrossed substantial attention due to unique properties such as high stability, high surface area, and abundance of functional groups over its surface [20–22].

The most interesting material among carbon nanomaterials is graphene, a two-dimensional sheet-like material having carbon atoms arranged in a hexagonal pattern forming a honeycomb-like structure of atomic thickness consisting sp^2 hybridized carbon [23]. Graphene became popular in 2004 when Geim and coworkers isolated monocrystalline graphitic film via mechanical exfoliation of graphite using scotch tape [24]. More interestingly what made this material a star in material science and technology is its unique physicochemical properties such as high intrinsic mobility, good electrical conductivity, thermal conductivity, ambipolar field effect, quantum Hall effect, high Young's modulus, large theoretical specific surface area, and optical transmittance. Due to these properties, graphene engrossed a huge amount of attention for its use in various applications and devices [25]. Graphene oxide (GO) is one of the forms of graphene that has been derived from graphite. The basic structure of GO is similar to that of graphene. However, there are differences such as the presence of both sp^2 and sp^3 hybridized carbon atoms and excess of oxygen-containing functional groups such as —OH (hydroxyl), C—O—C (alkoxy), C=O (carbonyl), and —COOH (carboxylic acid) on its surface [26]. Properties displayed by GO are high surface area, high mechanical strength, good chemical stability, and low electrical and thermal conductivities [27, 28]. Moreover, the presence of oxygen-containing functional groups on the surface of GO makes it a hydrophilic material. All other properties along with hydrophilicity, high surface area, and several functional groups on the surface allow various chemical modifications of GO to produce several nanocomposites of GO [29–32]. GO and its nanocomposites have been used for wide applications such as photocatalysis, catalysis, energy storage, environmental application, water treatment, sensors, biosensors, optical imaging and therapy, gene and drug delivery, theranostics, and other biomedical and biotechnological fields [27, 33–43]. Furthermore, in aqueous medium, stable suspension is obtained due to hydrophilic property of GO that makes it a superior material that can be used for water purification [44]. However, the aggregation of GO sheets in water and the problem of separating GO after adsorption of pollutants due to its high dispersibility are associated difficulties in water purification application. A most effective and practical approach to overcome this problem is immobilizing the GO sheets with NPs of metal and their oxides [45]. In this context, most preferred NPs are magnetic nanoparticles (MNPs) because of their biocompatibility, size and shape-dependent chemical and physical properties, and easy separation due to magnetism it possesses [46]. Therefore, bare MNPs have been extensively used for the removal of organic and inorganic contaminants from wastewater [47]. Despite the advantages of MNPs, disadvantages that are majorly associated with them are low dispersibility in an aqueous medium, sensitive to oxidation, and aggregation [48]. The combination of MNPs and GO to form a magnetic graphene oxide (MGO) composite is the way to overcome the problem of aggregation associated with both individuals and separation of GO from aqueous medium [45]. Nanosize, robust structure, chemical and thermal stabilities, high surface area, and magnetic nature are the salient features of this MGO [49]. This nanocomposite material can easily be separated by applying an external magnetic field and also will show superior properties as compared with bare MNPs and pristine GO [45, 50]. Compared with other separation methods such as filtration and centrifugation, magnetic separation is the most effective, and there is no loss of material during separation. Being a simple and cost-effective method, magnetic separation found to be advantageous in industrial applications [8].

This chapter is based on MGO and its composite nanomaterials for environmental applications. Herein, we have discussed the synthesis and application of MGO and its composite nanomaterials for environmental decontamination, by utilization of these nanomaterials for the removal of toxic contaminants (metal, organic, and other main contaminants) for sustainable water treatment.

2.2 Synthesis of Magnetic Graphene Oxide and Its Composite Nanomaterials

Binding of Fe_3O_4 NPs on GO for the synthesis of MGO nanocomposite can be achieved in two different methods such as *in situ* binding or *ex situ* binding. *Ex situ* binding involves mixing solutions of pre-synthesized Fe_3O_4 NPs and GO, whereas *in situ* binding involves the formation of Fe_3O_4 NPs in a solution having GO. In the following paragraph, we have discussed some methods used for the synthesis of MGO and its composite nanomaterials.

In synthesis of magnetic graphene oxide functionalized with cysteine (Cys-MGO), GO was synthesized using the modified Hummers method. However, using coprecipitation method, Fe_3O_4 NPs were synthesized, and then under sonication, Fe_3O_4 NPs were mixed to the solution of GO that resulted in MGO. Then Cys-MGO was obtained by sonicating mixture of MGO and cysteine in the presence of NaOH [51]. In the synthesis of Fe_3O_4@(PAH/GO–COOH)$_n$ nanocomposite, Fe_3O_4–COOH NPs were synthesized, and then in poly(allylamine) hydrochloride (PAH) solution, these were soaked. After this, the obtained particles were immersed in an aqueous solution containing carboxyl-functionalized graphene oxide (GO-COOH), and the resultant product was Fe_3O_4@(PAH/GO–COOH)$_1$ nanocomposite. Again soaking in PAH solution and then in GO-COOH solution results in Fe_3O_4@(PAH/GO–COOH)$_2$ nanocomposite. Furthermore, the number of time steps is repeated, and Fe_3O_4@(PAH/GO–COOH)$_n$ sandwiched nanocomposite is obtained using layer-by-layer self-assembly method [52]. Another example is Fe_3O_4@GO-Ag_xPt_{100-x} that was synthesized by deposition of 3-(2-aminoethylamino)propyltrimethoxysilane (APPTS)-modified Fe_3O_4 NPs that were synthesized solvothermally on GO through electrostatic interactions. Further on this MGO, by adjusting molar ratios of Ag and Pt, AgPt NPs were anchored to obtain Fe_3O_4@GO-Ag_xPt_{100-x} nanocomposites [53].

With diethylenetriaminepentaacetic acid-modified magnetic graphene oxide (DTPA/MGO), MGO composite was synthesized by *in situ* synthesis of Fe_3O_4 NPs on GO by coprecipitation method. Then by an amidation reaction between MGO and DTPA, DTPA/MGO nanocomposite was synthesized using diethylenetriamine as a cross-linker [54]. Magnetic graphene oxide modified with poly(allylamine) (PAA-MGO), a new composite, was synthesized by precipitating magnetic NPs via coprecipitation method onto poly(allylamine hydrochloride)-modified graphene oxide (PAA-GO) that was prepared separately [55]. Another composite is mGO/SiO_2@coPPy-Th, in which MNPs were *in situ* synthesized over GO by coprecipitation method to give mGO. Then SiO_2 layer was coated over this mGO, which further was modified with pyrrole–polythiophene copolymer via *in situ* polymerization for synthesizing mGO/SiO_2@coPPy-Th composite [56]. A different process was used for preparing MGO composite, in which MNPs on GO were synthesized *in situ* via solvothermal method. Afterward, this Fe_3O_4/graphene oxide composite was modified with ethylenediaminetetraacetic acid (EDTA) ligand by chemical method to obtain EDTA-mGO nanocomposite [57]. PMGO is another nanocomposite in which by coprecipitation method, Fe_3O_4 was *in situ* introduced onto GO for the synthesis of MGO. Furthermore, this MGO was modified with poly dimethyl diallyl ammonium chloride (PDMDAAC) for PMGO nanocomposite [58]. The different process was observed for graphene oxide/iron (III–VI) oxide nanocomposite that was synthesized in a one-step reaction by using ferrates as the oxidizing agent instead of $KMnO_4$ in GO synthesis [59]. The hydrothermal method has also been employed for the synthesis of MGO that on further blending in a ball mill with hexadecyltrimethylammonium bromide (CTAB) resulted in MAGO-CTAB nanocomposite [60].

2.3 Application of Magnetic Graphene Oxide and Its Composite Nanomaterials

Major pollutants such as heavy metal waste, radioactive metal wastes, organic dyes and pigments, pesticides and herbicides, inorganic anions, and other pollutants are present in wastewater, causing serious threat for the survival of living beings. Therefore researchers synthesized various MGOs and their composite nanomaterials and have used these materials for the removal of pollutants from aqueous solutions and wastewater samples. These MGOs and their composite nanomaterials served as either adsorbent or catalyst for the removal of water pollutants. In this section, a discussion has been made regarding the use of MGO and its composite nanomaterials for the removal of water contaminants (Figure 2.1) that have been divided into three categories: toxic metal contaminants, toxic organic contaminants, and other contaminants.

2.3.1 Removal of Toxic Metal Contaminants

Magnetic graphene oxide functionalized with cysteine (Cys-MGO) has been utilized for the removal of cadmium. Cysteine selectively binds cadmium in comparison with other metal ions. Cadmium removal process by Cys-MGO followed pseudo-second-order kinetic model, and Langmuir isotherm model fits best with the adsorption process data. For the removal of cadmium by Cys-MGO, adsorption capacities obtained were 24.39–30.30 mg/g at 298–328 K. Thermodynamic studies revealed the adsorption process was chemical in nature, spontaneous, and endothermic. Recyclability studies show that Cys-MGO can be reused several times without the loss of adsorption property for selective removal of

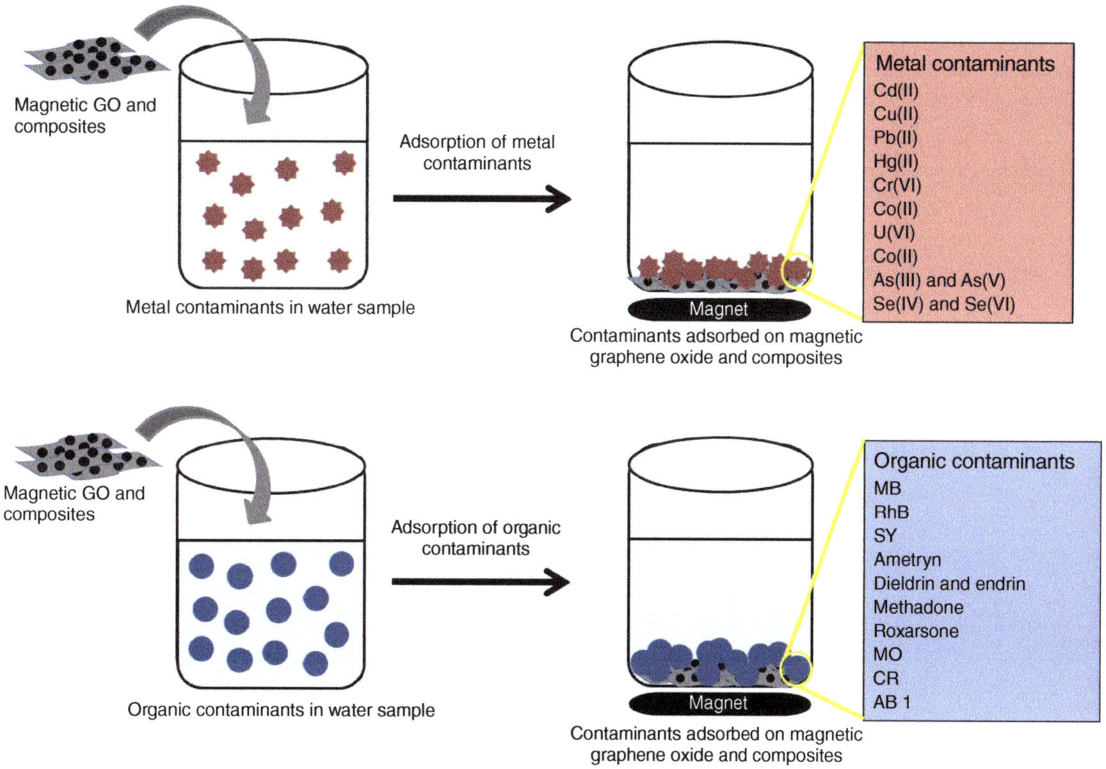

Figure 2.1 Graphical illustration of magnetic graphene oxide and its composites for adsorption of metal and organic contaminants from aqueous solution.

cadmium from aqueous solution [51]. Another study shows the removal of Cu(II), Cd(II), and Pb(II) ions using magnetic graphene oxide modified with diethylenetriamine-pentaacetic acid (DTPA/MGO) from aqueous solution under acidic conditions. The adsorption process of metal ions on DTPA/MGO followed pseudo-second-order kinetic model and Langmuir adsorption isotherm model. At pH 3.0, maximum adsorption capacities were 131.4, 286.5, and 387.6 mg/g for Cu(II), Cd(II), and Pb(II), respectively. Adsorption process was found to be spontaneous and endothermic as per thermodynamic studies. DTPA/MGO was found to be selective for Pb(II) adsorption from a mixture of various metal ions. Excellent reusability was exhibited by DTPA/MGO composite [54]. Acetylcysteine-functionalized magnetic graphene oxide (MGO-NH-SH) nanocomposite has been used as an adsorbent for the removal of Hg(II) from aqueous solution. Langmuir isotherm model fits well with the adsorption of Hg(II) on MGO-NH-SH, whereas adsorption process followed pseudo-second-order kinetic model. For the removal of Hg(II) from aqueous solution and wastewater, MGO-NH-SH exhibited higher adsorption capacity and reusability. The maximum adsorption capacity of MGO-NH-SH for Hg(II) removal was 208.8 mg/g [61].

Polystyrene and MGO composite, PS@Fe_3O_4@GO, was used as a nanosorbent for the removal of Pb(II) ions from the aqueous solution. The removal of Pb(II) ions using PS@Fe_3O_4@GO by adsorption follows first-order kinetics, and adsorption data fits Langmuir isotherm model well. The adsorption of Pb(II) over composite was found to be spontaneous. The estimated maximum adsorption capacity of PS@Fe_3O_4@GO was 73.52 mg/g. At slight acidic condition ($pH = 6$), the removal efficiency was reported as 93.78%. Studies revealed Pb(II) and PS@Fe_3O_4@GO have chemical interaction between them. The removal efficiency decreased to 40% at the end of fourth cycle for Pb(II) ions [62]. Selenium oxyanions removal has been studied using modified magnetic graphene oxide with poly(allylamine) composite (PAA-MGO). In the removal of Se(IV) and Se(VI), adsorption capacities of PAA-MGO were 120.1 and 83.7 mg/g at pH 5.8, respectively. Isotherms obtained for the adsorption of Se(IV) and Se(VI) fit well with Freundlich model. At 30 minutes, adsorption equilibrium for Se(IV) and Se(VI) on PAA-MGO reaches 90% and 93%. For the adsorption of selenium on PAA-MGO, acidic pH was beneficial, as maximum adsorption of Se(IV) and Se(VI) on PAA-MGO was achieved at pH 3.1. At alkaline pH, adsorbed selenium on PAA-MGO was stripped off. Ligand exchange between selenium oxyanions and –OH groups of iron oxide on PAA-MGO and electrostatic interactions between selenium oxyanions and amine were considered behind enhanced Se removal by adsorption on PAA-MGO. When solution was within 2 ppb for both Se oxyanions, 99.3% and 99.7% removals of Se(IV) and Se(VI) were obtained by using PAA-MGO [55]. There is another study in which polymer-modified nanocomposite has been used for water decontamination. In this, magnetic graphene oxide coated with SiO_2 was modified with pyrrole–polythiophene copolymer (mGO/SiO_2@coPPy-Th) has been applied for the removal of chromium, copper, zinc, lead, and cadmium from agricultural and water samples. For these metals, limit of detection ranged from 0.15 to 0.65 μg/l under optimized conditions. The mGO/SiO_2@coPPy-Th nanocomposite had maximum adsorption capacity of 98, 201, 125, 230, and 80 mg/g for Cr(III), Cu(II), Zn(II), Pb(II), and Cd(II), respectively. Study shows that at least for five cycles, this nanocomposite can be reused with no significant loss in adsorption property [56]. A homopolymer, PDMDAAC, has been used to functionalize MGO for improved adsorption of Cr(VI) from aqueous solution. Using Langmuir model, the maximum adsorption capacity of PMGO nanocomposite for Cr(VI) was 95.2 mg/g, and Langmuir model fits well to the adsorption isotherm of Cr(VI) adsorption on PMGO. The presence of quaternary amine and its electrostatic interaction with Cr(VI) resulted in higher adsorption capacity that was twice as compared with non-functionalized MGO. Pseudo-first-order kinetic model was followed by the adsorption process of Cr(VI) on PMGO. PMGO was found to be stable, as can be reused for five cycles with no significant loss in adsorption of Cr(VI) [58]. Beads of MGO and chitosan nanocomposite synthesized in a single step when applied for the removal of Co(II) from aqueous solution found the adsorption process to be endothermic. Studies on pH effect for cobalt adsorption show that pH in the range of 5–9 had slight effects, whereas acidic condition was less favorable for adsorption. Langmuir model fits well with the adsorption data of Co(II) on MGO/chitosan composite. Adsorption process followed pseudo-second-order kinetic model, and the maximum adsorption capacity of MGO/chitosan composite for Co(II) was 15.24 mg/g [63]. A novel MGO@PNB nanocomposite is synthesized by immobilizing Fe_3O_4 NPs and poly-(*N*-isopropylacrylamide-*co*-benzo-18-crown-6 acrylamide) microgels (PNB) on GO. This PNB is thermosensitive and selective for Pb(II) binding because it contains 18-crown-6 units, which is a host that selectively binds Pb(II) and forms a stable complex. Adsorption process of Pb(II) on MGO@PNB is endothermic and spontaneous as per thermodynamic studies. Langmuir isotherm model fits well with adsorption isotherm data, and the process followed pseudo-second-order kinetic model. The maximum adsorption capacity for MGO@ PNB toward Pb(II) is 11.76 mg/g [64].

A MGO nanocomposite in which Fe_3O_4/GO was modified using chelating group (EDTA) was used for the removal of U(VI) from aqueous solution. Fast removal of U(VI) was observed, and the maximum adsorption capacity of EDTA-mGO was 277.43 mg/g. The removal of U(VI) by EDTA-mGO was attributed to the presence of metal-chelating group (EDTA) and electrostatic interactions between oxygen-containing functional groups and U(VI). Adsorption process followed pseudo-second-order kinetic model, and this process was found to be spontaneous and endothermic in nature as per thermodynamic data. EDTA-mGO showed high stability and good reusability [57]. New efficient adsorbent material for U(VI) that can be used for the treatment of mine radioactive wastewater was designed using MGO. The polyamidoxime/polyethyleneimine magnetic graphene oxide (mGO-PP) nanocomposite showed 606.06 mg/g as maximum adsorption capacity toward U(VI) at pH 6.0 and 298 K. Adsorption process for U(VI) by mGO-PP fits Langmuir model and pseudo-second-order model well. The adsorption of U(VI) by mGO-PP was monolayer chemosorption, and this process was spontaneous and endothermic in nature. The removal rate of 93.68% was obtained for U(VI) by mGO-PP, and the distribution coefficient (K_d) in real mine radioactive wastewater was 444.73 l/g. Imine and hydroxyl of amidoxime groups in mGO-PP adsorbed U(VI) through complexation. This mGO-PP on reusability found to have 85.6% removal rate after five cycles [65]. To selectively remove cesium (Cs) from Cs-contaminated water, a nanocomposite of Prussian blue and pectin-stabilized magnetic graphene oxide (PSMGPB) was developed. This has good water dispersion that resulted in improved Cs removal. As compared with magnetic Prussian blue, magnetic pectin Prussian blue, and MGO Prussian blue nanocomposites, PSMGPB had improved adsorption capacity for the removal of Cs (1.609 mmol/g). Further pH 7.0 and temperature 30 °C were found optimum for the adsorption of Cs onto PSMGPB. Thermodynamic studies of adsorption revealed the process to be feasible, spontaneous, and exothermic. Cs adsorption process followed pseudo-second-order kinetics, and a study proposed that adsorption of Cs on PSMGPB is chemisorption. Moreover, experimental data of adsorption process fits Langmuir isotherm best [66]. Various types of MGO composite nanomaterials that were discussed above along with few other nanomaterials for the removal of toxic metal contaminants are given in Table 2.1.

Table 2.1 Magnetic graphene oxide and its composite nanomaterials for the removal of toxic metal contaminants.

S. no.	Magnetic composite	Contaminant	Adsorption isotherm	Kinetic model	Adsorption capacity	References
1	Cys-MGO	Cd(II)	Langmuir	Pseudo-second order	24.39–30.30 mg/g at 298–328 K	[51]
2	MGO	As(III) and As(V)	Freundlich	Pseudo-second order	54.18 and 26.76 mg/g	[67]
3	DTPA/MGO	Cu(II), Cd(II), and Pb(II)	Langmuir	Pseudo-second order	131.4, 286.5, and 387.6 mg/g	[54]
4	MGO-NH-SH	Hg(II)	Langmuir	Pseudo-second order	208.8 mg/g	[61]
5	PS@Fe_3O_4@GO	Pb(II)	Langmuir	First order	73.52 mg/g	[62]
6	PAA-MGO	Se(IV) and Se(VI)	Freundlich	—	120.1 and 83.7 mg/g	[55]
7	mGO/SiO_2@coPPy-Th	Cr(III), Cu(II), Zn(II), Pb(II), and Cd(II)	—	—	98, 201, 125, 230, and 80 mg/g	[56]
8	PMGO	Cr(VI)	Langmuir	Pseudo-first order	95.2 mg/g	[58]
9	Magnetic GO/chitosan	Co(II)	Langmuir	Pseudo-second order	15.24 mg/g	[63]
10	MGO@PNB	Pb(II)	Langmuir	Pseudo-second order	11.76 mg/g	[64]
11	EDTA-mGO	U(VI)	Langmuir	Pseudo-second order	277.43 mg/g	[57]
12	mGO-PP	U(VI)	Langmuir	Pseudo-second order	606.06 mg/g	[65]
13	RGO/poly-ethylenimine (PEI)/Fe_3O_4	Cr(VI)	—	—	266.6 mg/g	[68]
14	MCGO	U(VI)	Langmuir	Pseudo-second order	204.1 mg/g	[69]
15	MCGO	Pb(II)	Langmuir	Pseudo-second order	112.35 mg/g	[70]

(Continued)

Table 2.1 (Continued)

S. no.	Magnetic composite	Contaminant	Adsorption isotherm	Kinetic model	Adsorption capacity	References
16	SiO_2@ $CoFe_2O_4$-GO	Cr(VI)	Langmuir	Pseudo-second order	136.40 mg/g	[71]
17	Amino functionalized magnetic graphene oxide (AMGO)	U(VI)	Langmuir	Pseudo-second order	141.2 mg/g	[72]
18	MGO-poly (amidoamine) (PAMAM)-G3.0	Hg(II)	Langmuir	Pseudo-second order	113.71 mg/g	[73]
19	MGO-ionic liquid (IL)	As(III) As(V)	Langmuir Freundlich	Pseudo-second order	160.65 mg/g 104.125 mg/g	[74]
20	ternary hydrosulphonyl (THS)-deep eutectic solvent (DES)@MGO	Hg(II)	Langmuir	—	215.1 mg/g	[75]

2.3.2 Removal of Toxic Organic Contaminants

Fe_3O_4 deposited on sweet graphene nanosheets was used for the removal of methylene blue (MB), a cationic dye from aqueous solution. The sweet component is glucose moieties that have been attached to graphene nanosheets for binding Fe_3O_4 particles strongly. Along with this, the presence of these moieties proved to be advantageous in the removal of MB using this magnetic material. It is reported that 111 and 105 mg/g are maximum adsorption capacities for two different magnetic sweet graphene nanocomposites rGO-N_3-Glc-Fe_3O_4 and rGO-diazide-Glc-Fe_3O_4, respectively. The effect of pH was also studied and found increased adsorption rates with increasing pH of up to 7, whereas after pH 7, no significant change was observed [76]. Sandwiched MGO nanocomposite materials Fe_3O_4@(PAH/GO–COOH)$_1$ and Fe_3O_4@(PAH/GO–COOH)$_2$ were prepared depending upon the time they were treated during layer-by-layer self-assembly process. This material was used for the removal of rhodamine B (RhB) and MB. For the removal of RhB and MB, Fe_3O_4@(PAH/GO–COOH)$_2$ has efficiencies of 22.124 and 35.958 mg/g, respectively. However, efficiencies were 15.252 and 20.995 mg/g for RhB and MB removal by Fe_3O_4@(PAH/GO–COOH)$_1$, respectively. Fe_3O_4@(PAH/GO–COOH)$_2$ was found to have better removal efficiency by adsorption for both RhB and MB dye as compared with Fe_3O_4@(PAH/GO–COOH)$_1$ due to its sandwich structure. The reaction followed pseudo-second-order kinetic model. Reusability studies show, even after eight cycles, Fe_3O_4@(PAH/GO–COOH)$_2$ has 82% removal efficiency for MB dye [52]. Magnetic nanocomposite made from hydrogel of chitosan and graphene oxide embedded with iron oxide NPs (CSGOIO) was found effective for the removal of MB. Adsorption studies revealed electrostatic interactions between CSGOIO and MB because ionic strength and pH of medium influenced the adsorption of MB on CSGOIO. Studies revealed adsorption reaction was following pseudo-second-order kinetic model, and adsorption data fits Freundlich isotherm model well. For the removal of MB, the performance of CSGOIO was fast as the observed rate constant was 0.06 g/mg min, and after four consecutive cycles, from solution having different pH, excellent adsorption property was possessed by CSGOIO [77]. Magnetic graphene oxide composite formed with hexadecyltrimethylammonium bromide (MAGO-CTAB) was found to be fast, sensitive, and efficient for the removal of sunset yellow (SY). Studies show the removal of SY by MAGO-CTAB follows pseudo-second-order kinetic model. The maximum adsorption capacity of MAGO-CTAB for the removal of SY was 44.0 mg/g. Reusability studies show from the first to the fifth cycle, the removal efficiency of MAGO-CTAB was 98.4–95.7%, showing nanocomposite can be used up to five cycles without significant loss in adsorption capacity. While testing the MAGO-CTAB for the removal of SY from different soft drinks and food samples, the recovery values obtained were excellent (98–103%). Adsorption mechanism proposed the removal of SY by MAGO-CTAB via electrostatic interactions and π–π stacking interactions between nanocomposite and SY dye [60].

MGO that is Fe_3O_4/reduced graphene oxide has been used for the removal of five different harmful pesticides such as ametryn, simazine, prometryn, atrazine, and simeton. Adsorption capacity of Fe_3O_4/reduced graphene oxide is strongly related to the hydrophobic, electrostatic, and π–π interactions with pesticides. At 25 °C and pH 5, adsorption efficiencies for ametryn, simazine, prometryn, atrazine, and simeton were found to be 93.61%, 88.55%, 91.34%, 75.24%, and 81.22%, respectively. The reaction followed pseudo-second-order linear kinetic model. Also reported is 54.8 mg/g maximum adsorption capacity at pH 5, which in the presence of various ions (Na^+, Ca^{2+}, Mg^{2+}, and $SO_4{}^{2-}$) was enhanced, and the maximum capacity of 63.7 mg/g in seawater medium was found for ametryn adsorption. Even after seven cycles of reusability, adsorption efficiency was observed to be 88.66% for ametryn [8]. MGO nanocomposite has also been used for the stereoselective removal of isomeric pesticides endrin and dieldrin by adsorption. Conditions found for the best removal of pesticides were 30 mg of nanocomposite having contact time of 30 minutes at pH 4 at room temperature. For dieldrin and endrin, adsorption capacities were found to be 1 and 99 mg/g, respectively, with nanocomposite. The reaction for dieldrin and endrin was found to follow pseudo-second-order model and pseudo-first-order kinetic model, respectively. For the adsorption of pesticides on nanocomposite, a possible way is the interaction of pesticide oxygen entities with hydroxyl groups of nanocomposite. Also, both pesticides showed different adsorption behavior that is due to the endo- position of oxygen atoms in endrin structure and exo- position of oxygen atoms in dieldrin structure [78].

For catalytic reduction of 4-nitrophenol (4-NP), MGO supported with AgPt NPs with different ratios was used in the presence of $NaBH_4$. Fe_3O_4@GO-$Ag_{75}Pt_{25}$ was found to have the best activity for the reduction of 4-NP with rate constant of $140.6 \times 10^{-3}\ s^{-1}$. Moreover, within 30 seconds, reaction was completed. Reusability studies show Fe_3O_4@GO-$Ag_{75}Pt_{25}$ was able to reduce 4-NP for five consecutive runs [53]. MGO nanocomposite has also been used as an adsorbent for the removal of methadone. Using 0.0098 mg of adsorbent at pH 6.2 at 295.7 K was observed to be the ideal condition for the removal of methadone. The maximum concentration of methadone that was removed under observed ideal conditions was 87.23 mg/g, and the adsorption limit shown by Langmuir isotherm was 142.85 mg/g. It is also found that adsorption reaction followed the pseudo-second-order reaction mechanism [79]. For the removal of roxarsone (ROX), by attracting benzene groups and arsenate in ROX simultaneously, a nanocomposite material was designed that was made up of 3D graphene and nano-Fe_3O_4. Results revealed that toward ROX this nanomaterial had higher affinity, adsorption rate, and adsorption capacity. During adsorption and desorption, ROX was not degraded to inorganic arsenic. Adsorption rates of ROX on Fe_3O_4@RGO (reduced graphene oxide) were observed to be above 90% in solution that contains dissolved organic matter and swine manure lixivium. The π–π stacking, hydrogen bonding, and synergistic As–Fe coordination between Fe_3O_4@RGO and ROX attributed toward enhanced removal of ROX [80]. A nanocomposite material having composition as Fe_3O_4@rGO@ZnO whose morphology has been described as onion, further decorated with silver NPs Fe_3O_4@rGO@ZnO/Ag-NPs, was used as catalyst for the degradation of metformin under UV–visible irradiations. From this photocatalytic degradation process, metformin was completely degraded, and mineralization of 20 mg/l metformin was 60% within 60 minutes by using Fe_3O_4@rGO@ZnO/Ag-NPs as catalyst. The generation of $^{\cdot}O_2^-$ and $OH^{\cdot}$ radicals was considered as plausible mechanism for the degradation of metformin. During four successive runs, this catalyst was found to preserve its catalytic property toward metformin degradation [81].

Polydopamine and magnetic graphene oxide nanocomposite (M-GO-PDA) was applied for the sulfadiazine (SDZ) sonocatalytic degradation. Optimized conditions obtained were 0.16 mM of SDZ in the presence of 0.45 mg/l M-GO-PDA at pH 6.8 with ultrasonic power of 300 W, under which 93.1% sonocatalytic degradation was obtained for SDZ within 60 minutes. The chemical oxygen demand (COD) removal efficiency was 79.4%. Within 180 minutes sonocatalysis, the concentration of NH_4^+, SO_4^{2-}, and NO_3^- increased considerably that proved effective mineralization of SDZ [82]. Graphene oxide/iron (III–VI) oxide nanocomposites in combination with ferrate were applied for the removal of RhB and diclofenac (DCF). When using 20 mg/l concentration of GOFeO and 10 μM aqueous solution of RhB, 30% dye removal was obtained in 30 minutes. However, after five minutes, ~50% dye removal was obtained when 2 mM potassium ferrate is added along with GOFeO to an aqueous solution of RhB. Potassium ferrate (12 mM) in addition to GOFeO (625 mg/l) resulted in complete removal of RhB under lesser time (30–60 minutes). Furthermore, removal studies for DCF showed that after adding a mixture of potassium ferrate and GOFeO, the concentration of DCF reduced to 0.03 mg/l from 2.5 mg/l after 30 minutes (99% removal of DCF), whereas the removal was slightly lower than 97% when only GOFeO was used. This revealed the potential of potassium ferrate and GOFeO mixture in the water purification application [59]. Ciprofloxacin (CIP), an antibiotic present in water samples, is difficult to remove. Magnetic chitosan and graphene oxide composite (MCGO) was used as an absorbent for the removal of CIP. The adsorption process of CIP on MCGO followed pseudo-second-order kinetics while both Langmuir and Freundlich models were better at fitting adsorption data. Isotherm results revealed 282.9 mg/g as maximum adsorption capacity of MCGO. In CIP adsorption, ionic strength and pH of the initial solution had a significant effect. CIP adsorption was reduced significantly when Na and Ca are present. π–π electron interaction and electrostatic attraction were the reason behind the fast and efficient CIP removal by adsorption on MCGO. Even after four cycles, MCGO had 72% of the initial adsorption capacity showing MCGO can be used repeatedly [83]. Some of the various types of MGO composite nanomaterials that were discussed above along with few other nanomaterials for the removal of toxic organic contaminants are given in Table 2.2.

Table 2.2 Magnetic graphene oxide and its composite nanomaterials for the removal of toxic organic contaminants.

S. no.	Magnetic composite	Contaminant	Adsorption isotherm	Kinetic model	Adsorption capacity/removal	References
1	rGO-N_3-Glc-Fe_3O_4	MB	—	—	111 mg/g	[76]
	rGO-diazide-Glc-Fe_3O_4	MB	—	—	105 mg/g	
2	Fe_3O_4@(PAH/GO–COOH)$_2$	RhB and MB	—	Pseudo-second order	22.124 and 35.958 mg/g	[52]
	Fe_3O_4@(PAH/GO–COOH)$_1$				15.252 and 20.995 mg/g	
3	CSGOIO	MB	Freundlich	Pseudo-second order	74.93 mg/g	[77]
4	MAGO-CTAB	SY	Langmuir	Pseudo-second order	44.0 mg/g	[60]
5	Fe_3O_4/rGO	Ametryn	Langmuir	Pseudo-second order	54.8 mg/g (pH 5) and 63.7 mg/g (seawater medium)	[8]
6	GO-MNPs	Dieldrin and endrin	Langmuir	Pseudo-second order and pseudo-first order	1 and 99 mg/g	[78]
7	MGO	Methadone	Langmuir	Pseudo-second order	87.23 mg/g	[79]
8	Fe_3O_4@RGO	Roxarsone	Langmuir	Pseudo-second order	454.48 mg/g	[80]
9	M-GO-PDA	Sulfadiazine	—	Pseudo-first order	93.1%	[82]
10	MCGO	Ciprofloxacin (CIP)	Freundlich and Langmuir	Pseudo-second order	282.9 mg/g	[83]
11	MCGO	Methyl orange (MO)	Langmuir	Pseudo-second order	398.08 mg/g	[84]
12	Magnetic chitosan-GO	Bromophenol blue (BPB)	Freundlich	Pseudo-second order	79%	[85]
		Acid red-17 (AR-17)			97%	
13	Fe_3O_4@GO	MB, RhB, and MO	Langmuir	Pseudo-second order	131.10, 34.50, and 39.95 mg/g	[86]

(Continued)

Table 2.2 (Continued)

S. no.	Magnetic composite	Contaminant	Adsorption isotherm	Kinetic model	Adsorption capacity/removal	References
14	MGO	MB and Congo red (CR)	Langmuir	Pseudo-second order	546.45 and 628.93 mg/g	[87]
15	SiO_2@ $CoFe_2O_4$-GO	Acid black 1 (AB 1)	Langmuir	Pseudo-second order	130.74 mg/g	[71]
16	citric acid (CA)-magnetic graphene oxide coated corn straw (mGOCS)	MB	Freundlich	Pseudo-second order	315.5 mg/g	[88]
17	Chitosan (CS)-EDTA-mGO	RhB	Langmuir	Pseudo-second order	1085.3 mg/g	[89]
18	GC/MGO-SO_3H	Ibuprofen	Freundlich	Pseudo-second order	113.27–138.16 mg/g (298–313 K)	[90]
		Tetracycline			473.25–556.28 mg/g (298–313 K)	
19	Fe_3O_4/ persimmon tannin (PT)/GO	Malachite green (MG)	Langmuir	Pseudo-second order	560.58 mg/g (at 308 K and pH 6.0)	[91]

2.3.3 Removal of Other Contaminants

N-doped reduced graphene oxide and Fe_3O_4 nanocomposite (NrGO/Fe_3O_4) were utilized for the removal of phosphate from water under the acidic environment. Toward phosphate, it was found that NrGO/Fe_3O_4 nanocomposite exhibits high specific affinity and high adsorption capacity. Kinetic studies of phosphate adsorption revealed reaction followed pseudo-second-order. Maximum adsorption capacity of phosphate by NrGO/Fe_3O_4 nanocomposite was 169.7 mg/g, and Freundlich isotherm model fits phosphate adsorption better [92]. Nanocomposite composed of Fe_3O_4 and graphene oxide has been utilized for the wastewater treatment by removing and settling down the pollutants such as suspended solid particles in the presence of an external magnetic field. When, to the suspension of kaolinite solid particles, the result of the addition of a small amount of magnetic Fe_3O_4-GO reveals within 30 minutes, removal rate reaches nearly 90%, and little residue remains in the solution. In this process suspended solid particles are attracted by GO while Fe_3O_4 interacts with the applied external magnetic field. This combined effects of the nanocomposite components helps in settling of solid particles much faster than natural settling due to gravity. Langmuir and Freundlich models best fit with the adsorption behavior of Fe_3O_4-GO, and this nanocomposite has the maximum adsorption capacity of 58.46 mg/mg. Kinetic studies revealed that pseudo-second-order kinetic model is followed by adsorption process. Maximum initial adsorption rate observed was 0.0275 mg/mg min. This work shows the potential that magnetic graphene nanocomposite possesses toward the removal of suspended solid particles in the wastewater [93]. New magnetic material (IAO/GO) found effective in the removal of fluoride from aqueous solution had GO anchored with iron–aluminum oxide NPs. Fluoride removal studies after varying main factors showed effective removal in a wide pH range (3–9), and even in the presence of co-anions, IAO/GO had good selectivity for fluoride except for HPO_4^{2-} co-anion. The process of fluoride adsorption by IAO/GO followed pseudo-second-order kinetic model, and adsorption data fits Langmuir model well. As per Langmuir model, the maximum adsorption capacity for IAO/GO was calculated as 64.72 mg/g. Plausible mechanisms for fluoride adsorption on IAO/GO include inner-sphere complexation, electrostatic attraction, and anion exchange [94].

2.4 Conclusion

MGO and its composite nanomaterials are very advantageous for sustainable water treatment by efficient removal of water contaminants. Herein, we have discussed the synthesis and application of MGO and its composite nanomaterials that are applied for environmental decontamination, by utilization of these nanomaterials for removal of toxic contaminants (metal, organic, and other main contaminants) for sustainable water treatment. Important features like nanosize, surface functional groups, and high surface area are key advantages associated with MGO and its composites to exhibit higher adsorption capacities. After adsorption, these materials can be easily separated by the use of an external magnetic field. This magnetic separation method is most effective, and there is no loss of material during separation. The use of different moieties, functional groups, or molecules while synthesizing various nanocomposites of MGO can improve chemical interactions between material and contaminants for significant improvement in adsorption capacities and selectivity. Due to chemical and thermal stabilities and robust structure, these materials can be used for adsorption application at a wide pH range and have good reusability. Various nanocomposite materials synthesized from MGO exhibited better adsorption capacity compared with simple MGO. All properties together have proven that MGO and its composites are the best candidates for sustainable water treatment.

Important Websites

https://www.sciencedirect.com/science/article/pii/B9780128119891000014
http://edugreen.teri.res.in/explore/water/pollu.htm
https://link.springer.com/chapter/10.1007/978-981-15-0671-0_3

https://www.usgs.gov/special-topic/water-science-school/science/wastewater-treatment-water-use?qt-science_center_objects=0#qt-science_center_objects
https://link.springer.com/book/10.1007%2F978-3-319-58136-1#about
https://www.sciencedirect.com/science/article/abs/pii/S1385894716311287
https://books.google.co.in/books?id=4m8NDgAAQBAJ&pg=PA244&lpg=PA244&dq=wastewater+remediation+using+magnetic+graphene+oxide&source=bl&ots=EcON3DiJmB&sig=ACfU3U3IpFOYxNtFvnP4_n3d9MnfQyEbOA&hl=en&sa=X&ved=2ahUKEwjy4dCCy4zrAhVE9XMBHePbCF04KBDoATAFegQICxAB#v=onepage&q=wastewater%20remediation%20using%20magnetic%20graphene%20oxide&f=false
https://link.springer.com/book/10.1007%2F978-3-319-75484-0
https://link.springer.com/chapter/10.1007/978-3-030-05399-4_26

References

1 Lubchenco, J. (1998). Entering the century of the environment: a new social contract for science. *Science* 279 (5350): 491–497.
2 Halder, J. and Islam, N. (2015). Water pollution and its impact on the human health. *J. Environ. Hum.* 2 (1): 36–46.
3 Tangahu, B.V., Sheikh Abdullah, S.R., Basri, H. et al. (2011). A review on heavy metals (As, Pb, and Hg) uptake by plants through phytoremediation. *Int. J. Chem. Eng.* 2011: 1–31. https://doi.org/10.1155/2011/939161
4 Thakur, K. and Kandasubramanian, B. (2019). Graphene and graphene oxide-based composites for removal of organic pollutants: a review. *J. Chem. Eng. Data* 64 (3): 833–867.
5 Khan, M.A. and Ghouri, A.M. (2011). Environmental pollution: its effects on life and its remedies. *Res. World J. Arts Sci. Commerce* 2 (2): 276–285.
6 Srivastava, B., Jhelum, V., Basu, D.D., and Patanjali, P.K. (2009). Adsorbents for pesticide uptake from contaminated water: a review. *J. Sci. Ind. Res. (India)* 68 (10): 839–850.
7 Bandala, E.R. and Raichle, B.W. (2013). *Solar Driven Advanced Oxidation Processes for Water Decontamination and Disinfection*. CRC Press/Balkema.
8 Boruah, P.K., Sharma, B., Hussain, N., and Das, M.R. (2017). Magnetically recoverable Fe_3O_4/graphene nanocomposite towards efficient removal of triazine pesticides from aqueous solution: investigation of the adsorption phenomenon and specific ion effect. *Chemosphere* 168: 1058–1067.
9 Hussain, C.M. and Mishra, A.K. (eds.) (2018). *Nanotechnology in Environmental Science*, vol. 2. Weinheim: Wiley-VCH.
10 Hussain, C.M. (ed.) (2019). *Handbook of Environmental Materials Management*. Springer International Publishing.
11 Hussain, C.M. (ed.) (2018). *Handbook of Nanomaterials for Industrial Applications*. Elsevier.
12 Hussain, C.M. (ed.) (2020). *Handbook of Functionalized Nanomaterials for Industrial Applications*. Elsevier.
13 Hussain, C.M. (ed.) (2020). *Handbook of Nanomaterials for Manufacturing Applications*. Elsevier.

14 Hussain, C.M. and Kecili, R. (2019). *Modern Environmental Analysis Techniques for Pollutants*. Elsevier.

15 Hussain, C.M. and Kharisov, B. (eds.) (2016). *Advanced Environmental Analysis: Applications of Nanomaterials*. Royal Society of Chemistry.

16 Mallakpour, S. and Rashidimoghadam, S. (2018). Carbon nanotubes for dyes removal. In: *Composite Nanoadsorbents* (eds. G.Z. Kyzas and A.C. Mitropoulos) , 211–243. Elsevier.

17 Mallakpour, S. and Khadem, E. (2018). Carbon nanotubes for heavy metals removal. In: *Composite Nanoadsorbents* (eds. G.Z. Kyzas and A.C. Mitropoulos), 181–210. Elsevier.

18 Mallakpour, S. and Khadem, E. (2019). Linear and nonlinear behavior of crosslinked chitosan/N-doped graphene quantum dot nanocomposite films in cadmium cation uptake. *Sci. Total Environ.* 690: 1245–1253.

19 Mallakpour, S. and Hatami, M. (2019). An effective, low-cost and recyclable bio-adsorbent having amino acid intercalated LDH@Fe_3O_4/PVA magnetic nanocomposites for removal of methyl orange from aqueous solution. *Appl. Clay Sci.* 174: 127–137.

20 Liu, X., Wang, M., Zhang, S., and Pan, B. (2013). Application potential of carbon nanotubes in water treatment: a review. *J. Environ. Sci. (China)* 25 (7): 1263–1280.

21 Smith, S.C. and Rodrigues, D.F. (2015). Carbon-based nanomaterials for removal of chemical and biological contaminants from water: a review of mechanisms and applications. *Carbon (NY)* 91: 122–143.

22 Sweetman, M., May, S., Mebberson, N. et al. (2017). Activated carbon, carbon nanotubes and graphene: materials and composites for advanced water purification. *C* 3 (4): 18.

23 Eigler, S. and Hirsch, A. (2014). Chemistry with graphene and graphene oxide – challenges for synthetic chemists. *Angew. Chem. Int. Ed.* 53 (30): 7720–7738.

24 Novoselov, K.S., Geim, A.K., Morozov, S.V. et al. (2004). Electric field in atomically thin carbon films. *Science* 306 (5696): 666–669.

25 Zhu, Y., Murali, S., Cai, W. et al. (2010). Graphene and graphene oxide: synthesis, properties, and applications. *Adv. Mater.* 22 (35): 3906–3924.

26 Smith, A.T., LaChance, A.M., Zeng, S. et al. (2019). Synthesis, properties, and applications of graphene oxide/reduced graphene oxide and their nanocomposites. *Nano Mater. Sci.* 1 (1): 31–47.

27 Subodh, Mogha, N.K., Chaudhary, K. et al. (2018). Fur-imine-functionalized graphene oxide-immobilized copper oxide nanoparticle catalyst for the synthesis of xanthene derivatives. *ACS Omega* 3 (11): 16377–16385.

28 Gholami, P., Khataee, A., Soltani, R.D.C., and Bhatnagar, A. (2019). A review on carbon-based materials for heterogeneous sonocatalysis: fundamentals, properties and applications. *Ultrason. Sonochem.* 58: 104681.

29 Perreault, F., Fonseca De Faria, A., and Elimelech, M. (2015). Environmental applications of graphene-based nanomaterials. *Chem. Soc. Rev.* 44 (16): 5861–5896.

30 Compton, O.C. and Nguyen, S.T. (2010). Graphene oxide, highly reduced graphene oxide, and graphene: versatile building blocks for carbon-based materials. *Small* 6 (6): 711–723.

31 Abdolmaleki, A., Mallakpour, S., and Karshenas, A. (2017). Facile synthesis of glucose-functionalized reduced graphene oxide (GFRGO)/poly(vinyl alcohol) nanocomposites for improving thermal and mechanical properties. *Mater. Sci. Eng., B* 217: 26–35.

32 Mallakpour, S., Abdolmaleki, A., and Borandeh, S. (2014). Covalently functionalized graphene sheets with biocompatible natural amino acids. *Appl. Surf. Sci.* 307: 533–542.

33 Georgakilas, V., Tiwari, J.N., Kemp, K.C. et al. (2016). Noncovalent functionalization of graphene and graphene oxide for energy materials, biosensing, catalytic, and biomedical applications. *Chem. Rev.* 116 (9): 5464–5519.

34 Kumar, G., Mogha, N.K., Kumar, M. et al. (2020). NiO nanocomposites/rGO as a heterogeneous catalyst for imidazole scaffolds with applications in inhibiting the DNA binding activity. *Dalton Trans.* 49 (6): 1963–1974.

35 Chaudhary, K., Kumar, K., Venkatesu, P., and Masram, D. (2020). In depth understanding of nano-bio interface between lysozyme and Au NPs immobilized N-doped reduced graphene oxide 2-D scaffolds. *Nanoscale Adv.* 2 (5): 2146–2159.

36 Wang, Y., Li, Z., Wang, J. et al. (2011). Graphene and graphene oxide: biofunctionalization and applications in biotechnology. *Trends Biotechnol.* 29 (5): 205–212.

37 Chung, C., Kim, Y.K., Shin, D. et al. (2013). Biomedical applications of graphene and graphene oxide. *Acc. Chem. Res.* 46 (10): 2211–2224.

38 Chen, D., Feng, H., and Li, J. (2012). Graphene oxide: preparation, functionalization, and electrochemical applications. *Chem. Rev.* 112 (11): 6027–6053.

39 Mogha, N.K., Sahu, V., Sharma, M. et al. (2016). Biocompatible ZrO_2-reduced graphene oxide immobilized AChE biosensor for chlorpyrifos detection. *Mater. Des.* 111: 312–320.

40 Mogha, N.K., Kirti, S., and Masram, D.T. (2017). La_2O_3/reduced graphene oxide nanocomposite: a highly efficient, reusable heterogeneous catalyst for the synthesis of biologically important bis(indolyl)methanes under solvent free conditions. *J. Nanosci. Nanotechnol.* 17 (4): 2508–2514.

41 Mogha, N.K., Gosain, S., and Masram, D.T. (2017). Gold nanoworms immobilized graphene oxide polymer brush nanohybrid for catalytic degradation studies of organic dyes. *Appl. Surf. Sci.* 396: 1427–1434.

42 Mogha, N.K., Gosain, S., and Masram, D.T. (2020). Lanthanum oxide nanoparticles immobilized reduced graphene oxide polymer brush nanohybrid for environmental vitiation of organic dyes. *Arabian J. Chem.* 13 (1): 1367–1376.

43 Subodh, Chaudhary, K., Prakash, K., and Masram, D.T. (2020). TiO_2 nanoparticles immobilized organo-reduced graphene oxide hybrid nanoreactor for catalytic applications. *Appl. Surf. Sci.* 509: 144902.

44 Kyzas, G.Z., Deliyanni, E.A., and Matis, K.A. (2014). Graphene oxide and its application as an adsorbent for wastewater treatment. *J. Chem. Technol. Biotechnol.* 89 (2): 196–205.

45 Cao, Y. and Li, X. (2014). Adsorption of graphene for the removal of inorganic pollutants in water purification: a review. *Adsorption* 20: 713–727.

46 Indira, T.K. and Lakshmi, P.K. (2010). Magnetic nanoparticles: a review. *Int. J. Pharm. Sci. Nanotechnol.* 3 (3): 1035–1042.

47 Jiang, B., Lian, L., Xing, Y. et al. (2018). Advances of magnetic nanoparticles in environmental application: environmental remediation and (bio)sensors as case studies. *Environ. Sci. Pollut. Res.* 25 (31): 30863–30879.

48 Yao, Y., Miao, S., Liu, S. et al. (2012). Synthesis, characterization, and adsorption properties of magnetic Fe_3O_4@graphene nanocomposite. *Chem. Eng. J.* 184: 326–332.

49 Lingamdinne, L.P., Koduru, J.R., and Karri, R.R. (2019). A comprehensive review of applications of magnetic graphene oxide based nanocomposites for sustainable water purification. *J. Environ. Manage.* 231: 622–634.

50 Li, Y., Chu, J., Qi, J., and Li, X. (2011). An easy and novel approach for the decoration of graphene oxide by Fe_3O_4 nanoparticles. *Appl. Surf. Sci.* 257 (14): 6059–6062.

51 Banazadeh, A., Mozaffari, S., and Osoli, B. (2015). Facile synthesis of cysteine functionalized magnetic graphene oxide nanosheets: application in solid phase extraction of cadmium from environmental sample. *J. Environ. Chem. Eng.* 3 (4): 2801–2808.

52 Guo, R., Jiao, T., Li, R. et al. (2018). Sandwiched Fe_3O_4/carboxylate graphene oxide nanostructures constructed by layer-by-layer assembly for highly efficient and magnetically recyclable dye removal. *ACS Sustainable Chem. Eng.* 6 (1): 1279–1288.

53 Kohantorabi, M. and Gholami, M.R. (2017). AgPt nanoparticles supported on magnetic graphene oxide nanosheets for catalytic reduction of 4-nitrophenol: studies of kinetics and mechanism. *Appl. Organomet. Chem.* 31 (11): e3806.

54 Li, X., Wang, S., Liu, Y. et al. (2017). Adsorption of Cu(II), Pb(II), and Cd(II) ions from acidic aqueous solutions by diethylenetriaminepentaacetic acid-modified magnetic graphene oxide. *J. Chem. Eng. Data* 62 (1): 407–416.

55 Lu, Z., Yu, J., Zeng, H., and Liu, Q. (2017). Polyamine-modified magnetic graphene oxide nanocomposite for enhanced selenium removal. *Sep. Purif. Technol.* 183: 249–257.

56 Molaei, K., Bagheri, H., Asgharinezhad, A.A. et al. (2017). SiO_2-coated magnetic graphene oxide modified with polypyrrole–polythiophene: a novel and efficient nanocomposite for solid phase extraction of trace amounts of heavy metals. *Talanta* 167: 607–616.

57 Zhao, D., Zhang, Q., Xuan, H. et al. (2017). EDTA functionalized Fe_3O_4/graphene oxide for efficient removal of U(VI) from aqueous solutions. *J. Colloid Interface Sci.* 506: 300–307.

58 Li, N., Yue, Q., Gao, B. et al. (2018). Magnetic graphene oxide functionalized by poly dimethyl diallyl ammonium chloride for efficient removal of Cr(VI). *J. Taiwan Inst. Chem. Eng.* 91: 499–506.

59 Mura, S., Jiang, Y., Vassalini, I. et al. (2018). Graphene oxide/iron oxide nanocomposites for water remediation. *ACS Appl. Nano Mater.* 1 (12): 6724–6732.

60 Yakout, A.A. and Mahmoud, M.E. (2018). Fabrication of magnetite-functionalized-graphene oxide and hexadecyltrimethyl ammonium bromide nanocomposite for efficient nanosorption of sunset yellow. *Mater. Sci. Eng., C* 92: 287–296.

61 Kazemi, A., Bahramifar, N., Heydari, A., and Olsen, S.I. (2019). Synthesis and sustainable assessment of thiol-functionalization of magnetic graphene oxide and superparamagnetic $Fe_3O_4@SiO_2$ for Hg(II) removal from aqueous solution and petrochemical wastewater. *J. Taiwan Inst. Chem. Eng.* 95: 78–93.

62 Ravishankar, H., Wang, J., Shu, L., and Jegatheesan, V. (2016). Removal of Pb(II) ions using polymer based graphene oxide magnetic nano-sorbent. *Process Saf. Environ. Prot.* 104: 472–480.

63 Zhuang, S. and Wang, J. (2019). Removal of cobalt ion from aqueous solution using magnetic graphene oxide/chitosan composite. *Environ. Prog. Sustainable Energy* 38 (s1): S32–S41.

64 Pan, L., Zhai, G., Yang, X. et al. (2019). Thermosensitive microgels-decorated magnetic graphene oxides for specific recognition and adsorption of Pb(II) from aqueous solution. *ACS Omega* 4 (2): 3933–3945.

65 Dai, Z., Sun, Y., Zhang, H. et al. (2019). Highly efficient removal of uranium(VI) from wastewater by polyamidoxime/polyethyleneimine magnetic graphene oxide. *J. Chem. Eng. Data* 64 (12): 5797–5805.

66 Kadam, A.A., Jang, J., and Lee, D.S. (2016). Facile synthesis of pectin-stabilized magnetic graphene oxide Prussian blue nanocomposites for selective cesium removal from aqueous solution. *Bioresour. Technol.* 216: 391–398.

67 Yu, F., Sun, S., Ma, J., and Han, S. (2015). Enhanced removal performance of arsenate and arsenite by magnetic graphene oxide with high iron oxide loading. *Phys. Chem. Chem. Phys.* 17 (6): 4388–4397.

68 Wang, D., Zhang, G., Zhou, L. et al. (2017). Synthesis of a multifunctional graphene oxide-based magnetic nanocomposite for efficient removal of Cr(VI). *Langmuir* 33 (28): 7007–7014.

69 Huang, G., Peng, W., and Yang, S. (2018). Synthesis of magnetic chitosan/graphene oxide nanocomposites and its application for U(VI) adsorption from aqueous solution. *J. Radioanal. Nucl. Chem.* 317 (1): 337–344.

70 Samuel, M.S., Shah, S.S., Bhattacharya, J. et al. (2018). Adsorption of Pb(II) from aqueous solution using a magnetic chitosan/graphene oxide composite and its toxicity studies. *Int. J. Biol. Macromol.* 115: 1142–1150.

71 Santhosh, C., Daneshvar, E., Kollu, P. et al. (2017). Magnetic SiO_2@$CoFe_2O_4$ nanoparticles decorated on graphene oxide as efficient adsorbents for the removal of anionic pollutants from water. *Chem. Eng. J.* 322: 472–487.

72 Chen, L., Zhao, D., Chen, S. et al. (2016). One-step fabrication of amino functionalized magnetic graphene oxide composite for uranium(VI) removal. *J. Colloid Interface Sci.* 472: 99–107.

73 Ma, Y.X., Xing, D., Shao, W.J. et al. (2017). Preparation of polyamidoamine dendrimers functionalized magnetic graphene oxide for the adsorption of Hg(II) in aqueous solution. *J. Colloid Interface Sci.* 505: 352–363.

74 Zhang, M., Ma, X., Li, J. et al. (2019). Enhanced removal of As(III) and As(V) from aqueous solution using ionic liquid-modified magnetic graphene oxide. *Chemosphere* 234: 196–203.

75 Chen, J., Wang, Y., Wei, X. et al. (2018). Magnetic solid-phase extraction for the removal of mercury from water with ternary hydrosulphonyl-based deep eutectic solvent modified magnetic graphene oxide. *Talanta* 188: 454–462.

76 Namvari, M. and Namazi, H. (2016). Magnetic sweet graphene nanosheets: preparation, characterization and application in removal of methylene blue. *Int. J. Environ. Sci. Technol.* 13 (2): 599–606.

77 Singh, N., Riyajuddin, S., Ghosh, K. et al. (2019). Chitosan-graphene oxide hydrogels with embedded magnetic iron oxide nanoparticles for dye removal. *ACS Appl. Nano Mater.* 2 (11): 7379–7392.

78 Shrivas, K., Ghosale, A., Nirmalkar, N. et al. (2017). Removal of endrin and dieldrin isomeric pesticides through stereoselective adsorption behavior on the graphene oxide-magnetic nanoparticles. *Environ. Sci. Pollut. Res.* 24 (32): 24980–24988.

79 Kumar Gupta, V., Agarwal, S., Asif, M. et al. (2017). Application of response surface methodology to optimize the adsorption performance of a magnetic graphene oxide nanocomposite adsorbent for removal of methadone from the environment. *J. Colloid Interface Sci.* 497: 193–200.

80 Tian, C., Zhao, J., Zhang, J. et al. (2017). Enhanced removal of roxarsone by Fe_3O_4@3D graphene nanocomposites: synergistic adsorption and mechanism. *Environ. Sci. Nano* 4 (11): 2134–2143.

81 Cheshme Khavar, A.H., Moussavi, G., Mahjoub, A. et al. (2019). Novel magnetic Fe_3O_4@rGO@ZnO onion-like microspheres decorated with Ag nanoparticles for the efficient photocatalytic oxidation of metformin: toxicity evaluation and insights into the mechanisms. *Catal. Sci. Technol.* 9 (20): 5819–5837.

82 Gholami, P., Khataee, A., and Vahid, B. (2020). Integration of polydopamine and Fe_3O_4 nanoparticles with graphene oxide to fabricate an efficient recoverable catalyst for the degradation of sulfadiazine. *Ind. Eng. Chem. Res.* 59 (1): 183–193.

83 Wang, F., Yang, B., Wang, H. et al. (2016). Removal of ciprofloxacin from aqueous solution by a magnetic chitosan grafted graphene oxide composite. *J. Mol. Liq.* 222: 188–194.

84 Jiang, Y., Gong, J.L., Zeng, G.M. et al. (2016). Magnetic chitosan-graphene oxide composite for anti-microbial and dye removal applications. *Int. J. Biol. Macromol.* 82: 702–710.

85 Sohni, S., Gul, K., Ahmad, F. et al. (2018). Highly efficient removal of acid red-17 and bromophenol blue dyes from industrial wastewater using graphene oxide functionalized magnetic chitosan composite. *Polym. Compos.* 39 (9): 3317–3328.

86 Ganesan, V., Louis, C., and Damodaran, S.P. (2018). Graphene oxide-wrapped magnetite nanoclusters: a recyclable functional hybrid for fast and highly efficient removal of organic dyes from wastewater. *J. Environ. Chem. Eng.* 6 (2): 2176–2190.

87 Pu, S., Xue, S., Yang, Z. et al. (2018). In situ co-precipitation preparation of a superparamagnetic graphene oxide/Fe_3O_4 nanocomposite as an adsorbent for wastewater purification: synthesis, characterization, kinetics, and isotherm studies. *Environ. Sci. Pollut. Res.* 25 (18): 17310–17320.

88 Ge, H., Wang, C., Liu, S., and Huang, Z. (2016). Synthesis of citric acid functionalized magnetic graphene oxide coated corn straw for methylene blue adsorption. *Bioresour. Technol.* 221: 419–429.

89 Nekouei Marnani, N. and Shahbazi, A. (2019). A novel environmental-friendly nanobiocomposite synthesis by EDTA and chitosan functionalized magnetic graphene oxide for high removal of rhodamine B: adsorption mechanism and separation property. *Chemosphere* 218: 715–725.

90 Liu, Y., Liu, R., Li, M. et al. (2019). Removal of pharmaceuticals by novel magnetic genipin-crosslinked chitosan/graphene oxide-SO_3H composite. *Carbohydr. Polym.* 220: 141–148.

91 Gao, M., Wang, Z., Yang, C. et al. (2019). Novel magnetic graphene oxide decorated with persimmon tannins for efficient adsorption of malachite green from aqueous solutions. *Colloids Surf., A* 566: 48–57.

92 Akram, M.Y., Ahmed, S., Li, L. et al. (2019). N-doped reduced graphene oxide decorated with Fe_3O_4 composite: stable and magnetically separable adsorbent solution for high performance phosphate removal. *J. Environ. Chem. Eng.* 7 (3): 103137.

93 Wu, Y., Jiang, G., Cano, Z.P. et al. (2018). Highly efficient removal of suspended solid pollutants from wastewater by magnetic Fe_3O_4-graphene oxides nanocomposite. *ChemistrySelect* 3 (41): 11643–11648.

94 Liu, L., Cui, Z., Ma, Q. et al. (2016). One-step synthesis of magnetic iron-aluminum oxide/graphene oxide nanoparticles as a selective adsorbent for fluoride removal from aqueous solution. *RSC Adv.* 6 (13): 10783–10791.

3

Biomass- or Biowaste-Derived Carbon Nanoparticles as Promising Materials for Electrochemical Sensing Applications

Anila R. Cherian[1], Vinay S. Bhat[2], Anitha Varghese[1], and Gurumurthy Hegde[2]

[1]*CHRIST (Deemed to be University), Department of Chemistry, Hosur Road, Bengaluru, 560029, India*
[2]*B.M.S. College of Engineering, Centre for Nano-materials and Displays, Bull Temple Road, Bengaluru, 560019, India*

3.1 Introduction

Technological progression has had a pronounced effect on our lives. Advanced fabrication techniques resulting in miniaturization of devices has impacted all scientific domains, and the field of sensors is no exception. Headways in microelectronics, breakthroughs in computers, and advancement in instrumentation have greatly contributed to the design and development of newer, novel, and efficient sensors tapping onto the principles of physics and chemistry [1, 2]. Sensors are devices capable of detecting a given variant in input energy and converting it into same or different type of energy as output [3]. The sensor contains a recognition element that enables the selective response to a particular analyte or a group of analytes, thus minimizing interferences from other sample components [4]. Numerous methods have been used to classify sensors. Based on the technique adopted to gauge the effect while sensing, sensors can be classified as electrochemical [5, 6], mechanical [7], optical [8–11], thermometric [12], mass sensitive [13], magnetic [14–16], etc. They are also classified according to the application to determine a given analyte [3]. Examples include sensors to quantify biomolecules, environmental pollutants, metal ions, food adulterants, etc.

The use of nanotechnologies in improving the detection sensitivity and miniaturization of the devices in analytical procedures has advanced environmental analysis techniques. These techniques also help to achieve wide linear dynamic range, less interference from closely associated species, and very low detection limits [17]. Furthermore, such nanotechnology-based approaches promise an easier, faster, and cheaper processes in environmental monitoring and remediation [18–20].

Environmental Applications of Carbon Nanomaterials-Based Devices, First Edition.
Edited by Shadpour Mallakpour and Chaudhery Mustansar Hussain.

3.2 Electrochemical Sensors

Electrochemical sensors involve using electrodes to measure the electrochemical changes that occur when chemicals interact with a sensing surface. They recast the information, intrinsic of reaction between analyte and electrodes into qualitative or quantitative signals [21]. Electrochemical sensors contain a recognition element that selectively produces an electrical signal that is related to the concentration of the analyte being studied. The electroactive sensing material on the electrode acts as a catalyst and catalyzes the reaction of the chemical and biochemical molecules to obtain the output signals [4]. Figure 3.1 shows generalized steps involved in electrochemical sensing [22]. Electrochemical sensors can be redesigned for detecting wide range of analytes and also remain inexpensive unlike the other modes of detection, like chromatographic or spectroscopic. Furthermore, they can be integrated into a resilient, portable, miniaturized devices enabling tailor-made applications [23]. Distinct class of electrochemical sensors can be categorized depending on the electrical magnitude used for detection: conductometric (change of conductance), impedimetric (change of impedance), potentiometric (change of membrane potential), voltammetric, and amperometric (change of current with applied voltage) [5]. Operating principles are same for electrochemical sensors, irrespective of the type of sensing.

3.3 The Choice of Electrode Materials

The appearance of nanomaterials in the present-day world has changed the scenario of modern industries, and they created fascinating products [24–26]. This indeed has unlocked new avenues in designing the nanomaterial-modified electrodes for electrochemical sensor applications [27–31]. Newer novel materials that also involve graphene and graphene polymer composites have found to enhance mechanical properties of the materials [32–36]. Generally employed materials for the electrode and supporting substrate are metal/metal oxides, new hybrid and inorganic complexes [37], organic electroconductive polymers or salts [38, 39], metal–organic frameworks (MOFs) [40–45], and carbon-based materials [46, 47]. In this context, efforts have been directed to the modification of electrodes with novel nanomaterials with controlled assembly onto electrode surfaces for improved sensory performance. Among them, carbon-based materials have found widespread application in designing of electrodes used in electroanalytical chemistry because of their relatively wide potential window in aqueous media, low cost, low background current, and relative chemical inertness in most of the electrolyte solutions [4, 48].

A plethora of different forms of carbon-based material are subjected to research for their use in electroanalytical applications [49–52]. To enhance and magnify its sensitivity and selectivity, these materials are engineered with utmost skill to embed defects into them and also to graft functional groups. Exploitations and advancements over electrode modification have taken place over the recent years, owing to the introduction of graphene, carbon nanotubes (CNTs), and other nanoparticles. There was a rampant utilization of these particles that attributed to its high electronic conductivity, good catalytic activity, and very large surface area. Synthesizing such carbon materials generally involves multiple steps associated with expensive precursor feedstock that requires pretreatment [53, 54].

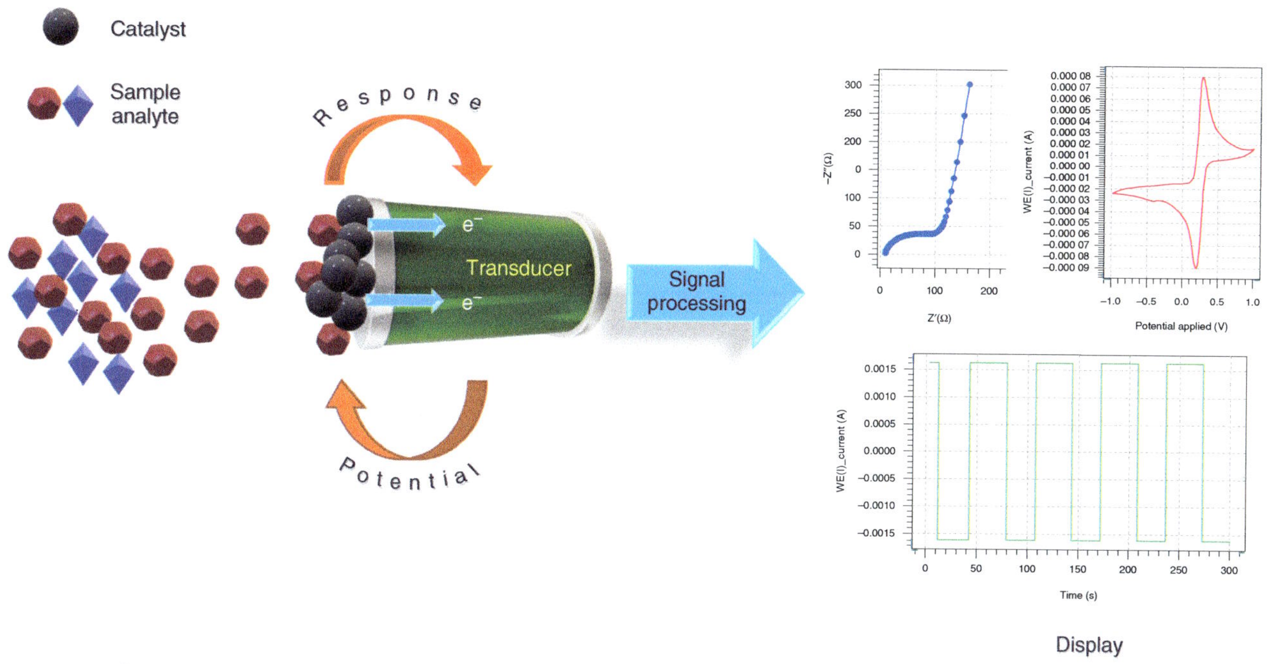

Figure 3.1 General operating principle of electrochemical sensors. Source: Omar et al. [22]. © 2016 Elsevier.

Commercial production of some of the carbon forms like carbon nanofibers, CNTs, and fullerenes involves expensive instrumentation [55–60]. Graphene, having larger surface area than that of single-walled CNTs, involves multiple steps of chemical and mechanical exfoliation, epitaxial growth, and chemical vapor deposition [59–64]. It is also poorly dispersed in aqueous media. To make a stable suspension, additional functionalization to graphene surface using dispersing agents such as amphiphilic polymers, alkylamines, and hydrophilic carboxyl groups becomes necessary [39]. Another major flaw these materials possessed is that they could form close-packed structures after they are assembled on electrode surface that would drastically reduce the surface area and obstruct mass transfer. Hence there was a need to engineer and develop new materials where graphene and CNTs bereft. Renewable lignocellulosic biomass, like agricultural and forest residues, is a promising resource that can be easily converted and processed into carbon materials. With this context, biomass-derived carbon materials for sensors are increasingly becoming popular due to the abundantly available bioprecursors and easier synthesis methods [65–67], which altogether reduces the cost in the development of novel electrochemical sensors.

3.4 Biomass-Derived Porous Carbons

In general, lignocellulosic biomass contains 30–50 wt% cellulose, 20–30 wt% hemicellulose, and 10–20 wt% lignin, the rest being organic components, oil and resins. The concentrations of these vary with biomass [68]. Thermochemical decomposition of biomass leads to non-graphitizable amorphous carbons. Based on the conversion method, the aromatization degree of carbon materials is lower and therefore has a higher reactivity and higher porosity [69]. Porous carbon materials with fairly high edge plane-like defective site density, functional groups containing oxygen, and large surface area showcased notable electrical conductivity and abundant sites for analyte detection. Also the incorporation of heteroatoms (such as N, B, and P) endows upon these structures higher and better sensor performance [70]. In this chapter, we restrict ourselves to some of the biomass-derived porous carbon material-modified electrodes and their application in the electrochemical sensing.

3.4.1 Synthesis

3.4.1.1 Hydrothermal Carbonization (HTC)

Bio-derived carbon materials are synthesized by pyrolysis and hydrothermal carbonization (HTC). HTC is carried at temperatures between 180 and 250 °C in a closed system using water as the reaction medium [71]. The type of carbon products obtained from the pyrolysis and HTC methods is different, owing to different decomposition pathways involved. In HTC, carbohydrates (starch, cellulose, hemicellulose, etc.) are hydrolyzed into simple sugars (glucose, fructose) that are dehydrated into furfurals, hydroxymethylfurfural, and organic acids. At long reaction times, furfurals and hydroxymethylfurfural can polymerize forming long-chained humins and secondary char. Lignin decomposition is slower, involving the cleavage of the β–O–4 linkage [72, 73]. A simplified reaction pathway of glucose/fructose during HTC is depicted in Figure 3.2. After analyzing

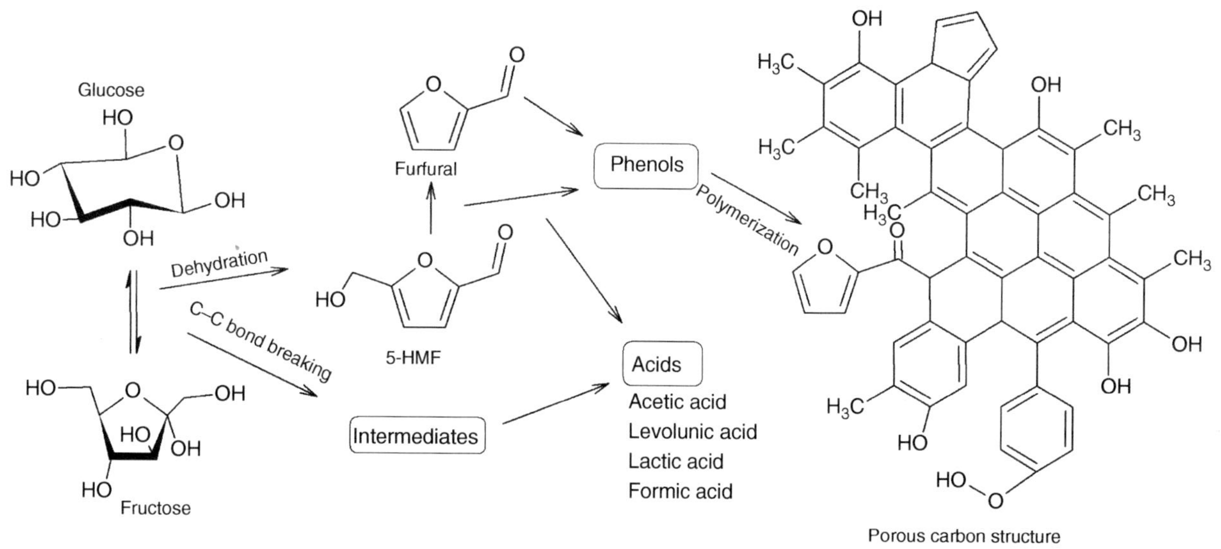

Figure 3.2 A simplified reaction route of biomass conversion during HTC. Source: Sharma et al. [74]. © 2020 Elsevier.

decomposition of cellulose, xylan, and lignin at different HTC temperatures, Kim et al. determined the optimum reaction temperatures for hemicellulose, cellulose, and lignin to be 180, 200, and 220 °C, respectively [75]. Repeated demonstrations also show that, after HTC, alkali metals are strongly leached to process water, whereas Ca, Mg, P, and other heavy metals remain in the solid matrix that may also effect the carbon performance on the targeted applications [68].

3.4.1.2 Pyrolysis

Slow pyrolysis is the most commonly used carbonization technique with temperatures ranging from 200 to 1000 °C [76–79]. During pyrolysis, condensable and non-condensable volatiles and pyrochar are produced by the complete decomposition of all the structural components of biomass. Depolymerization of cellulose leads into formation of low-molecular-weight compounds and levoglucosan [68]. The decomposition of hemicellulose also follows similar pathway as that of cellulose [80]. In contrast, lignin undergoes a physical transformation (softening and swelling) and later decomposes into oligomers, phenols, permanent gases, and char. Also, minerals influence pyrolysis reactions inside particles [68]. Possible pyrolytic carbonization routes of biomass components are represented in Figure 3.3.

Alternatively, carbon nanoparticles can also be synthesized based on templates. Li et al. synthesized mesoporous carbon nanospheres (CNSs) with diameters of c. 1.2 μm. Melamine-formaldehyde resin as carbon precursor and silica as the template were pyrolyzed at 800–1000 °C and treated with hydrogen fluoride (HF) to remove silica template. The obtained CNSs showed low degree of graphitization with high surface area.

3.4.2 Structure and Properties

Irrespective of the synthesis methods, most biomass-derived carbons present a poor graphitization degree characterized mainly by X-ray diffraction (XRD) and Raman spectroscopy. Figure 3.4 shows representative XRD and Raman analysis images associated with biomass-derived carbons. XRD patterns are characterized by a main diffraction peak at 25° corresponding to 002 graphite plane and a secondary peak at c. 44°, which could be assigned to the 100 graphitic planes (JCPDS-ICDD card no. 41-1487). The broadness of the peak indicates a lesser long-range structural order. Raman spectra presents two peaks at Raman shifts of c. 1355 and 1588 cm^{-1} commonly denoted as D-band and G-band, respectively. The D-band is associated with disorder in the carbon structure, while the G-band is associated with E_{2g} mode of graphite and related to vibrations of sp^2-bonded carbon atoms in two-dimensional hexagonal lattice similar to a graphitic layer indicating structural order. The relative intensities of D- and G-bands (I_D/I_G) are used as a tool to assess graphitic nature. Together with XRD, Raman analysis provides information regarding carbon structural order.

The porous nature of the biomass-derived carbons is generally assessed from N_2 adsorption desorption isotherms using Brunauer–Emmett–Teller (BET) method. According to the IUPAC classification, micropores have pore diameters less than 2 nm, and mesopores have pore diameters more than 2 nm. Pores with diameter more than 50 nm are called macropores. The presence of micropores imparts higher surface area, while the presence of

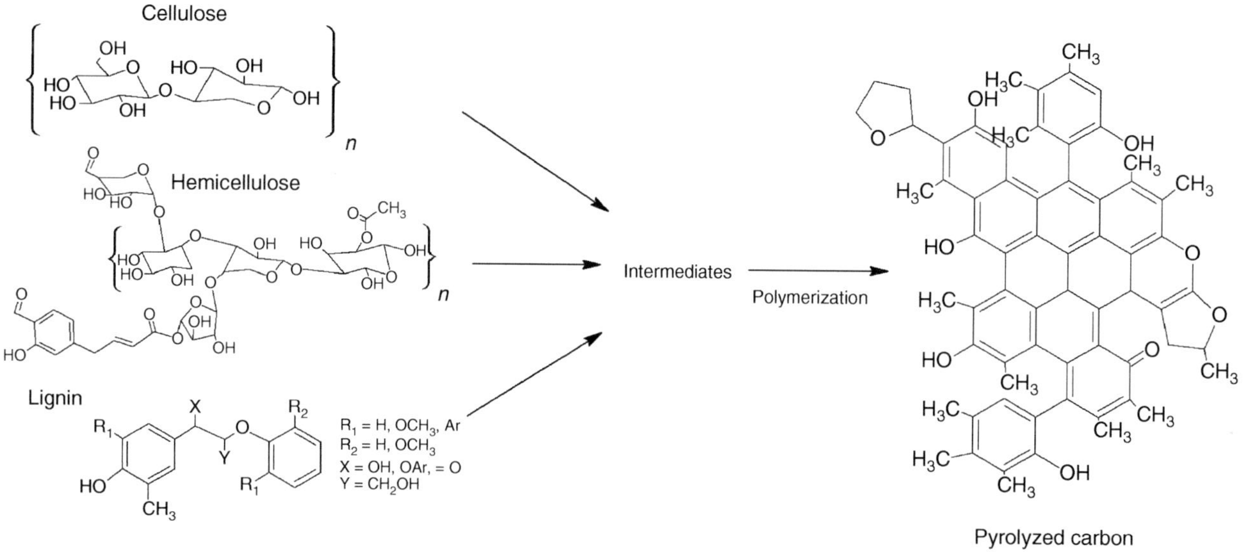

Figure 3.3 A simplified plausible pathway of biomass conversion during pyrolysis. Source: Deng et al. [81]. © 2016 American Chemical Society.

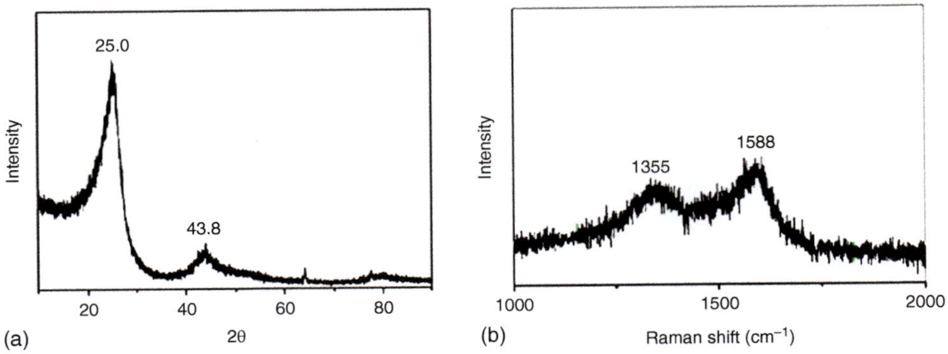

Figure 3.4 (a) A representative XRD pattern and (b) Raman spectrum associated with biomass-derived carbon materials. Source: Jin et al. [82]. © 2005 Elsevier.

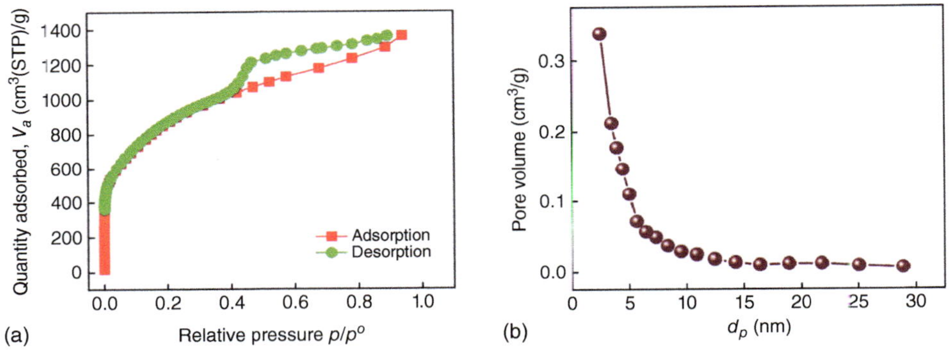

Figure 3.5 (a) A representative N_2 adsorption desorption isotherm obtained for onion peel-derived CNSs. (b) Pore size distribution of CNSs. Source: Supriya et al. [76]. © 2019 Springer Nature.

meso- and macropores provides for easier access to the electrolytes/analytes onto the surface. A representative N_2 adsorption desorption isotherm of onion peel-derived carbons done by Supriya et al. is provided in Figure 3.5 with pore size distribution calculated from Barrett, Joyner, and Halenda (BJH) method. BET surface area was found to be as high as 2961.9 m^2/g. Total pore volume was found to be 2.1 cm^3/g, and a mean pore diameter was 2.8 nm.

Supriya et al. demonstrated the effect of temperature on the transformation of biowaste to sustainable, porous CNSs. The biowaste was pyrolyzed at different consecutive temperatures from 200 to 1000 °C. Representative image showing morphology (scanning electron microscope [SEM] and transmission electron microscope [TEM]) of biomass-derived porous carbon materials at different temperatures is shown in Figure 3.6. The images confirm the formation of CNSs with uniform shape at higher temperatures. Also, the size of the particles decreased from 115 to 63 nm as the pyrolysis temperature was increased. This has been attributed to dissociation of solid clusters forming well-defined structures at higher temperatures.

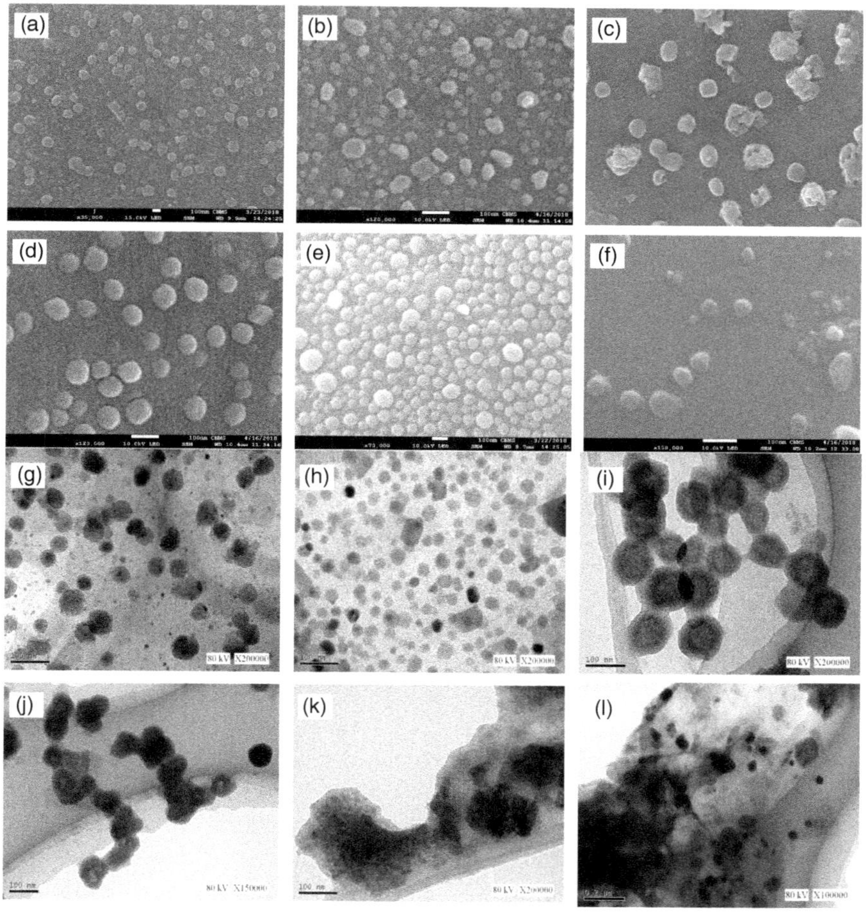

Figure 3.6 FESEM (a–f) and TEM (g–l) images of carbon nanospheres (ON5–ON10) obtained by pyrolysis of onion peel. Source: Reproduced with permission from reference Supriya et al. [76].

3.5 Biomass-Derived Carbons in Electrochemical Sensing

3.5.1 H_2O_2 Sensing from Okra-Derived Carbons

Okra (*Abelmoschus esculentus*), popularly known to common man as lady's finger, is an extensively consumed vegetable in the subtropical and tropical regions due to its excellent palatability [83]. Not only did okra enter the kitchen of every household, but also it managed to win over the hearts of electrochemists as well due to its richness in bioactive polysaccharides. Employment of green synthesis to engineer and develop heteroatom-doped porous carbon has garnered a lot of attention, mostly because of it being environmentally benign but also remarkably enhancing their performance and the synergistic effect contributed by the porosity and tuning structures of the heteroatoms. Zhang and workers had prepared N,P Co-doped carbon material pomelo [84] and cattle bones [85] as precursor.

However, nitrogen and phosphorous sources were additionally introduced in these works. Sen Liu et al. were successful in fabricating one such heteroatom material from mere thermal treatment of freeze-dried okra without the utilization of any chemicals and thereby creating nitrogen and phosphorous co-doped hierarchical micro/mesoporous (N,P-MMC) carbon material for non-enzymatic electrochemical H_2O_2 sensing (Figure 3.7) [86].

The freeze-dried okra was heat treated at 800 °C for three hours. A black carbon residue was obtained. The morphology of the carbon material was established from scanning transmission electron microscopy (STEM). The energy-dispersive X-ray spectroscopy (EDX) revealed that the N,P-MMC is consisting of C, N, P, and O elements. Figure 3.7 shows the STEM and elemental (C, N, P and O elements) mapped images of N, P-MMC. X-ray photoelectron spectroscopy (XPS) was utilized to examine the surface functionalities of carbon material. The C1s spectrum of carbon material (Figure 3.8a) exhibited four peaks centered at 284.6, 285.5, 286.0, and 286.6, which were attributed to the C=C/C—C, C=P, C=N, and C=O bonds, respectively [87]. The presence of elements other than carbon is derived from proteins and cellulose in okra, which is the same as the results of EDX. The as-synthesized material was used for the electrochemical detection of H_2O_2, and the sensor was constructed by drop-casting the porous carbon dispersion on glassy carbon electrode (GCE) with the aid of Nafion (binding agent).

Cyclic voltammetric studies was carried out to probe the reactions in 0.1 M PBS (phosphate buffer solution) at pH 7.0 in the presence and absence of 1.0 mM H_2O_2. Notably from the voltammograms shown in Figure 3.8b, no reduction peak was observed for bare GCE and N,P-MMC/GCE in the absence of H_2O_2. However, a strong reduction peak at −0.60 V for N,P-MMC/GCE was observed after the addition of 1 mM H_2O_2 for the electrochemical reduction of the same accounting for good electrochemical activity of N,P-MMC.

Coming to the amperometric investigation, the response of H_2O_2 at N,P-MMC/GCE was carried out in a N_2-saturated 0.1 M PBS 7.0 buffer at the applied potential of −0.60 V by adding H_2O_2. The steady-state response of the biomass-derived modified electrode to the successive injection of H_2O_2 is shown in Figure 3.8c. The linear detection range is from 0.1 to 10 mM, and the detection limit is estimated to be 6.8 μM. Additionally anti-interference study was also carried out to validate the strong selectivity and electrochemical catalytic activity of N,P-MMC. From the current–time curve as shown in Figure 3.8d, the addition of H_2O_2 leads to a strong current response, whereas no perceptible and apparent current response was noticed post the addition of interference, which included ascorbic acid (AA), citric acid (CA), glucose, and ethanol. In brief, one can draw a conclusion that the introduction of these heteroatoms has by and large multiplied the conductivity than it would have been for a bare mesoporous carbon material. This particular work not only showcases an effective method for the synthesis of heteroatom-doped porous carbon but also paves the way for the engineering and fabrication of novel non-enzymatic H_2O_2 sensors.

3.5.2 Acetaminophen (AM) Detection by Seaweed-Derived Carbons

Nowadays, researchers have gone ahead and are motivated to procure and utilize ocean resources as a precursor to convert them into porous carbon material. As an abundant seaweed material, kelp is one such material [88]. It is very easy to procure kelp in high quantity because of its cost efficiency and also because of its abundance in the oceans around the

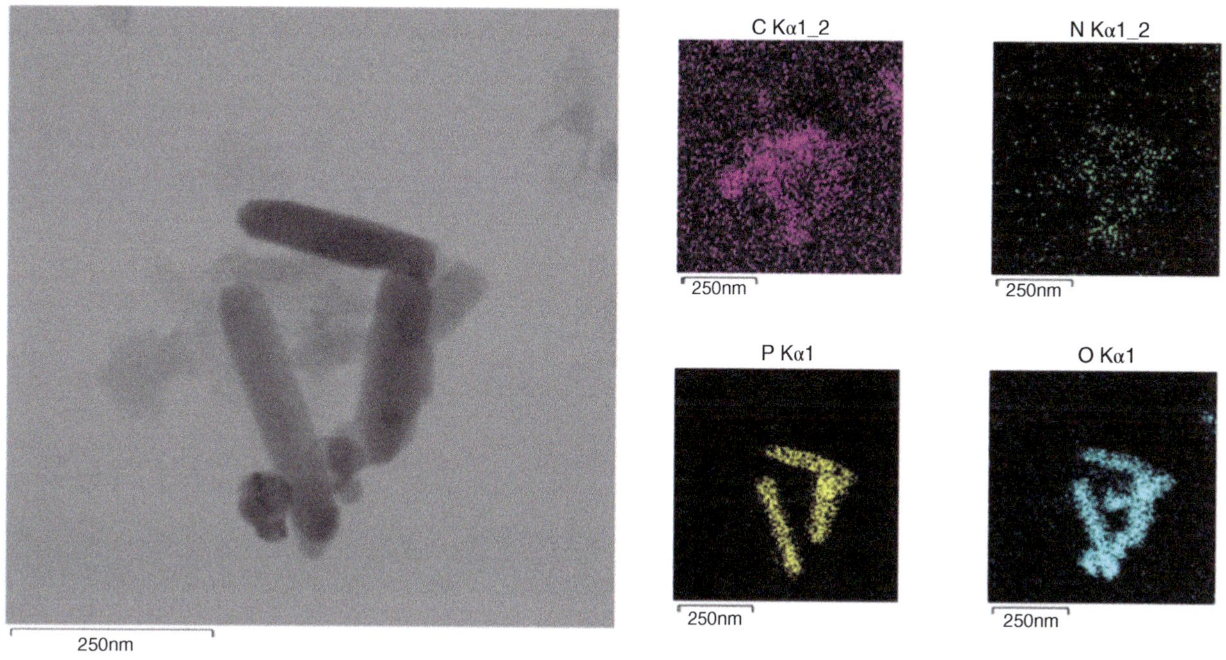

Figure 3.7 STEM image and the corresponding EDX images for C, N, P, O elements of N,P-MMC. Source: Rreproduced with permission from reference Liu et al. [86].

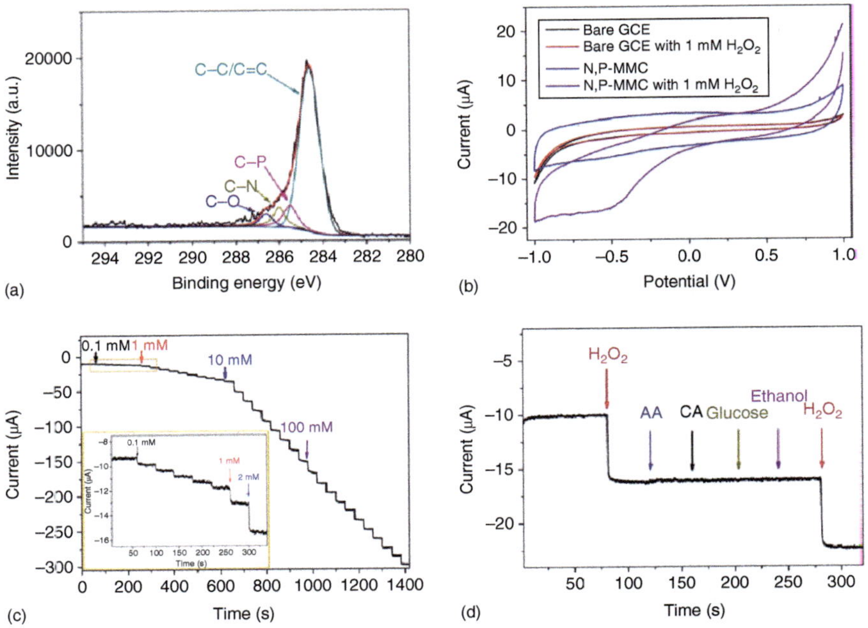

Figure 3.8 (a) C1s spectra of N,P-MMC sample. (b) Cyclic voltammetry (CVs) of a bare GCE and N,P-MMC/GCE in N_2 saturated 0.1 M PBS at pH 7.0 in the presence and absence of 1.0 mM H_2O_2 (scan rate of 0.05 V/s). (c) Typical steady-state response of the N,P-MMC/GCE to successive injection of H_2O_2 into the N_2-saturated PBS under stirring applied potential of −0.60 V and supporting electrolyte of 0.1 M PBS of pH 7.0. (d) Current–time curve for the N,P-MMC/GCE exposed to H_2O_2, AA, CA, glucose, and ethanol. Source: Liu et al. [86]. © 2018 John Wiley & Sons.

world, thereby making it an excellent and promising candidate for biomass-derived carbon material as a precursor for activated carbon. Monitoring and analyzing the concentration of therapeutic drugs in the human body is very relevant due to the drug safety and overall health of the people in the society. Acetaminophen (AM), popularly known at paracetamol, is one such drug, casually popped by people because of its pain-relieving, anti-inflammatory, and antipyretic effects. However, overconsumption of AM can tamper with the health of the humans and cause hepatoxicity [73].

Electrochemical method of sensing has bypassed the earlier conventional and expensive methods of detection of these compounds. Dongwon Kim et al. were successful in constructing a kelp powder-derived modified electrode for the sensitive determination of acetaminophen [89]. The electrode material for the sensing and determination of acetaminophen was prepared by employing not one but rather a two-step activation process. Under inert conditions of Ar atmosphere and 700 °C, kelp powder was carbonized, and the obtained kelp carbon (KC) was mixed with $ZnCl_2$ followed by pyrolysis at 700 °C. The obtained product was labeled (ZAKC), and this was subjected to second activation that involved the mixing of 1 equiv of ZAKC with one equivalent of KOH and pyrolyzed. The obtained product (ZKAKC) was used as the modified electrode material. ZKAKC slurry was drop-casted on GCE, and the electrochemical behavior of the bare and the electrode was tested in 5.0 mM $K_3Fe(CN)_6$ in 0.1 KCl solution. Classic redox-modified peaks were obtained at all electrodes as shown in Figure 3.9a. The effective surface area for each of the electrodes was calculated using the Randle Sevick's equation (3.1):

$$I_p = 2.69 \times 10^5 \times A \surd D \times n^{3/2} \times \surd vC \tag{3.1}$$

where A is the electrochemical active surface are (cm^2), D stands for diffusion coefficient of the molecule in the bulk solution, n is the number of electrons that have taken in the reaction, v is the scan rate (V/s), and C stands for the concentration of the bulk solution. The calculated active surface areas for ZAKC/GCE, KAKC/GCE, and ZKAKC/GCE were 0.088, 0.098, and 0.122 cm^2, respectively. It was observed that two-step activation significantly magnified the electroactive surface area. By employing cyclic voltammetry (CV) experiments, the performances of the bare and modified electrodes were examined for 0.3 mM AC in PBS 7.4. Negative potential shift from 0.43 V for bare to 0.37 V for ZKAKC/GCE was seen, and in Figure 3.9b, we can infer that with high current intensity, the modified electrode can bring about electrochemical oxidation better than GCE [9]. This is very much credible and can be attributed and owed to greater surface-area-to-volume ratio offered by ZKAKC/GCE. Anti-interference study of the modified electrode toward AC was tested along with AA and dopamine of concentration 0.1 mM and with 0.3 mM AC in 0.1 M PBS. However, AC portrayed a very notably high current as observed in Figure 3.9c exhibiting excellent electrochemical performance. To find out the detection limit, differential pulse voltammetry (DPV) technique with higher current sensitivity in comparison with CV was employed. Limit of detection (LOD) was calculated to be 0.004 μM.

3.5.3 4-Nitrophenol Detection from Mango Leave-Derived Carbons

Mango leaves (*Mangifera indica*) has been a subject of scrutiny and a focus of intense research due to its medicinal value in the treatment of various kinds of diseases. Particularly the leaves gained a lot of attention because the phytochemical screening of the leaves revealed the presence of phenols, saponins, tannins, steroids, flavonoids, anthraquinone, and glycosides [90]. The presence of these functional groups aides and enhances the stability of prepared carbon material and also can provide organic functionality of bioconjugation. The toxicity and the irreversible damage caused by 4-nitrophenol (4-NP) top the list of anthropogenic toxicants. 4-NP serves its purpose as an ingredient in pesticides and fertilizers, but little did the users realize that this would be a bane to the human health, and its damage can be irrevocable. Therefore, there was a need to adopt techniques to detect the same in environmental samples; subsequently electrochemical method of detection gained momentum due its advantages over the conventional techniques such as high-performance liquid chromatography (HPLC), liquid chromatography, fluorescence, and spectrophotometry.

Rajesh Madhu et al. took to developing a working electrode material by employing spherical carbon nanoparticle-decorated activated carbon (AC) materials from mango leaves (*M. indica*) by simple pyrolysis method at different temperatures (700–900 °C) [91]. Thus fabricated modified electrode was coded as ACx (x is the carbonization temperature ranging from 700 to 900 °C). To determine the electrocatalytic behavior of 4-NP at the modified electrode, AC was drop-casted on GCE. Figure 3.10 depicts the cyclic voltammograms

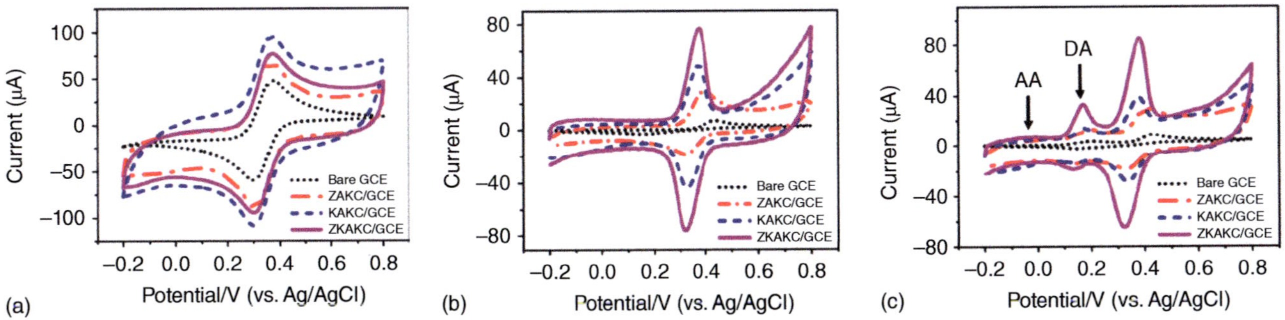

Figure 3.9 (a) CV curves of 5 mM $K_3Fe(CN)_6$ in 1 M KCl using bare and modified electrodes. (b) CV curves of 0.3 mM AC in 0.1 M PBS (pH = 7.4) using bare and the other modified electrodes. (c) CV curves of 0.3 mM AC in 0.1 M PBS with the presence of 0.1 mM of AA and 0.1 mM of DA using various modified electrodes. Source: Kim et al. [89]. © 2018 Elsevier.

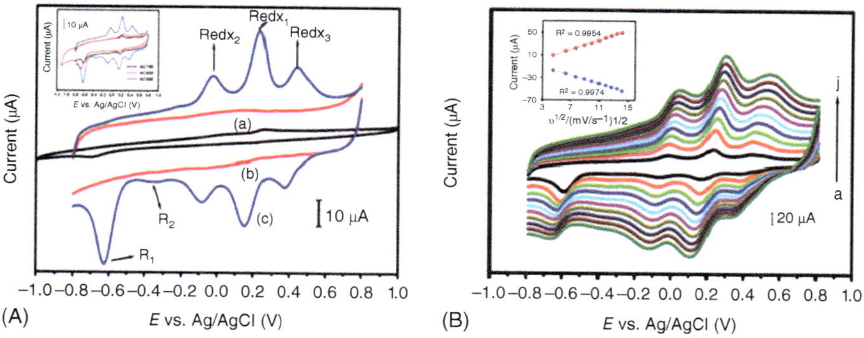

Figure 3.10 CV curves of (A) bare GCE and (c) AC-modified GCE in the presence of 200 μM 4-NP in 0.05 M acetate buffer solution (pH 5.0). Curve b is the profile in the absence of the analyte in the electrolyte solution. (B) CV curves of AC900-modified GCE in the presence of 200 μM 4-NP in 0.05 M buffer solution (pH 5.0) recorded at different scan rates ranging from 20 to 200 mV/s (curves a–j). Inset: Correlations between peak currents and square root of scan rate. Source: Madhu et al. [91]. © 2014 Royal Society of Chemistry.

for the bare and modified electrode with and without the addition and presence of 4-NP in acetate buffer. The CV curves of (a) and (c) correspond to bare GCE and AC-modified GCE in the presence 200 μM 4-NP in 0.05 M acetate buffer solution, and curve (b) is the voltammogram recorded in the absence of 4-NP. The inset in Figure 3.10A corresponds to the CV profiles obtained from different modified electrodes (AC700, AC800, AC900).

It can be clearly observed that three redox peaks ($Redx_1$, $Redx_2$, and $Redx_3$) along with two additional reduction peaks (R_1 and R_2) were clearly observed for the modified electrodes. The redox peaks reveal about reversible two-electron transfer oxidation–reduction reaction of 4-NP, and the additional reduction peaks indicate the reduction of the nitro group of 4-NP to hydroxylamine. From the inset in Figure 3.10B (with $R_2 = 0.99$), clearly the modified electrode with the highest carbonization temperature exhibited the best performance. This can be owed and attributed to the enhanced porosities and surface area; moreover the presence of oxygen surface functional groups and heteroatoms contributed to the excellent catalytic activity and reversible redox behavior. The CV curves of the effect of sweep rate on the modified electrode (AC900) are shown in Figure 3.10.

The effect of sweep rate on the AC900-modified electrode, with scan rates ranging from 20 to 200 mV, was recorded in the presence of 200 μM 4-NP in acetate buffer solution of pH 5. A diffusion-controlled process for the electrochemical reaction of 4-NP over AC-modified electrode was confirmed from the linear dependence between the $Redx_1$ and the square root of the sweep rate [12] (inset, Figure 3.10B). Linear sweep voltammetry (LSV) was further conducted to evaluate the detection limit and linear range (shown in Table 3.1). The practicability of the fabricated electrode was also tested for the detection of 4-NP in tap water and orange juice, and a fairly good electrochemical response was observed in both cases.

3.5.4 Bisphenol-A (BPA) Detection Using Bamboo Fungi-Derived Carbon

The synthesis of doped carbon material is not a green-friendly approach as they require expensive chemicals, and also the preparatory steps involved in the same are a laborious

Table 3.1 Some parameters deduced via electrochemical measurements for biomass-derived modified working electrodes.

Biomass	Electrode material	Analyte	Linear range (μM)	Limit (μM)	Sensitivity (μA/μM cm^2)	References
Okra	N,P-MMC	H_2O_2	100 to 10^4 20×10^3 to 200×10^3	6.8	—	[86]
Kelp	ZKAKC	Acetaminophen	0.01–20	0.004	—	[89]
Mango leaves	AC-modified GCE	4-nitropenol	1–500	0.16	5.810	[91]
Bamboo fungus	NDC/GCE	Bisphenol A	1–50	1.068	—	[92]
Areca nut	CNSs	Nitrite	0.2–400	0.015	28.91	[93]
Bougainvillea spectabilis	GPAC	Catechin	4–368	0.67	7.2	[94]
Onion peel	CNSs	Progesterone	39.39×10^{-5} to 25.0×10^{-5}	1.2×10^{-5}	19 590	[95]
Oil palm leaves	OPL-CNPs	Butein	10–100	7.6	—	[96]

process. This led to the sustainable making of heteroatom-doped material from biomass. Decades together, bamboo fungus has been considered as a delicacy and constitutes as a primary cuisine in China. However, what drew the attention of scientist is its enrichment in heteroatom nitrogen because of which it served as an easily available renewable source to functionalize carbon. In one such work, bamboo fungus-derived heteroatom-doped carbon material was synthesized for detection of bisphenol-A (BPA).

BPA is the most exhaustively utilized raw material and precursor on the industrial production of epoxy, unsaturated polyester resins, flame retardants, and polycarbonate. However, overtime exposure to (BPA) poses threat as endocrine disruptor chemical [97]. Therefore, there is a need to quantify the same to curb and clamp down its risk, and BPA being electrochemically active electrochemical sensing and determination of the same is undoubtedly surpasses all the other methods in the on-site determination. Though different materials like metal composites and conducting polymers were used, Yujuan Xu et al. developed a sustainable and environmentally benign biomass-derived porous nitrogen-doped carbon (NDC) material as a platform for the electrochemical sensing of BPA [92].

NDC material was prepared by simple hydrothermal and calcination process wherein bamboo fungi along with deionized water were taken in an autoclave and heated under 180 °C for 24 hours. The precipitate-obtained post-centrifugation was annealed in the presence of $ZnCl_2$, and the hence prepared material served as the modified working electrode material. This material upon drop-casting on GCE became the modified working electrode NDC/GCE. To gauge the charge transfer property of NDC/GCE, $[Fe(CN)_6]^{3-/4-}$ redox probe was utilized. In Figure 3.11A, we can see that the redox peak current of NDC/GCE is nearly twice than that of bare GCE, signifying that NDC can accelerate the charge transfer property of the electrode. This is attributed to the enhanced conductivity and surface area of NDC. To further verify the electron transfer properties of the bare and modified electrode, electrochemical impedance spectroscopy (EIS) was carried out in 0.1 M KCl in the presence of 5 mM $[Fe(CN)_6]^{3-/4-}$. The diameter of semicircle of the Nyquist plot in Figure 3.11B is directly related to the electron transfer resistance (R_{et}); the smaller diameter of the NDC/GCE resulted in the hastening of the electron transfer process between the electrode surface and the electrolyte.

Figure 3.11C depicts the CV profiles of 20 μM BPA at GCE (curve a) and NDC/GCE (curve b) in 0.1 M PBS (pH 8.0). Besides, the CV of NDC/GCE in 0.1 M PBS (pH 8.0) without BPA is presented as curve c (scan rate: 100 mV/s). Distinctively, the lack of the oxidation peak of curve c confirms that the peak at 0.44 V (a and b) is attributed to the oxidation of BPA [15]. DPV method was for the quantity determination of BPA at NDC/GCE. In Figure 3.11D, we can see that as the concentration of BPA increased, the anodic peak current increased linearly. The concentration was changed from 0.5 to 30 μM of BPA in 0.1 M PBS buffer (pH 8.0). The linear relationship between the anodic peak current and the BPA concentration in the range 1–50 μM is shown in Figure 3.11e.

3.5.5 Nitrite Ion Detection by Areca Nut-Derived Carbons

CNSs, a novel carbonaceous material, find its usage in many fields due to the diverse properties and advantages it had to offer. Areca nut, a very popular masticatory, is consumed by millions of people across the globe. It is the seed of areca palm fruit grown in most of the

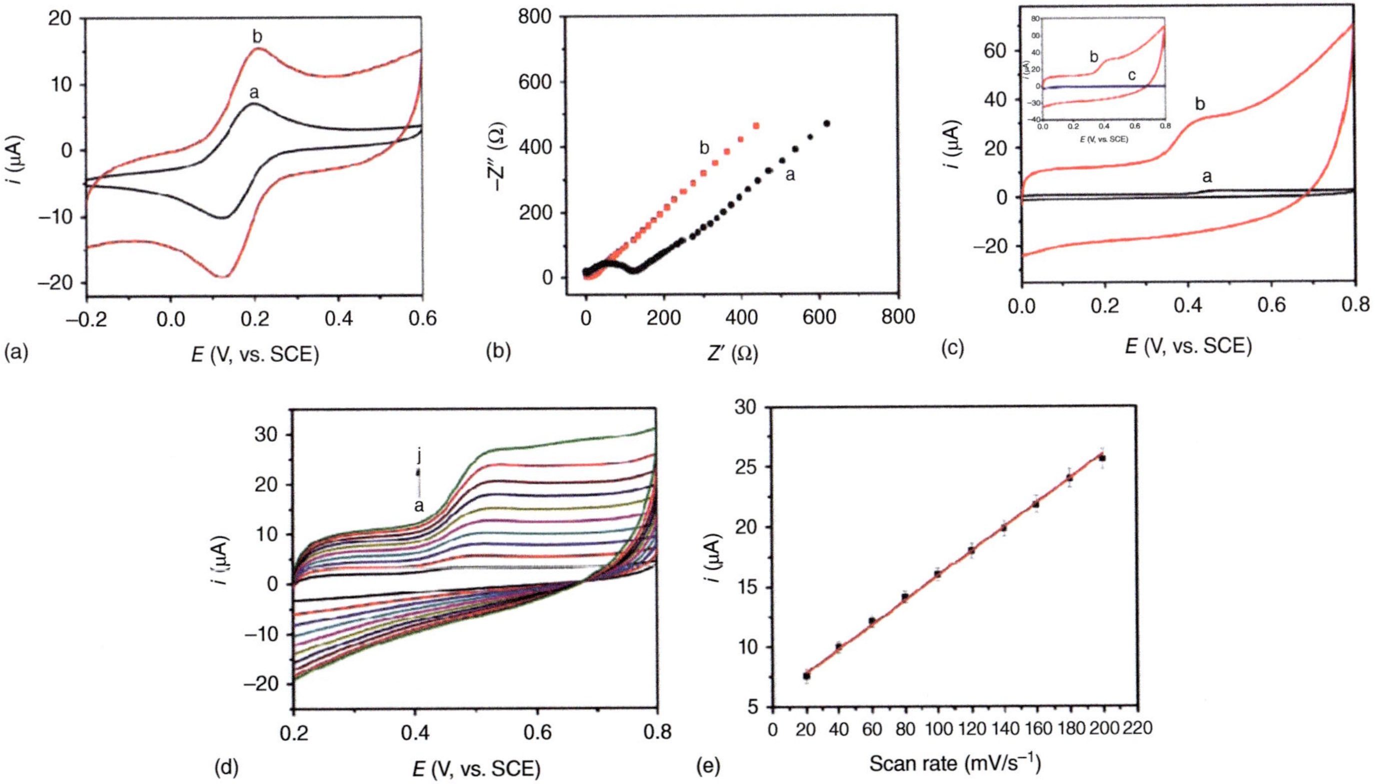

Figure 3.11 (A) CVs of the bare (curve a) and NDC/GCE (curve b) in 0.1 M KCl containing 0.5 mM $[Fe(CN)_6]^{3-/4-}$ (scan rate of 100 mV/s). (B) EIS of bare GCE (curve a) and NDC/GCE (curve b) in 0.1 M KCl with the presence of 5 mM $[Fe(CN)_6]^{3-/4-}$. (C) CVs of 20 μM BPA at GCE (curve a) and NDC/GCE (curve b) in 0.1 M PBS (pH 8.0) and CV of NDC/GCE (curve c) in 0.1 M PBS (pH 8.0) without BPA. Scan rate is 100 mV/s. (D) DPV curves of BPA at NDC/GCE in 0.1 M PBS (pH 8.0) with different concentrations. (E) The linear relationship between the anodic peak current and the BPA concentration in the range of 1–50 μM. Source: Xu et al. [92]. Licensed under CC BY 4.0.

tropical areas of Asia and Africa. Conversion of vast amount of areca nut waste into materials with practical application is meaningful and productive, considering the fact that it is economically and environmentally feasible. Previously, areca nut was carbonized thermally [98] to produce porous carbon as adsorbent for methylene blue dye, but its role as an electroactive material for sensing remained unexplored. Being used in fertilizers and food packaging, nitrite is an inorganic pollutant that is hazardous to the environment and poses threat to the human life. Bare GCE cannot be directly used for the nitrite ion sensing and due to its high overpotential for nitrite ion oxidation. Mesoporous CNSs were synthesized from carbon-rich areca nut seeds for nitrite ion detection [93]. CNSs were prepared by one-step pyrolysis at 550 °C for one hour under continuous N_2 gas flow. To conduct electrochemical experiments, the modified working electrode sensing platform was fabricated by drop-casting the CNS slurry on GCE and air-drying the same. Figure 3.12A depicts the cyclic voltammograms of CNSs/GCE and bare glassy carbon electrode (BGCE) electrodes at a scan rate of 50 mV/s in the potential range from −1.0 to +1.0 V in 0.1 M PBS (pH = 7.0) and 3 mM $NaNO_2$ solution. BGCE (curve a) and CNSs/GCE (curve b) did not show any response in the buffer solution. Upon the addition of 3 mM sodium nitrite, BGCE (curve c) showed an increase in the currents but absence of a peak potential in the potential range

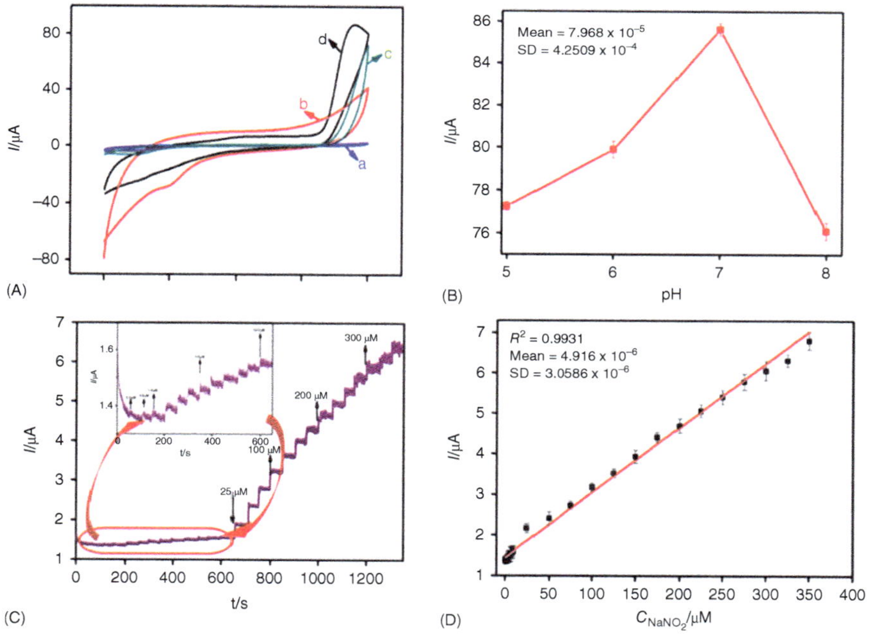

Figure 3.12 (A) Cyclic voltammograms for (a) BGCE in 0.1 M PBS (pH = 7), (b) CNSs/GCE 0.1 M PBS (pH = 7), (c) BGCE in 3 mM $NaNO_2$, and (d) CNSs/GCE in 3 mM $NaNO_2$. (B) Effect of buffer pH on the oxidation currents of 3 mM $NaNO_2$ at CNSs/GCE in 0.1 M PBS obtained from CVs. (C) The amperometric current to time response of CNSs/GCE with successive addition of different concentration of $NaNO_2$ in 0.1 M PBS at an applied of 0.85 V. Inset: Enlarged region between 0.2 and 10 μM nitrite. (D) Calibration plot of the amperometric current response at CNSs/GCE as a function of nitrite concentration. Source: Yallappa et al. [93]. © 2018 IOP Publishing Ltd.

studied. However, for CNSs/GCE (curve d), the currents are greater than those for BGCE, and a current peak appeared at 0.89 V with a negative shift compared with BGCE validating the catalytic activity of CNSs/GCE toward nitrite oxidation.

To determine the ideal pH for nitrite oxidation, the effect of pH at CNSs/GCE was conducted by recording the data in 0.1 M PBS of different pH values with $NaNO_2$ at a scan rate of 50 mV/s. In Figure 3.12B, we can see that the oxidation peak current increased from pH 5 to 7, after which it decreased, thereby making pH 7 the ideal pH at which further studies were carried out.

Upon successive addition of the nitrite solution to the buffer at pH 7 under constant stirring, the amperometric response of the sensor was monitored at 0.85 V. As illustrated in Figure 3.12C,D, there was a rapid response on the addition of nitrite ions, thus showcasing high sensitivity. A good linear relationship was observed in the sodium nitrite concentration range of 0.2–350 μM ($R = 0.9982$), with the detection limit being 0.0146 μM at a signal-to-noise ratio of 3 [17]. All these results indicate commendable catalytic activity of the electrode toward nitrite oxidation. This was assumed because of the increased adsorption sites and enhanced porosity of the electrode due to CNS, paving its way to electron transfer, thereby resulting in increased selectivity and sensitivity.

3.5.6 Catechin Sensing Using *Bougainvillea spectabilis*-Derived Carbons

There has been a penchant to synthesize activated carbon material from biomass as they surpass the need for the requirement of multiple preparation steps and expensive instrumental setup. Veeramani et al. used a biomass precursor (*Bougainvillea spectabilis*) to synthesize graphene sheet-like porous activated carbon (GPAC) for the sensing of catechin (CT) [94]. Catechin, which is a flavonoid, is used exuberantly in food, owing to its radical scavenging properties.

The electrode active material was prepared by preheating the flowers, after which it was ground into fine powder; it was activated by employing 10% $ZnCl_2$, followed by pyrolysis at 700, 800, and 900 °C. The pyrolyzed products were tagged as AC-700, GPAC, and AC-900. The hence prepared modified working electrode material was drop-casted on GCE for the electrochemical sensing of catechin by techniques like CV and DPV in 0.05 M phosphate buffer (PBS, pH 7). As expected the modified working electrode exhibited better profiles than the bare, with GPAC exhibiting a sharper peak compared with AC-700 and AC-900 as seen in Figure 3.13a. Figure 3.13 shows the effect of sweep rates that range 50–500 mV/s for GPAC electrode in the presence of analyte catechin (240 μM). As seen in Figure 3.13b, there was a progressive increase in the peak current on increasing the scan rate. To conclude that the process happening at the surface of the electrode is diffusion limited, we can see that the oxidation peak shifted toward the positive potential and the reduction peak shifted toward the negative potential, thereby substantiating the same. The linear regression equations (Eqs. (3.2) and (3.3))

$$y = 0.243x + 4.163 \tag{3.2}$$

and

$$y = -0.1036x - 4.95 \tag{3.3}$$

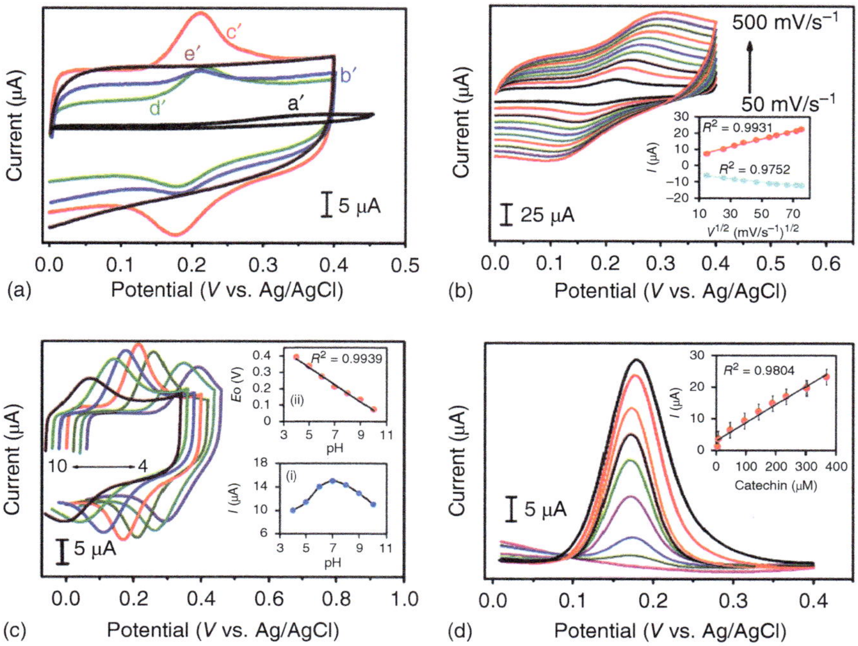

Figure 3.13 (a) CV profiles for bare (a′), AC-700 (b′), GPAC (c′), and AC-900 (d′) GCEs in the presence of 240 μM CA with a blank measurement (e′) at a scan rate of 50 mV/s. (b) CV curves at different scan rates (from 50 to 500 mV/s) for GPAC-modified GCE. The inset shows the corresponding peak current vs. square root of the scan rate. (c) Different pH studies from pH 4 to 10 for the GPAC electrode. Inset (i): Peak current vs. pH. Inset (ii): Correlation between the peak potential (E_0) vs. pH. (d) DPV curves for the various CA concentrations over GPAC-modified GCE. Inset: Anodic oxidation peak current (I_{pa}) vs. CA concentration (mM). Electrolyte solution, 0.05 M PBS (pH 7.0); scan rate, 50 mV/s. Source: Veeramani et al. [94]. Licensed under CC BY 3.0.

with regression coefficients, $R^2 = 0.9931$ and 0.9752 reaching unity, mathematically also proves that indeed catechin oxidation on the GPAC surface is a diffusion-controlled one [18]. The peak current has a direct correlation on the pH; hence it is critical to determine an optimized pH value. Figure 3.13c depicts the optimization of the pH value for GPAC-modified electrode in the presence of CA at a constant rate of 50 mV/s. In order to be commercially viable, the working electrode should be effective in physiological pH (pH = 7), and we can see that the maximum current value was obtained for pH 7. Beyond this pH value, the peak current decreased due to the prevention of absorption of CA by hydroxyl anions. DPV was further employed to procure high sensitivity of the fabricated electrode. A progressive increase in the oxidation peak on increasing the concentration in CA added is depicted by DPV as shown in Figure 3.13d. The linear range, sensitivity, and detection limit are shown in Table 3.1.

3.5.7 Progesterone Sensing by Onion Peel-Derived Carbons

The food waste value of natural food is often underestimated. Hence it is necessary to utilize it not only to recover its economic and nutritional value but also to employ in various

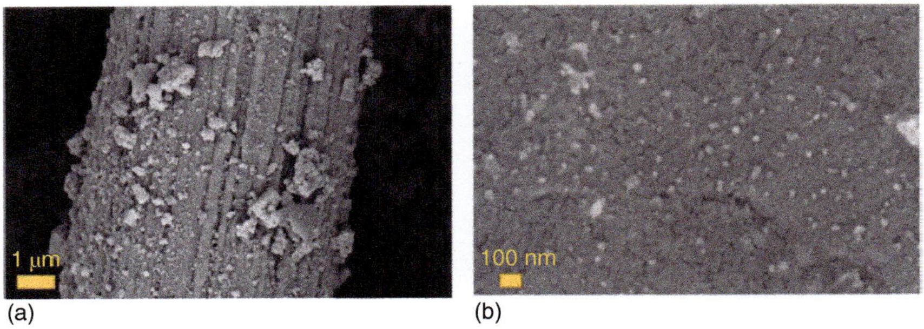

Figure 3.14 (a) and (b) SEM micrographs of CNS/CFP electrode. Source: Reproduced with permission from reference Akshaya et al. [95].

analytical applications. There are reports on how carbon source derived from onion peel synthesized by microwave-assisted method is used in speeding skin wound healing and living cell imaging [99]. K.B. Akshaya et al. synthesized highly porous CNSs from onion peels by a single-step pyrolysis method [95]. The hence prepared CNSs was coated on carbon fiber paper (CFP) that served as the working electrode (CNS/CFP) for the detection of progesterone [95].

This method bypassed the need of enzymes and steroid specific antibodies for the sensitive and selective determination of progesterone. From the SEM images rendered in Figure 3.14a,b of the CNS/CFP electrodes, one can notice agglomerated nanospheres seamlessly dispersed on the CFP substrate. The SEM images have displayed the porous nature of CNS and can be concluded that porosity plays a pivotal role in facilitating the electrolyte to sweep into the electrode.

Because of the porosity of CNSs, the authors were able to achieve detection of progesterone at an ultra-nanomolar level with a detection limit of 0.012 nM. From the voltammograms obtained as seen in Figure 3.15a, the modified electrode showed a better and higher current profile than the bare one for their response to the redox couple of 1 mM potassium ferrocyanide/ferricyanide in 1 M KCl. CVs of bare CFP and CNS/CFP in PBS in pH 7 for progesterone analysis are shown in Figure 3.15c,d. The impedimetric characterization in Figure 3.15b demonstrates that the modified CNS/CFP electrode exhibited significantly very low charge transfer resistance of 8.85 Ω with respect to the bare electrode that possessed a resistance of 93.68 Ω.

To gauge the kinetics of the process, the electrochemical response of 0.15 mM progesterone at CNS/CFP with varying scan rates from 0.01 to 0.1 V/s in PBS (pH 7.0) was recorded as seen in Figure 3.16a. In Figure 3.16b, we can see that i_{pa} is proportional to v, and hence the process is limited by the diffusion of species. The plot of logarithm of peak current against that of scan rate showcases a linear relationship. A logarithmic plot of anodic peak currents vs. scan rates is shown in Figure 3.16c.

$$\log i_{pa}(A) = -0.98806 \log v \text{ (mV/s)} + 3.03786 \ (R^2 = 0.9951) \tag{3.4}$$

From the above straight-line equation, the slope value approaches to 1, and this is indicative that electrode process was not diffusion controlled but rather an adsorption-controlled one. The developed sensor was successfully used in real-life samples to see the practicability of

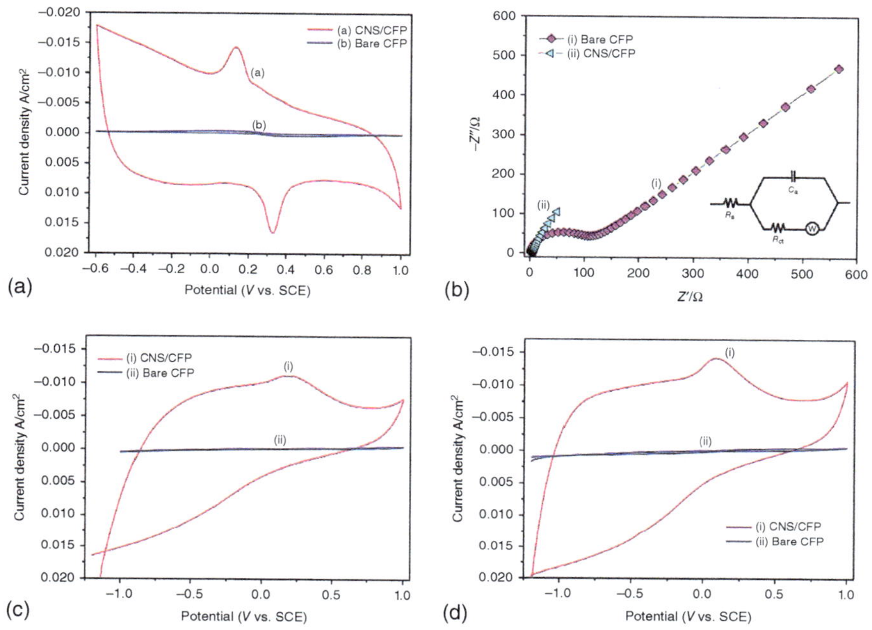

Figure 3.15 (a) Cyclic voltammograms of 1 mM potassium ferrocyanide/ferricyanide at CNS/CFP (curve a) and bare CFP (curve b) electrodes in 1 M KCl. (b) Nyquist plots of bare CFP (curve (i)) and CNS/CFP (curve (ii)) electrodes in 5 mM K4[Fe (CN)6]/K_3[Fe $(CN)_6$] solutions containing 0.1 M KCl. (c) Cyclic voltammograms of (i) CNS/CFP and (ii) bare CFP electrodes without progesterone in PBS (pH 7.0) at a scan rate of 0.05 V/s. (d) Cyclic voltammograms of 0.15 nM progesterone at (i) CNS/CFP and (ii) bare CFP electrodes in PBS (pH 7.0) at 0.05 V/s. Source: Akshaya et al. [95]. © 2019 IOP Publishing Ltd.

the electrode. DPV was conducted to find the capability of the sensor to discriminate the interferences, and Figure 3.16d shows the DPV response to progesterone in the presence of other interfering compounds. The applicability of the proposed method was evaluated by determining progesterone in pharmaceutical progesterone injections, spiked human blood serum, and cow milk samples.

3.5.8 Butein Detection from Oil Palm Leave-Derived Carbons

Inflammation is the state that results after activation of the immune system due to external attack or injury on the cells or tissues. The degree of inflammation is elevated to weeks or months in chronic inflammation. Chronic inflammation is often likely due to the foreign pathogens and continued production of inflammatory cytokines or genetically driven autoimmune disease processes. During the process, there is an increase in the biochemical markers of inflammation, resulting in a state of oxidative stress and elevation of inflammatory cytokines [100]. The drugs employed to ease inflammation-related indications proceed by restraining cyclooxygenase action or by blocking cytokine receptors. However, long-term usage of the drugs has a secondary effect. Non-synchronization of inflammatory response results in an increase in the concentration of inflammatory mediators, which

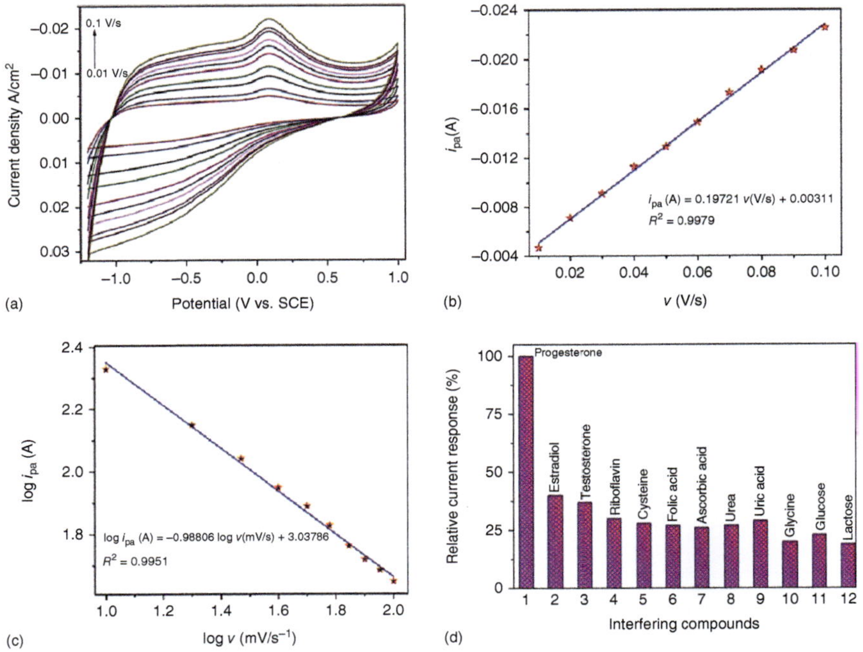

Figure 3.16 (a) Cyclic voltammograms of 0.15 nM progesterone at CNS/CFP electrode with varying scan rates from 0.01–0.1 V/s in PBS (pH 7.0). (b) Plot of anodic peak currents against varying scan rates and (c) plot of logarithm of the anodic peak currents vs. logarithm of scan rates. (d) DPV selective responses of CNS/CFP toward 0.15 nM of progesterone with 11 interfering compounds. Source: Akshaya et al. [95]. © 2019 IOP Publishing Ltd.

leads to the occurrence of numerous chronic diseases such as coronary diseases cancer, rheumatoid arthritis, and others [101]. It is recommended that inflammatory cytokines have potential as therapeutic targets to care for inflammatory disorders [102]. Nowadays, studying the consequence of flavonoids on inflammatory mediators, particularly by modulating cytokines, is significant in expanding alternative management for inflammation-related disorders [100]. Butein, a chalcone flavonoid, shows a wide range of pharmacological effects such as anti-inflammatory, anticancer, antioxidant, and antiangiogenic in diverse disease models [100]. Studying the direct electrochemical redox property of butein is useful in the pharmaceuticals. Rapid analysis of flavonoids based on the electrochemical fingerprinting is a novel area of interest that can complement the existing herbaceutical analysis methods. P. Kanagavalli et al. demonstrated non-enzymatic use of oil palm leave-derived carbon nanoparticles (OPL-CNPs) as electrode materials for butein sensing [96].

Oil palm leaves (OPL) were dried and pyrolyzed in an inert atmosphere at 600 °C for two hours, producing carbon residue that were treated with NaOH and washed with deionized water to get rid of impurities. Evaluation of electrochemical property of butein was done by using OPL-CNP as the modified working substrate on GCE. The CV revealed the redox nature of butein on OPL-CNP-modified GCE with phosphate buffer as supporting electrolyte. The cathodic and anodic peak ratio (I_{pc}/I_{pa}) was calculated to be 0.97 close to 1, indicating reversible redox reaction. The effect of pH on the redox nature of butein was understood by carrying out CV experiments in different pH ranges from 2 to 8. The results

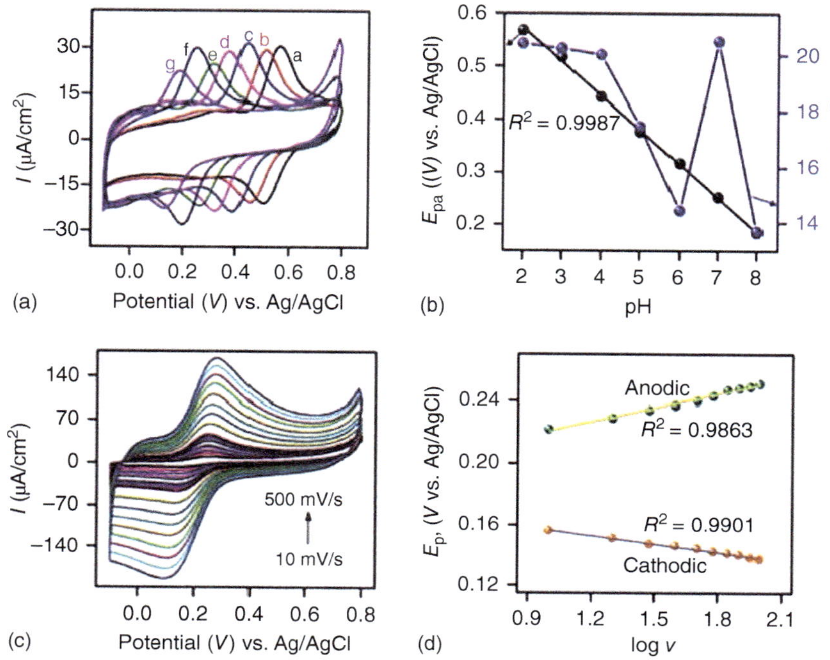

Figure 3.17 (A) CVs of OPL-CNP-modified GCE in 0.1 M PBS solution containing 0.05 mM of butein in different pHs – (a) 2.0, (b) 3.0, (c) 4.0, (d) 5.0, (e) 6.0, (f) 7.0, and (g) 8.0 – at a scan rate of 0.05 V/s. (B) Calibration plot obtained from CV responses at different pHs vs. anodic current density/anodic peak potential of butein (0.05 mM) in OPL-CNP-modified GCE. (C) CVs of OP-CNP-modified GCE recorded in PBS solution containing 0.05 mM of butein at different scan rates (0.01–0.5 V/s) and (D) its corresponding calibration plot of log scan rate vs. peak potential. Source: Kanagavalli et al. [96]. © 2020 John Wiley & Sons.

are shown in Figure 3.17a. This showed that at pH 7 the modified electrode exhibited better redox peak at lower potential. A calibration plot was obtained from CV responses at different pHs, against anodic current density/anodic peak potential of butein (0.05 mM) in OPL-CNPs modified GCE, which is depicted in Figure 3.17b. The effect of scan rate was studied by varying the scan rate from 10 to 500 mV/s (Figure 3.17c). A quasi-reversible redox behavior was observed at higher scan rates. This indicated reversible process, but at higher concentrations, the electrode process seemed to be adsorption/diffusion controlled. The corresponding calibration plot of log scan rate vs. peak potential is shown in Figure 3.17d.

The LOD and limit of quantification were reported to be 7.6 and 25.3 μM. Also for the first time, butein quantification was done by chronoamperometry (Figure 3.18a). The work demonstrated that distinct physical and chemical composition of OPL-CNPs can play significant role in electrochemical sensing of butein and also offering enhanced stability for continuous cycles. For practical applications of the system, real serum samples were utilized with butein spiking. Figure 3.18c–e illustrates the CVs of OPL-CNP-modified electrode measured against analyte in PBS and serum spiked analyte of known concentrations (60 and 80 μM). A histogram depicting the variations in the peak current density in comparison with pristine butein is also depicted in Figure 3.18f. The reported work suggested a new dimension for biowaste-derived CNPs toward electro-bioanalysis approach helpful for

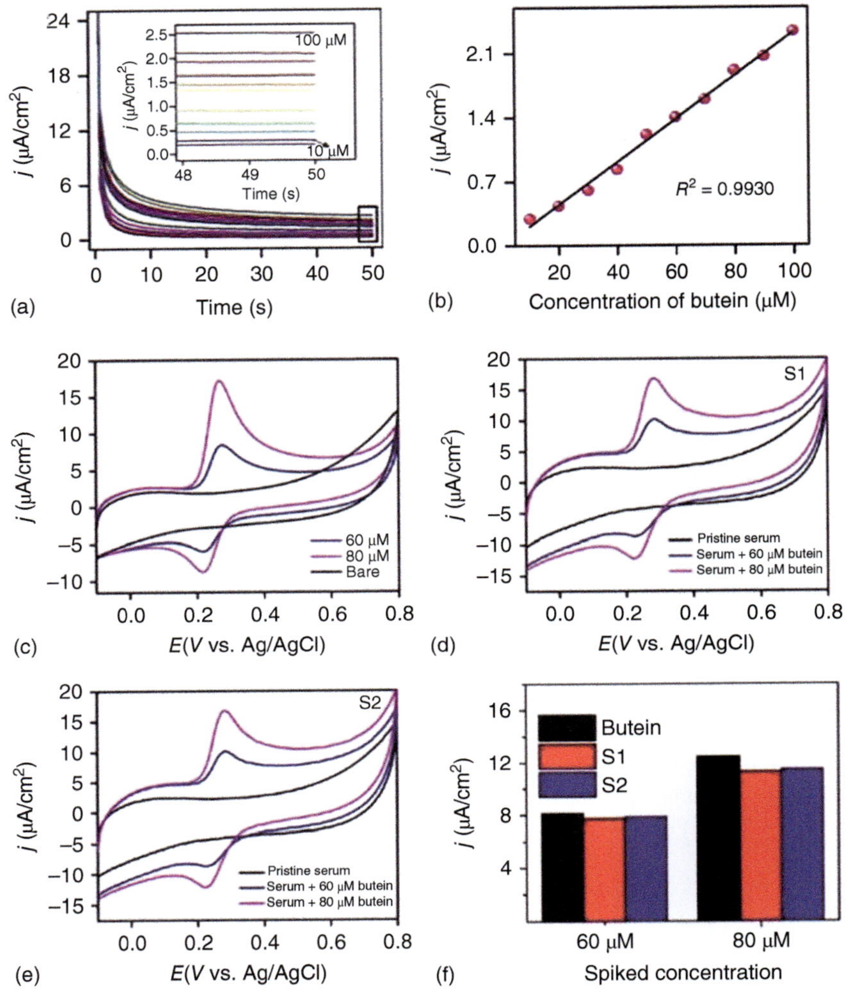

Figure 3.18 (a) Chronoamperometric results of OPL-modified GCE at fixed potential of +0.28 V in 0.1 M PBS in the presence of various concentrations of butein. (b) The plot between various concentrations of butein vs. current density (taken at 50 seconds) obtained from chronoamperometric results. (c) CVs of OPL-CNP-modified GCE with pristine butein samples in PBS electrolyte. (d) and (e) CVs of OPL-CNP-modified GCE electrodes in PBS containing patients' serum (S1 and S2) spiked with two different concentrations (60 and 80 μM) of butein. (f) Corresponding histogram depicting the variations in the peak current density in comparison with pristine butein. Source: Kanagavalli et al. [96]. © 2020 John Wiley & Sons.

herbaceutical fingerprinting. Also the methodology could also be tested for evaluating the binding affinity of flavonoids on selected biological targets like protein markers, which may find clinical significance in the drug macromolecular interactions.

Overall, modified electrodes enhance the analytical sensitivity in the detection and quantification of various molecules and biomolecules. Table 3.1 summarizes the determination of various compounds at different biomass/biowaste-derived carbon-modified electrodes.

3.6 Conclusion and Future Perspective

To conclude, biomass/biowaste carbonization produced porous materials rich in carbon with high surface area. BET surface area for one of the discussed carbon materials derived from onion peels had very high surface area of 2961 m^2/g. The presence of large surface area has been believed to provide an efficient platform for the electrochemical reaction to occur, thereby enhancing the adsorption (surface)-controlled mechanism of electrode–analyte interaction. Most of the discussed porous carbon structures were either microporous ($d < 2$ nm), mesoporous ($d > 2$ nm to 50 nm), or macroporous ($d > 50$ nm) or hierarchical in nature that facilitates easier diffusion of molecules and ions, enhancing surface diffusion phenomena [47, 103–106]. As a noble material, biomass-derived carbon nanoparticles have extensively been applied to detect various molecules by electrochemical methods. These carbon nanoparticles have been utilized by the researchers effectively for the determination of various biomolecules due to an increase in sensitivity, avoidance of the fouling effect, and better peak resolution. The simultaneous determination of biomolecules in a mixture of analytes is still found to be in an incipient phase. Hence, it is important that a novel method for the fabrication of carbon nanoparticle-modified electrodes are developed for sensitive determination of analytes in a mixture. Although there are few obstacles in using the biomass-derived carbon as an electrode material, we anticipate them to be a superior electrode material due to its biocompatibility, wide potential window, chemical and electrochemical inertness, abundance, low cost, lack of metallic impurities, high surface area, and enhanced interaction with analyte materials via π–π interaction. The electrocatalysis and the sensitivity of the modified electrodes toward several analytes are related to (i) the electrostatic interactions between the structural defects of the carbon planes and the analytes and (ii) the π–π interaction between the carbon and the aromatic moieties of analytes (biomolecules).

However, there are few obstacles that need to be overcome for effective use of biomass-derived carbon nanoparticles as an electrode. Besides, the synthesis of carbon nanomaterial and the size and uniformity should be controlled to obtain reproducible results, which are important when designing a sensor. The modification of electrodes with carbon nanomaterials is a challenging task. Though drop-cast electrodes show superior electrocatalytic activity and have low detection limits toward various analytes and biomolecules, the reproducibility and the stability of the electrode still face challenges. Another important concern is the defects in carbon nanomaterials. The exact relation between the defects of basal plane and the edge plane and the electrocatalytic activity of carbon nanoparticles should be studied in detail.

Acknowledgment

Dr. Gurumurthy Hegde acknowledges DST-Nanomission, Government of India, for providing research grant on "Biowaste based porous nano materials for efficient low-cost energy storage devices" with file no SR/NM/NT-1026/2017.

Important Websites

www.electrochem.org.
https://www.sciencedirect.com/topics/engineering/electrochemical-sensor.
https://www.eia.gov/energyexplained/biomass.
https://www.reenergyholdings.com.

References

1 Patel, S.V., Mlsna, T.E., Fruhberger, B. et al. (2003). Chemicapacitive microsensors for volatile organic compound detection. *Sens. Actuators, B* 96 (3): 541–553.
2 Peixoto, A.C. and Silva, A.F. (2017). Smart devices: micro- and nanosensors. In: *Bioinspired Materials for Medical Applications*, Chapter 11 (eds. L. Rodrigues and M. Mota), 297–329. Woodhead Publishing.
3 Hulanicki, A., Glab, S., and Ingman, F. (1991). Chemical sensors: definitions and classification. *Pure Appl. Chem.* 63 (9): 1247.
4 Raj, M.A. and John, S.A. (2019). Graphene-modified electrochemical sensors. In: *Graphene-Based Electrochemical Sensors for Biomolecules*, Chapter 1 (eds. A. Pandikumar and P. Rameshkumar), 1–41. Elsevier.
5 Antuña-Jiménez, D., Díaz-Díaz, G., Blanco-López, M.C. et al. (2012). Molecularly imprinted electrochemical sensors: past, present, and future. In: *Molecularly Imprinted Sensors*, Chapter 1 (eds. S. Li, Y. Ge, S.A. Piletsky and J. Lunec), 1–34. Amsterdam: Elsevier.
6 Tokonami, S., Shiigi, H., and Nagaoka, T. (2012). Molecularly imprinted overoxidized polypyrrole films for sensor applications from enantiorecognition to trace analysis. In: *Molecularly Imprinted Sensors*, Chapter 4 (eds. S. Li, Y. Ge, S.A. Piletsky and J. Lunec), 73–89. Amsterdam: Elsevier.
7 Zhang, J.X.J. and Hoshino, K. (2019). Mechanical transducers: cantilevers, acoustic wave sensors, and thermal sensors. In: *Molecular Sensors and Nanodevices*, Chapter 6, 2e (eds. J.X.J. Zhang and K. Hoshino), 311–412. Academic Press.
8 Baker, B.A., Zhang, H., Cha, T.G., and Choi, J.H. (2013). Carbon nanotube solar cells. In: *Carbon Nanotubes and Graphene for Photonic Applications*, Chapter 9 (eds. S. Yamashita, Y. Saito and J.H. Choi), 241–269. Woodhead Publishing.
9 Peng, L.M., Wang, S., and Zhang, Z. (2013). Carbon nanotube-based photovoltaic and light-emitting diodes. In: *Carbon Nanotubes and Graphene for Photonic Applications*, Chapter 11 (eds. S. Yamashita, Y. Saito and J.H. Choi), 298–318. Woodhead Publishing.
10 Walden-Newman, W. and Strauf, S. (2013). Quantum light sources based on individual carbon nanotubes. In: *Carbon Nanotubes and Graphene for Photonic Applications*, Chapter 13 (eds. S. Yamashita, Y. Saito and J.H. Choi), 346–385e. Woodhead Publishing.
11 Wilkinson, T. and Butt, H. (2013). Hybrid carbon nanotube–liquid crystal nanophotonic devices. In: *Carbon Nanotubes and Graphene for Photonic Applications*, Chapter 12 (eds. S. Yamashita, Y. Saito and J.H. Choi), 319–350e. Woodhead Publishing.
12 Bársony, I., Dücső, C., and Fürjes, P. (2009). Thermometric gas sensing. In: *Solid State Gas Sensing* (eds. E. Comini, G. Faglia and G. Sberveglieri), 1–24. Boston, MA: Springer.

13 Mujahid, A. and Dickert, F.L. (2012). Molecularly imprinted polymers for sensors: comparison of optical and mass-sensitive detection. In: *Molecularly Imprinted Sensors*, Chapter 6 (eds. S. Li, Y. Ge, S.A. Piletsky and J. Lunec), 125–159. Amsterdam: Elsevier.

14 Reyne, G., Delamare, J., and Cugat, O. (2005). Magnetic micro-actuators (MAGMAS). In: *Encyclopedia of Materials: Science and Technology* (eds. K.H.J. Buschow, R.W. Cahn, M.C. Flemings, et al.), 1–5. Oxford: Elsevier.

15 Ripka, P. (2007). Magnetic sensors: principles and applications. In: *Encyclopedia of Materials: Science and Technology* (eds. K.H.J. Buschow, R.W. Cahn, M.C. Flemings, et al.), 1–11. Oxford: Elsevier.

16 Wang, M.L. and Wang, G. (2014). Electromagnetic sensors for assessing and monitoring civil infrastructures. In: *Sensor Technologies for Civil Infrastructures*, Chapter 9 (eds. M.L. Wang, J.P. Lynch and H. Sohn), 238–264. Woodhead Publishing.

17 Sharma, D., Kanchi, S., Bisetty, K., and Nuthalapati, V.N. (2017). Chapter 1. In: *Perspective on Analytical Sciences and Nanotechnology, Advanced Environmental Analysis: Applications of Nanomaterials* (eds. Kharisov, B. and Hussain, C.M.), vol. 1, 1–34: The Royal Society of Chemistry.

18 Mishra, A.K. and Hussain, C.M. (2018). *Nanotechnology in Environmental Science*, 1e. Wiley-VCH.

19 Hussain, C.M. (2019). *Handbook of Environmental Materials Management*, 1e. Springer International Publishing.

20 Keçili, R. and Hussain, C.M. (2019). *Modern Environmental Analysis Techniques for Pollutants*, 1e. Elsevier.

21 Coyle, S., Curto, V.F., Benito-Lopez, F. et al. (2014). Wearable bio and chemical sensors. In: *Wearable Sensors*, Chapter 2.1 (eds. E. Sazonov and M.R. Neuman), 65–83. Oxford: Academic Press.

22 Omar, F.S., Duraisamy, N., Ramesh, K., and Ramesh, S. (2016). Conducting polymer and its composite materials based electrochemical sensor for nicotinamide adenine dinucleotide (NADH). *Biosens. Bioelectron.* 79: 763–775.

23 Kimmel, D.W., LeBlanc, G., Meschievitz, M.E., and Cliffel, D.E. (2012). Electrochemical sensors and biosensors. *Anal. Chem.* 84 (2): 685–707.

24 Hussain, C.M. (2018). *Handbook of Nanomaterials for Industrial Applications*, 1e. Elsevier.

25 Hussain, C.M. (2020). *Handbook of Functionalized Nanomaterials for Industrial Applications*, 1e. Elsevier.

26 Hussain, C.M. (2020). *Handbook of Nanomaterials for Manufacturing Applications*, 1e. Elsevier.

27 Daniel, M.-C. and Astruc, D. (2004). Gold nanoparticles: assembly, supramolecular chemistry, quantum-size-related properties, and applications toward biology, catalysis, and nanotechnology. *Chem. Rev.* 104 (1): 293–346.

28 Katz, E., Willner, I., and Wang, J. (2004). Electroanalytical and bioelectroanalytical systems based on metal and semiconductor nanoparticles. 16 (1–2): 19–44.

29 McCreery, R.L. (2008). Advanced carbon electrode materials for molecular electrochemistry. *Chem. Rev.* 108 (7): 2646–2687.

30 Wang, Y., Li, Y., Tang, L. et al. (2009). Application of graphene-modified electrode for selective detection of dopamine. *Electrochem. Commun.* 11 (4): 889–892.

31 Xu, C.-X., Huang, K.-J., Fan, Y. et al. (2012). Simultaneous electrochemical determination of dopamine and tryptophan using a TiO_2-graphene/poly(4-aminobenzenesulfonic acid) composite film based platform. *Mater. Sci. Eng., C* 32 (4): 969–974.

32 Abdolmaleki, A., Mallakpour, S., and Karshenas, A. (2017). Facile synthesis of glucose-functionalized reduced graphene oxide (GFRGO)/poly(vinyl alcohol) nanocomposites for improving thermal and mechanical properties. *Mater. Sci. Eng., B* 217: 26–35.

33 Abdolmaleki, A., Mallakpour, S., and Karshenas, A. (2017). Synthesis and characterization of new nanocomposites films using alanine-Cu-functionalized graphene oxide as nanofiller and PVA as polymeric matrix for improving of their properties. *J. Solid State Chem.* 253: 398–405.

34 Abdolmaleki, A., Mallakpour, S., and Borandeh, S. (2016). Improving interfacial interaction of l-phenylalanine-functionalized graphene nanofiller and poly(vinyl alcohol) nanocomposites for obtaining significant membrane properties: morphology, thermal, and mechanical studies. *Polym. Compos.* 37 (6): 1924–1935.

35 Mallakpour, S., Abdolmaleki, A., and Borandeh, S. (2014). Covalently functionalized graphene sheets with biocompatible natural amino acids. *Appl. Surf. Sci.* 307: 533–542.

36 Mallakpour, S. and Rashidimoghadam, S. (2018). Poly(vinyl alcohol)/carbon nanotube nanocomposites: challenges and opportunities. In: *Biodegradable and Biocompatible Polymer Composites*, Chapter 10 (ed. N.G. Shimpi), 297–315. Woodhead Publishing.

37 Matos-Peralta, Y. and Antuch, M. (2020). Review—Prussian blue and its analogs as appealing materials for electrochemical sensing and biosensing. 167 (3): 037510.

38 George, J.M., Antony, A., and Mathew, B.J.M.A. (2018). Metal oxide nanoparticles in electrochemical sensing and biosensing: a review. 185 (7): 358.

39 Robert Săndulescu, M.T., Cristea, C., and Bodoki, E. (2015). New materials for the construction of electrochemical biosensors, biosensors – micro and nanoscale applications. In: *Biosensors – Micro and Nanoscale Applications* (ed. T. Rinken), 1. London, UK: IntechOpen.

40 Kumar, P., Deep, A., and Kim, K.-H. (2015). Metal organic frameworks for sensing applications. *TrAC, Trends Anal. Chem.* 73: 39–53.

41 Achmann, S., Hagen, G., Kita, J. et al. (2009). Metal-organic frameworks for sensing applications in the gas phase. 9 (3): 1574–1589.

42 Meek, S.T., Greathouse, J.A., and Allendorf, M.D. (2011). Metal-organic frameworks: a rapidly growing class of versatile nanoporous materials. 23 (2): 249–267.

43 Cui, Y., Li, B., He, H. et al. (2016). Metal–organic frameworks as platforms for functional materials. *Acc. Chem. Res.* 49 (3): 483–493.

44 He, L., Liu, Y., Liu, J. et al. (2013). Core–Shell Noble-metal@metal-organic-framework nanoparticles with highly selective sensing property. 52 (13): 3741–3745.

45 Fleker, O., Borenstein, A., Lavi, R. et al. (2016). Preparation and properties of metal organic framework/activated carbon composite materials. *Langmuir* 32 (19): 4935–4944.

46 Llobet, E. (2013). Gas sensors using carbon nanomaterials: a review. *Sens. Actuators, B* 179: 32–45.

47 Kumar, A., Hegde, G., Manaf, S.A.B.A. et al. (2014). Catalyst free silica templated porous carbon nanoparticles from bio-waste materials. *Chem. Commun.* 50 (84): 12702–12705.

48 Mallakpour, S., Abdolmaleki, A., and Karshenas, A. (2017). Graphene oxide supported copper coordinated amino acids as novel heterogeneous catalysts for epoxidation of norbornene. *Catal. Commun.* 92: 109–113.

49 Mallakpour, S., Abdolmaleki, A., Mahmoudian, M. et al. (2017). Synergetic effect of synthesized sulfonated polyaniline/quaternized graphene and its application as a high-performance supercapacitor electrode. *J. Mater. Sci.* 52 (16): 9683–9695.

50 Bhat, V.S., Supriya, S., Jayeoye, T.J. et al. (2020). Influence of surface properties on electro-chemical supercapacitors utilizing *Callerya atropurpurea* pod derived porous nanocarbons: structure property relationship between porous structures to energy storage devices. *Nano Select* 1 (2): 226–243.

51 Rison, S., Akshaya, K.B., Bhat, V.S. et al. MnO_2 nanoclusters decorated on graphene modified pencil graphite electrode for non-enzymatic determination of cholesterol. *Electroanalysis* https://doi.org/10.1002/elan.202000049.

52 Chaithra, K.P., Bhat, V.S., Akshaya, K.B. et al. (2020). Unique host matrix to disperse Pd nanoparticles for electrochemical sensing of morin: sustainable engineering approach. *ACS Biomater. Sci. Eng.* 6 (9): 5264–5273.

53 Tibbetts, G.G. (1983). Carbon fibers produced by pyrolysis of natural gas in stainless steel tubes. 42 (8): 666–668.

54 Tibbetts, G.G. (1990). Vapor-grown carbon fibers. In: *Carbon Fibers Filaments and Composites* (eds. J.L. Figueiredo, C.A. Bernardo, R.T.K. Baker and K.J. Hüttinger), 73–94. Dordrecht: Springer Netherlands.

55 Shao, Y., El-Kady, M.F., Wang, L.J. et al. (2015). Graphene-based materials for flexible supercapacitors. *Chem. Soc. Rev.* 44 (11): 3639–3665.

56 Pandikumar, A., Soon How, G.T., See, T.P. et al. (2014). Graphene and its nanocomposite material based electrochemical sensor platform for dopamine. *RSC Adv.* 4 (108): 63296–63323.

57 Li, S.-J., Qian, C., Wang, K. et al. (2012). Application of thermally reduced graphene oxide modified electrode in simultaneous determination of dihydroxybenzene isomers. *Sens. Actuators, B* 174: 441–448.

58 El-Kady, M.F., Strong, V., Dubin, S., and Kaner, R.B. (2012). Laser scribing of high-performance and flexible graphene-based electrochemical capacitors. *Science* 335 (6074): 1326.

59 Park, S. and Ruoff, R.S. (2009). Chemical methods for the production of graphenes. *Nat. Nanotechnol.* 4: 217.

60 Seyller, T., Bostwick, A., Emtsev, K.V. et al. (2008). Epitaxial graphene: a new material. 245 (7): 1436–1446.

61 Kim, K.S., Zhao, Y., Jang, H. et al. (2009). Large-scale pattern growth of graphene films for stretchable transparent electrodes. *Nature* 457 (7230): 706–710.

62 Li, X., Cai, W., An, J. et al. (2009). Large-area synthesis of high-quality and uniform graphene films on copper foils. *Science* 324 (5932): 1312.

63 Newson, R.W., Dean, J., Schmidt, B., and Driel, H.M.V. (2009). Ultrafast carrier kinetics in exfoliated graphene and thin graphite films. *Opt. Express* 17 (4): 2326–2333.

64 Pumera, M., Ambrosi, A., Bonanni, A. et al. (2010). Graphene for electrochemical sensing and biosensing. *TrAC, Trends Anal. Chem.* 29 (9): 954–965.

65 Iyer, D., Manaf, S., Yallappa, S. et al. (2016). Low cost, high performance supercapacitor electrode using coconut wastes: eco-friendly approach. *J. Energy Chem.* 25 (5): 880–887.

66 Hegde, G., Abdul Manaf, S.A., Kumar, A. et al. (2015). Biowaste sago bark based catalyst free carbon nanospheres: waste to wealth approach. *ACS Sustainable Chem. Eng.* 3 (9): 2247–2253.

67 Xing, Z., Ju, Z., Zhao, Y. et al. (2016). One-pot hydrothermal synthesis of nitrogen-doped graphene as high-performance anode materials for lithium ion batteries. *Sci. Rep.* 6 (1): 26146.

68 Rodriguez Correa, C., Hehr, T., Voglhuber-Slavinsky, A. et al. (2019). Pyrolysis vs. hydrothermal carbonization: understanding the effect of biomass structural components and inorganic compounds on the char properties. *J. Anal. Appl. Pyrolysis* 140: 137–147.

69 Iyer, D. and Hegde, G. (2015). Activated carbon nanospheres derived from bio-waste materials for supercapacitor applications – a review. *RSC Adv.* 5 (107): 88339–88352.

70 Takahashi, H., Li, B., Sasaki, T. et al. (2000). Catalytic activity in organic solvents and stability of immobilized enzymes depend on the pore size and surface characteristics of mesoporous silica. *Chem. Mater.* 12 (11): 3301–3305.

71 Kruse, A., Funke, A., and Titirici, M.-M. (2013). Hydrothermal conversion of biomass to fuels and energetic materials. *Curr. Opin. Chem. Biol.* 17 (3): 515–521.

72 Wikberg, H., Ohra-aho, T., Pileidis, F., and Titirici, M.-M. (2015). Structural and morphological changes in Kraft lignin during hydrothermal carbonization. *ACS Sustainable Chem. Eng.* 3 (11): 2737–2745.

73 Bhat, V.S., Supriya, S., and Hegde, G. (2019). Biomass derived carbon materials for electrochemical sensors. *J. Electrochem. Soc.* 167 (3): 037526.

74 Sharma, H.B., Sarmah, A.K., and Dubey, B. (2020). Hydrothermal carbonization of renewable waste biomass for solid biofuel production: a discussion on process mechanism, the influence of process parameters, environmental performance and fuel properties of hydrochar. *Renewable Sustainable Energy Rev.* 123: 109761.

75 Kim, D., Lee, K., and Park, K.Y. (2016). Upgrading the characteristics of biochar from cellulose, lignin, and xylan for solid biofuel production from biomass by hydrothermal carbonization. *J. Ind. Eng. Chem.* 42: 95–100.

76 Supriya, S., Sriram, G., Ngaini, Z. et al. (2019). The role of temperature on physical–chemical properties of green synthesized porous carbon nanoparticles. *Waste Biomass Valorization*.

77 Ali, G.A.M., Divyashree, A., Supriya, S. et al. (2017). Carbon nanospheres derived from Lablab purpureus for high performance supercapacitor electrodes: a green approach. *Dalton Trans.* 46 (40): 14034–14044.

78 Ali, G.A.M., Manaf, S.A.A., Iyer, D. et al. (2016). Superior supercapacitive performance in porous nanocarbons. *J. Energy Chem.* 25 (4): 734–739.

79 Ali, G.A.M., Manaf, S.A.B.A., Kumar, A. et al. (2014). High performance supercapacitor using catalysis free porous carbon nanoparticles. *J. Phys. D: Appl. Phys.* 47 (49): 495307.

80 Patwardhan, P.R., Brown, R.C., and Shanks, B.H. (2011). Product distribution from the fast pyrolysis of hemicellulose. *ChemSusChem* 4 (5): 636–643.

81 Deng, J., Xiong, T., Wang, H. et al. (2016). Effects of cellulose, hemicellulose, and lignin on the structure and morphology of porous carbons. *ACS Sustainable Chem. Eng.* 4 (7): 3750–3756.

82 Jin, Y.Z., Gao, C., Hsu, W.K. et al. (2005). Large-scale synthesis and characterization of carbon spheres prepared by direct pyrolysis of hydrocarbons. *Carbon* 43 (9): 1944–1953.

83 Liu, J., Zhao, Y., Wu, Q. et al. (2018). Structure characterisation of polysaccharides in vegetable "okra" and evaluation of hypoglycemic activity. *Food Chem.* 242: 211–216.

84 Xu, D., Tong, Y., Yan, T. et al. (2017). N,P-Codoped meso−/microporous carbon derived from biomass materials via a dual-activation strategy as high-performance electrodes for deionization capacitors. *ACS Sustainable Chem. Eng.* 5 (7): 5810–5819.

85 Zan, Y., Zhang, Z., Liu, H. et al. (2017). Nitrogen and phosphorus co-doped hierarchically porous carbons derived from cattle bones as efficient metal-free electrocatalysts for the oxygen reduction reaction. *J. Mater. Chem. A* 5 (46): 24329–24334.

86 Liu, S., Han, T., Wang, Z. et al. (2019). Biomass-derived nitrogen and phosphorus co-doped hierarchical micro/mesoporous carbon materials for high-performance non-enzymatic H_2O_2 sensing. *Electroanalysis* 31 (3): 527–534.

87 Wu, J., Jin, C., Yang, Z. et al. (2015). Synthesis of phosphorus-doped carbon hollow spheres as efficient metal-free electrocatalysts for oxygen reduction. *Carbon* 82: 562–571.

88 Guan, J., Fang, Y., Zhang, T. et al. (2020). Kelp-derived activated porous carbon for the detection of heavy metal ions via square wave anodic stripping voltammetry. *Electrocatalysis* 11 (1): 59–67.

89 Kim, D., Kim, J.M., Jeon, Y. et al. (2018). Novel two-step activation of biomass-derived carbon for highly sensitive electrochemical determination of acetaminophen. *Sens. Actuators, B* 259: 50–58.

90 Jhaumeer Laulloo, S., Bhowon, M.G., Soyfoo, S., and Chua, L.S. (2018). Nutritional and biological evaluation of leaves of *Mangifera indica* from Mauritius. *J. Chem.* 2018: 6869294.

91 Madhu, R., Karuppiah, C., Chen, S.-M. et al. (2014). Electrochemical detection of 4-nitrophenol based on biomass derived activated carbons. *Anal. Methods* 6 (14): 5274–5280.

92 Xu, Y., Lei, W., Zhang, Y. et al. (2016). Bamboo fungus-derived porous nitrogen-doped carbon for the fast, sensitive determination of bisphenol a. *J. Electrochem. Soc.* 164 (5): B3043–B3048.

93 Yallappa, S., Shivakumar, M., Nagashree, K.L. et al. (2018). Electrochemical determination of nitrite using catalyst free mesoporous carbon nanoparticles from bio renewable areca nut seeds. *J. Electrochem. Soc.* 165 (10): H614–H619.

94 Veeramani, V., Sivakumar, M., Chen, S.-M. et al. (2017). Lignocellulosic biomass-derived, graphene sheet-like porous activated carbon for electrochemical supercapacitor and catechin sensing. *RSC Adv.* 7 (72): 45668–45675.

95 Akshaya, K.B., Bhat, V.S., Varghese, A. et al. (2019). Non-enzymatic electrochemical determination of progesterone using carbon nanospheres from onion peels coated on carbon fiber paper. *J. Electrochem. Soc.* 166 (13): B1097–B1106.

96 Kanagavalli, P., Radhakrishnan, S., Pandey, G. et al. (2020). Electrochemical tracing of butein using carbon nanoparticles interfaced electrode processed from biowaste. *Electroanalysis* 32 (6): 1220–1225.

97 Hiroi, H., Tsutsumi, O., Takeuchi, T. et al. (2004). Differences in serum bisphenol a concentrations in premenopausal normal women and women with endometrial hyperplasia. *Endocr. J.* 51 (6): 595–600.

98 Yallappa, S., Manaf, S.A.A., Shiddiky, M.J.A. et al. (2017). Synthesis of carbon nanospheres through carbonization of areca nut. *J. Nanosci. Nanotechnol.* 17 (4): 2837–2842.

99 Bankoti, K., Rameshbabu, A.P., Datta, S. et al. (2017). Onion derived carbon nanodots for live cell imaging and accelerated skin wound healing. *J. Mater. Chem. B* 5 (32): 6579–6592.

100 Shukla, R., Pandey, V., Vadnere, G.P., and Lodhi, S. (2019). Role of flavonoids in management of inflammatory disorders. In: *Bioactive Food as Dietary Interventions for Arthritis and Related Inflammatory Diseases*, Chapter 18, 2e (eds. R.R. Watson and V.R. Preedy), 293–322. Academic Press.

101 Baecklund, E., Iliadou, A., Askling, J. et al. (2006). Association of chronic inflammation, not its treatment, with increased lymphoma risk in rheumatoid arthritis. 54 (3): 692–701.

102 Siebert, S., Tsoukas, A., Robertson, J., and McInnes, I. (2015). Cytokines as therapeutic targets in rheumatoid arthritis and other inflammatory diseases. *Pharmacol. Rev.* 67 (2): 280.

103 Borchardt, L., Oschatz, M., and Kaskel, S. (2014). Tailoring porosity in carbon materials for supercapacitor applications. *Mater. Horiz.* 1 (2): 157–168.

104 Lv, Y., Zhang, F., Dou, Y. et al. (2012). A comprehensive study on KOH activation of ordered mesoporous carbons and their supercapacitor application. *J. Mater. Chem.* 22 (1): 93–99.

105 Rodriguez Correa, C., Otto, T., and Kruse, A. (2017). Influence of the biomass components on the pore formation of activated carbon. *Biomass Bioenergy* 97: 53–64.

106 Zhu, Z., Jiang, H., Guo, S. et al. (2015). Dual tuning of biomass-derived hierarchical carbon nanostructures for supercapacitors: the role of balanced meso/microporosity and graphene. *Sci. Rep.* 5: 15936.

4

Applications of Carbon-Based Nanomaterials for Wastewater Treatment

Ramesh K. Guduru, Anurag A. Gupta, Parwathi Pillai, and Swapnil Dharaskar

Pandit Deendayal Petroleum University, Raisan, Gandhinagar, 382010, Gujarat, India

4.1 Introduction

Explosive growth of global industrialization and population followed by urbanization has resulted in a tremendous rise in the waste production and pollution levels in almost all the countries. This has therefore led to numerous problems in our ecosystem including the lack of fresh and clean water supply. Despite the imposition of several legislations, industrialization has always resulted in the release of harmful products, such as dyes, heavy metals, and toxic compounds, into the environment affecting the air, soil, and groundwater quality. On the other hand, there is a shortage of low-cost and efficient practicable solutions for purification of wastewater. Thus, there has been a great demand for the solutions or materials that are safe to deal with and provide efficient and economical performance in the detection and the removal of contaminants from the wastewater. At this juncture, several low dimensional materials, i.e. nanomaterials, with large surface area have been found to exhibit superior performance in the detection of the contaminants and the wastewater treatment for the removal of different types of pollutants. Among those, porous carbon or activated carbon materials have long been in use for the wastewater treatment applications. However, the latest materials like carbon nanotubes (CNTs) and graphene and their derivatives have been found to show far superior characteristics with enhanced selectivity for large number of contaminants because of their extremely large surface area and excellent adsorption and disinfection characteristics. While considering a huge progress in the past couple of decades, herein we review and present the recent advances and current progress in the wastewater treatment using the latest carbonaceous nanomaterials for the removal of various contaminants from the wastewater. At the same time, we will also review morphological aspects and multifunctional characteristics of the carbonaceous nanomaterials. This chapter is intended to provide insights into the performance of graphene and CNTs and their derivatives including composites for wastewater treatment applications following different mechanisms.

Environmental Applications of Carbon Nanomaterials-Based Devices, First Edition.
Edited by Shadpour Mallakpour and Chaudhery Mustansar Hussain.

4.2 Wastewater

Irrespective of the developed or developing or under developed nations, the wastewater generation has become a big challenge to deal with in all the countries across the globe. In fact, the rate of wastewater production is more in the developing and developed nations. The discharges from the municipal and industrial treatment plants are some of the major contributors to the aquatic pollution. In most of the developing nations, the domestic and industrial waste is directly released into the water streams without any treatment [1]. Even in the countries like China, which is a highly industrialized nation, approximately 55% of sewage is discharged without any treatment [2]. Thus the wastewater generation in any nation is done by three major sources [3, 4]:

- Improper drainage, disposal of sewage, and human waste in the low income areas and highly urbanized cities.
- Release of effluence, dyes, medical waste, and toxic chemicals from urban-based hospitals, motor garages, mining, and industries.
- Agricultural emissions: In the process of maximizing yields of farming, the usage of fertilizers and pesticides has increased irrespective of the large- and small-scale farming, which eventually started to impact the environment by polluting the soil and groundwaters.

The other sources of wastewater generation are attributed to underground leakages, marine dumping, atmospheric deposition, global warming, and oil pollution [5]. The contamination levels and the types of contaminants in the wastewater vary depending on the types of sources and the processes of contamination. The contaminants in the wastewater could vary from physical to chemical and biological in nature [6]. Table 4.1 shows an exemplary list of types of contaminants found in the wastewater and their impact on our lives. The untreated wastewater poses several environmental problems when it enters into the regular water streams. For example, (i) the organic matter waste in the wastewater will

Table 4.1 Common types of pollutants/contaminants found in the wastewater and their impact on our lives.

Contaminant	Reason of importance
Pathogenic organisms	Found in wastewater can cause infectious diseases
Suspended solid(s)	They lead to development of sludge deposits and anaerobic conditions when untreated wastewater is discharged to the aquatic environment
Refractory organics	They tend to resist conventional wastewater treatment including surfactants, phenols, and agricultural pesticides
Heavy metals	They are highly toxic and environmentally hazardous species
Priority pollutants	Including organic and inorganic compounds
Dissolved inorganic constituent	Such as calcium, sodium, and sulfate are often initially added to domestic water supplies and may have to be removed for wastewater reuse

consume the dissolved oxygen for biochemical oxygen demand (BOD) and deplete the dissolved oxygen required by other aquatic lives; (ii) the pathogenic or disease causing microorganisms and toxic compounds can dwell in the human intestinal tract and pose risks to human health; (iii) the nutrients in the wastewater can lead to eutrophication of the ponds and streams by stimulating the growth of other aquatic lives, such as plants and algal blooms; and (iv) the decomposed organic compounds in the wastewater can lead malodorous gases. Therefore, the wastewater treatment is a must before it enters the natural water bodies. At the same time, it should be noted that the wastewater treatment and following resultant water produced could be highly dependent on the processes or the technologies being implemented and their efficiencies.

4.3 Wastewater Treatment Methods

The scarcity of fresh water has been impacting the whole world in different aspects. The policy makers across the globe have been trying to solve the water scarcity issues while building the dams, recharging the groundwater, seeding the clouds, desalinating, treating the wastewater, and developing large-scale water transfer projects [7–9]. In the twentieth century, the world population has grown by three times, whereas the water usage increased by sixfold [10, 11]. Therefore, the continuous shortage of freshwater resources motivated the researchers to find radical and cheap solutions, especially by solving the problems associated with the wastewater production and treatment. There are basically four fundamental steps involved in the wastewater treatment processes, which are shown in Table 4.2. The initial physical methods focus on removing insoluble solids without any chemical interventions. Following that, the chemical methods undertake the removal of dissolved contents and ions using different chemical approaches including chlorination, ozonation, adsorption, ion exchange, etc. In the process of chlorination and ozonation, the bacteria also get killed. The biological methods then use different processes to break down the organic matter. In the final stage, the leftover micro contents of the wastewater are removed to obtain higher purity levels, and at the same time, for quality correction, required dosing will be done to the water. For a detailed overview of all the above methods, the reader is suggested to refer to the review articles [12–14].

Table 4.2 Wastewater treatment methods.

Preliminary treatment	• *Physical*: Sedimentation, screening, aeration, particle filtration (≥1 μm), floatation and skimming, degasification, equalization
Primary treatment	• *Chemical*: Chlorination, ozonation, neutralization, coagulation, adsorption, ion exchange
Secondary treatment	• *Biological*: Aerobic – activated sludge treatment method, trickling filtration, oxidation pond, lagoons, aerobic digestion. Anaerobic – anaerobic digestion, septic tank, lagoons
Tertiary treatment	• *Final treatment*: Disinfection, oxidation, chemical dosing for water quality correction, chemically aided settling, filtration, softening, activated carbon treatment, ion exchange, membrane process

In spite of so many advances in the conventional wastewater treatment technologies, the removal of some of the emerging micro contaminants, such as endocrine-disrupting compounds (EDCs) in the wastewater, has become a challenge. The discharge of some of these complex compounds into the water streams has affected all the living organisms and the environment. The emerging complex chemicals from the pharmaceuticals, surfactants, industrial additives, personal care products, and many numerous chemicals rendered the conventional treatment methods quite inefficient as they could not meet the environmental standards during the purification process. The traditional materials/technologies like activated carbon, nanofiltration (NF), oxidation, activated sludge, and reverse osmosis (RO) membranes were observed to be ineffective in treating the complex pollutants. Thus many of the conventional water treatment technologies are not able to address the efficient removal of a wide range of toxic chemicals and pathogenic microorganisms from the wastewater. Although dilution of micropollutants was thought to reduce the toxic effects in the downstream waters [15, 16], their existence even in micro- or nanograms per liter is quite hazardous. On the other hand, biological treatment systems, such as biological trickling filters and activated sludge, are not able to support the filtration of a wide range of soluble contaminants [17–19]. Similarly, the physicochemical methods like flocculation, lime softening, or coagulation turned out to be ineffective in removing the pharmaceutical compounds and EDCs [20–22]. Chlorination against the regrowth of pathogens and bacteria [23, 24] resulted in undesirable tastes and odors and formation of disinfection by-products (DBPs) [25–28]. As an alternative to chlorination, ozonation is considered to be less attractive due to higher costs and shorter lifetime. Ultraviolet (UV) photolysis and ion exchange are advanced treatment techniques; however, they are not plausible alternatives for the removal of micropollutants [29]. The pressure-driven membrane-based microfiltration, ultrafiltration, NF, and RO, are considered to be effective processes in removing the organic micropollutants [29–34], but for wastewater treatment, these techniques are not cost-efficient solutions [35]. Thus, above limitations in different aspects of wastewater treatment certainly project a huge demand for the development of advanced materials and treatment methods that are cost-effective and can treat newly emerging micropollutants. The new inventions need to be stable, economical, and more efficient than the already existing techniques. Following that, the traditional technologies need to be modernized/updated/modified/replaced with the new material systems and technologies that are efficient, cost-effective, and reliable. This is very crucial in today's world as the water scarcity challenges are ramping up, and hence achieving a considerable quality in the treated wastewater by reusing it and saving the potable water will help tackle the worsening quality of the groundwater and drinking water. This is where the nanotechnology has been found to be quite effective in solving the wastewater problems, especially from the quality point of view while dealing with large quantities [36].

4.4 Nanomaterials

There are different types of nanomaterials including organic and inorganic in nature with different size, shape, morphology, and dimensionality in their structures. They could either exist in the form of nanoparticles or nanolayers or nanostructured materials. The nanostructured materials usually have internal features (e.g. grains) with size range between 1 and 100 nm; on the other hand, the nanophase/nanoparticle materials exist as individual particles of 1–100 nm size [37, 38]. The nanolayer materials exist in 2D with thickness in the nano regime. Thus, the nanomaterials can be divided into different categories based on the dimension, and their size could have an effect on its resultant properties, especially 0D, 1D, 2D, and 3D nanostructures [39]. The nanomaterials exhibit unique and enhanced properties when compared with the bulk materials because of their high surface area-to-volume ratio. The improvement in their properties is achieved through small feature size with large percentage of atoms on the surface or along the grain boundaries. The characteristics of nanomaterials are quite sensitive to their sizes [40–42] due to a large percentage of surface atoms and active sites. The large surface area of nanomaterials can be exploited for many practical/industrial applications, such as filtration processes and catalytic processing because of their enhanced chemical reactivity [43–45]. They exhibit excellent specific functionality and selectivity. Nanoscale adsorbents have a great affinity to adsorb various molecules and substances. High porosity and large active surface area enable the nanomaterials sequester the contaminants with hydrophilic/hydrophobic characteristics, varying molecular size, and speciation behavior. Nano-adsorbents show rapid kinetics with superior binding capacities compared with the micro- and macroscale adsorbents because of their large exposed active surface area. On the other hand, regeneration of nanomaterials is also easier after their exhaustion [46]. These salient features of nanomaterials can be very effectively utilized in the wastewater treatment processes, especially in the chemical, biological, and final treatment stages, where the reactive and functionalization nature of nanomaterials can be thoroughly exploited.

There is a plenty of research literature existing on the uses of various nanomaterials for wastewater treatment applications ranging from pure metallic nanoparticles to metal oxides, carbonaceous nanomaterials, and nanocomposites [47–53]. As mentioned earlier, the nanomaterials have strong adsorption capacities and reactivity with more active sites exposed on their surface. Importantly, their mobility is also high in solutions [50]. Therefore, utilizing different characteristics of nanomaterials, organic pollutants [51], inorganic anions [52], heavy metals [53], and bacteria have been reported to be successfully removed from the wastewater. On the basis of numerous studies, nanomaterials showed a great promise for applications in the wastewater treatment. Among different types of nanomaterials, carbon-based systems have garnered a major attention due to the long history of activated carbon uses in the water filtration applications. Also, the recent discovery of graphene and its capabilities drove the research toward realizing its potentials for filtration technologies, including the wastewater treatment. At the same time, carbonaceous materials, such as activated carbon, CNTs, and graphene showed high adsorption capacities and thermal stability [54–57]. They are capable of removing many pollutants, including inorganic and organic nature. Since the discovery of CNTs, and recently graphene, these materials have been extensively investigated for their adsorption and filtration characteristics [58].

In this chapter, the focus will be given to the types of carbonaceous nanomaterials and their structures, properties, and functionalities with regard to the wastewater treatment applications. Finally, commercial prospects, along with challenges and a summary, will be provided toward the end.

4.5 Carbon-Based Nanomaterials

Carbon-based nanomaterials (CNMs) have captured a great interest due to their unique structures and excellent properties. They exhibit superior and distinctive physical and chemical properties due to their unique ability to establish diverse hybridization states (sp, sp^2, and sp^3) that result in different varieties of structures, including its allotropes and nanostructures [59]. Figure 4.1 shows the types of CNMs based on their dimensionality.

Based on hybridization, the CNMs can be divided into sp^2 and sp^3 hybridization materials. The sp^2 hybridization materials with graphenic nanostructures are densely packed with hexagonal honeycomb crystal lattice, and they also have sp^3 hybridization at the defects and edges. The CNMs that fall under this category are graphene (2D), CNTs (1D), onion-like carbon nanospheres, and carbon dots (0D). These materials exhibit 2D lattice structure with simple hexagons. The simplest basic representative of this group is a single-atom-thick graphene sheet, which is the building block for other nanostructures, such as fullerene (0D), CNT (1D), and multilayer-stacked carbon nanosheets or graphite (2D). The complex carbon nanohorn structures are also made of the same building block, i.e. the graphene sheet. On the other hand, the CNMs with sp^3 hybridization contain both sp^3

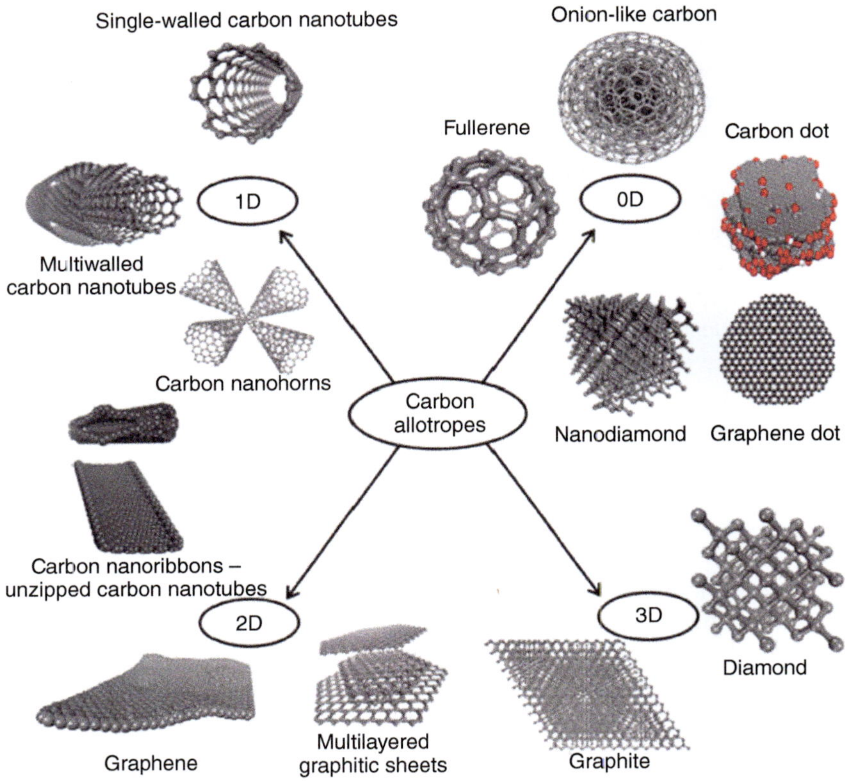

Figure 4.1 Classification of CNMs based on their dimensionality. Source: Georgakilas et al. [59]. © 2015 American Chemical Society.

and sp^2 hybridizations at various sites of the bonding. They can also possess amorphous and graphitic regions but predominantly sp^3 carbon atoms. Nanodiamond and certain types of carbon dot structures fall under this paradigm. The distinction of these materials is that they are not built upon the single atom or monolayer of graphene building block. For more details on the CNM classification and their properties, one can refer to the review articles [59, 60].

Depending on the structures and varieties, the CNMs have attracted a great attention in various applications ranging from filtration to catalysis and many more. With several advances in the nanotechnologies, the widespread role of CNMs in the water treatment applications has seen a rapid growth in the recent past, especially after the realizing the potentials of graphene. CNMs are attractive materials for fundamental research as well as diverse applications, especially in sorption processes. Their advantages for wastewater treatment include (i) a great capacity to adsorb a wide range of contaminants due to their large specific surface area with more functional sites on the surface, (ii) fast kinetics, and (iii) selectivity toward aromatics [60]. Among all the new-generation porous nanomaterials, the CNMs, such as the CNTs and graphene, have been studied extensively for wastewater treatment applications. They are mesoporous structured with good chemical stability and are considerably different from the conventional adsorbent materials. The CNTs are the rolled up sheets of graphene into the cylindrical format with diameter as small as 1 nm [50]. They exist as single wall and multiwall. The single-walled carbon nanotubes (SWCNTs) consist only a single graphene sheet, and the multiwalled carbon nanotubes (MWCNTs) have multiple graphene sheets rolled up coaxially. The CNTs exhibit exceptional adsorption properties with high adsorption efficiencies for various contaminants, such as dichlorobenzene, ethyl benzene [61], Zn^{2+} [62], Pb^{2+}, Cu^{2+}, Cd^{2+} [63], and dyes [64]. Similarly, the graphene and graphene oxide (GO) nanosheets were employed as the adsorbents for the removal of various metallic (including heavy metals), nonmetallic, organic, and inorganic pollutants from the wastewater. They showed excellent water purification capabilities while manipulating their structures and functionalization and varying the pH conditions [65, 66]. Tables 4.3 and 4.4 provide a list of contaminants removed from wastewater using CNTs and graphene and their derivatives as adsorbents.

4.6 Adsorption Mechanisms of CNTs and Graphene

4.6.1 Adsorption Through Physical and Chemical Methods

The adsorption behavior varies based on the physicochemical characteristics of the adsorbents, including their specific surface area, surface functional groups, hydrophobicity, and chemical bonds (e.g. electrostatic, π–π bond, hydrogen bond) [99]. Both CNTs and graphene have sp^2 hybridization in which three of the four valance electrons of carbon atoms are covalently shared in a 2D plane forming a strong σ bond. Remaining one valence electron is then delocalized among all the carbon atoms and results in the formation of a weak π bond in the third dimension [100, 101]. This is the reason why CNTs and graphene could easily form π bonds, secondary interactions, such van der Waals forces, hydrogen bonds, and electrostatic interactions with certain organics and make them attached without any physical or chemical or property changes [102]. Also, the chemical bonding of

Table 4.3 List of contaminants removed using CNTs and its derivatives as adsorbing agents.

Adsorbent	Adsorbate	Maximum adsorption capacity	Suitable adsorption model	Optimized conditions	Remarks	References
HNO_3 oxidized MWCNTs	Ni^{2+}	6.346×10^5 mol/g	Langmuir	Contact time, pH, foreign ions	i) The adsorption of Ni(II) on oxidized MWCNTs is dependent on foreign ions at low pH values and independent of foreign ions at high pH values ii) The adsorption of Ni(II) on oxidized MWCNTs increased with increasing pH iii) Kinetic adsorption of Ni(II) on oxidized MWCNTs can be demonstrated by a pseudo-second-order model well	[67]
Oxidized MWCNTs	Ni^{2+}	2.7099 mg/g	Freundlich	pH, contact time, initial concentration MWCNT dosage	i) The adsorption of Ni(II) on oxidized MWCNTs increases with increasing pH ii) Oxidized MWCNTs is a promising material for pre-concentration and solidification of Ni(II) ion from large volume of solutions	[68]

Microwave-assisted MWCNTs	Cd^{2+}	88.62 mg/g	Freundlich	pH, MWCNTs, dosage, contact time	i) Kinetic study of the adsorption process obeys pseudo-second order ii) MWCNTs emerge as a better promising adsorbent in the removal of Cd(II) as the removal percentage is more than 98%	[66]
Oxidized CNTs	Cd^{2+}	34.36 mg/g	Langmuir	pH, CNT dosage, contact time, agitation speed	i) Cadmium removal in acidic condition (the lower the pH value, the higher the cadmium removal) ii) The highest percentage removal of cadmium obtained from the adsorption experiment is 19.14%	[69]
Oxidized CNTs	Cd^{2+}, Pb^{2+}, Cu^{2+}	10.86, 97.08, 28.49 mg/g	Langmuir	pH, CNT dosage	i) The competitive adsorption studies showed that the affinity order of three metal ions adsorbed by CNTs is $Pb^{2+} > Cu^{2+} > Cd^{2+}$ ii) Higher dosage of CNTs increases the adoption of metal ions	[70]

(Continued)

Table 4.3 (Continued)

Adsorbent	Adsorbate	Maximum adsorption capacity	Suitable adsorption model	Optimized conditions	Remarks	References
CNTs	Cd^{2+}	1.1 mg/g	Langmuir	pH, CNT dosage, contact time	i) The high efficiency of adsorption capacity with increase CNT dosage 0.2 g per 100 ml gives increase in a doption capacity of 3.5 mg/g ii) Adsorption of cadmium(II) by CNTs was strongly pH dependent	[67]
Oxidized CNTs	Cu^{2+}, Cd^{2+}, Zn^{2+}, Ni^{2+}	0.0404, 0.0260, 0.0152, 0.0311 mmol/g of Cu^{2+}, Cd^{2+}, Zn^{2+}, Ni^{2+}	Langmuir	pH, initial concentration, contact time	i) The adsorption was more complex. The order of amount adsorbed at the same concentration was $Cu^{2+} > Cd^{2+} > Zn^{2+} > Ni^{2+}$ ii) The mechanism of adsorption is governed by the surface features, ion exchange process, and electrochemical potential	[71]
CNTs	Cu^{2+}, Co^{2+}, Pb^{2+}	3.49, 2.60, 2.96 mg/g of Cu^{2+}, Co^{2+}, Pb^{2+}, respectively	Freundlich	pH, initial concentration, contact time, CNTs dosage	i) The decrease of pH leads to neutralization of surface charge which decreases the adsorption of cations ii) The metal ions onto the CNTs followed the orderof $Cu^{2+} > Pb^{2+} > Co^{2+} >$	[72]

As-grown and HNO_3 oxidized MWCNTs	Pb^{2+}, Cu^{2+}, and Cd^{2+}	Pb^{2+}: 63.29 mg/g Cu^{2+}: 28.33 mg/g Cd^{2+}: 11.01 mg/g	Langmuir	Competitive ions	i) Pseudo-second-order rate model fits the adsorption data well ii) Adsorption on oxidized CNTs followed an order of $Pb^{2+} > Cu^{2+} > Cd^{2+}$	[73]
MWCNTs	Pb^{2+}	162.16 m^2/g	Langmuir	pH, contact time, CNT dosage	i) The acidified MWCNTs can be efficiently regenerated by adjusting the solution to pH 2 ii) The equilibrium time needed for adsorption of Pb^{2+} about 20 min	[74]
Microwave-assisted MWCNTs	Pb^{2+}	104.2 mg/g	Freundlich	pH, MWCNTs, contact time, agitation speed	i) The MWCNTs emerge as a promising adsorbent in the removal of 99% Pb(II) from aqueous solution ii) The adoption kinetics obeyed pseudo-second-order rate model	[66]
Oxidized MWCNTs	Pb^{2+}	9.92×10^{-6} mol/g	Langmuir	pH, contact time, CNT dosage	i) Oxidized MWCNTs is able to provide a higher adsorption capacity than raw CNTs, and the adsorption capacity is strongly dependent on the pH values while being independent of ionic strength and the type of foreign ions ii) The kinetic sorption of Pb^{2+} onto the oxidized MWCNTs was well described by the pseudo-second-order rate equation	[75]

(Continued)

Table 4.3 (Continued)

Adsorbent	Adsorbate	Maximum adsorption capacity	Suitable adsorption model	Optimized conditions	Remarks	References
MnO_2/CNTs	Pb^{2+}	78.74 mg/g	Langmuir	pH, contact time, MnO_2/CNT dosage	i) The adsorption of Pb^{2+} is endothermic nature ii) The Pb(II) removal capacity of MnO_2/CNTs decreased with the decrease of pH iii) The equilibrium and the kinetic adsorption can be described by the pseudo-second-order rate	[76]
Oxidized MWCNTs	Pb^{2+}	91 mg/g	Langmuir	pH, CNT dosage	Pb(II) adsorbed on the acidified MWCNTs reacts with the oxygenous functional groups	[74]
Oxidized with CNT-DES	Pb^{2+}	288.4 mg/g	Langmuir	pH, contact time, CNT dosage	i) The maximum adsorption capacity of 288.4 mg/g for CNTs functionalized with $KMnO_4$ ii) The Pb(II) adsorption system fitted excellently to a pseudo-second-order kinetics model	[77]
Microwave-induced MWCNTs	Cu^{2+}	99 mg/g	Freundlich	pH, MWCNTs, dosage, contact time, agitated speed	i) The adsorption process obeys the pseudo-second-order kinetics ii) The adsorption of Cu^{2+} was feasible, spontaneous, and endothermic in nature	[66]

CNTs	Cu^{2+}	8.25 mg/g	Langmuir	pH, ionic strength, CNT dosage	i) The adsorption capacity of Cu^{2+} onto the surface of CNTs increased with the system pH from acidic to alkaline solution ii) The adsorption of Cu^{2+} onto CNTs was endothermic in nature	[78]
HNO_3 oxidized MWCNTs	Th^{4+}	57.14, 65. 9, and 74.07 μmol/g at 293, 303, and 323 K	Langmuir	Initial concentration of Th^{4+} ions	i) Adsorption data fits pseudo-second-order very well ii) Adsorption of Th^{4+} onto oxidized MWCNTs is endothermic iii) Adsorption of Th^{4+} is reversible, and these ions can be desorbed from the surface of oxidized MWCNTs by manipulating the pH values of the solution	[79]

(Continued)

Table 4.3 (Continued)

Adsorbent	Adsorbate	Maximum adsorption capacity	Suitable adsorption model	Optimized conditions	Remarks	References
Raw and 30 ml HNO_3 + 10 ml H_2SO_4 (3N + 1S), HNO_3, $KMnO_4$, and NaClO oxidized MWCNTs	Zn^{2+}	10.21, 18.14, 27.20, 28.01, and 32.68 mg/g in respect of raw MWCNTs and oxidized MWCNTs of (3N + 1S), (HNO_3), ($KMnO_4$), (NaClO)	Langmuir	Different oxidations of MWCNTs	i) Adsorption capacity found highest for MWCNTs (NaClO), followed by MWCNTs ($KMnO_4$), MWCNTs (HNO_3), MWCNTs (3N + 1S), and raw MWCNTs ii) The kinetics of Zn^{2+} sorption onto MWCNTs (NaClO) was found to follow the pseudo-second-order rate law iii) Functionalization process increased the carbon content in MWCNTs, oxygen-containing functional group, and negatively charged carbon surfaces	[80]
Oxidized MWCNTs	Zn^{2+}	1.05 mg/l	Langmuir and Freundlich	pH, MWCNTs dosage, contact time, agitation speed	i) Oxidized CNTs proved high efficiency of removal of Zn^{2+} (99%) from aqueous solution ii) The adsorption kinetics follows pseudo-second order	[81]

$NaClO_3$ oxidized SWCNTs and MWCNTs	Ni^{2+}	51.28 and 40.00 mg/g for oxidized SWCNTs and MWCNTs, respectively, at 45 °C	Langmuir	Contact time and temperature	i) The amount of Ni^{2+} sorbed onto SWCNTs and MWCNTs increased with a rise in temperature ii) SWCNTs and MWCNTs possess more sorption capacities of Ni^{2+} than granular activated carbon (GAC) iii) Sorption kinetics follows the pseudo-second-order rate law while the sorption thermodynamics indicates the endothermic and spontaneous nature. iv) The sorption affinity between Zn^{2+} and CNT's surfaces is stronger than that between Ni^{2+} and CNT's surface	[82]
Oxidized CNTs	Ni^{2+}	49.261 mg/g	Langmuir	pH, CNT dosage, adsorption time	i) At pH 6, the adsorption uptake reaches the maximum due to the effect of the functional groups formed on the CNTs surface during oxidation ii) Oxidized CNTs have greater adsorption capacity for nickel ion removal	[83]

(Continued)

Table 4.3 (Continued)

Adsorbent	Adsorbate	Maximum adsorption capacity	Suitable adsorption model	Optimized conditions	Remarks	References
MWCNTs/iron oxide	Ni^{2+}	9.18 mg/g		pH, CNT dosage	i) Adsorption of Ni^{2+} ions depends on ionic strength and pH of solution ii) The adsorption capacity of the magnetic composites is much higher than that of MWCNTs and iron oxide	[84]
Oxidized MWCNTs	Ni^{2+}	38.46 mg/g	Langmuir	pH, CNT dosage, agitation speed	i) The oxidized MWCNTs have shorter equilibrium time and better Ni^{2+} sorption ii) The amount of Ni^{2+} sorbed onto oxidized CNTs increase with increase in agitation speed	[82]
MWCNTs	Cr^{6+}	34.9 mg/g at 310 K	Brunauer–Emmett–Teller (BET)	Adsorbent dosage, initial Cr^{6+} concentration, initial pH, contact time and temperature	i) Low pH, low initial concentrations of Cr^{6+}, increasing contact time, and high temperature were found as optimal conditions ii) The pseudo-second-order kinetics model was found to correlate with the experiment data as well iii) Thermodynamic parameters were estimated, and result suggests that the adsorption process is spontaneous, physical, and endothermic	[85]

MWCNTs	Cr^{6+}	34.9 mg/g	Langmuir	pH, contact time, temperature, initial concentration	i) The adsorption kinetic studies proved that pseudo-second-order rate equation ii) The removal of Cr^{6+} (95%) was obtained at an optimal condition of low pH, low initial concentrations of Cr(VI), increasing contact time, and high temperature	[85]
Oxidized MWCNTs	Cr^{6+}	2.46 mg/g	Langmuir	pH, initial concentration of Cr(VI), MWCNT content, contact time, and ionic strength	i) The kinetic adsorption of Cr^{6+} can be described well by pseudo-second-order models ii) The removal was favored at low pH with maximum removal at $pH < 2$	[86]
(CeO_2/ACNTs)	Cr^{6+}	30.2 mg/g	Langmuir	pH, CNT dosage	i) Employing ceria nanoparticles supported on aligned carbon nanotubes (CeO_2/ACNTs) tend to be an outstanding adsorbent as well by obtaining a high adsorption capacity of CeO_2/ACNTs reaching 30.2 mg/g at an equilibrium Cr(VI) concentration of 35.3 mg/l at pH 7.0 ii) CeO_2/ACNTs provided an adsorption capacity that is 1.5 times higher than activated carbon, 2.0 times higher than Al_2O_3, and 1.8 times higher than ball-milled ACNTs	[63]

(Continued)

Table 4.3 (Continued)

Adsorbent	Adsorbate	Maximum adsorption capacity	Suitable adsorption model	Optimized conditions	Remarks	References
Activated carbon (AC) and AC–CNT	Cr^{6+}	8.25 and 9 mg/g of AC and AC–CNT, respectively	Langmuir	pH, contact time, agitation speed	i) The adsorption capacity due to the higher surface area on AC–CNT-coated adsorbent ii) The particle size plays vital role in adsorption of Cr^{6+} from aqueous solution	[87]
Oxidized CNTs	CrO_4^{2-}	249 μmol/g	Langmuir	pH, CNT dosage, initial concentration of CrO_4^{2-}	i) The adsorption isotherm of CrO_4^{2-} onto functionalized CNTs was well demonstrated by Langmuir isotherm model ii) The functional groups has a higher affinity toward the chromate ion due to the ion exchange occurring between the chromate ion and the acidic surface OH groups on the surface of functionalized CNTs	[73]
Oxidized MWCNTs	Sn^{2+}	0.459 mg/g	Langmuir	pH, MWCNTs dosage, agitation speed, contact time	i) Adsorption kinetic obeyed pseudo-second-order kinetic model ii) The functionalized MWCNTs is a better adsorbent for removal of 93% of Sn^{2+} from aqueous solution	[81]

CeO_2-coated aligned CNTs (CeO_2/ACNTs)	Cr^{6+}	30.2 mg/g	Langmuir and Freundlich	pH of Cr^{6+} solution	i) Nanoparticles of CeO_2 supported on ACNTs can be prepared by chemical reaction of $CeCl_3$ and NaOH in ACNTs solution, followed by heat treatment ii) Adsorption capacity of CeO_2/ACNTs 1.5 times higher than activated carbon, 2.0 times higher than Al_2O_3, and 1.8 times higher than ball-milled ACNTs	[63]
Al_2O_3 supported on CNTs (Al_2O_3/CNTs)	F^-	14.9 mg/g	Freundlich	pH of F^- solution	i) Al_2O_3/CNTs exhibit high adsorption efficiency from water in pH values of 5.0–9.0 ii) Adsorption capacity of Al_2O_3/CNTs is 13.5 times higher than AlC-300 carbon and four times higher than $\gamma - Al_2O_3$ and IRA-410 polymeric resin	[88]

(Continued)

Table 4.3 (Continued)

Adsorbent	Adsorbate	Maximum adsorption capacity	Suitable adsorption model	Optimized conditions	Remarks	References
CeO_2-coated CNTs (CeO_2/CNTs)	AsO_4^{3-} in function of Ca^{2+} and Mg^{2+}	81.9 and 78.8 mg/g in respect of Ca^{2+} and Mg^{2+}	Freundlich	pH of AsO_4^{3-} solution	i) CeO_2/CNTs have higher zeta potentials than oxidized CNTs and can be expected to reduce the repulsion between the particles and negatively charged substances and enhance the adsorption capacity ii) The AsO_4^{3-}-loaded adsorbent can be regenerated by diluted NaOH and high regeneration efficiency of 94% achieved using 0.1 mol/l NaOH	[89]
Thermally treated MWCNTs	natural organic matter (NOM)	26.138 mg/g	Langmuir	pH, ionic strength, contact time, initial concentration	i) The amount of NOM adsorbed onto the CNTs increased with a rise in initial NOM concentration and solution ionic strength but decreased with a rise in solution pH ii) Heat treatment leads to increase in graphitized structure and decrease in surface functional groups and negative charges. Adsorption capacity of both dissolved organic carbon (DOC) and assimilable organic carbon (AOC) increased with initial concentration and ionic strength but decreased with initial pH	[90]

$HNO_3 + H_2SO_4$ oxidized CNTs	Trihalomethanes (THM) consists of $CHCl_3$, $CHCl_2Br$, $CHClBr_2$, and $CHBr_3$	2.41, 1.23, 1.08, and 0.92 mg/g of $CHCl_3$, $CHCl_2Br$, $CHClBr_2$, and $CHBr_3$, respectively	Langmuir and Freundlich	Contact time and solution's pH	i) Adsorption of THMs onto CNTs fluctuates very little in the pH range of 3–7 but decreases with pH value as pH exceeds 7 ii) $CHCl_3$ being the smallest molecule most preferentially adsorbed onto CNTs followed by $CHBrCl_2$, $CHBr_2Cl$, and then $CHBr_3$ iii) The purity, structure, and nature of the surface of CNTs can be improved by treatment with acid	[62]
NH_3 and HNO_3 oxidized MWCNTs	Chlorophenols	110.3 and 73.5 mg/g for NH_3 and HNO_3, respectively	N/A	pH, shaking time	i) The specific surface area provides a larger area for the interactions of chlorophenols (CPs) ions and the surface of CNTs ii) The stronger hydrophobic interactions can be another reason for the highest adsorption capacity of NH_3–MWCNTs	[91]
As-grown and graphitized CNTs	Dichloro-benzene (DCB)	30.8 and 28.7 mg/g of as-grown and graphitized CNTs, respectively	Freundlich	Contact time and DCB solution's pH	i) As-grown CNTs are better for adsorption of DCB compared with graphitized CNTs ii) Adsorption increased in the pH range of 3–10 and decreases beyond pH 10 due to the ionization of the oxygen groups on the surface and formation of water cluster on these groups	[61]

Table 4.4 List of contaminants removed using graphene and its derivatives as adsorbing agents.

Magnetic reduced graphene oxide (M-RGO)	As(III) and As(V)	5.83 and 13.10 mg/g, 2.28, and 3.79 mg/g in respect of As(V) and As(III) for Langmuir and Freundlich	Langmuir and Freundlich	Initial arsenic concentration	i) The temperature of the maximum adsorption of arsenic ions is at 30 °C ii) The M-RGO composites that were dispersed in water solution (1 mg/ml) can be separated from water by using a magnet to prove the magnetic properties	[92]
Polypyrrole–reduced graphene oxide (PPy-RGO)	Hg^{2+}	400 and 980 mg/g in respect of PPy and PPy-RGO	Langmuir and Freundlich	Weight of GO loaded on PPy	i) The enhanced adsorption capacity of the PPy-RGO composite originates due to enhanced adsorption sites for mercury ions in the presence of graphene sheets ii) The desorption capacities of PPy for 50 ml HCl at pH 3, 2, and 1 are 54%, 65%, and 86%, respectively, while the corresponding capacities of PPy-RGO composite are 45%, 63%, and 92.3%, respectively	[92]
Magnetic chitosan/graphene oxide (MCGO)	Pb^{2+}	76.94 mg/g	Langmuir	Effects of pH, contact time, and concentration of Pb^{2+} ions	i) MCGO composite was synthesized via a facile and fast process that provided a composite with a larger surface area ii) MCGO demonstrated an extremely high desorption capacity up to 90.3%	[93]

Reduced graphene oxide, RGO-MnO_2, and RGO-Ag	Hg^{2+}	More than 10 l/g in respect of RGO-MnO_2 and RGO-Ag	Distribution coefficient, K_d	The amount of loading of MnO_2 and Ag on RGO	i) The immobilized RGO composites were carried by utilizing none external acid, which made the whole method simple and ecofriendly ii) The size of the nanoparticles formed can be tuned by controlling the precursor concentration added iii) The predicted model kinetic parameters demonstrate that the pseudo-first-order model is more suitable to predict the experimental data	[94]
Graphene oxide (GO) aerogel	Cu^{2+}	17.73, 19.65, and 29.59 mg/g at 10, 25, and 40 °C	Langmuir	pH of Cu^{2+} solution	i) A fast adsorption rate of Cu^{2+} (15 min) was attained due to the interconnected pore structure of GO aerogels, which is vital for the diffusion of metal ions in the adsorption sites on GO aerogels ii) Cu^{2+} adsorption data fitted well on pseudo-second-order kinetics iii) The adsorption of Cu^{2+} on GO aerogel increased with the increase in pH, which indicates an ion exchange mechanism	[95]

(Continued)

Table 4.4 (Continued)

Graphene oxide nanosheet	Cd^{2+} and Co^{2+}	106.3 and 68.2 mg/g in respect of Cd^{2+} and Co^{2+}	Langmuir	Effect of humic acid, pH, and ionic strength	i) The result indicates that Cd^{2+} and Co^{2+} sorption on graphene oxide nanosheets was strongly dependent on pH and weakly dependent on ionic strength ii) The presence of humic acid reduced Cd^{2+} and Co^{2+} sorption on graphene oxide nanosheets for $pH < 8$ iii) Thermodynamic parameters suggest that the sorption of Cd^{2+} Co^{2+} on graphene oxide nanosheets was endothermic and spontaneous process	[79]
Graphene nanosheets (GNSs)	Pb^{2+}	22.42, 35.21, and 35.46 mg/g in respect of GNS, GNS 500, and GNS 700	Langmuir	Initial concentration of Pb^{2+} solution and different temperature on heat treatment on GNS	i) Adsorption against Pb^{2+} ions is enhanced by heat treatment that is described by the Lewis basicity and electrostatic attraction of graphene ii) Adsorption of Pb^{2+} onto GNSs increased as the pH value of solution increases. It is well defined that the Lewis basicity of GNSs is improved by heat treatment under a high vacuum, in favor of simultaneous adsorption of Pb^{2+} and protons onto GNSs	[96]

Functionalized graphene sheets (f-HEG)	Na, As(V), and As(III)	0.216 52, 0.350 73, and 0.653 41 mg/g with respect of Na, As(V), and As(III)	Freundlich	Initial concentration of arsenic and sodium solution	i) Functionalization of highly exfoliated graphene (HEG) provides better contact between electrodes and water, resulting in better performance of metal removal from aqueous solution and seawater ii) Adsorption kinetics was found to fit better for intraparticle diffusion as compared with Elovich model iii) High removal efficiency and good cyclic repeatability of electrodes demonstrate for a high concentration of both arsenic ions containing aqueous solution and seawater	[97]
SiO_2/graphene	Pb^{2+}	113.6 mg/g	Langmuir	Effect of pH, contact time, and ionic strength	i) SiO_2/graphene composites reduce the serious stacking of graphene sheets and prevent the agglomeration of SiO_2 nanoparticles ii) The soluble graphene oxide-based sheets were produced by complete exfoliation of graphite oxide as an entry into SiO_2/graphene composite	[98]

CNT/graphene with organics is possible via functional groups or dangling bonds (e.g. carboxyl or hydroxyl groups), existing on the basal surfaces of CNT/graphene; see in Figure 4.2. However, the number of chemical interactions and formation of chemical bonds or chemisorption depends on the amount of functional groups available on the surfaces of CNTs/graphene. The near-perfect structure of graphene/CNTs mostly results in physical adsorption [104] with very little chemisorption phenomenon.

In addition, the other driving forces, such as hydrophobic/hydrophilic nature, surface area, and porous structure, also play an important role in the adsorption characteristics. The graphene/CNTs having sp^2 hybridized carbon are hydrophobic in nature and favor the adsorption of hydrocarbons at high efficiency. Nonpolar and hydrophobic organic contaminants interact with graphene/CNTs and reduced graphene oxide (rGO) quite significantly [105]. These interactions come under π–π interactions. On the other hand, through surface functionalization or modification with dangling bonds or the diversified modification, the CNTs/graphene can be turned partially or completely into hydrophilic in nature. This would be quite effective in adsorbing organic molecules of both hydrophilic and hydrophobic natures. Thus, the π–π interactions or π–π electron donor–acceptor (π–π EDA) interactions can grow stronger in the presence of functional groups as the specific surface area increases as it provides more active sites and enhances the adsorption capacity [106]. Extensive efforts have been put forth to understand the hydrophilic nature and other properties of CNTs and graphene along with their adsorption mechanisms. To induce the hydrophilic nature, oxidation is the most widely used method for graphene and CNTs [107]. During the oxidation process, the hydroxyl, carboxyl, or carbonyl groups are introduced onto CNTs/graphene, which increase the hydrophilic properties and consequently enhance their adsorption capacity for certain contaminants [108]. For CNTs, the acid treatment is the most widely used oxidation method.

The MWCNTs could be oxidized by treating with H_2SO_4/H_2O_2 mixture and result in abundant functional groups of —OH, —COOH, —C—O, —C—C, and —C=O on the surface of MWCNTs. These functional groups exhibit van der Waals forces, hydrogen bonds, and π–π interactions [109]. The MWCNTs oxidized by H_2SO_4/HNO_3 mixture exhibited efficient adsorption of Cu(II) and pefloxacin (PEF) (1 g) due to the π–π interactions and hydrogen

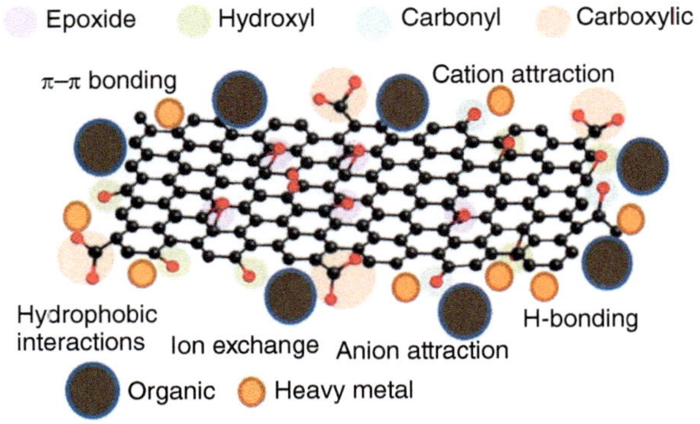

Figure 4.2 Adsorption mechanisms between graphene and different kinds of contaminants. Source: Kim et al. [103]. © 2018 Elsevier.

bonds. Nitric acid and citric acid were also observed to increase the carboxylic functional groups on MWCNTs, which enhance the hydrophilicity and adsorption of aniline, phenols, and other chemicals due to the formation of hydrogen bonds [110]. $KMnO_4$ was also used to oxidize CNTs and graphene [50]. GO is the most common product of graphene oxidation. The GO synthesized by Hummers method [111, 112] showed enhanced adsorption of ciprofloxacin (CIP) through electrostatic interactions and sulfamethoxazole (SMX) due to π–π EDA attraction. The functionalization of GO with ethylenediaminetetraacetic acid (EDTA) groups helped enhance its capacity for adsorption of toxic heavy metals [113] due to the increased hydrophilic nature with large amounts of −OH and −COOH groups. Interestingly, the −OH and −COOH groups can also form a surface complex with contaminants and increase the adsorption capacity. Despite so many alterations and enhancements, the oxidized CNTs and graphene could yet have limitations in the removal of complicated and recalcitrant organics [114].

The adsorption capacity of graphene/CNTs is also influenced by the functional groups of the adsorbates. For example, if there are chemicals with more groups of aniline or phenol substitution, then the adsorption of those molecules would have higher affinity with MWCNTs. The adsorption efficiency of various adsorbates varies with different functional groups in the following sequence: methyl group < chloride group < nitro group. Similarly, at a given equilibrium concentration, the adsorption affinity of MWCNTs decreases in the following sequence for various functional groups on the adsorbates: pyrene > phenanthrene > naphthalene, which corroborates very well with their hydrophobic behavior [115]. Theoretically, there is a linear relationship between the adsorption capacity and the specific surface area for CNM-based adsorbents as the surface area and micropore volume increase, and this is valid if adsorption follows a monolayer mechanism. It was practically observed for SWCNTs in the adsorption of phenanthrene. However, sometimes, it may not work out in the same way because of the multilayer adsorption. For example, the adsorbed volume of phenanthrene on SWCNTs and MWCNTs was higher than the total capacity predicted for the monolayer adsorption based on the volume capacity measured by nitrogen volumetric adsorption measurements. Thus, the multilayer adsorption could also become predominant, or the mesopore filling could occur [116, 117].

From the structure point of view, graphene is a 2D flat material and can adsorb on both directions of the flat platelet structure, whereas CNTs are 1D tubular material and only external surface and hollow core allow for adsorption. The arrangement of these two materials in a 3D fashion would also provide the formation of a stacked pore structure. The agglomerates of CNT and graphene always provide a sufficient number of micro-, meso-, and macropore structures that can significantly enhance their adsorption capacity and possibly enable the containment of organics that may be heavier than the adsorbates [118]. The capillary filling of CNTs and pore structures also promotes the adsorption as it combines the effects of pore structure, surface area, and curvature. Sometimes, the agglomerates of CNT could exhibit flexible and stacked pore structures of wedge or slit shape between the nanotubes. The hollow hydrophobic 3D pores can adsorb large quantities of aromatic pollutants, allowing the friction-free passage of water without requiring much energy, which has been observed to be quite faster and easier when compared with the adsorption by activated carbon [119].

The diameter of SWCNTs also affects the adsorption process. For example, the SWCNTs with large diameter and small curvature can act similar to a planar graphene sheet. It can allow the access of other chemicals as well [120], and this is very similar to that of face-to-face adsorption. In another case, with a smaller diameter of a curved SWCNT, the surface contact area for polycyclic aromatic hydrocarbons decreased, and hence adsorption decreased. This reflected how the adsorption capacity decreased with increasing the curvature of CNTs. This behavior also explains how the complicated adsorption of organics in the 3D stacking pores of CNTs and graphene agglomerates varies with curvature.

4.6.2 Adsorption Through Biological Methods

The other methods to enhance the specific adsorption of CNMs utilize biological molecules. In this approach, CNMs are functionalized or decorated with biological molecules to obtain specific selectivity. For example, the decoration of functionalized MWCNTs with glycine and β-cyclodextrin helps remove various organic dyes from the wastewater [121] through formation of compounds between the dyes and the hydrophobic cavity of β-cyclodextrin (β-CD). Following this method, the sequence of saturated absorption follows MWCNTs < MWCNTs/Gly < MWCNTs/Gly/β-CD, where MWCNTs/Gly/β-CD has the highest adsorption capacity due to a large number of –OH groups present on the CNTs. In another similar approach, the magnetic multi-walled carbon nanotubes (m-MWCNTs) were functionalized by amino groups and immobilized a recombinant cyanate hydratase (rTl-Cyn) [122]. These modified MWCNTs (m-MWCNT-rTl-Cyn) were able to remove cyanate (>84%) and heavy metals (>29%) simultaneously from a lab-prepared wastewater sample. However, as the adsorption of heavy metals increased, the adsorption of cyanate decreased due to the inhibition effects of metals adsorbed, which hindered by occupying the binding sites of cyanate and rTl-Cyn.

4.6.3 Adsorption Using Deep Eutectic Solvents (DESs)

The deep eutectic solvents (DESs) help void the aggregation of nanomaterials like CNTs or graphene through functionalization process. They are prepared by combining various hydrogen bond donors (HBDs) with different inorganic and organic salts [123]. The functionalization of MWCNTs using different DES systems showed enhanced adsorption capacity of arsenic ions. The specific surface area CNTs or their composites can be increased using different DESs. The DES solutions help enhance the dispersion, as well as easy and low-cost functionalization of CNMs (e.g. CNTs) [124].

4.6.4 CNT- and Graphene-Based Composite Adsorbents

Composites of CNMs with polymers can develop different adsorption attributes along with superior pore structures [125, 126]. Usually, the π–π stacking forces and weak van der Waals interactions help wrap the polymer around the CNTs. Graphene oxide-polyaniline (GO-PANI) composite is an example for enhanced adsorption capacity, and it was utilized to remove Zn ions [127]. The addition of PANI to GO increased the adsorption of Zn ions because of entrapment of Zn(II) in the PANI pores when compared with the limited adsorption of Zn^{2+} by GO through electrostatic interactions. In another study [128], oleophobic–superhydrophilic composite was developed with excellent adsorption capacity for aromatics and methyl blue (MB) using strong π–π interaction. They were fabricated using a mixture of aqueous poly(diallyldimethylammonium) chloride and nanoparticles, and graphene was used as a plug in these composites. Following this approach, multifunctional composite membranes were prepared to separate hexadecane from the wastewater with MB and also to decolor the MB solution. Metal sulfide-based composites were also prepared using CNMs [129]. For example, Fe nanoparticle-attached MWCNTs were treated with sulfur to form CNTs/FeS Fenton-like adsorbents, which are then used for the adsorption of ciprofloxacin. The capability of these composites depended on the unbounded −OH generated in a Fenton-like process, and they showed good adsorption for pH 3–8 up to a capacity > 75%. Nevertheless, these needed to be improved further due to their poor regeneration. Sarkar et al. [130] utilized GO-CNT-based composites for simultaneous functionalization, reduction, and hybridization of GO in GO-MWCNT hybrids while using *p*-phenylenediamine (PPD). These composites developed a new adsorbent (rGO-CNT-PPD), which was able to adsorb MO and methyl violet (MV) selectively from their mixture under different pH values.

4.7 Membrane-Based Filtration of Contaminants Using CNTs and Graphene-Based Materials

Membrane technology is a physical process, and it functions based on the movement of particles as well as the concentration gradients. The performance of a membrane is usually determined by its selectivity and permeability. The membranes used in microfiltration/ultrafiltration/NF or even RO are usually made of polymers, and their hydrophobic nature leads to irreversible adsorption of organic, inorganic, and biological substances, resulting in membrane fouling and reduced flux [131]. Thus, the performance of a membrane is controlled by the materials it is made of, and this is where incorporation of CNMs into membrane technologies provided excellent opportunities to enhance their permeability, stability, and fouling resistance. Incorporation of CNTs into membranes, such as polymers and composites, helped reduce the membrane fouling and biofouling as well. The CNTs are hydrophobic in nature; however, they were turned into hydrophilic through oxidation/acid treatment and incorporated into polysulfone [132] and polyethersulfone (PES) membranes [133] to enhance their fouling resistance using the hydrophilic and carboxylic groups on the CNTs. The hydrophilic isophthaloyl chloride groups [134] and amphiphilic polymer groups were introduced onto CNTs to impart protein resistance to the membranes [135]. The CNT-incorporated polymer membranes demonstrated high water flux at reasonably low pressures [136]. Wang et al. [137] developed a high flux filtration membrane by introducing surface-oxidized MWCNTs in the coating layer. Increasing the MWCNT dosage improved their water flux while retaining the total organic rejection, and this improvement was attributed to the generation of more effective hydrophilic nanochannels for water passage in the composite membranes [133].

Celik et al. [133] reported that MWCNTs and PES-blended membranes were more hydrophilic in nature and helped achieve higher pure water flux than the pure PES membranes alone. The properties of these membranes, such as morphology and permeation, were influenced by the amount of MWCNTs added to the blended membranes. They displayed very high flux and slow fouling rates than the usual PES membranes. However, in another study, the aligned CNTs resulted in high fouling rates due to the hydrophobic nature of the pore walls with CNTs [138]. These results reflected an increase in the fluxes with increasing the hydrophilicity of the membranes, whereas their hydrophobic nature always displayed a higher decline in their fluxes when compared with the hydrophilic membranes [139].

Another important contribution of CNTs to the polymer composite membranes is enhancing their mechanical strength and rigidity to sustain high pressures during the filtration process. According to [140], MWCNT–polymer (polyamide [PA]) composite membranes showed superior strength and mechanical properties while controlling their selectivity with varying concentration polymer matrix, i.e. PA for increased salt (NaCl) rejection. This was quite superior compared with the commercial PA membranes. With increasing MWCNTs, the structural compactness of the PA nanocomposite membranes gets better with stronger interactions between the MWCNTs and the polymer PA matrix, leading to higher permeability and higher salt rejection [141]. Thus, the addition of MWCNTs to PA not only enhanced their mechanical rigidity but also improved the purification process. With superior mechanical properties of CNTs, the tensile strength of polyacrylonitrile – 2% MWCNT [112] and chitosan – 2% MWCNT composite membranes [142] increased by 97% and 90%, respectively, compared with the pure polymer membranes.

Interestingly, the effect of CNT diameter was also demonstrated on the fouling resistance. Ajmani et al. [143] showed more effectiveness of larger diameter CNTs compared with the smaller-diameter CNTs on membrane fouling, and it was attributed to the removal of larger organic foulants. Biofouling was also reduced with incorporation of CNTs onto the polymer membranes by utilizing their antibacterial properties [144]. Tiraferri et al. [144] reported that SWCNTs were successfully attached to the surface of ethylenediamine thin film membranes (TFMs) and their incorporation was entirely based on the activation of the carboxylate groups on both the membrane and CNTs to maximize their interaction. These SWCNT-coated TFMs were able to achieve up to 60% inactivation of *Escherichia coli* bacteria within an hour of the contact time. In the case of ozonized SWCNTs, the inactivation of *E. coli* exceeded 95%, which was quite higher than the purified only CNTs. Brady-Estévez et al. showed a high bacterial retention using SWCNTs and high viral removal with MWCNTs under low-pressure conditions [145]. The combination of SWCNT and MWCNT in hybrid filters also achieved both bacterial inactivation and viral retention at low pressures quite efficiently [146]. In combination with silver nanomaterials [147], CNT-incorporated filters showed even enhanced antibacterial abilities. Srivastava et al. [148] showed the application of monolithic CNT-based hollow cylinders with radially aligned CNT walls for filtration of bacteria (*E. coli*) with low blockage. These CNT filters were found to be superior over the conventional membrane filters because of their easy regeneration or cleaning through simple ultrasonication and autoclaving. CNTs were used as additives for polymer membrane antifouling applications. Thus, the CNTs with excellent merits, along with antimicrobial activity, higher water flux, tunable surface chemistry, and pore size, are very promising materials for filtration/separation applications. Table 4.5 lists the summary of CNT-incorporated membrane performance in the water treatment applications.

Table 4.5 List of contaminants filtered using CNM-based membranes.

Nanomaterials membrane	Production method of nanomaterials membrane	Contaminated removed	Capacity of contaminated removed	Remarks	References
MWCNTs/polyethersulfone (C/P)	Phase inversion	Total organic carbon	58.81, 5.64, 11.61, and 41.57 mg/m^2 in respect of desorbed NOM, hydrophobic NOM, transphilic NOM, and hydrophilic NOM	i) Increase in the MWCNTs content in the membranes increased the membrane roughness, hydrophilicity, porosity, and pure water flux ii) Morphology and separation of the C/P blend depend on MWCNTs content iii) C/P membranes displayed 42% less flux decline than PES membranes	[133]
MWCNTs/aromatic polyamide (PA)	Polymer grafting process	NaCl and humic acid	15 mg/g of MWCNTs in a 10% PA casting solution exhibits a rejection of NaCl and humic acid by factors of 3.17 and 1.67, respectively	i) Increase in the MWCNTs content leads to the increase in the Young's modulus, toughness, and tensile strength of the membrane ii) An increase in the PA concentration in membrane casting solution leads to a higher salt rejection from 24% to 64.5% accompanied by a decrease in both permeability and specific flux that dropped from 0.76 to 0.28 l/m^2 h bar and from 32 to 11 l/m^2 h, respectively	[140]

(Continued)

Table 4.5 (Continued)

Nanomaterials membrane	Production method of nanomaterials membrane	Contaminated removed	Capacity of contaminated removed	Remarks	References
new type of thin-film nanocomposite (*n*-TFN) membrane consists of *n*Ag in thin film layer and modified MWCNTs in support layer	Interfacial polymerization	*Pseudomonas aeruginosa*, PAO1	96% of PAO1 bacteria attached on the membrane	i) Membrane permeability was enhanced in the *n*-TFN membrane compared with a non-CNT matrix through the diffusive tunnel effects of MWCNTs ii) MWCNTs at 5.0 wt% in the support layer and *n*Ag particles at 10 wt% in the thin film layer enhanced the pure water permeability of the *n*-TFN membrane by 23% and 20%, respectively iii) PAO1 batch test indicates greater anti-adhesive and antibacterial properties of the *n*-TFN membranes than *n*Ag	[149]

Nanomaterials membrane	Production method of nanomaterials membrane	Contaminated removed	Capacity of contaminated removed	Remarks	References
Sub-2-nm CNT membranes fabricated by conformal deposition of silicon nitride	Etching process	$Fe(CN)_6^{3-}$ and KCl	98% achieved at low ionic strength for $Fe(CN)_6^{3-}$ while a KCl rejection of 50% at 0.3 mM and decreased to almost zero at 10 mM	i) Percentage rejection that is sensitive to solution pH and electrostatic length suggests that electrostatic interactions dominate over steric effects in governing ion rejection ii) Biological pore channels share a number of structural and functional features with CNTs that made CNT nanofluidic platforms ideal candidates for the realization of a robust biomimetic system that could exploit the fast transport, selectivity, and gating properties of biological pores	[150]
SWCNTs covalent bonded to TFM	Covalent binding	*E. coli*	95% loss of *E. coli* viability for purified and oxidized SWCNTs	i) SWCNT membranes were able to achieve up to 60% inactivation of bacterial attached to the membrane within 1 h of contact time ii) The cooperation of SWCNTs into the TFM was entirely aqueous based and encompasses activation of the carboxylate groups of both membrane and nanomaterials to maximize reaction with ethylenediamine	[144]

(Continued)

Table 4.5 (Continued)

Nanomaterials membrane	Production method of nanomaterials membrane	Contaminated removed	Capacity of contaminated removed	Remarks	References
MWCNTs/polysulfone (PSf)	Phase inversion process	polyethylene oxide (PEO) 100 000	Over 95% of rejection as the pressure increases	i) Surface of MWCNTs/PSf membranes was found to be more hydrophilic than just PSf membranes ii) The PSf membrane with 4.0% MWCNTs showed a higher flux and rejection than PSf membrane without MWCNTs iii) When polyvinylpyrrolidone (PVP) 55 000 was used, a solute of the feed solution, overall flux of membranes was improved, and the overall rejection decreased compared with PEO 100 000	[132]
CNTs	Molecular dynamic simulation	Salt, NaCl	100% salt rejection for (5,5) size membrane	i) The energy barriers at the entrance and exit of the tubes mean that the transport is almost frictionless once molecules are inside. The permeation of ions into nonpolar interior of CNTs is controlled by the energy to remove the water molecules i) Efficiency gained of 5–1000 times can be achieved if membranes comprise less than 1 pore in 100 over the size of 10 Angstrom in diameters	[151]

4.8 Use of CNTs and Derivative Materials as Disinfecting Agents for Water Purification

Disinfection is another way of removing contaminants from the wastewater. The removal of pathogens from the wastewater could be challenging sometimes due to the fluctuating concentrations and the types of pathogens present. On the other hand, the release of toxins by the microbial could also impact the purification or quality of the treated water. For example, the bacteria like cyanobacteria releases toxic microcystins during the water treatment [152]. In addition, the formation of toxic DBPs is also considered as a challenge during the wastewater treatment. Chlorine and ozone disinfectants are capable of forming DBPs, and they can be easily formed when the oxidants are present in the water treatment [153]. However, some of the pathogens, such as *Cryptosporidium* and *Giardia*, show a high resistance toward disinfectants, and it is considered as a major problem. Therefore, to avoid the formation of DBPs and improve the quality of the treated water, nanotechnology can be adopted as an alternative [154]. Different types of engineered nanomaterials were observed to present excellent disinfectant characteristics, including nanosilver [155], fullerol [156], aqueous fullerene nanoparticles n-C_{60} [157], and CNTs [158]. Most of these antimicrobial nanomaterials are relatively inert in the water and behave as weak oxidants and excellent disinfectants. Thus they have secured important position even in the disinfection stage of wastewater treatment.

CNTs have been investigated for the removal of bacteria and viruses in the water. They exert an acute pulmonary toxicity and cytotoxicity onto the mammalian cells [159] in the following severity sequence SWCNTs > MWCNTs. CNTs were implemented as inhibitor for microbial attachment and biofouling formation [160]. Dispersion, stability, and bioavailability of the CNTs enhance their effectiveness as a disinfectant [154]. According to Kang et al. [158], highly purified SWCNTs display a strong antimicrobial activity and cause an extreme damage to cell membranes during the direct contact. However, it should be remembered that the difficulty of dispersion of CNTs in water makes them quite inefficient in the disinfection process. Interestingly, functionalization and coating of CNTs can also increase their efficiency as a disinfectant. Kang and coworkers increased the efficiency of SWCNTs by immobilizing them on a membrane filter surface and demonstrated a reasonable eradication of *E. coli* (87%). In another study [136], a thin layer coating of SWCNTs on a microporous polyvinylidene difluoride (PVDF) membrane showed inactivation of *E. coli* and the removal of bacteriophage MS2 (Emesvirus zinderi). Brunet et al. [161] also observed inactivation of *E. coli* and poliovirus by CNTs. On another note, Lilly et al. [162] showed a higher sporicidal effect of SWCNTs in the presence of oxidizing antimicrobial agents like H_2O_2 and NaOCl on *Bacillus antharacis* in comparison with the absence of oxidizing antimicrobial chemicals. The combination of H_2O_2 and NaOCl had more enhancement compared with the individual chemicals. Arias and Yang [163] also investigated the antimicrobial effects of both SWCNTs and MWCNTs with –OH and –COOH functional groups and found the antimicrobial activity of SWCNTs to be dependent on the buffer and its concentration. They exhibited stronger antimicrobial activity with –OH and –COOH functional groups when compared with $-NH_2$ on both the Gram-positive and Gram-negative bacterial cells. Thus, many studies have clearly indicated that the activity of bacteria or viruses is terminated when they come in contact with the CNTs with and without functional groups, as well as depending on the surrounding chemicals. Compared with SWCNTs, the MWCNTs were found to have slightly lower antimicrobial characteristics due to their low toxicity levels.

4.9 Commercial Use of CNMs in Wastewater Treatment

The nanotechnology research in the context of wastewater treatment should focus on improving not only the properties of nanomaterials but also their viability from the efficiency and cost point of view in the practical applications. Nanomaterials have plenty of potentials to penetrate into consumer markets with superior characteristics [164–167]. However, they may impact human health and the ecosystems. The toxicological studies on nanomaterials are still at nascent stage, and a lot of work needs to be done to ensure accurate conclusions about their short- and long-term effects on the human lives and the environment. However, the nanomaterials possess other salient features that could make them generally acceptable such as their ability to detect and prevent contamination. Due to their size and other properties, they can act as good sensors for targeted contaminants in the wastewater with extreme selectivity and accuracy. This characteristic could be very well exploited even for detection of pathogens or emerging pollutants. However, there are three main challenges in the commercial implementation of nanomaterials: scalability, cost-effectiveness, and impact on human life and the environment [168].

With respect to the applications of CNTs and graphene for wastewater treatment, they showed superb adsorbing characteristics, and their mechanisms are very well understood. Even pilot plant studies were also successfully conducted. However, their commercial exploitation is greatly dictated by the cost when compared with the existing commercial technologies. Implementing CNT- or graphene-based adsorbents into practical or commercial applications is feasible only when their performance is superior by multifold and with longer lifetime while the running costs are much lower than that of the existing commercial adsorbents. Incidentally, the energy costs incurred for regeneration of CNTs/graphene adsorbents is also much lower than that of the existing technologies (e.g. activated carbon-based adsorbents), which makes them commercially attractive, and their practical implementation renders a plausibility [101]. On the other hand, both CNT and graphene have also shown excellent potentials for membrane technology in the water purification. However, the synthesis of their membranes and implementation in the wastewater are not thoroughly investigated. The major challenge with CNTs in membrane technologies is their alignment. The irregular alignment of CNTs affects the membrane properties, such as flux and salt rejection. Further research is needed to address this challenge in order to enhance the membrane performance.

4.10 Conclusions

Nanomaterials provide unprecedented possibilities, and the ongoing research in the arena effectively lays solid foundation for their diverse applications. Carbon nanomaterials, such as CNTs and graphene, show great potential for wastewater treatment apart from their varied applications. Breakthrough research and other detailed studies have demonstrated the flexibility of CNTs and graphene in various applications. However, probably owing to the limited supply of CNTs and graphene, their commercial application was only realized in a limited way. CNMs still require sustainable development, including scaling up of functionalized CNTs and their composites with good reproducibility and acceptable costs. The selection of important and difficult applications that could not be accomplished by using conventional adsorbents, such as zeolite, activated carbon, and clay, may provide an opportunity for the rapid acceptance of CNTs and graphene by the industry. The cost–performance ratio is a critical factor too. On the other hand, it is generally realized that, even though the applications are still few, the combination of new adsorption materials and traditional adsorption and desorption equipment and technologies should greatly shorten the industrialization process of these new materials especially in the area of wastewater treatment. In the light of this, exploring and promoting practical application of new CNMs is likely to bring about new vitality in the field of wastewater treatment. Finally, the importance of safety while using CNMs cannot be ignored. The monodispersed state of CNTs and graphene is very small and light, resulting in the easy loss in water and also intake via the respiratory systems of people. However, it is relatively safe to use composites on a macroscopic scale because the interaction of CNTs and graphene with binders is very strong.

Important Websites

The following websites may be referred for constant updates on wastewater treatment and related activities recommended by Environmental Protection Agency (USA), the UN, and the worldwide bodies:

https://www.epa.gov/sites/production/files/2019-02/documents/emerging-tech-wastewater-treatment-management.pdf
https://sustainabledevelopment.un.org/topics/water/unsgab/wastewater
http://www.oas.org/DSD/publications/Unit/oea59e/ch25.htm

References

1 Dhote, J., Ingole, S., and Chavhan, A. (2012). Review on waste water treatment technologies. *Int. J. Eng. Res. Technol.* 1: 1–10.
2 The People's Daily. (2001). World Bank supports China's wastewater treatment. *The People's Daily*. Beijing.
3 UN Water. https://www.unwater.org/water-facts/quality-and-wastewater/ (accessed 15–30 July 2020).
4 Almasi, A. (2018). Wastewater treatment technologies for processing plants. https://www.watertechonline.com/wastewater/article/15550541/wastewater-treatment-technologies-for-processing-plants (accessed 15–30 July 2020).
5 Rizzo, L., Manaia, C., Merlin, C. et al. (2013). Urban wastewater treatment plants as hotspots for antibiotic resistant bacteria and genes spread into the environment: a review. *Sci. Total Environ.* 447: 345–360.

6 Schwarzenbach, R.P., Escher, B.I., Fenner, K. et al. (2006). The challenge of micropollutants in aquatic systems. *Science* 313: 1072–1077.

7 Hutchinson, C.F., Varady, R.G., and Drake, S. (2010). Old and new: changing paradigms in arid lands water management. In: *Water and Sustainability in Arid Regions* (eds. G. Schneier-Madanes, M. F. Courel), 311–332. Springer.

8 Abdel-Shafy, H.I. and Aly, R.O. (2007). Wastewater management in Egypt. In: *Wastewater Reuse–Risk Assessment, Decision-Making and Environmental Security* (ed. M. K. Zaidi), 375–382. Springer.

9 Amer, M., Attia, M., Fahmy, H., and Tawfik, M. (2005). Water Policy Issues of Egypt-Country Policy Support Programme (CPSP) Sustainable Economic Development Department, National Policy Environment Division, the Govt of the Netherlands New Delhi: International Commission on Irrigation and Drainage (ICID).

10 Engelman, R. and LeRoy, P. (1993). *Sustaining Water. Population and the Future of Renewable Water Supplies*. Washington, DC: Population Action International Population and Environment Program.

11 Cosgrove, W.J. and Rijsberman, F.R. (2000). World Water Vision: Making Water Everybody's Business. Earthscan Publications Ltd. ISBN: 9781315071763, https://doi.org/10.4324/9781315071763.

12 Amin, M.T., Alazba, A.A., and Manzoor, U. (2014). A review of removal of pollutants from water/wastewater using different types of nanomaterials review. *Adv. Mater. Sci. Eng.* 2014: 825910, 24 pp.

13 Yin, Z., Cui, C., Chen, H. et al. (2020) The application of arbon nanotube/graphene-based nanomaterials in wastewater treatment. *Small*, 16 (15): 1902301. doi: https://doi.org/10.1002/smll.201902301

14 Seow, T.W., Lim, C.K., Nor, M.H.M. et al. (2016). Review on wastewater treatment technologies. *Int. J. Appl. Environ. Sci.* ISSN: 0973-6077 11 (1): 111–126.

15 Bowman, J.C., Zhou, J.L., and Readman, J.W. (2002). Sediment-water interactions of natural oestrogens under estuarine conditions. *Mar. Chem.* 77 (4): 263–276.

16 Verstraeten, I.M., Heberer, T., Vogel, J.R. et al. (2003). Occurrence of endocrine-disrupting and other wastewater compounds during water treatment with case studies from Lincoln, Nebraska and Berlin, Germany. *Pract. Period. Hazard. Toxic Radioact. Waste Manage* 7 (4): 253–263.

17 Ozaki, H. (2004). Rejection of micropollutants by membrane filtration. In: *Regional Symposium on Membrane Science and Technology* 2004, 21–25 April 2004, Puteri Pan Pacific Hotel, Johor Bahru, Johor, Malaysia.

18 Servos, M.R., Bennie, D.T., Burnison, B.K. et al. (2005). Distribution of estrogens, 17β-estradiol and estrone, in Canadian municipal wastewater treatment plants. *Sci. Total Environ.* 336 (1–3): 155–170.

19 Urase, T. and Kikuta, T. (2005). Separate estimation of adsorption and degradation of pharmaceutical substances and estrogens in the activated sludge process. *Water Res.* 39 (7): 1289–1300.

20 Petrovic, M., Diaz, A., Ventura, F., and Barcel´o, D. (2003). Occurrence and removal of estrogenic short-chain ethoxy nonylphenolic compounds and their halogenated derivatives during drinking water production. *Environ. Sci. Technol.* 37 (19): 4442–4448.

21 Westerhoff, P., Yoon, Y., Snyder, S., and Wert, E. (2005). Fate of endocrine-disruptor, pharmaceutical, and personal care product chemicals during simulated drinking water treatment processes. *Environ. Sci. Technol.* 39 (17): 6649–6663.

22 Vieno, N., Tuhkanen, T., and Kronberg, L. (2006). Removal of pharmaceuticals in drinking water treatment: effect of chemical coagulation. *Environ. Technol.* 27 (2): 183–192.

23 Szewzyk, U., Szewzyk, R., Manz, W., and Schleifer, K.-H. (2000). Microbiogical safety of drinking water. *Annu. Rev. Microbiol.* 54: 81–127.

24 Zhang, W. and DiGiano, F.A. (2002). Comparison of bacterial regrowth in distribution systems using free chlorine and chloramine: a statistical study of causative factors. *Water Res.* 36 (6): 1469–1482.

25 Becher, G. (1999). Drinking water chlorination and health. *Acta Hydrochim. Hydrobiol.* 27 (2): 100–102.

26 Hozalski, R.M., Zhang, L., and Arnold, W.A. (2001). Reduction of haloacetic acids by Fe^0: implications for treatment and fate. *Environ. Sci. Technol.* 35 (11): 2258–2263.

27 Sadiq, R. and Rodriguez, M.J. (2004). Disinfection by-products (DBPs) in drinking water and predictive models for their occurrence: a review. *Sci. Total Environ.* 321 (1–3): 21–46.

28 Gopal, K., Tripathy, S.S., Bersillon, J.L., and Dubey, S.P. (2007). Chlorination byproducts, their toxicodynamics and removal from drinking water. *J. Hazard. Mater.* 140 (1–2): 1–6.

29 Adams, C., Wang, Y., Loftin, K., and Meyer, M. (2002). Removal of antibiotics from surface and distilled water in conventional water treatment processes. *J. Environ. Eng.* 128 (3): 253–260.

30 Strathmann, H. (2001). Membrane separation processes: current relevance and future opportunities. *AICHE J.* 47 (5): 1077–1087.

31 Van der Bruggen, B. and Vandecasteele, C. (2002). Distillation vs. membrane filtration: overview of process evolutions in seawater desalination. *Desalination* 143 (3): 207–218.

32 Ahmad, A.L., Ooi, B.S., Mohammad, A.W., and Choudhury, J.P. (2004). Development of a highly hydrophilic nanofiltrationmembrane for desalination and water treatment. *Desalination* 168 (1–3): 215–221.

33 Qin, J.-J., Oo, M.H., and Kekre, K.A. (2007). Nanofiltration for recovering wastewater from a specific dyeing facility. *Sep. Purif. Technol.* 56 (2): 199–203.

34 Bodzek, M., Dudziak, M., and Luks-Betlej, K. (2004). Application of membrane techniques to water purification. Removal of phthalates. *Desalination* 162 (1–3): 121–128.

35 Owen, G., Bandi, M., Howell, J.A., and Churchouse, S.J. (1995). Economic assessment of membrane processes for water and waste water treatment. *J. Membr. Sci.* 102 (1–3): 77–91.

36 Walha, K., Amar, R.B., Firdaous, L. et al. (2007). Brackish groundwater treatment by nanofiltration, reverse osmosis and electrodialysis in Tunisia: performance and cost comparison. *Desalination* 207 (1–3): 95–106.

37 Mulder, M. (1994). The use of membrane processes in environmental problems. an introduction. In: *Membrane Processes in Separation and Purification* (eds. J.G. Crespo and K.W. Böddeker), 229–262. Springer.

38 Kiso, Y., Kon, T., Kitao, T., and Nishimura, K. (2001). Rejection properties of alkyl phthalates with nanofiltration membranes. *J. Membr. Sci.* 182 (1–2): 205–214.

39 Bottero, J.-Y., Rose, J., and Wiesner, M.R. (2006). Nanotechnologies: tools for sustainability in a new wave of water treatment processes. *Integr. Environ. Assess. Manag.* 2 (4): 391–395.

40 Edelstein, A.S. (ed.) (1999). *Nanomaterials: Synthesis, Properties and Applications*. Bristol: IOP publishing.

41 Zhang, Z.Y. and Lagally, M.G. (1997). Atomistic processes in the early stages of thin-film growth. *Science* 276: 377.

42 Malhotra, B.D. and Ali, A. (2018). Chapter 1: Nanomaterials in biosensors: fundamentals and applications. In: *Nanomaterials for Biosensors, Fundamentals and Applications (eds.* B. D. Malhotra and M. A. Ali), 1–74. Micro and Nano Technologies. https://doi.org/10.1016/B978-0-323-44923-6.00001-7.

43 Weertman, J.R. and Averback, R.S. (1996) Mechanical properties. In: *Nanomaterials: Synthesis, Properties and Applications* (eds. A.S. Edelstein and R.C. Cammarata), 323. London: IOP publishing.

44 Koch, C.C. (ed.) (2007). *Nanostructured Materials; Processing, Properties and Applications*, 2e. USA: William Andrew Inc, ISBN: 9780815515340.

45 Decker, S. and Klabunde, K.J. (1996). Enhancing effect of Fe_2O_3 on the ability of nanocrystalline calcium oxide to adsorb SO_2. *J. Am. Chem. Soc.* 118: 12465.

46 Yang, K. and Xing, B. (2007). Desorption of polycyclic aromatic hydrocarbons from carbon nanomaterials in water. *Environ. Pollut.* 145: 529–537.

47 Mallakpour, S. and Khadem, E. (2019). Chapter 8: Carbon nanotubes for heavy metals removal. In: *Composite Nanoadsorbents* (eds. G. Kyzas and A.C. Mitrpoulos), 181–210. Amsterdam, Netherlands: Elsevier. https://doi.org/10.1016/B978-0-12-814132-8.00009-5; eBook ISBN: 9780128141335.

48 Mallakpour, S. and Rashidimoghadam, S. (2019). Chapter 9: Carbon nanotubes for dyes removal. In: *Composite Nanoadsorbents* (eds. G. Kyzas and A.C. Mitrpoulos), 211–244. Amsterdam, Netherlands: Elsevier. https://doi.org/10.1016/B978-0-12-814132-8.00010-1; eBook ISBN: 9780128141335.

49 Mallakpour, S. and Rashidimoghadam, S. Chapter 29: Utilization of starch and starch/carbonaceous nanocomposites for removal of pollutants from wastewater. In: *Handbook of Polymer Nanocomposites for Industrial Applications* (ed. C.M. Hussain), Elsevier, ISBN: 9780128214978.

50 Khin, M.M., Nair, A.S., Babu, V.J. et al. (2012). A review on nanomaterials for environmental remediation. *Energy Environ. Sci.* 5 (8): 8075–8109.

51 Yan, J., Han, L., Gao, W. et al. (2015). Biochar supported nanoscale zerovalent iron composite used as persulfate activator for removing trichloroethylene. *Bioresour. Technol.* 175: 269–274.

52 Liu, F., Yang, J.H., Zuo, J. et al. (2014). Graphene-supported nanoscale zero-valent iron: removal of phosphorus from aqueous solution and mechanistic study. *J. Environ. Sci.* 26 (8): 1751–1762.

53 Tang, W.-W., Zeng, G.-M., Gong, J.-L. et al. (2014). Impact of humic/fulvic acid on the removal of heavy metals from aqueous solutions using nanomaterials: a review. *Sci. Total Environ.* 468–469: 1014–1027.

54 Ren, X., Chen, C., Nagatsu, M., and Wang, X. (2011). Carbon nanotubes as adsorbents in environmental pollution management: a review. *Chem. Eng. J.* 170: 395–410.

55 Rao, G.P., Lu, C., and Su, F. (2007). Sorption of divalent metal ions from aqueous solution by carbon nanotubes: a review. *Sep. Purif. Technol.* 58: 224–231.

56 Seymour, M.B., Su, C., Gao, Y. et al. (2012). Characterization of carbon nano-onions for heavy metal ion remediation. *J. Nanopart. Res.* 14: 1–13.

57 Wang, S., Ng, C.W., Wang, W. et al. (2012). A comparative study on the adsorption of acid and reactive dyes on multiwall carbon nanotubes in single and binary dye systems. *J. Chem. Eng. Data* 57: 1563–1569.

58 Yang, K., Zhu, L., and Xing, B. (2006). Adsorption of polycyclic aromatic hydrocarbons by carbon. *Environ. Sci. Technol.* 40 (6): 1855–1861.

59 Georgakilas, V., Perman, J.A., Tucek, J., and Zboril, R. (2015). Broad family of carbon nanoallotropes: classification, chemistry, and applications of fullerenes, carbon dots, nanotubes, graphene, nanodiamonds, and combined superstructures. *Chem. Rev.* 115: 4744–4822.

60 Barhoum, A., Shalan, A.E., El-Hout, S. I. et al. (2019). A broad family of carbon nanomaterials: classification, properties, synthesis, and emerging applications. In: *Handbook of Nanofibers* (eds. A. Barhoum, M. Bechelany and A. Makhlouf). Cham: Springer. https://doi.org/10.1007/978-3-319-42789-8_59-1.

61 Chatterjee, A. and Deopura, B.L. (2002). Carbon nanotubes and nanofibre: an overview. *Fibers Polym.* 3 (4): 134–139.

62 Peng, X., Li, Y., Luan, Z. et al. (2003). Adsorption of 1,2-dichlorobenzene from water to carbon nanotubes. *Chem. Phys. Lett.* 376 (1–2): 154–158.

63 Lu, C., Su, F.S., and Hu, S. (2008). Surface modification of carbon nanotubes for enhancing BTEX adsorption from aqueous solutions. *Appl. Surf. Sci.* 254 (21): 7035–7041.

64 Cho, H.-H., Wepasnick, K., Smith, B.A. et al. (2010). Sorption of aqueous Zn[II] and Cd[II] by multiwall carbon nanotubes: the relative roles of oxygen-containing functional groups and graphenic carbon. *Langmuir* 26 (2): 967–981.

65 Mallakpour, S. and Soltanian, S. (2016). Surface functionalization of carbon nanotubes: fabrication and applications. *RSC Adv.* 6: 109916–109935.

66 Kalhapure, R.S., Sonawane, S.J., Sikwal, D.R. et al. (2015). Solid lipid nanoparticles of clotrimazole silver complex: an efficient nano antibacterial against *Staphylococcus aureus* and MRSA. *Colloids Surf., B* 136: 651–658.

67 Li, Y.H., Di, Z., Ding, J. et al. (2005). Adsorption thermodynamic, kinetic and desorption studies of Pb^{2+} on carbon nanotubes. *Water Res.* 39 (4): 605–609. https://doi.org/10.1016/j.watres.2004.11.004.

68 Yang, S., Li, J., Shao, D. et al. (2009). Adsorption of Ni(II) on oxidized multi-walled carbon nanotubes: effect of contact time, pH, foreign ions and PAA. *J. Hazard. Mater.* 166 (1): 109–116. https://doi.org/10.1016/j.jhazmat.2008.11.003.

69 Kabbashi, N.A., Daoud, J.I., Isam, Y.Q. et al. (2011). Statistical analysis for removal of cadmium from aqueous solution at high pH. *Aust. J. Basic Appl. Sci.* 5 (6): 440–446.

70 Li, Y.H., Ding, J., Luan, Z. et al. (2003). Competitive adsorption of Pb^{2+}, Cu^{2+} and Cd^{2+} ions from aqueous solutions by multiwalled carbon nanotubes. *Carbon (NY)* 41 (14): 2787–2792. https://doi.org/10.1016/S0008-6223(03)00392-0.

71 Gao, Z., Bandosz, T.J., Zhao, Z. et al. (2009). Investigation of factors affecting adsorption of transition metals on oxidized carbon nanotubes. *J. Hazard. Mater.* 167 (1–3): 357–365. https://doi.org/10.1016/j.jhazmat.2009.01.050.

72 Stafiej, A. and Pyrzynska, K. (2007). Adsorption of heavy metal ions with carbon nanotubes. *Sep. Purif. Technol.* 58 (1): 49–52. https://doi.org/10.1016/j. seppur.2007.07.008.

73 Xu, Y.J., Arrigo, R., Liu, X., and Su, D.S. (2011). Characterization and use of functionalized carbon nanotubes for the adsorption of heavy metal anions. *New Carbon Mater.* 26 (1): 57–62. https://doi.org/10.1016/S1872-5805(11)60066-8.

74 Wang, H.J., Zhou, A.L., Peng, F. et al. (2007). Adsorption characteristic of acidified carbon nanotubes for heavy metal Pb(II) in aqueous solution. *Mater. Sci. Eng., A* 466 (1–2): 201–206. https://doi.org/10.1016/j.msea.2007.02.097.

75 Xu, D., Tan, X., Chen, C., and Wang, X. (2008). Removal of Pb(II) from aqueous solution by oxidized multiwalled carbon nanotubes. *J. Hazard. Mater.* 154 (1–3): 407–416. https://doi.org/10.1016/j.jhazmat.2007.10.059.

76 Wang, S.G., Gong, W.X., Liu, X.W. et al. (2007). Removal of lead(II) from aqueous solution by adsorption onto manganese oxide-coated carbon nanotubes. *Sep. Purif. Technol.* 58 (1): 17–23. https://doi.org/10.1016/j.seppur.2007.07.006.

77 AlOmar, M.K., Alsaadi, M.A., Hayyan, M. et al. (2016). Lead removal from water by choline chloride based deep eutectic solvents functionalized carbon nanotubes. *J. Mol. Liq.* 222: 883–894. https://doi.org/10.1016/j.molliq.2016.07.074.

78 Wu, C.H. (2007). Studies of the equilibrium and thermodynamics of the adsorption of Cu^{2+} onto as-produced and modified carbon nanotubes. *J. Colloid Interface Sci.* 311 (2): 338–346. https://doi.org/10.1016/j.jcis.2007.02.077.

79 Chen, C., Li, X., Zhao, D. et al. (2007). Adsorption kinetic, thermodynamic and desorption studies of Th(IV) on oxidized multi-wall carbon nanotubes. *Colloids Surfaces A Physicochem. Eng. Asp.* 302 (1–3): 449–454. https://doi.org/10.1016/j.colsurfa.2007.03.007.

80 Goswami, S. and Ghosh, U.C. (2005). Studies on adsorption behaviour of Cr(VI) onto synthetic hydrous stannic oxide. *Water SA* 31 (4): 597–602. https://doi.org/10.4314/wsa.v31i4.5150.

81 Mubarak, N.M., Alicia, R.F., Abdullah, E.C. et al. (2013). Statistical optimization and kinetic studies on removal of Zn^{2+} using functionalized carbon nanotubes and magnetic biochar. *J. Environ. Chem. Eng.* 1 (3): 486–495. https://doi.org/10.1016/j.jece.2013.06.011.

82 Lu, C., Liu, C., and Su, F. (2009). Sorption kinetics, thermodynamics and competition of Ni^{2+} from aqueous solutions onto surface oxidized carbon nanotubes. *Desalination* 249 (1): 18–23. https://doi.org/10.1016/j.desal.2009.06.009.

83 Kandah, M.I. and Meunier, J.L. (2007). Removal of nickel ions from water by multi-walled carbon nanotubes. *J. Hazard. Mater.* 146 (1–2): 283–288. https://doi.org/10.1016/j.jhazmat.2006.12.019.

84 Chen, C., Hu, J., Shao, D. et al. (2009). Adsorption behavior of multiwall carbon nanotube/iron oxide magnetic composites for Ni(II) and Sr(II). *J. Hazard. Mater.* 164 (2–3): 923–928. https://doi.org/10.1016/j.jhazmat.2008.08.089.

85 Gholipour, M. and Hashemipour, H. (2012). Procenjivanje Performansi Višeslojnih Ugljeničnih Nanocevi U Adsorpciji I Desorpciji Šestovalentnog Hroma. *Chem. Ind. Chem. Eng. Q.* 18 (41): 509–523. https://doi.org/10.2298/CICEQ111104025G.

86 Hu, J., Chen, C., Zhu, X., and Wang, X. (2009). Removal of chromium from aqueous solution by using oxidized multiwalled carbon nanotubes. *J. Hazard. Mater.* 162 (2–3): 1542–1550. https://doi.org/10.1016/j.jhazmat.2008.06.058.

87 Ali Atieh, M. (2011). Removal of chromium(VI) from polluted water using carbon nanotubes supported with activated carbon. *Procedia Environ. Sci.* 4: 281–293. https://doi.org/10.1016/j.proenv.2011.03.033.

88 Li, Y.H., Wang, S., Cao, A. et al. (2001). Adsorption of fluoride from water by amorphous alumina supported on carbon nanotubes. *Chem. Phys. Lett.* 350 (5–6): 412–416. https://doi.org/10.1016/S0009-2614(01)01351-3.

89 Peng, X., Luan, Z., Ding, J. et al. (2005). Ceria nanoparticles supported on carbon nanotubes for the removal of arsenate from water. *Mater. Lett.* 59 (4): 399–403. https://doi.org/10.1016/j.matlet.2004.05.090.

90 Lu, C. and Su, F. (2007). Adsorption of natural organic matter by carbon nanotubes. *Sep. Purif. Technol.* 58 (1): 113–121. https://doi.org/10.1016/j.seppur.2007.07.036.

91 Liao, Q., Sun, J., and Gao, L. (2008). Adsorption of chlorophenols by multi-walled carbon nanotubes treated with HNO_3 and NH_3. *Carbon (NY)* 46 (3): 553–555. https://doi.org/10.1016/j.carbon.2007.12.009.

92 Chandra, V. and Kim, K.S. (2011). Highly selective adsorption of Hg^{2+} by a polypyrrole–reduced graphene oxide composite. *Chem. Commun.* 47 (13): 3942–3944. https://doi.org/10.1039/c1cc00005e.

93 Fan, L., Luo, C., Sun, M. et al. (2013). Highly selective adsorption of lead ions by water-dispersible magnetic chitosan/graphene oxide composites. *Colloids Surf., B.* 103: 523–529. https://doi.org/10.1016/j.colsurfb.2012.11.006.

94 Sreeprasad, T.S., Maliyekkal, S.M., Lisha, K.P., and Pradeep, T. (2011). Reduced graphene oxide-metal/metal oxide composites: facile synthesis and application in water purification. *J. Hazard. Mater.* 186 (1): 921–931. https://doi.org/10.1016/j.jhazmat.2010.11.100.

95 Mi, X., Huang, G., Xie, W. et al. (2012). Preparation of graphene oxide aerogel and its adsorption for Cu^{2+} ions. *Carbon (NY)* 50 (13): 4856–4864. https://doi.org/10.1016/j.carbon.2012.06.013.

96 Huang, Z., Zheng, X., Lv, W. et al. (2011). *Langmuir* 27(12): 7558–7562.

97 Mishra, A.K. and Ramaprabhu, S. (2011). Functionalized graphene sheets for arsenic removal and desalination of sea water. *Desalination* 282: 39–45. https://doi.org/10.1016/j.desal.2011.01.038.

98 Hao, L., Song, H., Zhang, L. et al. (2012). SiO_2/graphene composite for highly selective adsorption of Pb(II) ion. *J. Colloid Interface Sci.* 369 (1): 381–387. https://doi.org/10.1016/j.jcis.2011.12.023.

99 Li, Y.-H., Ding, J., Luan, Z. et al. (2003). Competitive adsorption of Pb^{2+}, Cu^{2+} and Cd^{2+} ions from aqueous solutions by multiwalled carbon nanotubes. *Carbon* 41 (14): 2787–2792.

100 Madrakian, T., Afkhami, A., Ahmadi, M., and Bagheri, H. (2011). Removal of some cationic dyes from aqueous solutions using magnetic-modified multi-walled carbon nanotubes. *J. Hazard. Mater.* 196: 109–114.

101 Thines, R.K., Mubarak, N.M., Nizamuddin, S. et al. (2017). Application potential of carbon nanomaterials in water and wastewater treatment: a review. *J. Taiwan Inst. Chem. Eng.* 72: 116–133.

102 Zhang, D., Pan, B., Wu, M. et al. (2011). Adsorption of sulfamethoxazole on functionalized carbon nanotubes as affected by cations and anions. *Environ. Pollut.* 159: 2616–2621.

103 Ajayan, P.M. (1999). Nanotubes from carbon. *Chem. Rev.* 99: 1787.

104 Apul, O.G. and Karanfil, T. (2015). Adsorption of synthetic organic contaminants by carbon nanotubes: a critical review. *Water Res.* 68: 34.

105 Teow, Y.H. and Mohammad, A.W. (2019). New generation nanomaterials for water desalination: a review. *Desalination* 451: 2.

106 Chen, G.C., Shan, X.Q., Zhou, Y.Q. et al. (2009). Adsorption kinetics, isotherms and thermodynamics of atrazine on surface oxidized multiwalled carbon nanotubes. *J. Hazard. Mater.* 169: 912.

107 Kim, S., Park, C.M., Jang, M. et al. (2018). Aqueous removal of inorganic and organic contaminants by graphene-based nanoadsorbents: a review. *Chemosphere* 212: 1104.

108 Ersan, G., Apul, O.G., Perreault, F., and Karanfil, T. (2017). Adsorption of organic contaminants by graphene nanosheets: a review. *Water Res.* 126: 385.

109 Li, Y.H., Zhao, Y.M., Hu, W.B. et al. (2007). Carbon nanotubes - the promising adsorbent in wastewater treatment. *J. Phys. Conf. Ser.* 61: 698.

110 Esumi, K., Ishigami, M., Nakajima, A. et al. (1996). Chemical treatment of carbon nanotubes. *Carbon* 34: 279.

111 Chen, H., Gao, B., and Li, H. (2015). Removal of sulfamethoxazole and ciprofloxacin from aqueous solutions by graphene oxide. *J. Hazard. Mater.* 282: 201.

112 Babaei, A.A., Lima, E.C., Takdastan, A. et al. (2016). Removal of tetracycline antibiotic from contaminated water media by multi-walled carbon nanotubes: operational variables, kinetics, and equilibrium studies. *Water Sci. Technol.* 74: 1202.

113 Xie, X., Gao, L., and Sun, J. (2007). Thermodynamic study on aniline adsorption on chemical modified multi-walled carbon nanotubes. *Colloids Surf., A* 308: 54.

114 Wu, L., Liu, L., Gao, B. et al. (2013). Aggregation kinetics of graphene oxides in aqueous solutions: experiments, mechanisms, and modeling. *Langmuir* 29: 15174.

115 Madadrang, C.J., Kim, H.Y., Gao, G. et al. (2012). Adsorption behavior of EDTA-graphene oxide for Pb (II) removal. *ACS Appl. Mater. Interfaces* 4: 1186.

116 Lee, K.M., Wong, C.P.P., Tan, T.L., and Lai, C.W. (2018). Functionalized carbon nanotubes for adsorptive removal of water pollutants. *Mater. Sci. Eng., B* 236–237, 61.

117 Yang, K., Zhu, L., and Xing, B. (2006). Adsorption of polycyclic aromatic hydrocarbons by carbon nanomaterials. *Environ. Sci. Technol.* 40: 1855.

118 Yang, K. and Xing, B. (2007). Desorption of polycyclic aromatic hydrocarbons from carbon nanomaterials in water. *Environ. Pollut.* 145: 529.

119 Anjum, H., Johari, K., Gnanasundaram, N. et al. (2019). A review on adsorptive removal of oil pollutants (BTEX) from wastewater using carbon nanotubes. *J. Mol. Liq.* 277: 1005.

120 Akinpelu, A.A., Ali, M.E., Johan, M.R. et al. (2019). Polycyclic aromatic hydrocarbons extraction and removal from wastewater by carbon nanotubes: A review of the current technologies, challenges and prospects. *Process Saf. Environ. Prot.* 122: 68.

121 Rajarajeswari, M., Iyakutti, K., and Kawazoe, Y. (2011). Effect of chirality and curvature of single-walled carbon nanotubes on the adsorption of uracil. *Phys. Status Solidi B* 248: 1431.

122 O'Dea, A.R., Smart, R.S.C., and Gerson, A.R. (1999). Molecular modelling of the adsorption of aromatic and aromatic sulfonate molecules from aqueous solutions onto graphite. *Carbon* 37: 1133.

123 Gotovac, S., Honda, H., Hattori, Y. et al. (2007). Effect of nanoscale curvature of single-walled carbon nanotubes on adsorption of polycyclic aromatic hydrocarbons. *Nano Lett.* 7: 583.

124 Gu, Z., Yang, Z., Chong, Y. et al. (2015). Surface curvature relation to protein adsorption for carbon-based nanomaterials. *Sci. Rep.* 5: 10886.

125 Mohammadi, A. and Veisi, P. (2018). High adsorption performance of β-cyclodextrin-functionalized multi-walled carbon nanotubes for the removal of organic dyes from water and industrial wastewater. *J. Environ. Chem. Eng.* 6: 4634.

126 Ranjan, B., Pillai, S., Permaul, K., and Singh, S. (2019). Simultaneous removal of heavy metals and cyanate in a wastewater sample using immobilized cyanate hydratase on magnetic-multiwall carbon nanotubes. *J. Hazard. Mater.* 363: 73.

127 Zhang, Q., De Oliveira Vigier, K., Royer, S., and Jerome, F. (2012). Deep eutectic solvents: syntheses, properties and applications. *Chem. Soc. Rev.* 41: 7108.

128 AlOmar, M.K., Alsaadi, M.A., Hayyan, M. et al. (2016). Functionalization of CNTs surface with phosphonuim based deep eutectic solvents for arsenic removal from water. *Appl. Surf. Sci.* 389: 216.

129 Ibrahim, R.K., El-Shafie, A., Hin, L.S. et al. (2019). A clean approach for functionalized carbon nanotubes by deep eutectic solvents and their performance in the adsorption of methyl orange from aqueous solution. *J. Environ. Manag.* 235: 521.

130 Sarkar, C., Bora, C. and Dolui, S. K. et al. (2014). Selective dye adsorption by pH modulation on amine functionalized reduced graphene oxide–carbon nanotube hybrid, *Ind. Eng. Chem. Res.* 53 (42): 16148–16155.

131 Zhang, Z., Sun, D., Li, G. et al. (2019). Calcined products of Mg–Al layered double hydroxides/single-walled carbon nanotubes nanocomposites for expeditious removal of phenol and 4-chlorophenol from aqueous solutions. *Colloids Surf., A* 565: 143.

132 Choi, J.H., Jegal, J., and Kim, W.N. (2006). Fabrication and characterization, of multi-walled carbon nanotubes/polymer blend membranes. *J. Membr. Sci.* 284 (1–2): 406–415.

133 Celik, E., Park, H., Choi, H., and Choi, H. (2011). Carbon nanotube blended polyethersulfone membranes for fouling control in water treatment. *Water Res.* 45 (1): 274–282.

134 Qiu, S., Wu, L.G., Pan, X.J. et al. (2009). Preparation and properties of functionalized carbon nanotube/PSF blend ultrafiltration membranes. *J. Membr. Sci.* 342 (1–2): 165–172.

135 Liu, Y.L., Chang, Y., Chang, Y.H., and Shih, Y.J. (2010). Preparation of amphiphilic polymer-functionalized carbon nanotubes for low-protein-adsorption surfaces and protein-resistant membranes. *ACS Appl. Mater. Interfaces* 2 (12): 3642–3647.

136 Brady-Estévez, A.S., Kang, S., and Elimelech, M. (2008). A single-walled-carbon-nanotube filter for removal of viral and bacterial pathogens. *Small* 4 (4): 481–484.

137 Wang, X.F., Chen, X.M., Yoon, K. et al. (2005). High flux filtration medium based on nano fibrous substrate with hydrophilic nanocomposite coating. *Environ. Sci. Technol.* 39 (19): 7684–7691.

138 Kim, J. and Van der Bruggen, B. (2010). The use of nanoparticles in polymeric and ceramic membrane structures: review of manufacturing procedures and performance improvement for water treatment. *Environ. Pollut.* 158 (7): 2335–2349.

139 Fan, L., Harris, J.L., Roddick, F.A., and Booker, N.A. (2001). Influence of the characteristics of natural organic matter on the fouling of microfiltration membranes. *Water Res.* 35 (18): 4455–4463.

140 Shawky, H.A., Chae, S.-R., Lin, S., and Wiesner, M.R. (2011). Synthesis and characterization of a carbon nanotube/polymer nanocomposite membrane for water treatment. *Desalination* 272 (1): 46–50.

141 Lee, S.Y., Kim, H.J., Patel, R. et al. (2007). Silver nanoparticles immobilized on thin film composite polyamide membrane: characterization, nanofiltration, antifouling properties. *Polym. Adv. Technol.* 18 (7): 562–568.

142 Tang, C.Y., Zhang, Q., Wang, K. et al. (2009). Water transport behavior of chitosan porous membranes containing multi-walled carbon nanotubes (MWNTs). *J. Membr. Sci.* 337 (1–2): 240–247.

143 Ajmani, G.S., Goodwin, D., Marsh, K. et al. (2012). Modification of low pressure membranes with carbon nanotube layers for fouling control. *Water Res.* 46 (17): 5645–5654.

144 Tiraferri, A., Vecitis, C.D., and Elimelech, M. (2011). Covalent binding of single-walled carbon nanotubes to polyamide membranes for antimicrobial surface properties. *ACS Appl. Mater. Interfaces* 3 (8): 2869–2877.

145 Brady-Estévez, A.S., Schnoor, M.H., Vecitis, C.D. et al. (2010). Multiwalled carbon nanotube filter: Improving viral removal at low pressure. *Langmuir* 26 (18): 14975–14982.

146 Brady-Estévez, A.S., Schnoor, M.H., Kang, S., and Elimelech, M. (2010). SWNT-MWNT hybrid filter attains high viral removal and bacterial inactivation. *Langmuir* 26 (24): 19153–19158.

147 Akhavan, O., Abdolahad, M., Abdi, Y., and Mohajerzadeh, S. (2011). Silver nanoparticles within vertically aligned multi-wall carbon nanotubes with open tips for antibacterial purposes. *J. Mater. Chem.* 21 (2): 387–393.

148 Srivastava, A., Srivastava, O.N., Talapatra, S. et al. (2004). Carbon nanotube filters. *Nat. Mater.* 3 (9): 610–614.

149 Kim, E.S., Hwang, G., Gamal El-Din, M., and Liu, Y. (2012). Development of nanosilver and multi-walled carbon nanotubes thin-film nanocomposite membrane for enhanced water treatment. *J. Membr. Sci.* 394–395: 37–48. https://doi.org/10.1016/j.memsci.2011.11.041.

150 Fornasiero, F., Park, H.G., Holt, J.K. et al. (2008). Grigoropoulos-Ion exclusion by sub-2-nm carbon nanotube pores. *Proc. Natl. Acad. Sci. U.S.A.* 105 (45): 17250–17255. http://www.pnas.org/content/105/45/17250.full.pdf.

151 Corry, B. (2008). Designing carbon nanotube membranes for efficient water desalination. *J. Phys. Chem. B* 112 (5): 1427–1434. https://doi.org/10.1021/jp709845u.

152 Upadhyayula, V.K.K., Deng, S., Mitchell, M.C., and Smith, G.B. (2009). Application of carbon nanotube technology for removal of contaminants in drinking water: a review. *Sci. Total Environ.* 408 (1): 1–13.

153 Krasner, S.W., Weinberg, H.S., Richardson, S.D. et al. (2006). Occurrence of a new generation of disinfection byproducts. *Environ. Sci. Technol.* 40 (23): 7175–7185.

154 Li, Q., Mahendra, S., Lyon, D.Y. et al. (2008). Antimicrobial nanomaterials for water disinfection and microbial control: potential applications and implications. *Water Res.* 42 (18): 4591–4602.

155 Morones, J.R., Elechiguerra, J.L., Camacho, A. et al. (2005). The bactericidal effect of silver nanoparticles. *Nanotechnology* 16 (10): 2346.

156 Badireddy, A.R., Hotze, E.M., Chellam, S. et al. (2007). Inactivation of bacteriophages via photosensitization of fullerol nanoparticles. *Environ. Sci. Technol.* 41 (18): 6627–6632.

157 Lyon, D.Y., Adams, L.K., Falkner, J.C., and Alvarez, P.J.J. (2006). Antibacterial activity of fullerene water suspensions: effects of preparation method and particle size. *Environ. Sci. Technol.* 40 (14): 4360–4366.

158 Kang, S., Pinault, M., Pfefferle, L.D., and Elimelech, M. (2007). Single-walled carbon nanotubes exhibit strong antimicrobial activity. *Langmuir* 23 (17): 8670–8673.

159 Wick, P., Manser, P., Limbach, L.K. et al. (2007). The degree and kind of agglomeration affect carbon nanotube cytotoxicity. *Toxicol. Lett.* 168 (2): 121–131.

160 Jia, Z., Shen, D., and Xu, W. (2001). Synthesis and antibacterial activities of quaternary ammonium salt of chitosan. *Carbohydr. Res.* 333 (1): 1–6.

161 Brunet, L., Lyon, D., Zodrow, K. et al. (2008). Properties of membranes containing semi-dispersed carbon nanotubes. *Environ. Eng. Sci.* 25 (4): 565–576.

162 Lilly, M., Dong, X., McCoy, E., and Yang, L. (2012). Inactivation of *Bacillus anthracis* spores by single-walled carbon nanotubes coupled with oxidizing antimicrobial chemicals. *Environ. Sci. Technol.* 46 (24): 13417–13424.

163 Arias, L.R. and Yang, L. (2009). Inactivation of bacterial pathogens by carbon nanotubes in suspensions. *Langmuir* 25 (5): 3003–3012.

164 Hussain, C.M. (2018). *Handbook of Nanomaterials for Industrial Applications*. Elsevier.

165 Hussain, C.M. and Mishra, A.K. (2019). *Nanotechnology in Environmental Science, 2 Volumes*. Wiley.

166 Hussain, C.M. (2020). *Handbook of Functionalized Nanomaterials for Industrial Applications*. Elsevier.

167 Hussain, C.M. (2020). *Handbook of Manufacturing Applications of Nanomaterials*. Elsevier.

168 Zekić, E., Vuković, Ž., and Halkijević, I. (2018). Application of nanotechnology in wastewater treatment. *Građevinar* 70 (4): 315–323. https://doi.org/10.14256/JCE.2165.2017.

5

Electrochemical Determination of Indigotine Based on Poly(Gibberellic Acid)-Modified Carbon Nanotube Paste Electrode

Girish Tigari, Jamballi G. Manjunatha, and Chenthattil Raril

Mangalore University Constituent College, FMKMC College, Department of Chemistry, Madikeri, Karnataka, India

5.1 Introduction

Indigotine (IT) is a blue synthetic coloring agent widely used in the food and cosmetic industries. Recent research shows that IT is toxic to human health and can cause injury in cornea and conjunctiva and even tumors. Being azo dye, it is eco-toxic and causes water pollution [1–3]. So simple and effective IT detection in a water sample is important for human health and environmental pollution issues. To date, a lot of analytical methods were applied for detection and estimation of IT such as high-performance liquid chromatography [4], spectrophotometry [5], and flow injection analysis [6]. However, these needs time, sophisticated laboratory, handling experts and are also expensive. Compared with the abovementioned methods, the electroanalytical tools are cheap and rapid with high sensitivity [7, 8]. Few voltammetric methods were previously used for the determination of IT [9, 10]. Still, a more sensitive and stable electrochemical sensor device is required for quantization of IT in tracer levels. Recently different types of polymer-coated electrodes are used as sensing surfaces for determination of bioactive molecules such as poly(nigrosine) [11], poly(threonine) [12], poly(arginine) [13], 3-amino-1,2,4-triazole-5-thiol [14], etc. Polymer-based electrodes are used frequently due to their biocompatibility, film-forming ability, and stability. These modified surfaces contain active groups such as –COOH, –NH, and –OH. The oxygen functionalities are responsible for the sensing performance of electrodes [15–17]. Carbon nanotubes (CNTs) composite and functionalized nanomaterials find many applications in biomedical, environmental, electronic, and food and pharmaceutical fields. These extensive utility is due to the electrical conductivity, mechanical strength, elasticity, durability, and optical, thermal, and magnetic properties of the materials [18–33]. The present work describes the fabrication of poly(gibberellic acid)-electropolymerized carbon nanotube paste electrode for determination of IT in the water sample.

Environmental Applications of Carbon Nanomaterials-Based Devices, First Edition.
Edited by Shadpour Mallakpour and Chaudhery Mustansar Hussain.

5.2 Experimental

All electroanalytical measurements were performed with model CHI-6038 E. It is connected to conventional three-electrodes and private computer. Poly(gibberellic acid) film-modified carbon nanotube paste electrode (PGAMCNTPE) is used as a working electrode, Pt wire as an auxiliary, and $Hg/HgCl_2$ (calomel electrode) as reference electrode. All electrochemical measurements were performed at laboratory temperature.

5.2.1 Chemicals

IT and gibberellic acid (GA) were purchased from Molychem, India. CNTs was obtained from Sisco Laboratories Ltd., Mumbai. The remaining chemicals are of analytical grade. IT (0.5×10^{-5} M) and GA (1×10^{-4} M) were prepared using double-distilled water. The phosphate buffers (PBS) of different pH values were prepared by using a suitable volume of 0.1 M sodium dihydrogen phosphate and disodium hydrogen phosphate.

5.2.2 Bare Carbon Nanotube Paste Electrode (BCNTPE) Preparation

CNTs and silicone oil were taken in a mortar with proportion 60 : 40 and mixed well with a pestle for 20 minutes. After thorough mixing, the uniform paste is obtained. Paste obtained was filled in the cavity of Teflon rod that is attached to copper wire. Wire is inserted into Teflon rod to provide an electrical connection. Paste-packed cavity was polished with smooth tissue to get a shiny surface.

5.3 Results and Discussion

5.3.1 Electropolymerization of BCNTPE with GA

GA electrodeposition was performed on BCNTPE in potential range from −0.2 to 1.2 V using 0.1 mM GA with 0.1 V/s sweep rate as shown in Figure 5.1a. It can be analyzed in Figure 5.1a that as cyclic voltammetry (CV) cycling period increases, the polymerization

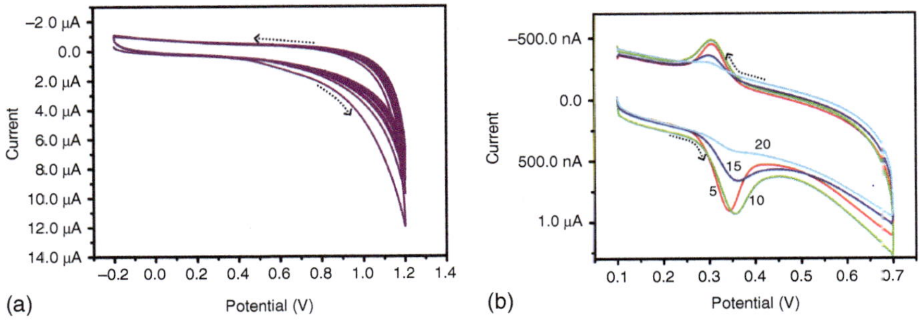

Figure 5.1 (a) Electropolymerization of 0.1 mM GA on BCNTPE in 0.1 PBS (pH 7.5). (b) Effect of polymerization cycles from 5 to 20 on IT (0.5×10^{-5} M) response.

curves get descended at the positive potential. This shows the formation of polymer layer on the surface of electrode. Also, a number of CV scans were calibrated for IT detection as depicted in Figure 5.1b. The 10 electropolymerization scans were optimized for entire IT measurements.

5.3.2 FE-SEM Characterization of BCNTPE and PGAMCNTPE

Field emission scanning electron microscope (FE-SEM) obtained for bare carbon nanotube paste electrode (BCNTPE) and PGAMCNTPE was shown in Figure 5.2a,b. Bare CNTPE exhibits asymmetric morphology with a random arrangement of CNTs. However, in the case of PGAMCNTPE, it is symmetric and smooth morphology. This provides that the bare electrode is covered with a polymer layer.

5.3.3 EIS Characterization for PGAMCNTPE and BCNTPE

Electrochemical impedance spectroscopy (EIS) studies are performed to check the electronic conductance of modified and bare electrodes. EIS study for BCNTPE (curve a) and PGAMCNTPE (curve b) was performed in 0.1 M KCl containing 1 mM $[Fe(CN)_6]^{3-/4-}$ redox probe as shown in Figure 5.3. BCNTPE shows a bigger semicircle with higher electron transfer resistance ($R_{ct} = 23.47\,k\Omega$), but PGAMCNTPE displays a very small semicircle with lesser electron transfer resistance ($R_{ct} = 14.77\,k\Omega$). These results show that PGAMCNTPE facilitates more electronic conduction than BCNTPE.

5.3.4 CV Behavior of IT at PGAMCNTPE and BCNTPE

Redox behavior of IT (0.5×10^{-5} M) was examined at PGAMCNTPE (curve c) and BCNTPE (curve a) in phosphate buffers (PBS) (0.1 M, pH 7.0) with a scan rate of 0.1 V/s using CV as depicted in Figure 5.4. BCNTPE (curve a) traces IT oxidation at 0.411 V and reduction at 0.351 V with the diminished current response. PGAMCNTPE detects IT anodic peak at 0.355 V and cathodic at 0.305 V with improved current signals by lowering overpotential. PGAMCNTPE also checked for blank (0.1 M PBS, pH 7.0), and it has no interruption in IT behavior (curve b).

5.3.5 Variation of IT Behavior at Different pHs

Selection of pH is significant for electroanalysis because pH has a marked effect on analyte response and electrode performance. By optimization of pH, we can boost up the electrode and analyte interface. PGAMCNTPE and IT (0.5×10^{-5} M) interface was tested in 0.1 M PBS at different pHs ranging from 5.5 to 8.0 with a sweep rate of 0.1 V/s as shown in Figure 5.5a. It can be analyzed in Figure 5.5a that as pH increases from acidic to basic (5.5–8.0), oxidation peak current also increases into pH 7.0 and after pH 7.0, the anodic peak current decreases. IT and PGAMCNTPE interface at pH 7.0 gives good current sensitivity. So pH 7.0 was considered for electrochemical measurement of IT. The impact of pH on oxidation peak potential was examined by plotting E_{pa} vs. pH. The plot E_{pa} vs. pH (Figure 5.5b) shows as the pH increased from 5.5 to 8.0, the anodic potential signal gets moved towards

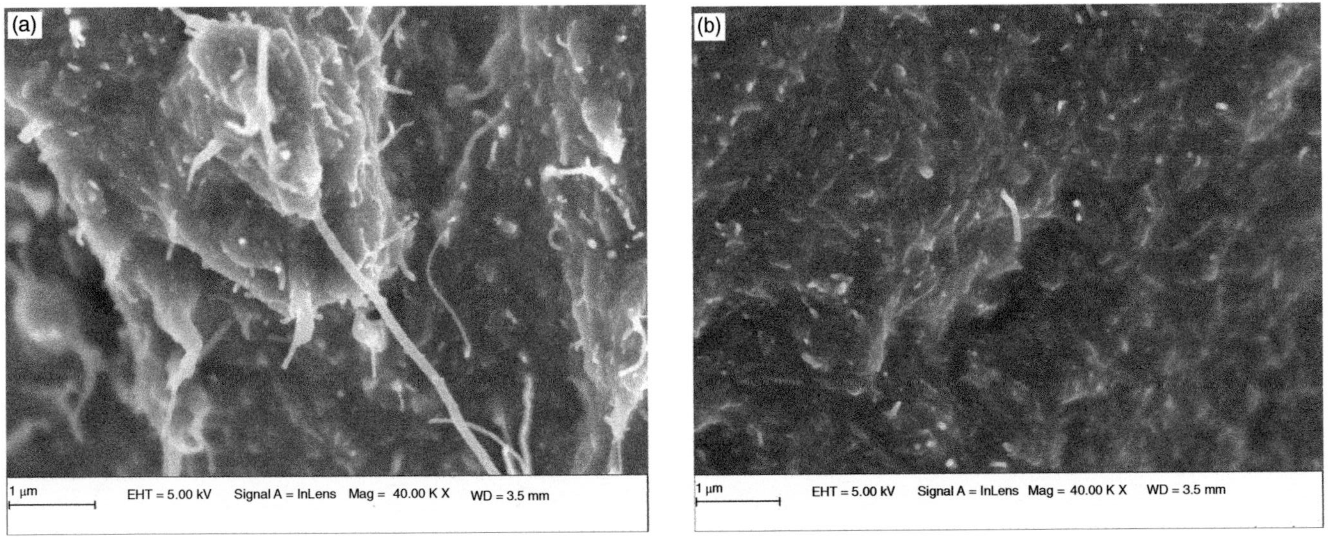

Figure 5.2 FE-SEM characterization for BCNTPE (a) and PGAMCNTPE (b).

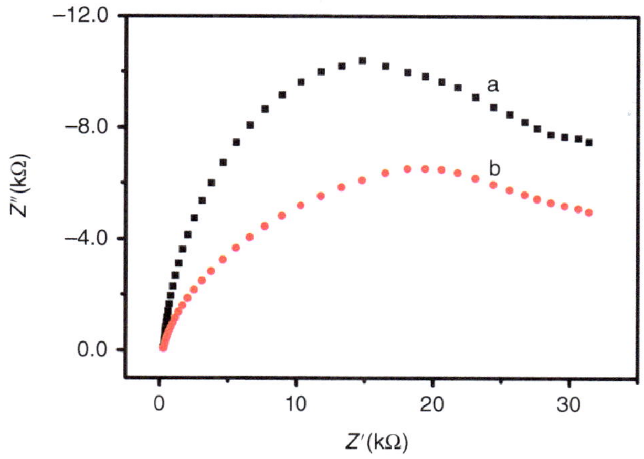

Figure 5.3 EIS curves for BCNTPE (a) and PGAMCNTPE (b).

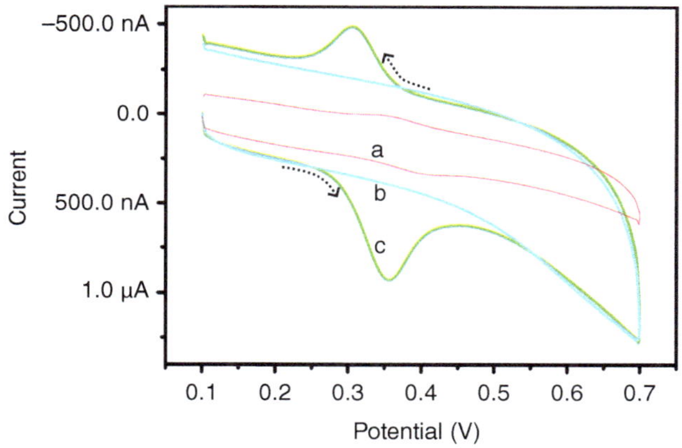

Figure 5.4 Electrochemical behavior of IT (0.5×10^{-5} M) at PGAMCNTPE (curve c) and BCNTPE (curve a) in PBS (0.1 M, pH 7.0) with scan rate of 0.1 V/s and curve b represents PGAMCNTPE response in blank (without analyte).

the negative side. It is explained by linear regression equation E_{pa} (V) = 0.052 pH + 0.713 ($r = 0.99$). Slope 0.052 V/pH indicates that the IT oxidation reaction involves an equal number of electrons and protons as shown in Scheme 5.1.

5.3.6 Effect of Voltage Sweep Rate

The impact of scan rate studies is important in voltammetry to know whether the electrode process is diffusion or adsorption controlled. Hence, the effect of voltage scan rates from 0.1 to 0.3 V/s for IT detection was carried in 0.1 M PBS (pH 7.0) as shown in Figure 5.6a. As the potential scan rate increases from 0.1 to 0.3 V/s, the peak currents also increases, and peak potential values related to oxidation and reduction get shifted opposite to each other.

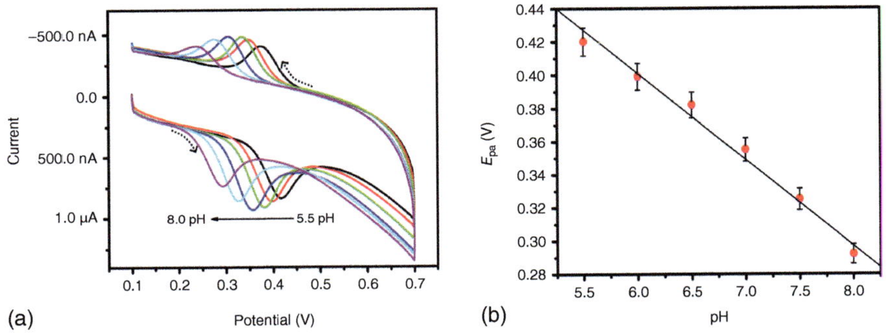

Figure 5.5 (a) Impact of pH (5.5–8.0) on electrochemical behavior of IT in PBS (0.1) with scan rate of 0.1 V/s. (b) Plot E_{pa} vs. pH.

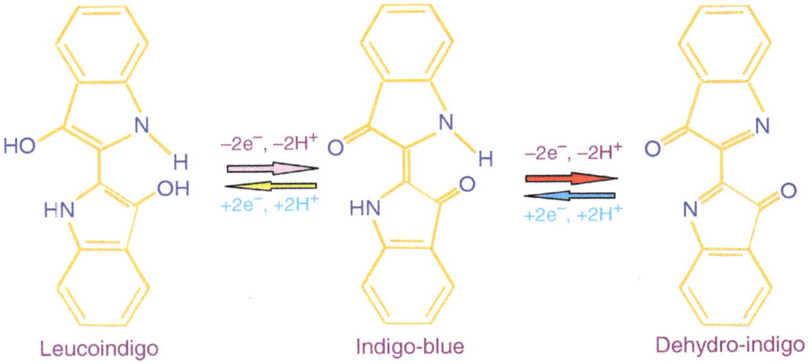

Scheme 5.1 Electron transfer mechanism of IT.

So IT reduction at PGAMCNTPE is a quasi-reversible process. Figure 5.6b gives a plot between the variation of scan rate and peak current with correlation coefficient of 0.99. This indicates the IT reduction at PGAMCNTPE is controlled by adsorption.

5.3.7 Calibration Curve

Differential pulse voltammetry (DPV) was used to develop a calibration plot for IT from 1.0 to 10 µM in PBS (0.1 M, pH 7.0) with a sweep rate of 0.05 V/s as shown in Figure 5.7a,b. The plot of concentration of IT vs. peak current shows for each addition of IT from 1.0 to 10 µM results in increases of peak current linearly. The linearity of the standard curve is described by the equation I_{pa} (A) = $8.96 \times 10^{-7} + 0.043$ (M). The limit of detection (LOD) and limit of quantification (LOQ) were evaluated as $3\sigma/b$ and $10\sigma/b$ [34]. Here, σ is standard deviation of DPV measurement in blank, and b is the slope of calibration curve. The LOD and LOQ of 4.11×10^{-8} and 1.37×10^{-7} M were obtained, respectively, for IT determination. Table 5.1 [35–39] gives the comparison of IT sensors. The proposed sensor provides lower LOD with high sensitivity.

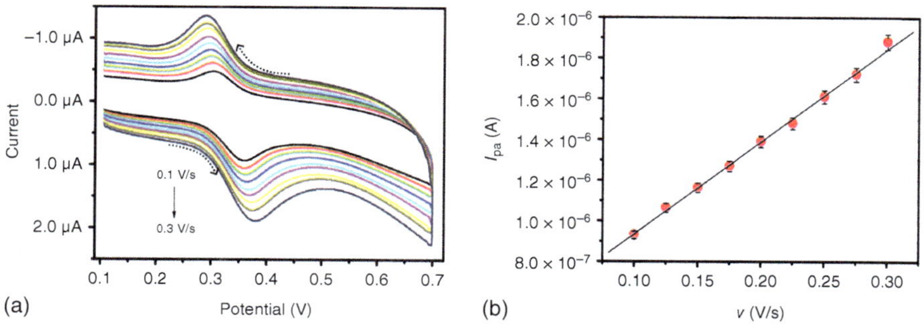

Figure 5.6 (a) Influence of scan rate (0.1–0.3 V/s) on IT behavior in 0.1 PBS, pH 7.0. (b) Graphical representation of *v* vs. I_{pa}.

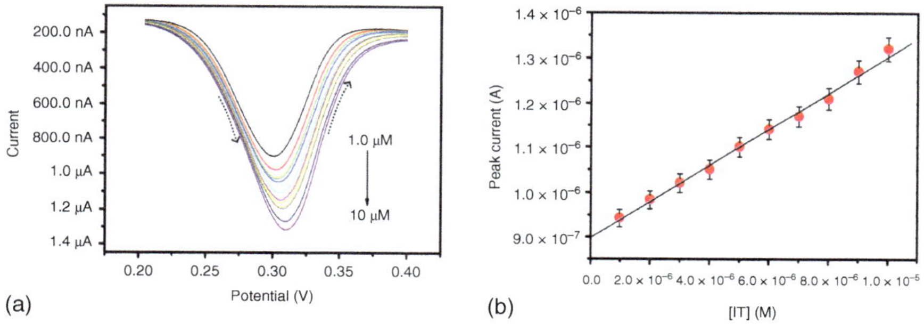

Figure 5.7 (a) DPV curves for IT at different concentrations in PBS (0.1 M, pH 7.0) with scan rate of 0.05 V/s. (b) Calibration plot for IT.

5.3.8 Reproducible and Stable Sensor

Good reproducibility for IT (0.5×10^{-5} M) detection was analyzed at PGAMCNTPE with relative standard deviation (RSD) 1.58% using CV in 0.1 PBS (pH 7.0) with a scan rate of 0.1 V/s. Stability of PGAMCNTPE was assessed by running continuous 20 CVs for IT (0.5×10^{-5} M) determination, and 87.25% peak current retention was observed. So the prepared sensor device is a stable and reproducible sensor for IT estimation.

5.3.9 Interference Analysis

PGAMCNTPE was tested for IT analysis in the presence of organic molecules such as anthrone, riboflavin, tartrazine, hydroquinone, nigrosine, sudan I, methyl orange, and isatin (0.1 mM) under calibrated conditions, and there is no significant interference observed for the detection of IT. This shows the high selectivity of the proposed electrode.

5.3.10 Water Sample Analysis

Under optimized circumstances, PGAMCNTPE was used for determination of IT in the tap water sample as shown in Figure 5.8. The collected tap water is diluted with PBS

Table 5.1 Comparison of proposed sensor with existing sensors for IT determination.

Electrode	Method	Linear range (μM)	LOD (μM)	References
TX-100MCPE	CV	1.5–20	0.16	[35]
TX-100MGPE	CV	3.5–15	0.24	[36]
PGMCPE	CV	2–60	0.11	[37]
SPCE	CV	0.5–100	0.19	[38]
PAMCNTPE	CV	8–130	0.27	[39]
PGAMCNTPE	DPV	1.0–10	0.0418	Present work

TX-100, triton X100; MGPE, modified graphene paste electrode; PGMCPE, poly(glycine)-modified carbon paste electrode; CV, cyclic voltammetry; SPCE, screen-printed carbon electrode; PAMCNTPE, poly(adenine)-modified carbon nanotube paste electrode; MCPE, modified carbon paste electrode. Sources: Raril et al. [35], Pushpanjali et al. [39].

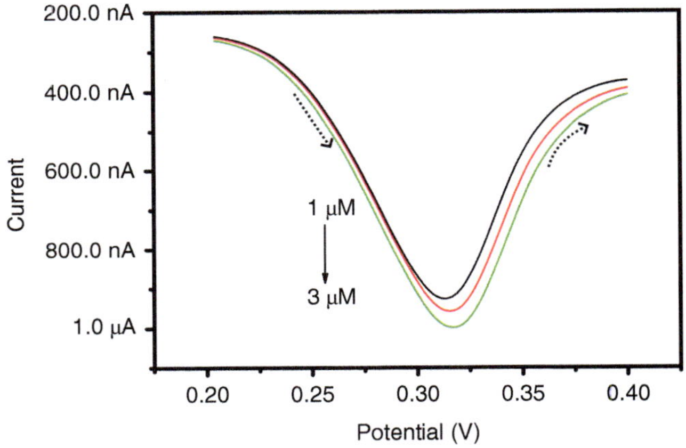

Figure 5.8 DPV curves for IT (1–3 μM) determination in tap water in PBS (0.1 M, pH 7.0) with scan rate of 0.05 V/s.

(0.1 M, pH 7.0), analysis is performed, and desirable peak for IT was not observed for a diluted tap water sample. So IT detection was performed by standard addition method with recoveries between the ranges 96.6% and 100.1%.

5.4 Conclusion

In this work, simple and rapid electrochemically modified electrode was presented for IT determination. The poly(gibberellic acid) formation on BCNTPE was confirmed by FE-SEM and EIS studies. The fabricated sensor showed a high electrocatalytic effect toward IT as compared with BCNTPE. This was attributed to active sites produced on the electrode during electrodeposition of gibberellic acid. The developed electrode yields lower detection and quantification limits as compared with recent reports. Moreover, the electrode provides good stability, reproducibility, and anti-interference with organic molecules. So the proposed sensor surface can be routinely utilized for quantification IT in different water samples.

Acknowledgment

We gratefully acknowledge the financial support from VGST, Bangalore, under Research Project No. KSTePS/VGST-KFIST (L1)2016-2017/GRD-559/2017-18/126/333, 21/11/2017.

Important Websites for Reference

https://www.sciencedirect.com/topics/agricultural-and-biological-sciences/indigo-carmine
http://www.inchem.org/documents/jecfa/jecmono/v46aje14.htm
https://hazard.com//msds/mf/baker/baker/files/i1440.htm
https://www.sciencedirect.com/topics/chemistry/voltammetric-method
https://www.sciencedirect.com/topics/materials-science/carbon-nanotubes
https://www.sciencedirect.com/topics/materials-science/electropolymerization

References

1 De Sa, I.C., Feiteira, F.N., and Pacheco, W.F. (2020). Quantification of the food dye indigo carmine in candies using digital image analysis in a polyurethane foam support. *Food Anal. Methods* 13: 962–969. https://doi.org/10.1007/s12161-020-01715-5.
2 Lakshmi, U.R., Srivastava, V.C., Mall, I.D., and Lataye, D.H. (2009). Rice husk ash as an effective adsorbent: evaluation of adsorptive characteristics for indigo carmine dye. *J. Environ. Manage.* 90 (2): 710–720. https://doi.org/10.1016/j.jenvman.2008.01.002.
3 Segura, E.G., Solache-Rios, M., and Colin Cruz, A. (2009). Sorption of indigo carmine by a Fe-zeolitic tuff and carbonaceous material from pyrolyzed sewage sludge. *J. Hazard. Mater.* 170 (2–3): 1227–1235. https://doi.org/10.1016/j.jhazmat.2009.05.102.
4 Trandafir, I., Nour, V., and Ionica, M.E. (2007). Development and evaluation of an HPLC-DAD method for determination of indigotine in soft drinks. *Sci. Study Res.* 7 (4): 391–396.
5 Guray, T. (2019). A novel method for simultaneous analysis of tartrazine and indigo carmine by cloud point extraction using spectrophotometric technique. *Int. J. Chem. Stud.* 7 (6): 17–23.
6 Derocoa, P.B., Medeiros, R.A., Rocha-Filho, R.C., and Fatibello-Filho, O. (2018). Selective and simultaneous determination of indigo carmine and allura red in candy samples at the nano-concentration range by flow injection analysis with multiple pulse amperometric detection. *Food Chem.* 247: 66–72. https://doi.org/10.1016/j.foodchem.2017.12.006.

7 Manjunatha, J.G., Deraman, M., Basri, N.H. et al. (2014). Sodium dodecyl sulfate modified carbon nanotubes paste electrode as a novel sensor for the simultaneous determination of dopamine, ascorbic acid, and uric acid. *C.R. Chim.* 17 (5): 465–476. https://doi.org/10.1016/j.crci.2013.09.016.

8 Manjunatha, J.G., Deraman, M., Basri, N.H., and Talib, I.A. (2018). Fabrication of poly(Solid Red A) modified carbon nano tube paste electrode and its application for simultaneous determination of epinephrine, uric acid and ascorbic acid. *Arabian J. Chem.* 11 (2): 149–158. https://doi.org/10.1016/j.arabjc.2014.10.009.

9 Arvand, M., Saberi, M., Ardaki, M.S., and Mohammadi, A. (2017). Mediated electrochemical method for the determination of indigo carmine levels in food products. *Talanta* 173: 60–68. https://doi.org/10.1016/j.talanta.2017.05.062.

10 Shucheng, M., Jianmin, N., Hua, M., and Chen, L. (1992). Voltammetric determination of indigo carmine and amaranth on a silver-based mercury film electrode. *Anal. Lett.* 25 (5): 899–909. https://doi.org/10.1080/00032719208020044.

11 Manjunatha, J.G. (2016). Poly(nigrosine) modified electrochemical sensor for the determination of dopamine and uric acid: a voltammetry study. *Int. J. Chemtech. Res.* 9: 136–146.

12 Raril, C. and Manjunatha, J.G. (2019). Electrogenarated thin poly(threonine) film modified carbon nanotube paste electrode for the simultaneous resolution of dopamine and uric acid. *Curr. Top. Electrochem.* 21: 93–105.

13 D'Souza, E.S., Manjunatha, J.G., Raril, C. et al. (2020). Arginine electropolymerized carbon nanotube paste electrode as sensitive and selective sensor for electrochemical determination of vanillin. *J. Mater. Environ. Sci.* 11 (3): 512–521.

14 Calam, T.T. and Uzun, D. (2019). Rapid and selective determination of vanillin in the presence of caffeine, its electrochemical behavior on an Au electrode electropolymerized with 3-amino-1,2,4-triazole-5-thiol. *Electroanalysis* 31 (12): 2347–2358. https://doi.org/10.1002/elan.201900328.

15 Tigari, G. and Manjunatha, J.G. (2019). Electrochemical preparation of poly(arginine)-modified carbon nanotube paste electrode and its application for the determination of pyridoxine in the presence of riboflavin: an electroanalytical approach. *J. Anal. Test.* 3: 331–340. https://doi.org/10.1007/s41664-019-00116-w.

16 D'Souza, E.S., Manjunatha, J.G., Raril, C. et al. (2020). Polymer modified carbon paste electrode as a sensitive sensor for the electrochemical determination of riboflavin and its application in pharmaceutical and biological samples. *Anal. Bioanal. Chem. Res.* 7 (4): 461–472.

17 Pushpanjali, P.A., Manjunatha, J.G., Tigari, G., and Fattepur, S. (2020). Poly(niacin) based carbon nanotube sensor for the sensitive and selective voltammetric detection of vanillin with caffeine. *Anal. Bioanal. Electrochem.* 12 (4): 553–568.

18 Hussain, C.M. and Kharisov, B. (2017). *Advanced Environmental Analysis: Application of Nanomaterials*. RSC. https://doi.org/10.1039/9781782623625.

19 Hussain, C.M. (2018). *Handbook of Nanomaterials for Industrial Applications*. Elsevier.

20 Hussain, C.M. and Mishra, A.K. (2019). *Nanotechnology in Environmental Science*. Wiley. https://doi.org/10.1002/9783527808854.

21 Hussain, C.M. (2019). *Handbook of Environmental Materials Management*, 1e. Springer International Publishing.

22 Hussain, C.M. and Keçili, R. (2019). *Modern Environmental Analysis Techniques for Pollutants*, 1e. Elsevier.

23 Hussain, C.M. (2020). *Handbook of Functionalized Nanomaterials for Industrial Applications*. Elsevier.

24 Hussain, C.M. (2020). *Handbook of Manufacturing Applications of Nanomaterials*. Elsevier.

25 Mallakpour, S., Abdolmaleki, A., and Azimi, F. (2017). Ultrasonic-assisted biosurface modification of multi-walled carbon nanotubes with thiamine and its influence on the properties of PVC/Tm-MWCNTs nanocomposite films. *Ultrason. Sonochem.* 39: 589–596. https://doi.org/10.1016/j.ultsonch.2017.05.028.

26 Mallakpour, S. and Azimi, F. (2019). Using sonochemistry for the production of poly(vinyl alcohol)/MWCNT–vitamin B1 nanocomposites: exploration of morphology, thermal and mechanical properties. *sNew J. Chem.* 43 (19): 7502–7510. https://doi.org/10.1039/C9NJ00116F.

27 Mallakpour, S. and Ezhieh, A.N. (2018). Preparation of polystyrene/MWCNT-valine composites: investigation of optical, morphological, thermal, and electrical conductivity properties. *Polym. Adv. Technol.* 29 (3): 1182–1190. https://doi.org/10.1002/pat.4229.

28 Mallakpour, S. and Khodadadzadeh, L. (2018). Fructose functionalized MWCNT as a filler for starch nanocomposites: fabrication and characterizations. *Prog. Org. Coat.* 114: 244–249. https://doi.org/10.1016/j.porgcoat.2017.08.013.

29 Mallakpour, S. and Khodadadzadeh, L. (2018). Ultrasonic-assisted fabrication of starch/MWCNT-glucose nanocomposites for drug delivery. *Ultrason. Sonochem.* 40: 402–409. https://doi.org/10.1016/j.ultsonch.2017.07.033.

30 Mallakpour, S. and Rashidimoghadam, S. (2020). Preparation, characterization, and in vitro bioactivity study of glutaraldehyde crosslinked chitosan/poly(vinyl alcohol)/ascorbic acid-MWCNTs bionanocomposites. *Int. J. Biol. Macromol.* 144: 389–402. https://doi.org/10.1016/j.ijbiomac.2019.12.073.

31 Mallakpour, S. and Rashidimoghadam, S. (2017). Starch/MWCNT-vitamin C nanocomposites: electrical, thermal properties and their utilization for removal of methyl orange. *Carbohydr. Polym.* 169: 23–32. https://doi.org/10.1016/j.carbpol.2017.03.081.

32 Mallakpour, S. and Ezhieh, A.N. (2017). Effect of starch-MWCNT@ valine nanocomposite on the optical, morphological, thermal, and adsorption properties of chitosan. *J. Polym. Environ.* 25: 875–883. https://doi.org/10.1007/s10924-016-0874-4.

33 Mallakpour, S. and Soltanian, S. (2016). Efficient surface modification of MWCNTs with vitamin B1 and production of poly(ester-imide)/MWCNTs nanocomposites containing l-phenylalanine moiety: thermal and microscopic study. *eXPRESS Polym. Lett.* 10 (1): 54–64. https://doi.org/10.3144/expresspolymlett.2016.6.

34 Raril, C. and Manjunatha, J.G. (2018). Sensitive electrochemical analysis of resorcinol using polymer modified carbon paste electrode. A cyclic voltammetric study. *Anal. Bioanal. Electrochem.* 10 (4): 488–498.

35 Raril, C., Manjunatha, J.G., Nanjundaswamy, L. et al. (2018). Surfactant immobilized electrochemical sensor for the detection of indigotine. *Anal. Bioanal. Electrochem.* 10 (11): 1479–1490.

36 Raril, C. and Manjunatha, J.G. (2019). A sensitive and selective procedure for the voltammetric determination of brilliant indigo and acid yellow 23 at surfactant modified graphene paste electrode. *J. Mater. Environ. Sci.* 10 (6): 510–519.

37 Manjunatha, J.G. (2018). A novel poly(glycine) biosensor towards the detection of indigo carmine: a voltammetric study. *J. Food Drug Anal.* 26 (1): 292–299. https://doi.org/s10.1016/j.jfda.2017.05.002.

38 Diaz-Gonzalez, M., Fernandez-Sanchez, C., and Costa-Garcia, A. (2002). Comparative voltammetric behavior of indigo carmine at screen printed carbon electrodes. *Electroanalysis* 14 (10): 665–670. https://doi.org/10.1002/1521-4109(200205)14:10<665::AID-ELAN665>3.0.CO;2-C.

39 Pushpanjali, P.A., Manjunatha, J.G., Raril, C., and Ravishankar, D.K. (2019). Determination of indigo carmine at poly(adenine) modified carbon nanotube paste electrode. *RJLBPCS* 5 (1): 820–831.

6

Toxicity of Carbon Nanomaterials

Arpita Adhikari[1] and Joydip Sengupta[2]

[1]*Affiliated to Maulana Abul Kalam Azad University of Technology, Techno Main Salt Lake, Department of Electronics and Communication Engineering, Kolkata, 700091, India*
[2]*Affiliated to University of Calcutta, Jogesh Chandra Chaudhuri College, Department of Electronic Science, Prince Anwar Shah Road, Kolkata, 700033, India*

6.1 Introduction

In 1959, the idea of manipulating and controlling things on a small scale was first introduced by Richard P. Feynman [1], and later in 1974, Norio Taniguchi, a professor of Tokyo University of Science, first introduced the term "nanotechnology" [2]. Since then nanotechnology finds its application in all commercial sectors like medical [3, 4], fire safety [5], food industry [6], transportation [7], agriculture [8], electrical transformers [9], oil and gas industry [10], renewable energies [11], etc. According to a study performed by Vance et al. [12], the use of nanotechnology-based products has increased rapidly in recent times and is expected to elevate in future in the perspective of more discoveries associated with the unique features of new nanomaterials [13] (Figure 6.1).

However, as the famous proverb says "*Every light has its shadow*," likewise, the excessive use of nanotechnology enhances the probability of exposure of nanomaterials toward mankind. There exist several potential routes of human exposure [15] to nanomaterials such as ingestion, inhalation, injection, and skin absorption [16] (Figure 6.2a). This enforces the scientists around the globe to research upon the effect of exposure of nanomaterials since way back in the 1990s [19] that consequently coined new terms like nanotoxicology [20] and nanotoxicity [21] to be evolved in 2004 and 2005, respectively. Till now the nanotoxicity of different materials [22–26] had been examined to explore the impact of their physicochemical characteristics on the nanoparticle–biological system (Figure 6.2b).

Among different nanomaterials [27], carbon nanomaterials are revolutionary due to the unique and significant utilization of each of the allotrope. Thus within the scope of the book, the toxicity of carbon nanomaterials will be discussed in this chapter.

Environmental Applications of Carbon Nanomaterials-Based Devices, First Edition.
Edited by Shadpour Mallakpour and Chaudhery Mustansar Hussain.

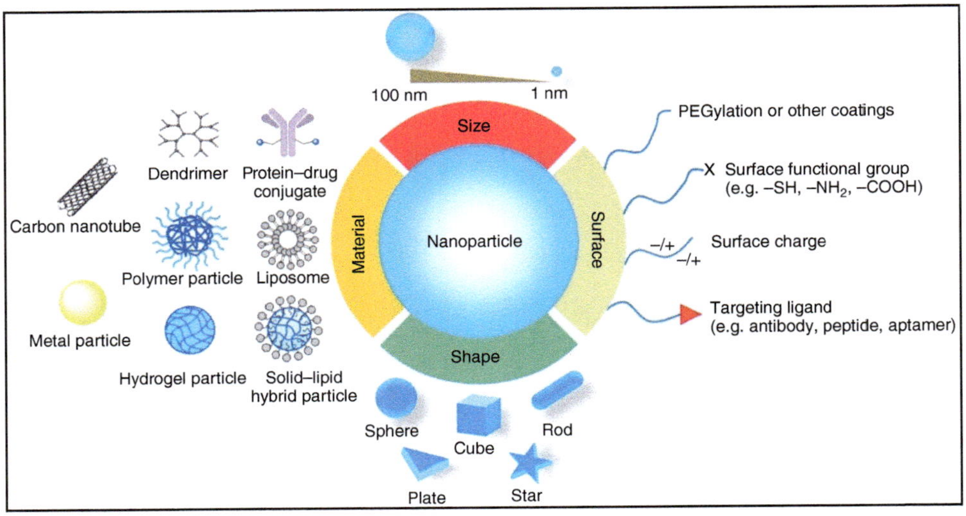

Figure 6.1 Material, dimension, and surface properties of nanoparticles govern their characteristics and applications. Source: Sun et al. [14]. © 2014 John Wiley & Sons.

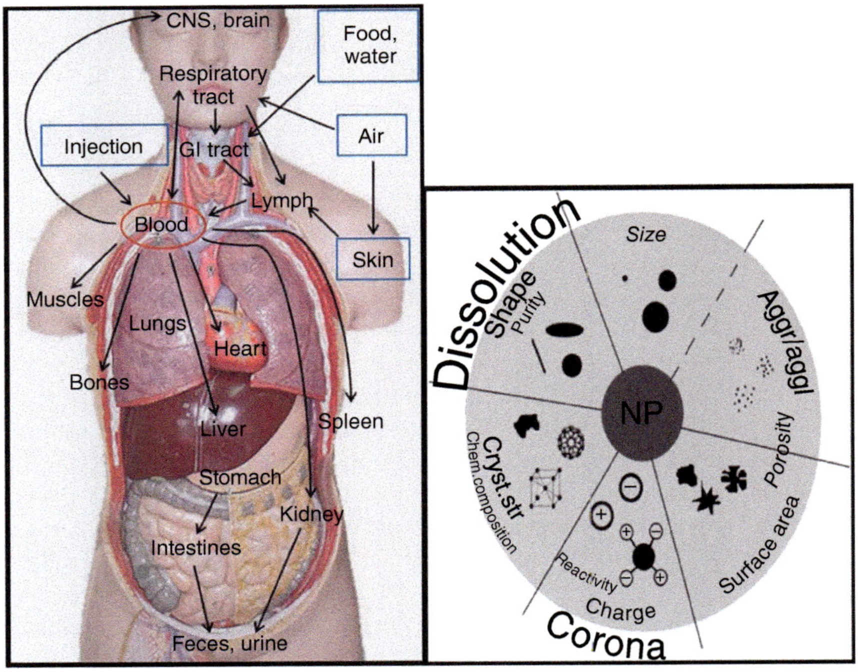

(a) (b)

Figure 6.2 (a) Potential routes of exposure, translocation, and deposition sites of nanomaterials. Source: Teow et al. [17]. © 2011 Royal Society of Chemistry. (b) Impact of physicochemical characteristics on nanoparticle–biological system interaction. Source: Luyts et al. [18]. © 2012 Royal Society of Chemistry.

6.2 Carbon Nanomaterials

Carbon, placed at group 14 (IV A), can form sp^3 and sp^2 hybrids and stable multiple π and σ bonds that in turn help it to configure its nanoallotropes like fullerene, carbon nanotube (CNT) [28, 29], and graphene (Figure 6.3a,b). All of these nanoallotropes along with their composites [32–34] possess great promises in finding applications in different potential sectors (Figure 6.3c) [35].

6.2.1 Fullerene

In 1985, Kroto et al. [36] discovered a nearly spherical closed-cage molecule having 60 numbers of carbon atoms and named then buckminsterfullerene. Structurally, C_{60} is a type of polygon called truncated icosahedron, composed of 60 vertices and 32 faces (12 pentagons and 20 hexagons) [37]. The C_{60} molecule possesses high symmetry with numerous transformations, each one of which maps the molecule onto itself [38]. A class of water-soluble, oxidized fullerenes are termed as fullerols or fullerenols. Fullerene can be synthesized by laser vaporization method [39], resistive and inductive heating [40], arc discharge [41], pyrolysis [42], or combustion/flame synthesis [43]. There are various prospective application areas of fullerene such as fuel cells [44], biology [45] and medicine [46], organic photovoltaics/solar cells [47], etc.

6.2.2 Carbon Nanotube

Radushkevich and Lukyanovich [48] observed a tubular carbon nanostructure in 1952 that was later termed as the carbon nanotube [49, 50]. However, this observation remained unnoticed by the researchers for a few decades. In 1991, Iijima [51] reported the observation of CNTs in the *Nature* journal, which created global interest and excitement. Structurally, CNT is a rolled-up sheet of carbon atoms that forms the tube structure where the number of sheets may be one (single-walled carbon nanotube [SWCNT]) or more (multiwalled carbon nanotube [MWCNT]) [52]. The difference in roll-up direction gives rise to different structures of SWCNTs such as zigzag, armchair, or chiral [53]. MWCNT can be viewed as either a single sheet of carbon rolled in around itself multiple times (parchment model) or multiple layers of graphitic sheets rolled in on themselves (Swiss roll). CNTs can be produced primarily via laser ablation [54], arc discharge [55], or chemical vapor deposition [56] methods. There are several possible application areas of CNT such as wearable electronics [57], flat-panel display [58], solar cell [59], sensor [60], textile industry [61], Li-ion battery [62], and even stimuli-responsive material [63].

6.2.3 Graphene and Graphene Derivatives

In 1962, first experimental observation of single layer of graphite was reported by Boehm et al. [64], and in 1994, he named the single layer of graphite as "graphene" [65]. In 2004, the report of the observation of graphene by Novoselov and coworkers [66] in Science journal fostered a huge interest in graphene among the global researchers. Structurally, graphene is one-atom-thick two-dimensional allotrope of carbon, consisting of a single sheet

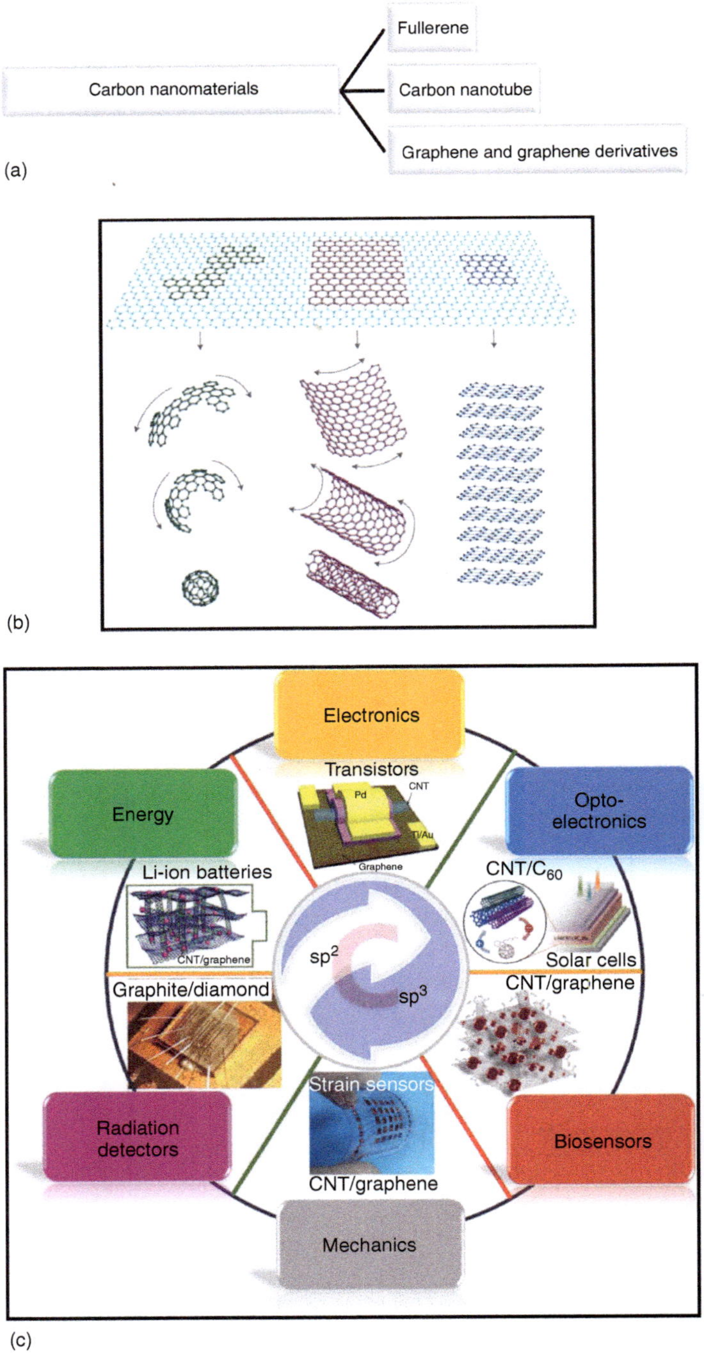

Figure 6.3 (a) Different carbon nanomaterials and (b) their structures. Source: Geim and Novoselov [30]. © 2007 Springer Nature. (c) Different potential application sectors of carbon nanomaterials. Source: Yuan et al. [31]. Licensed under CC BY 4.0.

of sp^2-hybridized carbon atoms. Graphene can be synthesized using exfoliation methods [67, 68], pyrolysis [69], arc discharge [70], epitaxial growth on SiC [71], or chemical vapor deposition [72]. Graphene oxide (GO) is produced by the exfoliation of graphite oxide and can be chemically reduced at a later stage to synthesize graphene. As complete reduction cannot be achieved, the end material is called "reduced graphene oxide" (rGO). Graphene has a gigantic possibility for applications in different sectors such as energy storage [73], flexible electronics [74], aerospace [75], and biomedical [76] applications.

6.3 Nanotoxicology and Resulting Cytotoxicity or Cellular Toxicity

Nanotoxicology is the branch of biological science that aims to measure the degree to which nanomaterials can cause a hazard to human health (Figure 6.4). The large surface-area-to-volume ratio and tiny size (which gives rise to quantum size effects) of nanomaterials along with environmental factors [78] collectively result in biological changes that widely differ from those caused by their larger counterparts and ultimately cause toxicity. Nanotoxicity can be measured in various ways; however, the primary measurement that needs to be performed is cytotoxicity.

Cytotoxicity or cellular toxicity is defined as the potential of a material to cause cell death. For an exact evaluation of cytotoxicity, information about cell origin, cell concentration, drug

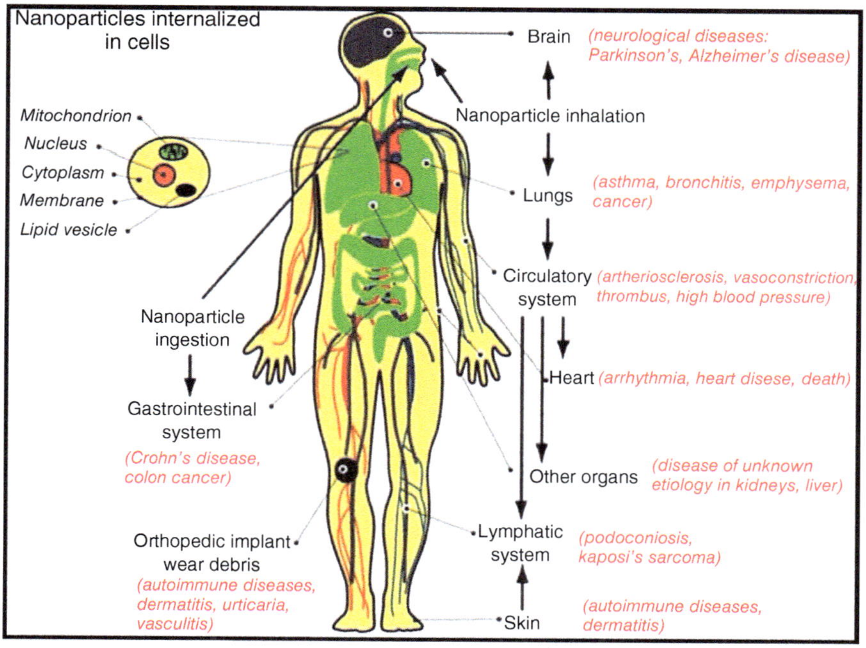

Figure 6.4 Pathways of exposure of the human body to nanoparticles, affected organs, and associated diseases. Source: Buzea et al. [77]. © 2007 AIP Publishing.

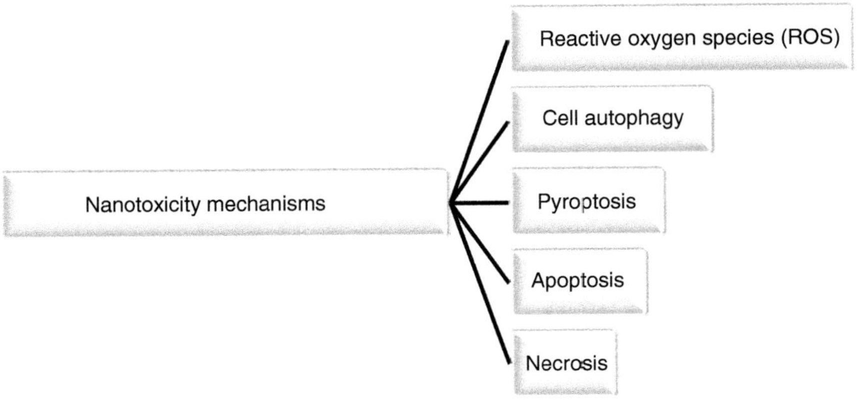

Figure 6.5 Various underlying mechanisms of cytotoxicity.

concentration, and treatment tenure require special attention. There are various underlying mechanisms of cytotoxicity, namely, reactive oxygen species [79] (ROS), cell autophagy [80], pyroptosis [81], apoptosis [82], and necrosis [83] (Figure 6.5).

ROS is a group of oxidative species (Table 6.1) that are produced intrinsically or extrinsically within the cell. These are mostly produced as by-products at the time of mitochondrial electron transport. Additionally, metal-catalyzed oxidation reactions also generate ROS as an essential intermediate. Abundant ROS have the potential to impair biological responses, leading to oxidative stress. Excessive toxic levels of oxidative stress can cause electron transfer chain dysfunction along with mitochondrial membrane damage, resulting in cell death (Figure 6.6a).

Table 6.1 Production pathways of reactive oxygen species [84].

Reactive oxygen species	Mechanism of generation
Superoxide (O_2^-)	Reduction of molecular oxygen in the electron transport chain of mitochondria and other enzymatic routes: monooxygenase, reduced form of nicotinamide adenine dinucleotide (NADPH) oxidase, and xanthine oxidase
Hydrogen peroxide (H_2O_2)	Converted from O_2^- by enzyme superoxide dismutase (SOD)
Hydroxyl radical ($^{\cdot}OH$)	Produced in Haber–Weiss reaction from O_2^- and H_2O_2
Singlet oxygen ($^1O_2$)	Produced in the reaction of hypochlorous acid (HOCl) and H_2O_2
Peroxynitrite ($ONOO^-$)	Produced in the reaction of nitric oxide (NO) and (O_2^-)

Source: Patel et al. [84]. Licensed under CC BY 3.0.

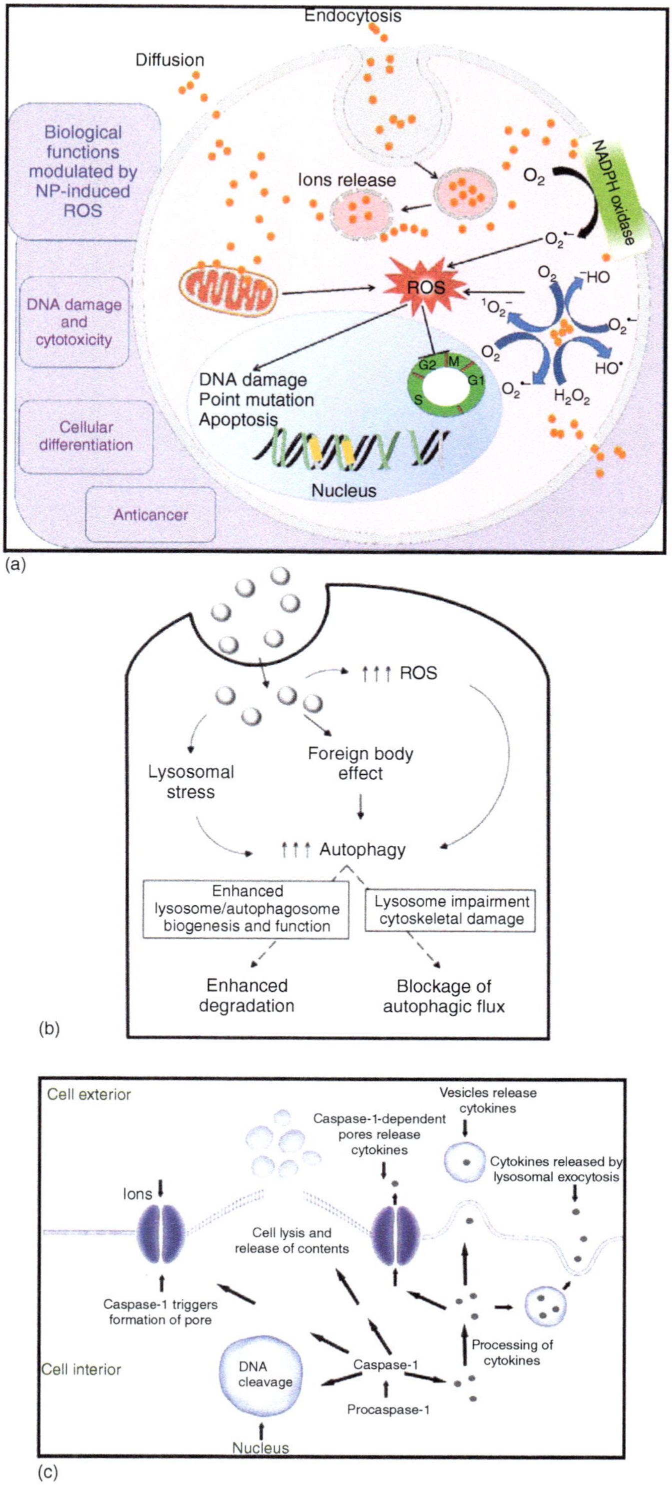
Endocytosis
Diffusion
Biological functions modulated by NP-induced ROS
DNA damage and cytotoxicity
Cellular differentiation
Anticancer
Ions release
NADPH oxidase
ROS
DNA damage
Point mutation
Apoptosis
Nucleus
(a)
ROS
Lysosomal stress
Foreign body effect
Autophagy
Enhanced lysosome/autophagosome biogenesis and function
Lysosome impairment cytoskeletal damage
Enhanced degradation
Blockage of autophagic flux
(b)
Cell exterior
Vesicles release cytokines
Caspase-1-dependent pores release cytokines
Cytokines released by lysosomal exocytosis
Ions
Cell lysis and release of contents
Caspase-1 triggers formation of pore
Processing of cytokines
Cell interior
DNA cleavage
Caspase-1
Procaspase-1
Nucleus
(c)

(Continued)

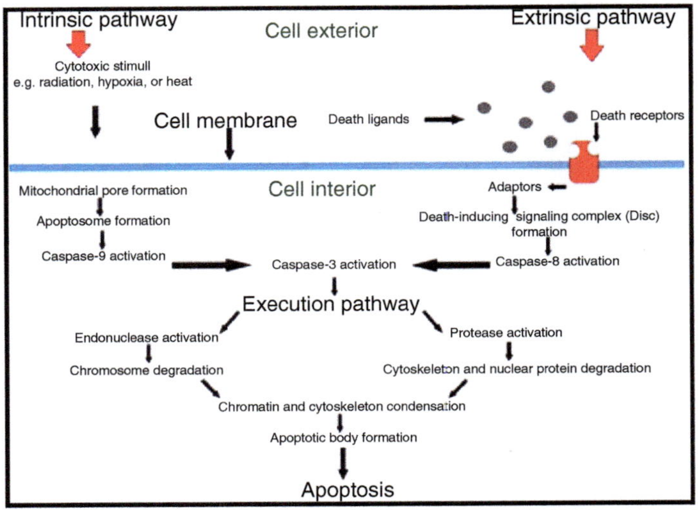

(d)

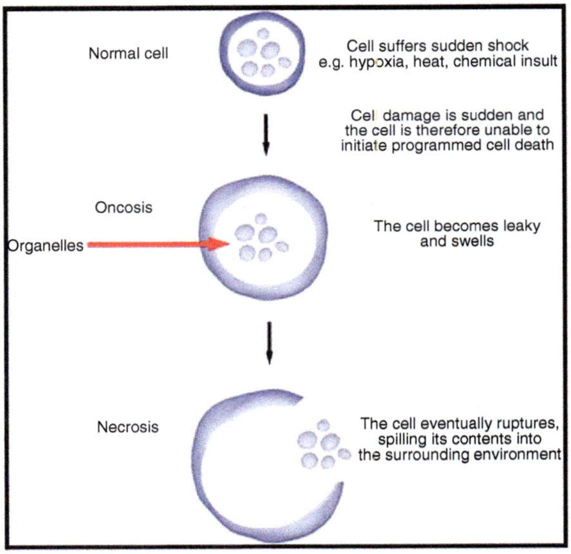

(e)

Figure 6.6 (a) Effect of reactive oxygen species in cells. Source: Dayem et al. [85]. Licensed under CC BY 4.0. (b) Mechanisms of nanomaterial-induced autophagy activation. Source: Popp and Segatori [86]. © 2015 Elsevier. Pathways leading to cell death: (c) pyroptosis, (d) apoptosis, and (e) necrosis. Source: D'Arcy [87]. © 2019 John Wiley & Sons.

Autophagy constitutes the processes of autophagosome synthesis and lysosomal degradation (Figure 6.6b). Under stressed conditions, reduced lysosomal activity may increase autophagosome synthesis; thus processing of excessive autophagosomes will be reduced due to the rate-limiting lysosomal activity. The formation/accumulation of uncontrolled autophagosomes, which are unfused to lysosomes, leads to autophagy and directly induces cellular toxicity.

The term "pyroptosis" originated from the Greek words "pyro" and "ptosis" that mean "fever/inflammation" and "to fall," respectively. It is a kind of regulated cell death caused by inflammatory stimuli, transduced via inflammatory caspases, and executed by gasdermins. Inflammatory caspases include critical assessment of protein structure prediction (CASP)-1, CASP-4/-5 (human), and CASP-11 (murine) (Figure 6.6c).

In multicellular organisms, the unwanted cells (not needed or threat to the organism) are abolished by a controlled cell suicide procedure known as apoptosis (originated from the Greek word that means "falling off," as leaves from a tree). Proteolytic enzymes (caspases) mediate apoptosis and initiate cell death via cleaving specific proteins in the cytoplasm and nucleus. The members of B-cell lymphoma 2 and inhibitors of apoptosis protein families regulate the activation process of caspase (Figure 6.6d).

Necrosis (originated from the Greek word that means "death") is a type of cell injury that causes premature cell death in living tissue via autolysis (Figure 6.6e). It is caused by external (to cell or tissue) factors, namely, trauma, infection, or toxins, and leads to the unregulated digestion of cell components.

The cytotoxicity of carbon-based nanomaterials is influenced by physical and chemical properties of carbon nanomaterials [88] (Figure 6.7), surface modification of nanomaterials [91], and metal impurities [92].

6.4 Assessment of Nanocytotoxicity

Now cytotoxicity can be assessed by examining the toxicity in different organs of the human body. This section aims to review the cytotoxic effects of carbon nanomaterials in different organs of the human body (Figure 6.8).

6.4.1 Respiratory or Pulmonary Toxicity

An aggregated form of underivatized C_{60} and a fully derivatized, highly water-soluble derivative of C_{60} was used to study the pulmonary toxicity on rats by Sayes et al. [93]. However, no adverse effects on lung tissue were reported. Two different-sized (nano and micro) fullerene aggregates were used by Sayers et al. [94] to study respiratory toxicity in mice, and lung inflammatory responses were observed. Morimoto et al. [95] also studied inhalation exposure and intratracheal instillation effects of fullerene on rats, and their result showed that well-dispersed fullerenes did not cause any neutrophil inflammation.

CNT-mediated pulmonary toxicity can occur via inflammation, injury, fibrosis, and pulmonary tumors. Park et al. [96] treated mice with MWCNTs by intratracheal instillation and observed allergic and pro-inflammatory responses, owing to B-cell (B-lymphocyte) activation and production of immunoglobulin E (IgE). Inoue et al. [97] examined *in vivo* and *in vitro* effects of pulmonary exposure of MWCNT on mice and showed that MWCNT can intensify allergic airway inflammation with increased humoral immunity and thus has the potential to cause allergic asthma. The effect of short-term inhalation of the rodlike CNT on mice was studied by Rydman et al. [98], and they found that novel natural immunity-mediated allergic-like airway inflammation was induced in mice by the said exposure. Heish et al. [99] studied the potential toxic effects of SWCNT exposure on the respiratory

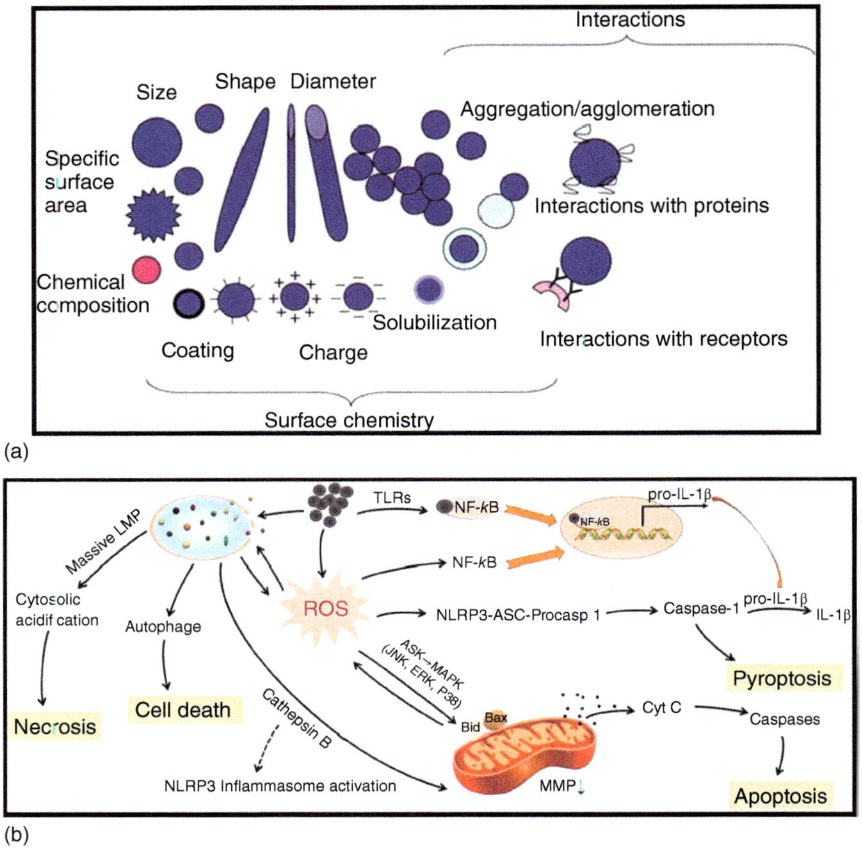

Figure 6.7 (a) Different physicochemical characteristics of nanomaterials, responsible for their toxicological activity. Source: Marano et al. [89]. © 2016 Springer Nature. (b) Cytotoxicity initiation mechanisms of a nanoparticle. Source: Yuan et al. [90]. Licensed under CC BY 4.0.

system of mice and found that intratracheal instillation of SWCNTs could result in an airflow obstruction and airway hyperreactivity that could consequently lead to irreversible obstructive airway disease. Xu et al. [100] reported that the intrapulmonary administration of MWCNTs can induce mesothelial proliferation that is closely associated with mesothelioma development. It was reported that myofibroblasts and lung fibrosis can be caused by CNT exposure [101]. Chang et al. [102] reported that epithelial–mesenchymal transition results in SWCNT-induced pulmonary fibrosis. The investigation of Luanpitpong et al. [103] concluded that chronic (six-month) exposure of human lung epithelial cells to SWCNTs can cause lung cancer. Inhalation exposure to MWCNTs can also promote lung adenocarcinoma [104] and lung carcinogenicity [105].

The pulmonary toxicities of different-sized graphite nanoplates were explored upon exposure to mice by Roberts et al. [106] that revealed that graphite nanoplates may induce lung inflammation and injury in lavage fluid depending on their sizes. Mao et al. [107] analyzed the toxicity of few-layer graphene in mice after intratracheal instillation, which

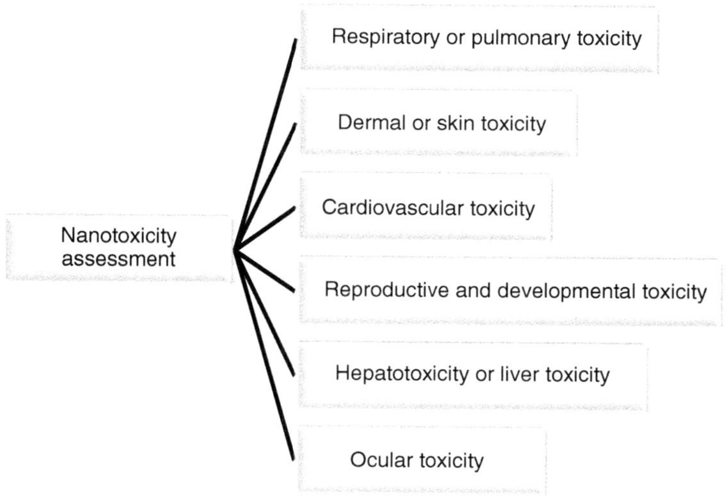

Figure 6.8 Nanotoxicity assessment pathways.

proved that it can cause acute lung injury and pulmonary edema. Ma et al. [108] found that the activation of macrophages and the stimulation of pro-inflammatory responses are governed by the lateral size of GO. Hu et al. [109] performed theoretical studies to demonstrate that inhaled GO nanosheets can severely damage pulmonary surfactant film and thereby induce toxicity. Zhang et al. [110] reported that GO, which is deposited on the lungs, has long retention time and thereby can cause pulmonary edema. Park et al. [111] demonstrated that intratracheally instilled graphene nanoplatelets can remain in lung for long duration and can cause cytoskeletal damage in the lung. Intratracheally instilled functionalized graphene nanoplatelets result in significant acute neutrophilic inflammation as reported by Lee et al. [112]. Nanoscale GO sheets were intratracheally instilled in mice by Li et al. [113], and the study revealed that the incorporation of GO causes acute lung injury and chronic pulmonary fibrosis.

6.4.2 Dermal or Skin Toxicity

Human epidermal keratinocytes were exposed to functionalized fullerene by Rouse et al. [114], which decreased cell viability and initiated a pro-inflammatory response. Saathoff et al. [115] reported that high dose of $C_{60}(OH)_{32}$ (42.5 μg/ml) decreased the viability of human epidermal keratinocytes.

Patlolla et al. [116] applied different doses of purified MWCNT in normal human dermal fibroblast cells to study the cellular response. They found that MWCNTs are very toxic at sufficiently high concentrations. Mice skin was exposed to SWCNT by Murray et al. [117], which resulted in dermal toxicity via enhancement of dermal cell numbers and skin thickening. Zhang et al. [118] studied the effect of the exposure of functionalized SWCNT on human epidermal keratinocytes and conclude that lower concentration of SWCNT can induce mild cytotoxicity, while higher concentration of SWCNT demonstrated an irritation response. Human epidermal keratinocytes were used as a target for the exposure of unrefined SWCNT, and this caused dermal toxicity caused by accelerated oxidative stress as reported by Shvedova et al. [119]. Palmer et al. [120] examined the effect of exposure of carboxylated MWCNT on mice, and the study revealed that functionalized MWCNT can increase epidermal thickness and gives rise to allergic skin conditions.

As reported by Liao et al. [121], mammalian fibroblasts can be damaged via the exposure to graphene though the exposure environment and mode of interaction with cells are to be considered as crucial parameters. Pelin et al. [122] investigated differential cytotoxic effects of graphene and GO on skin keratinocytes and found that these materials are cytotoxic and can cause significant cellular damage. In a follow-up study by the same group, they demonstrated that ROS production in human skin keratinocytes could be induced by graphene and GO, which is responsible for their cytotoxic effects.

6.4.3 Cardiovascular Toxicity

To examine the potential vascular toxicity of fullerene in humans, Yamawaki and Iwai [123] investigated the effects of water-soluble fullerene on vascular endothelial cells and reported that fullerene can cause cytotoxic injury. Thompson et al. [124] found that exposure of fullerene to rats may enhance cardiac ischemia/reperfusion injury via narrowing the coronary artery. Adverse effects of fullerenes on the human umbilical vein endothelial cells were reported by Gelderman et al. [125]. Vesterdal et al. [126] observed that even low doses of fullerene exposure to mice can cause modest vasomotor dysfunction and different degree of atherosclerosis.

Radomski et al. [127] reported that CNT is capable of stimulating platelet aggregation, thereby increasing the rate of vascular thrombosis. The cardiac effects of exposure of CNTs in rat were evaluated by Hosseinpour et al. [128], and their study showed that nanotube exposure can induce increment of heart rate. Li et al. [129] demonstrated that intrapharyngeal instillation of SWCNT in mice is capable to cause cardiovascular diseases. Both systolic and diastolic blood pressures were increased, and heart rate was decreased on inhalation of MWCNT as reported by Zheng et al. [130].

Singh et al. [131] reported that atomically thin GO sheets exerted strong aggregatory reaction in platelets depicting its prothrombotic property. Myocardial cells were used to check the cytotoxicity of rGO by Contreras-Torres et al. [132] and reported that rGO is very toxic to myocardial cells. Bangeppagari et al. [133] exposed zebrafish to GO and detected several cardiovascular defects due to the exposure. Cardiotoxicity of GO was evaluated by Arbo et al. [134] via mitochondrial disturbances.

6.4.4 Reproductive and Developmental Toxicity

Tsuchiya et al. [135] exposed mouse embryos to fullerene, and after an exposure of 18 hours, all embryos died. Fullerene can cause oxidative stress in embryonic zebrafish, leading to mortality as reported by Usenko et al. [136]; however, reduced light during exposure can reduce the mortality rate [137]. Zhu et al. [138] studied the toxicity effects of fullerene aggregates on zebrafish embryos and found that the exposure hindered the development of zebrafish embryo and larva, reducing the survival and hatching rates, and resulted in pericardial edema. Fullerene aggregates were exposed to the Chinese hamster ovary by Han and Karim [139] and found that cell mortality increases with the increase in the fullerene concentration and incubation time.

Pietroiusti et al. [140] reported that the application of even a low dose of SWCNT to embryonic stem cell of the mouse can affect embryonic development. Chicken embryos were exposed to SWCNT by Roman et al. [141], and they found that most of the embryos died before incubation. Fujitani et al. [142] studied the teratogenicity of MWCNT in mice and observed various types of fetuses with skeletal malformation. Cheng and Cheng [143] assessed the exposure effect of MWCNT on zebrafish embryos and demonstrated that the developmental toxicity is dependent on the length of MWCNT. The effect of functionalized MWCNTs on pregnant mice was studied by Huang et al. [144], and the study revealed that large MWCNT could restrict the fetus development and also could induce brain deformity.

Xu et al. [145] injected mouse dams with rGO nanosheets at different doses and time (pre- or postfertilization). rGO nanosheets injected at a late gestational stage are reported to cause abortions and death. GO can induce developmental malformation of the cardiac/yolk sac edema, eye, and tail flexure and reduction in heart rate when it is introduced during zebrafish embryogenesis, as reported by Chen et al. [146]. Fu et al. [147] reported that GO can impart many negative effects on the development of mice in the lactation period. Akhavan et al. [148] intravenously injected nanoscale GO sheets into male mice that caused a reduction in sperm viability and motility.

6.4.5 Hepatotoxicity or Liver Toxicity

Kamat et al. [149] introduced fullerene in rat liver microsomes while exposed to light, which resulted in significant lipid peroxidation and other forms of oxidative damage causing liver injury. Cytotoxic effects of hydroxylated fullerenes on isolated rat hepatocytes were studied by Nakagawa et al. [150] and concluded that the exposure can cause cell death via mitochondrial dysfunction. Shipkowski et al. [151] intratracheally instilled fullerene in rats and found that fullerenes are mostly concentrated in the liver and the concentration remains nearly unchanged, resulting in detrimental health effects.

Differential cytotoxicity of water-soluble fullerenes on human liver carcinoma cells (HepG2) was investigated by Sayes et al. [152] and found that cytotoxicity of water-soluble fullerene species is governed by surface derivatization. Patlolla et al. [153] demonstrated that the introduction of functionalized SWCNTs in mice causes oxidative stress in the liver followed by morphological alterations of the liver. Awasthi et al. [154] orally administered mice with MWCNTs, which resulted in cellular swelling, inflammation, and blood coagulation in the liver. Systemic distribution of SWCNT in mice was studied by Principi et al. [155], and the study revealed that the SWCNTs get accumulated in the liver causing liver damage.

Patlolla et al. [156] exposed rats to GO and found that GO cause damage to the liver tissue via ROS induction. To study the biological effects of GOs on parenchymal hepatocytes, Zhang et al. [157] exposed mouse to GOs and concluded that large GOs cause ROS production and thereby can damage the liver. Mohamed et al. [158] injected GO nanosheets in male albino mice and observed that GO has hepatotoxic effects on liver cells via the creation of oxidative stress. The effect of exposure of GO on zebrafish was analyzed by Xiong et al. [159], and they reported that GO could induce liver dysfunction to trigger the oxidative stress along with inflammatory responses to cause liver injury.

6.4.6 Ocular Toxicity

Aoshima et al. [160] observed that highly purified fullerenes can cause corneal epithelial defects and conjunctival redness upon exposure. Cytotoxicity of fullerol in human lens epithelial cells was examined by Roberts et al. [161], and their study revealed that exposure to fullerol along with the presence of sunlight may cause early cataractogenesis. In a similar study performed by Wielgus et al. [162], it was evidenced that, in the presence of sunlight, ocular exposure to fullerol can result in retinal damage.

Ema et al. investigated ocular toxicity on fullerene [163] and CNTs [164]. The study indicated that both fullerene and CNT can cause acute irritation of the eyes. Rabbits were exposed to CNTs to assess their ocular toxicity by Kishore et al. [165], and upon exposure, reversible conjunctival redness and discharge from the eye had been reported. Liu's group performed ocular toxicity studies for both SWCNTs [166] and MWCNTs [167] and concluded that CNTs are toxic to human ocular cells. Ocular exposure to SWCNTs and MWCNTs can decrease in the cell survival rate and initiate cell apoptosis via ROS generation.

To assess the toxicity of GO exposure to the eye, Wu et al. [168] performed short-term GO exposure on human epithelium cells, which showed significant cytotoxicity with increased intracellular ROS. Wu et al. [169] also examined the ocular toxicity of polyethylene glycol (PEG)-GO samples having different oxidation levels and found that oxidative stress-induced cytotoxicity of the PEG-GO samples depends on their oxidation levels. An et al. [170] found that short-term repeated GO exposure can cause intraocular inflammation and in turn can initiate cell apoptosis in the cornea, thus depicting significant cytotoxicity in rat epithelial cells.

6.5 Conclusions

Nanotechnology is progressing rapidly, as well as its commercial applications. The toxicity of the resulting exposure has to be assessed with utmost care in saving mankind. Among different nanomaterials, carbon nanomaterials stand far ahead because of their tremendous application potential. Thus many researchers around the globe are diligently trying to rightly assess the toxic nature of carbon nanomaterials. The researchers are using different methodologies and standards for the evaluation of the toxicity of carbon nanomaterials, which possesses difficulty in making a comparative statement based on their work. However, it can be stated that the physical structure and the chemical nature of the nanomaterials are responsible for their nature of toxicity level. Finally, it is the prerequisite for the industries to properly assess the risk factors of any carbon nanomaterials in terms of associated toxicity and accordingly develop proper regulations for their safe applications in concerned industries and healthcare sectors.

Important Websites

www.bp.com
https://copublications.greenfacts.org/en/nanotechnologies/l-2/6-ealth-effects-nanoparticles.htm
https://www.cdc.gov/niosh/topics/nanotech/default.html
https://www.sigmaaldrich.com/technical-documents/articles/material-matters/strategy-for-toxicity.html

References

1 Feynman, R.P. (1992). There's plenty of room at the bottom [data storage]. *J. Microelectromech. Syst.* 1 (1): 60–66. https://doi.org/10.1109/84.128057.
2 Taniguchi, N. (1974). On the basic concept of 'nano-technology'. In: *Proceedings of the International Conference on Production Engineering, Tokyo, Japan*, vol. 2, 18–23. Japan Society of Precision Engineering.
3 Viswanathan, V.K., Manoharan, S.R.R., Subramanian, S. et al. (2019). Nanotechnology in spine surgery: a current update and critical review of the literature. *World Neurosurg.* 123: 142–155. https://doi.org/10.1016/j.wneu.2018.11.035.
4 Sun, Y., Ma, W., Yang, Y. et al. (2019). Cancer nanotechnology: enhancing tumor cell response to chemotherapy for hepatocellular carcinoma therapy. *Asian J. Pharm. Sci.* 14 (6): 581–594. https://doi.org/10.1016/j.ajps.2019.04.005.
5 Olawoyin, R. (2018). Nanotechnology: the future of fire safety. *Saf. Sci.* 110: 214–221. https://doi.org/10.1016/j.ssci.2018.08.016.
6 de Francisco, E.V. and García-Estepa, R.M. (2018). Nanotechnology in the agrofood industry. *J. Food Eng.* 238: 1–11. https://doi.org/10.1016/j.jfoodeng.2018.05.024.
7 Mathew, J., Joy, J., and George, S.C. (2019). Potential applications of nanotechnology in transportation: a review. *J. King Saud Univ. Sci.* 31 (4): 586–594. https://doi.org/10.1016/j.jksus.2018.03.015.
8 Usman, M. et al. (2020). Nanotechnology in agriculture: current status, challenges and future opportunities. *Sci. Total Environ.* 721: 137778. https://doi.org/10.1016/j.scitotenv.2020.137778.
9 Contreras, J.E., Rodriguez, E.A., and Taha-Tijerina, J. (2017). Nanotechnology applications for electrical transformers – a review. *Electr. Power Syst. Res.* 143: 573–584. https://doi.org/10.1016/j.epsr.2016.10.058.
10 Fakoya, M.F. and Shah, S.N. (2017). Emergence of nanotechnology in the oil and gas industry: emphasis on the application of silica nanoparticles. *Petroleum* 3 (4): 391–405. https://doi.org/10.1016/j.petlm.2017.03.001.
11 Hussein, A.K. (2015). Applications of nanotechnology in renewable energies – a comprehensive overview and understanding. *Renewable Sustainable Energy Rev.* 42: 460–476. https://doi.org/10.1016/j.rser.2014.10.027.
12 Vance, M.E. et al. (2015). Nanotechnology in the real world: redeveloping the nanomaterial consumer products inventory. *Beilstein J. Nanotechnol.* 6 (1): 1769–1780. https://doi.org/10.3762/bjnano.6.181.

13 Hussain, C.M. (ed.) (2020). *Handbook of Functionalized Nanomaterials for Industrial Applications*. Elsevier. https://doi.org/10.1016/C2018-0-00341-2.

14 Sun, T. et al. (2014). Engineered nanoparticles for drug delivery in cancer therapy. *Angew. Chem. Int. Ed.* 53 (46): 12320–12364. https://doi.org/10.1002/anie.201403036.

15 Hussain, C.M. and Keçili, R. (eds.) (2020). *Modern Environmental Analysis Techniques for Pollutants*. Elsevier. https://doi.org/10.1016/C2018-0-01639-4.

16 Oberdörster, G. et al. (2005). Principles for characterizing the potential human health effects from exposure to nanomaterials: elements of a screening strategy. *Part. Fibre Toxicol.* 2 (1): 8. https://doi.org/10.1186/1743-8977-2-8.

17 Teow, Y. et al. (2011). Health impact and safety of engineered nanomaterials. *Chem. Commun.* 47 (25): 7025–7038. https://doi.org/10.1039/C0CC05271J.

18 Luyts, K. et al. (2012). How physico-chemical characteristics of nanoparticles cause their toxicity: complex and unresolved interrelations. *Environ. Sci. Processes Impacts* 15 (1): 23–38. https://doi.org/10.1039/C2EM30237C.

19 Ferin, J. et al. (1990). Increased pulmonary toxicity of ultrafine particles? I. Particle clearance, translocation, morphology. *J. Aerosol Sci.* 21 (3): 381–384. https://doi.org/10.1016/0021-8502(90)90064-5.

20 Donaldson, K. et al. (2004). Nanotoxicology. *Occup. Environ. Med.* 61 (9): 727–728. https://doi.org/10.1136/oem.2004.013243.

21 Wood, J. (2005). Profiling the cellular response to nanomaterials: toxicology. *Mater. Today* 8 (12, Supplement 1): 13. https://doi.org/10.1016/S1369-7021(05)71270-0.

22 Espanani, H.R. et al. (2015). Toxic effect of nano-zinc oxide. *Bratisl. Med. J.* 116 (10): 616–620. https://doi.org/10.4149/BLL_2015_119.

23 Grande, F. and Tucci, P. (2016). Titanium dioxide nanoparticles: a risk for human health? *Mini Rev. Med. Chem.* 16 (9): 762–769. http://www.eurekaselect.com/140530/article.

24 Kustov, L.M. et al. (2018). Silicon nanoparticles: characterization and toxicity studies. *Environ. Sci.: Nano* 5 (12): 2945–2951. https://doi.org/10.1039/C8EN00934A.

25 Carnovale, C. et al. (2019). Identifying trends in gold nanoparticle toxicity and uptake: size, shape, capping ligand, and biological corona. *ACS Omega* 4 (1): 242–256. https://doi.org/10.1021/acsomega.8b03227.

26 Hussain, C.M. (ed.) (2019). *Handbook of Environmental Materials Management*. Springer International Publishing. https://www.springer.com/gp/book/9783319736440.

27 Hussain, C.M. and Kharisov, B. (eds.) (2016). *Advanced Environmental Analysis: Application of Nanomaterials*, vol. 1. Royal Society of Chemistry. https://doi.org/10.1039/9781782623625.

28 Mallakpour, S. and Khodadadzadeh, L. (2018). Biocompatible and biodegradable chitosan nanocomposites loaded with carbon nanotubes. In: *Biodegradable and Biocompatible Polymer Composites*, Woodhead Publishing Series in Composites Science and Engineering (ed. N.G. Shimpi), 187–221. Woodhead Publishing. https://doi.org/10.1016/B978-0-08-100970-3.00007-9.

29 Mallakpour, S. and Rashidimoghadam, S. (2019). Chapter 9: Carbon nanotubes for dyes removal. In: *Composite Nanoadsorbents*, Micro and Nano Technologies (eds. G.Z. Kyzas and A.C. Mitropoulos), 211–243. Elsevier. https://doi.org/10.1016/B978-0-12-814132-8.00010-1.

30 Geim, A.K. and Novoselov, K.S. (2007). The rise of graphene. *Nat. Mater.* 6: 183–191. https://doi.org/10.1038/nmat1849.

31 Yuan, Q., Lin, C.-T., and Chee, K.W.A. (2019). All-carbon devices based on sp^2-on-sp^3 configuration. *APL Mater.* 7 (3). https://aip.scitation.org/doi/10.1063/1.5082767.

32 Mallakpour, S., Abdolmaleki, A., and Borandeh, S. (2015). Fabrication of amino acid-based graphene-zinc oxide (ZnO) hybrid and its application for poly(ester–amide)/graphene-ZnO nanocomposite synthesis. *J. Thermoplast. Compos. Mater.* https://doi.org/10.1177/0892705715598365.

33 Mallakpour, S. and Behranvand, V. (2018). Synthesis of mesoporous recycled poly(ethylene terephthalate)/MWNT/carbon quantum dot nanocomposite from sustainable materials using ultrasonic waves: application for methylene blue removal. *J. Cleaner Prod.* 190: 525–537. https://doi.org/10.1016/j.jclepro.2018.04.120.

34 Mallakpour, S., Behranvand, V., and Mallakpour, F. (2019). Synthesis of alginate/carbon nanotube/carbon dot/fluoroapatite/TiO_2 beads for dye photocatalytic degradation under ultraviolet light. *Carbohydr. Polym.* 224: 115138. https://doi.org/10.1016/j.carbpol.2019.115138.

35 Sengupta, J. (2020). Chapter 17: Application of carbon nanomaterials in the electronic industry. In: *Handbook of Nanomaterials for Manufacturing Applications*, Micro and Nano Technologies (ed. C.M. Hussain), 421–450. Elsevier. https://doi.org/10.1016/B978-0-12-821381-0.00017-X.

36 Kroto, H.W. et al. (1985). C_{60}: buckminsterfullerene. *Nature* 318 (6042): 162–163. https://doi.org/10.1038/318162a0.

37 David, W.I.F., Ibberson, R.M., Matthewman, J.C. et al. (1991). Crystal structure and bonding of ordered C_{60}. *Nature* 353: 147–149. https://doi.org/10.1038/353147a0.

38 Curl, R.F. and Smalley, R.E. (1991). Fullerenes. *Sci. Am.* 265 (4): 54–63.

39 Lieber, C.M. and Chen, C.-C. (1994). Preparation of fullerenes and fullerene-based materials. *Solid State Phys.* 48 (1986): 109–148. https://doi.org/10.1016/S0081-1947(08)60578-0.

40 Fostiropoulos, K., Huffman, D.R., Krätschmer, W., and Lamb, L.D. (1990). Solid C_{60}: a new form of carbon. *Nature* 347: 354–358. https://doi.org/10.1038/347354a0.

41 Haufler, R.E., Conceicao, J., Chibante, L.P.F. et al. (1990). Efficient production of C_{60} (buckminsterfullerene), $C_{60}H_{36}$, and the solvated buckide ion. *J. Phys. Chem.* 94 (8): 8634–8636. https://doi.org/10.1021/j100387a005.

42 Taylor, R. et al. (1993). Formation of C_{60} by pyrolysis of naphthalene. *Nature* 366 (6457): 728–731. https://doi.org/10.1038/366728a0.

43 Howard, J.B. et al. (1991). Fullerenes C_{60} and C_{70} in flames. *Nature* 352 (6331): 139–141. https://doi.org/10.1038/352139a0.

44 Lee, G. et al. (2009). Monodisperse Pt and PtRu/C_{60} hybrid nanoparticles for fuel cell anode catalysts. *Chem. Commun.* (33): 5036–5038. https://doi.org/10.1039/b911068b.

45 Castro, E. et al. (2017). Fullerenes in biology and medicine. *J. Mater. Chem. B* 5 (32): 6523–6535. https://doi.org/10.1039/C7TB00855D.

46 Bakry, R. et al. (2007). Medicinal applications of fullerenes. *Int. J. Nanomed.* 2 (4): 639–649. https://www.ncbi.nlm.nih.gov/pmc/articles/PMC2676811/.

47 Ganesamoorthy, R., Sathiyan, G., and Sakthivel, P. (2017). Review: fullerene based acceptors for efficient bulk heterojunction organic solar cell applications. *Sol. Energy Mater. Sol. Cells* 161: 102–148. https://doi.org/10.1016/j.solmat.2016.11.024.

48 Radushkevich, L.V. and Lukyanovich, V.M. (1952). The structure of carbon forming in thermal decomposition of carbon monoxide on an iron catalyst. *Russ. J. Phys. Chem.* 26: 88–95.

49 Mallakpour, S. and Khadem, E. (2019). Chapter 8: Carbon nanotubes for heavy metals removal. In: *Composite Nanoadsorbents*, Micro and Nano Technologies (eds. G.Z. Kyzas and A.C. Mitropoulos), 181–210. Elsevier. https://doi.org/10.1016/B978-0-12-814132-8.00009-5.

50 Mallakpour, S. and Khadem, E. (2016). Carbon nanotube–metal oxide nanocomposites: fabrication, properties and applications. *Chem. Eng. J.* 302: 344–367. https://doi.org/10.1016/j.cej.2016.05.038.

51 Iijima, S. (1991). Helical microtubules of graphitic carbon. *Nature* 354: 56–58. https://doi.org/10.1038/354056a0.

52 Amelinckx, S., Bernaerts, D., Zhang, X.B. et al. (1995). A structure model and growth mechanism for multishell carbon nanotubes. *Science* 267: 1334–1338. https://doi.org/10.1126/science.267.5202.1334.

53 Sisto, T.J. et al. (2016). Towards pi-extended cycloparaphenylenes as seeds for CNT growth: investigating strain relieving ring-openings and rearrangements. *Chem. Sci.* 7 (6): 3681–3688. https://doi.org/10.1039/C5SC04218F.

54 Guo, T. et al. (1995). Catalytic growth of single-walled nanotubes by laser vaporization. *Chem. Phys. Lett.* 243: 49–54. https://doi.org/10.1016/0009-2614(95)00825-O.

55 Iijima, S. and Ichihashi, T. (1993). Single-shell carbon nanotubes of 1-nm diameter. *Nature* 363: 603–605. https://doi.org/10.1038/363603a0.

56 Sengupta, J. (2018). Chapter 9: Carbon nanotube fabrication at industrial scale: opportunities and challenges. In: *Handbook of Nanomaterials for Industrial Applications*, Micro and Nano Technologies (ed. C.M. Hussain), 172–194. Elsevier. https://doi.org/10.1016/B978-0-12-813351-4.00010-9.

57 Kim, J.H. et al. (2018). Simple and cost-effective method of highly conductive and elastic carbon nanotube/polydimethylsiloxane composite for wearable electronics. *Sci. Rep.* 8 (1): 1375 (1–11). https://doi.org/10.1038/s41598-017-18209-w.

58 Kim, J. et al. (2006). Flat panel display using carbon nanotube. In: *International Symposium on Discharges and Electrical Insulation in Vacuum*, 888–889. IEEE. https://doi.org/10.1109/DEIV.2006.357448.

59 Kumar, U. et al. (2016). Carbon nanotube: synthesis and application in solar cell. *J. Inorg. Organomet. Polym. Mater.* 26 (6): 1231–1242. https://doi.org/10.1007/s10904-016-0401-z.

60 Zaporotskova, I.V. et al. (2016). Carbon nanotubes: sensor properties. A review. *Mod. Electron. Mater.* 2 (4): 95–105. https://doi.org/10.1016/j.moem.2017.02.002.

61 Shahidi, S. and Moazzenchi, B. (2018). Carbon nanotube and its applications in textile industry – a review. *J. Text. Inst.*: 1–14. https://doi.org/10.1080/00405000.2018.1437114.

62 Sehrawat, P., Julien, C., and Islam, S.S. (2016). Carbon nanotubes in Li-ion batteries: a review. *Mater. Sci. Eng., B* 213: 12–40. https://doi.org/10.1016/j.mseb.2016.06.013.

63 Bafkary, R. and Khoee, S. (2016). Carbon nanotube-based stimuli-responsive nanocarriers for drug delivery. *RSC Adv.* 6 (86): 82553–82565. https://doi.org/10.1039/C6RA12463A.

64 Boehm, H.P. et al. (1962). Dünnste Kohlenstoff-Folien. *Z. Naturforsch., B: Chem. Sci.* 17 (3): 150–153. https://doi.org/10.1515/znb-1962-0302.

65 Boehm, H.P., Setton, R., and Stumpp, E. (1994). Nomenclature and terminology of graphite intercalation compounds (IUPAC recommendations 1994). *Pure Appl. Chem.* 66 (9): 1893–1901. https://doi.org/10.1351/pac199466091893.

66 Zhang, Y., Novoselov, K.S., Geim, A.K. et al. (2004). Electric field effect in atomically thin carbon films. *Science* 306 (5696): 666–669. https://doi.org/10.1126/science.1102896.

67 Shmavonyan, G.S., Sevoyan, G., and Aroutiounian, V.M. (2013). Enlarging the surface area of monolayer graphene synthesized by mechanical exfoliation. *Am. J. Phys* 6 (1): 1–6.

68 Abdelkader, A.M. et al. (2015). How to get between the sheets: a review of recent works on the electrochemical exfoliation of graphene materials from bulk graphite. *Nanoscale* 7 (16): 6944–6956. https://doi.org/10.1039/C4NR06942K.

69 Choucair, M., Thordarson, P., and Stride, J.A. (2009). Gram-scale production of graphene based on solvothermal synthesis and sonication. *Nat. Nanotechnol.* 4 (1): 30–33. https://doi.org/10.1038/nnano.2008.365.

70 Subrahmanyam, K.S. et al. (2009). Simple method of preparing graphene flakes by an arc-discharge method. *J. Phys. Chem. C* 113: 4257–4259. https://doi.org/10.1021/jp900791y.

71 Van Bommel, A.J., Crombeen, J.E., and Van Tooren, A. (1975). LEED and auger electron observations of the SiC(0001) surface. *Surf. Sci.* 48 (2): 463–472. https://doi.org/10.1016/0039-6028(75)90419-7.

72 Sengupta, J. et al. (2019). Substrate free synthesis of graphene nanoflakes by atmospheric pressure chemical vapour deposition using Ni powder as a catalyst. *Bull. Mater. Sci.* 42 (4): 136. https://doi.org/10.1007/s12034-019-1818-0.

73 Liu, J. (2014). Charging graphene for energy. *Nat. Nanotechnol.* 9 (10): 739–741. https://doi.org/10.1038/nnano.2014.233.

74 Ahn, J.H. and Hong, B.H. (2014). Graphene for displays that bend. *Nat. Nanotechnol.* 9 (10): 737–738. https://doi.org/10.1038/nnano.2014.226.

75 Siochi, E.J. (2014). Graphene in the sky and beyond. *Nat. Nanotechnol.* 9 (10): 745–747. https://doi.org/10.1038/nnano.2014.231.

76 Kostarelos, K. and Novoselov, K.S. (2014). Graphene devices for life. *Nat. Nanotechnol.* 9 (10): 744–745. https://doi.org/10.1038/nnano.2014.224.

77 Buzea, C., Pacheco, I.I., and Robbie, K. (2007). Nanomaterials and nanoparticles: sources and toxicity. *Biointerphases* 2 (4): MR17–MR71. https://doi.org/10.1116/1.2815690.

78 Hussain, C.M. and Mishra, A.K. (eds.) (2018). *Nanotechnology in Environmental Science*. Wiley. https://www.dovepress.com/medicinal-applications-of-fullerenes-peer-reviewed-article-IJN.

79 Manke, A., Wang, L., and Rojanasakul, Y. (2013). Mechanisms of nanoparticle-induced oxidative stress and toxicity. *BioMed Res. Int.* (Hindawi, 2013). https://doi.org/10.1155/2013/942916.

80 Chen, R.-J. et al. (2020). The current understanding of autophagy in nanomaterial toxicity and its implementation in safety assessment-related alternative testing strategies. *Int. J. Mol. Sci.* 21 (7): 2387. https://doi.org/10.3390/ijms21072387.

81 Reisetter, A.C. et al. (2011). Induction of inflammasome-dependent pyroptosis by carbon black nanoparticles. *J. Biol. Chem.* 286 (24): 21844–21852. https://doi.org/10.1074/jbc.M111.238519.

82 Liang, C., Wu, L.-Y., and Yang, W.-X. (2018). Nanoparticles induce apoptosis via mediating diverse cellular pathways. *Nanomedicine* 13 (22): 2939–2955. https://doi.org/10.2217/nnm-2018-0167.

83 Hurtley, S.M. (2016). Apoptosis, necrosis, and pyroptosis. *Science* 352 (6281): 48–50. https://doi.org/10.1126/science.352.6281.48-j.

84 Patel, R. et al. (2017). Reactive oxygen species: the good and the bad. In: *Reactive Oxygen Species (ROS) in Living Cells*. https://doi.org/10.5772/intechopen.71547.

85 Dayem, A.A. et al. (2017). The role of reactive oxygen species (ROS) in the biological activities of metallic nanoparticles. *Int. J. Mol. Sci.* 18 (1): 120. https://doi.org/10.3390/ijms18010120.

86 Popp, L. and Segatori, L. (2015). Differential autophagic responses to nano-sized materials. *Curr. Opin. Biotechnol.*, Pathway engineering 36: 129–136. https://doi.org/10.1016/j.copbio.2015.08.016.

87 D'Arcy, M.S. (2019). Cell death: a review of the major forms of apoptosis, necrosis and autophagy. *Cell Biol. Int.* 43 (6): 582–592. https://doi.org/10.1002/cbin.11137.

88 Sukhanova, A. et al. (2018). Dependence of nanoparticle toxicity on their physical and chemical properties. *Nanoscale Res. Lett.* 13 (1): 44. https://doi.org/10.1186/s11671-018-2457-x.

89 Marano, F. et al. (2016). Cellular mechanisms of nanoparticle toxicity. In: *Encyclopedia of Nanotechnology* (ed. B. Bhushan), 498–505. Dordrecht: Springer Netherlands. https://doi.org/10.1007/978-94-017-9780-1_175.

90 Yuan, X. et al. (2019). Cellular toxicity and immunological effects of carbon-based nanomaterials. *Part. Fibre Toxicol.* 16 (1): 18. https://doi.org/10.1186/s12989-019-0299-z.

91 Rybak-Smith, M.J. (2012). Effect of surface modification on toxicity of nanoparticles. In: *Encyclopedia of Nanotechnology* (ed. B. Bhushan), 645–652. Dordrecht: Springer Netherlands. https://doi.org/10.1007/978-90-481-9751-4_174.

92 Ge, C. et al. (2012). The contributions of metal impurities and tube structure to the toxicity of carbon nanotube materials. *NPG Asia Mater.* 4 (12): e32. https://doi.org/10.1038/am.2012.60.

93 Sayes, C.M. et al. (2007). Comparative pulmonary toxicity assessments of C_{60} water suspensions in rats: few differences in fullerene toxicity in vivo in contrast to in vitro profiles. *Nano Lett.* 7 (8): 2399–2406. https://doi.org/10.1021/nl0710710.

94 Sayers, B.C. et al. (2016). Respiratory toxicity and immunotoxicity evaluations of microparticle and nanoparticle C_{60} fullerene aggregates in mice and rats following nose-only inhalation for 13 weeks. *Nanotoxicology* 10 (10): 1458–1468. https://doi.org/10.1080/17435390.2016.1235737.

95 Morimoto, Y. et al. (2010). Inflammogenic effect of well-characterized fullerenes in inhalation and intratracheal instillation studies. *Part. Fibre Toxicol.* 7 (1): 4. https://doi.org/10.1186/1743-8977-7-4.

96 Park, E.-J. et al. (2009). Pro-inflammatory and potential allergic responses resulting from B cell activation in mice treated with multi-walled carbon nanotubes by intratracheal instillation. *Toxicology* 259 (3): 113–121. https://doi.org/10.1016/j.tox.2009.02.009.

97 Inoue, K.-i. et al. (2009). Effects of multi-walled carbon nanotubes on a murine allergic airway inflammation model. *Toxicol. Appl. Pharmacol.* 237 (3): 306–316. https://doi.org/10.1016/j.taap.2009.04.003.

98 Rydman, E.M. et al. (2014). Inhalation of rod-like carbon nanotubes causes unconventional allergic airway inflammation. *Part. Fibre Toxicol.* 11 (1): 48. https://doi.org/10.1186/s12989-014-0048-2.

99 Hsieh, W.-Y. et al. (2012). Single-walled carbon nanotubes induce airway hyperreactivity and parenchymal injury in mice. *Am. J. Respir. Cell Mol. Biol.* 46 (2): 257–267. https://doi.org/10.1165/rcmb.2011-0010OC.

100 Xu, J. et al. (2012). Multi-walled carbon nanotubes translocate into the pleural cavity and induce visceral mesothelial proliferation in rats. *Cancer Sci.* 103 (12): 2045–2050. https://doi.org/10.1111/cas.12005.

101 Dong, J. and Ma, Q. (2016). Myofibroblasts and lung fibrosis induced by carbon nanotube exposure. *Part. Fibre Toxicol.* 13 (1): 60. https://doi.org/10.1186/s12989-016-0172-2.

102 Chang, C.-C. et al. (2012). Epithelial-mesenchymal transition contributes to SWCNT-induced pulmonary fibrosis. *Nanotoxicology* 6 (6): 600–610. https://doi.org/10.3109/17435390.2011.594913.

103 Luanpitpong, S. et al. (2014). Caveolin-1 regulates lung cancer stem-like cell induction and p53 inactivation in carbon nanotube-driven tumorigenesis. *Oncotarget* 5 (11): 1. https://doi.org/10.18632/oncotarget.1956.

104 Sargent, L.M. et al. (2014). Promotion of lung adenocarcinoma following inhalation exposure to multi-walled carbon nanotubes. *Part. Fibre Toxicol.* 11 (1): 3. https://doi.org/10.1186/1743-8977-11-3.

105 Kasai, T. et al. (2016). Lung carcinogenicity of inhaled multi-walled carbon nanotube in rats. *Part. Fibre Toxicol.* 13 (1): 53. https://doi.org/10.1186/s12989-016-0164-2.

106 Roberts, J.R. et al. (2016). Evaluation of pulmonary and systemic toxicity following lung exposure to graphite nanoplates: a member of the graphene-based nanomaterial family. *Part. Fibre Toxicol.* 13 (1): 34. https://doi.org/10.1186/s12989-016-0145-5.

107 Mao, L. et al. (2016). Biodistribution and toxicity of radio-labeled few layer graphene in mice after intratracheal instillation. *Part. Fibre Toxicol.* 13 (1): 7. https://doi.org/10.1186/s12989-016-0120-1.

108 Ma, J. et al. (2015). Crucial role of lateral size for graphene oxide in activating macrophages and stimulating pro-inflammatory responses in cells and animals. *ACS Nano* 9 (10): 10498–10515. https://doi.org/10.1021/acsnano.5b04751.

109 Hu, Q. et al. (2015). Effects of graphene oxide nanosheets on the ultrastructure and biophysical properties of the pulmonary surfactant film. *Nanoscale* 7 (43): 18025–18029. https://doi.org/10.1039/C5NR05401J.

110 Zhang, X. et al. (2011). Distribution and biocompatibility studies of graphene oxide in mice after intravenous administration. *Carbon* 49 (3): 986–995. https://doi.org/10.1016/j.carbon.2010.11.005.

111 Park, E.-J. et al. (2017). Pulmonary persistence of graphene nanoplatelets may disturb physiological and immunological homeostasis. *J. Appl. Toxicol.* 37 (3): 296–309. https://doi.org/10.1002/jat.3361.

112 Lee, J.K. et al. (2017). The role of surface functionalization on the pulmonary inflammogenicity and translocation into mediastinal lymph nodes of graphene nanoplatelets in rats. *Arch. Toxicol.* 91 (2): 667–676. https://doi.org/10.1007/s00204-016-1706-y.

113 Bo, L. et al. (2013). Biodistribution and pulmonary toxicity of intratracheally instilled graphene oxide in mice. *NPG Asia Mater.* 5 (4): e44. https://doi.org/10.1038/am.2013.7.

114 Rouse, J.G. et al. (2006). Fullerene-based amino acid nanoparticle interactions with human epidermal keratinocytes. *Toxicol. in Vitro* 20 (8): 1313–1320. https://doi.org/10.1016/j.tiv.2006.04.004.

115 Saathoff, J.G. et al. (2011). In vitro toxicity assessment of three hydroxylated fullerenes in human skin cells. *Toxicol. in Vitro* 25 (8): 2105–2112. https://doi.org/10.1016/j.tiv.2011.09.013.

116 Patlolla, A., Knighten, B., and Tchounwou, P. (2010). Multi-walled carbon nanotubes induce cytotoxicity, genotoxicity and apoptosis in normal human dermal fibroblast cells. *Ethn. Dis.* 20 (1 Suppl 1): S1-65–72.

117 Murray, A.R. et al. (2009). Oxidative stress and inflammatory response in dermal toxicity of single-walled carbon nanotubes. *Toxicology* 257 (3): 161–171. https://doi.org/10.1016/j.tox.2008.12.023.

118 Zhang, L.W. et al. (2016). Biological interactions of functionalized single-wall carbon nanotubes in human epidermal keratinocytes. *Int. J. Toxicol.* https://doi.org/10.1080/10915810701225133.

119 Shvedova, A. et al. (2003). Exposure to carbon nanotube material: assessment of nanotube cytotoxicity using human keratinocyte cells. *J. Toxicol. Environ. Health Part A* 66 (20): 1909–1926. https://doi.org/10.1080/713853956.

120 Palmer, B.C., Phelan-Dickenson, S.J., and DeLouise, L.A. (2019). Multi-walled carbon nanotube oxidation dependent keratinocyte cytotoxicity and skin inflammation. *Part. Fibre Toxicol.* 16 (1): 3. https://doi.org/10.1186/s12989-018-0285-x.

121 Liao, K.-H. et al. (2011). Cytotoxicity of graphene oxide and graphene in human erythrocytes and skin fibroblasts. *ACS Appl. Mater. Interfaces* 3 (7): 2607–2615. https://doi.org/10.1021/am200428v.

122 Pelin, M. et al. (2017). Differential cytotoxic effects of graphene and graphene oxide on skin keratinocytes. *Sci. Rep.* 7 (1): 1–12. https://doi.org/10.1038/srep40572.

123 Yamawaki, H. and Iwai, N. (2006). Cytotoxicity of water-soluble fullerene in vascular endothelial cells. *Am. J. Physiol. Cell Physiol.* 290 (6): C1495–C1502. https://doi.org/10.1152/ajpcell.00481.2005.

124 Thompson, L.C. et al. (2014). C_{60} exposure augments cardiac ischemia/reperfusion injury and coronary artery contraction in Sprague Dawley rats. *Toxicol. Sci.* 138 (2): 365–378. https://doi.org/10.1093/toxsci/kfu008.

125 Gelderman, M.P. et al. (2008). Adverse effects of fullerenes on endothelial cells: fullerenol $C_{60}(OH)_{24}$ induced tissue factor and ICAM-1 membrane expression and apoptosis in vitro. *Int. J. Nanomed.* 3 (1): 59–68. https://doi.org/10.2147/IJN.S1680.

126 Vesterdal, L.K. et al. (2009). Modest vasomotor dysfunction induced by low doses of C_{60} fullerenes in apolipoprotein E knockout mice with different degree of atherosclerosis. *Part. Fibre Toxicol.* 6 (1): 5. https://doi.org/10.1186/1743-8977-6-5.

127 Radomski, A. et al. (2005). Nanoparticle-induced platelet aggregation and vascular thrombosis. *Br. J. Pharmacol.* 146 (6): 882–893. https://doi.org/10.1038/sj.bjp.0706386.

128 Hosseinpour, M. et al. (2016). The cardiac effects of carbon nanotubes in rat. *BioImpacts* 6 (2): 79–84. https://doi.org/10.15171/bi.2016.11.

129 Li, Z. et al. (2007). Cardiovascular effects of pulmonary exposure to single-wall carbon nanotubes. *Environ. Health Perspect.* 115 (3): 377–382. https://doi.org/10.1289/ehp.9688.

130 Zheng, W. et al. (2018). The effects of inhaled multi-walled carbon nanotubes on blood pressure and cardiac function. *Nanoscale Res. Lett.* 13 (1): 189. https://doi.org/10.1186/s11671-018-2603-5.

131 Singh, S.K. et al. (2011). Thrombus inducing property of atomically thin graphene oxide sheets. *ACS Nano* 5 (6): 4987–4996. https://doi.org/10.1021/nn201092p.

132 Contreras-Torres, F.F. et al. (2017). Differential cytotoxicity and internalization of graphene family nanomaterials in myocardial cells. *Mater. Sci. Eng., C* 73: 633–642. https://doi.org/10.1016/j.msec.2016.12.080.

133 Bangeppagari, M. et al. (2019). Graphene oxide induces cardiovascular defects in developing zebrafish (*Danio rerio*) embryo model: in-vivo toxicity assessment. *Sci. Total Environ.* 673: 810–820. https://doi.org/10.1016/j.scitotenv.2019.04.082.

134 Arbo, M.D. et al. (2019). *In vitro* cardiotoxicity evaluation of graphene oxide. *Mutat. Res. Genet. Toxicol. Environ. Mutagen.* 841: 8–13. https://doi.org/10.1016/j.mrgentox.2019.03.004.

135 Tsuchiya, T. et al. (1996). Novel harmful effects of [60]fullerene on mouse embryos in vitro and in vivo. *FEBS Lett.* 393 (1): 139–145. https://doi.org/10.1016/0014-5793(96)00812-5.

136 Usenko, C.Y., Harper, S.L., and Tanguay, R.L. (2007). In vivo evaluation of carbon fullerene toxicity using embryonic zebrafish. *Carbon* 45 (9): 1891–1898. https://doi.org/10.1016/j.carbon.2007.04.021.

137 Usenko, C.Y., Harper, S.L., and Tanguay, R.L. (2008). Fullerene C_{60} exposure elicits an oxidative stress response in embryonic zebrafish. *Toxicol. Appl. Pharmacol.* 229 (1): 44–55. https://doi.org/10.1016/j.taap.2007.12.030.

138 Zhu, X. et al. (2007). Developmental toxicity in zebrafish (*Danio rerio*) embryos after exposure to manufactured nanomaterials: buckminsterfullerene aggregates (nC_{60}) and fullerol. *Environ. Toxicol. Chem.* 26 (5): 976–979. https://doi.org/10.1897/06-583.1.

139 Han, B. and Nazmul Karim, M. (2008). Cytotoxicity of aggregated fullerene C_{60} particles on CHO and MDCK cells. *Scanning* 30 (2): 213–220. https://doi.org/10.1002/sca.20081.

140 Pietroiusti, A. et al. (2011). Low doses of pristine and oxidized single-wall carbon nanotubes affect mammalian embryonic development. *ACS Nano* 5 (6): 4624–4633. https://doi.org/10.1021/nn200372g.

141 Roman, D. et al. (2013). Significant toxic role for single-walled carbon nanotubes during normal embryogenesis. *Nanomed. Nanotechnol. Biol. Med.* 9 (7): 945–950. https://doi.org/10.1016/j.nano.2013.03.010.

142 Fujitani, T. et al. (2012). Teratogenicity of multi-wall carbon nanotube (MWCNT) in ICR mice. *J. Toxicol. Sci.* 37 (1): 81–89. https://doi.org/10.2131/jts.37.81.

143 Cheng, J. and Cheng, S.H. (2012). Influence of carbon nanotube length on toxicity to zebrafish embryos. *Int. J. Nanomed.* 7: 3731–3739. https://doi.org/10.2147/IJN.S30459.

144 Huang, X. et al. (2014). The genotype-dependent influence of functionalized multiwalled carbon nanotubes on fetal development. *Biomaterials* 35 (2): 856–865. https://doi.org/10.1016/j.biomaterials.2013.10.027.

145 Xu, S., Zhang, Z., and Chu, M. (2015). Long-term toxicity of reduced graphene oxide nanosheets: effects on female mouse reproductive ability and offspring development. *Biomaterials* 54: 188–200. https://doi.org/10.1016/j.biomaterials.2015.03.015.

146 Chen, Y. et al. (2016). Specific nanotoxicity of graphene oxide during zebrafish embryogenesis. *Nanotoxicology* 10 (1): 42–52. https://doi.org/10.3109/17435390.2015.1005032.

147 Fu, C. et al. (2015). Effects of graphene oxide on the development of offspring mice in lactation period. *Biomaterials* 40: 23–31. https://doi.org/10.1016/j.biomaterials.2014.11.014.

148 Akhavan, O. et al. (2015). Dose-dependent effects of nanoscale graphene oxide on reproduction capability of mammals. *Carbon* 95: 309–317. https://doi.org/10.1016/j.carbon.2015.08.017.

149 Kamat, J.P. et al. (1998). Oxidative damage induced by the fullerene C_{60} on photosensitization in rat liver microsomes. *Chem. Biol. Interact.* 114 (3): 145–159. https://doi.org/10.1016/S0009-2797(98)00047-7.

150 Nakagawa, Y. et al. (2011). Cytotoxic effects of hydroxylated fullerenes on isolated rat hepatocytes via mitochondrial dysfunction. *Arch. Toxicol.* 85 (11): 1429–1440. https://doi.org/10.1007/s00204-011-0688-z.

151 Shipkowski, K.A. et al. (2019). Disposition of fullerene C_{60} in rats following intratracheal or intravenous administration. *Xenobiotica* 49 (9): 1078–1085. https://doi.org/10.1080/00498254.2018.1528646.

152 Sayes, C.M. et al. (2004). The differential cytotoxicity of water-soluble fullerenes. *Nano Lett.* 4 (10): 1881–1887. https://doi.org/10.1021/nl0489586.

153 Patlolla, A., McGinnis, B., and Tchounwou, P. (2011). Biochemical and histopathological evaluation of functionalized single-walled carbon nanotubes in Swiss–Webster mice. *J. Appl. Toxicol.* 31 (1): 75–83. https://doi.org/10.1002/jat.1579.

154 Awasthi, K.K. et al. (2013). Multi walled carbon nano tubes induced hepatotoxicity in Swiss albino mice. *Micron* 44: 359–364. https://doi.org/10.1016/j.micron.2012.08.008.

155 Principi, E. et al. (2016). Systemic distribution of single-walled carbon nanotubes in a novel model: alteration of biochemical parameters, metabolic functions, liver accumulation, and inflammation in vivo. *Int. J. Nanomed.* 11: 4299–4316. https://doi.org/10.2147/IJN.S109950.

156 Patlolla, A.K., Rondalph, J., and Tchounwou, P.B. (2017). Biochemical and histopathological evaluation of graphene oxide in Sprague-Dawley rats. *Austin J. Environ. Toxicol.* 3 (1) https://www.ncbi.nlm.nih.gov/pmc/articles/PMC5831506.

157 Zhang, Y. et al. (2020). Large-sized graphene oxide synergistically enhances parenchymal hepatocyte IL-6 expression monitored by dynamic imaging. *Nanoscale* 12 (15): 8147–8158. https://doi.org/10.1039/C9NR10713D.

158 Mohamed, W.N.A.E.M. et al. (2017). Evaluation of genotoxic and hepatotoxic effects of graphene oxide nanosheets in male albino mice. *Egypt. J. Exp. Biol. (Zoo.)* 13 (1): 43–53.

159 Xiong, G. et al. (2020). Graphene oxide nanoparticles induce hepatic dysfunction through the regulation of innate immune signaling in zebrafish (*Danio rerio*). *Nanotoxicology* 14 (5): 1–16. https://doi.org/10.1080/17435390.2020.1735552.

160 Aoshima, H. et al. (2009). Safety evaluation of highly purified fullerenes (HPFs): based on screening of eye and skin damage. *J. Toxicol. Sci.* 34 (5): 555–562. https://doi.org/10.2131/jts.34.555.

161 Roberts, J.E. et al. (2008). Phototoxicity and cytotoxicity of fullerol in human lens epithelial cells. *Toxicol. Appl. Pharmacol.* 228 (1): 49–58. https://doi.org/10.1016/j.taap.2007.12.010.

162 Wielgus, A.R. et al. (2010). Phototoxicity and cytotoxicity of fullerol in human retinal pigment epithelial cells. *Toxicol. Appl. Pharmacol.* 242 (1): 79–90. https://doi.org/10.1016/j.taap.2009.09.021.

163 Ema, M. et al. (2013). Dermal and ocular irritation and skin sensitization studies of fullerene C_{60} nanoparticles. *Cutan. Ocul. Toxicol.* 32 (2): 128–134. https://doi.org/10.3109/15569527.2012.727937.

164 Ema, M. et al. (2011). Evaluation of dermal and eye irritation and skin sensitization due to carbon nanotubes. *Regul. Toxicol. Pharm.* 61 (3): 276–281. https://doi.org/10.1016/j.yrtph.2011.08.007.

165 Sairam Kishore, A., Surekha, P., and Balakrishna Murthy, P. (2009). Assessment of the dermal and ocular irritation potential of multi-walled carbon nanotubes by using in vitro and in vivo methods. *Toxicol. Lett.* 191 (2): 268–274. https://doi.org/10.1016/j.toxlet.2009.09.007.

166 Yan, L. et al. (2011). Cytotoxicity of single-walled carbon nanotubes with human ocular cells. In: *Advanced Materials Research* (ed. J. Bu, P. Wang, L. Ai, X. Sang, Y. Li), Applications of Engineering Materials. Trans Tech Publications Ltd. http://www.scientific.net/AMR.287-290.32.

167 Yan, L. et al. (2013). Cytotoxicity and genotoxicity of multi-walled carbon nanotubes with human ocular cells. *Chin. Sci. Bull.* 58 (19): 2347–2352. https://doi.org/10.1007/s11434-013-5800-8.

168 Wu, W. et al. (2016). Evaluation of the toxicity of graphene oxide exposure to the eye. *Nanotoxicology* 10 (9): 1329–1340. https://doi.org/10.1080/17435390.2016.1210692.

169 Wu, W. et al. (2018). Investigating oxidation state-induced toxicity of PEGylated graphene oxide in ocular tissue using gene expression profiles. *Nanotoxicology* 12 (8): 819–835. https://doi.org/10.1080/17435390.2018.1480813.

170 An, W. et al. (2018). Ocular toxicity of reduced graphene oxide or graphene oxide exposure in mouse eyes. *Exp. Eye Res.* 174: 59–69. https://doi.org/10.1016/j.exer.2018.05.024.

7

Fundamentals of Functionalized Carbon Nanomaterials (CNMs) for Environmental Devices and Techniques

Kiran Soni[1] *and Rekha Yadav*[2]

[1] *University of Delhi, Maitreyi College, Department of Chemistry, Bapudham, Chanakyapuri, New Delhi, 110021, India*
[2] *University of Delhi, Sri Venkateswara College, Department of Chemistry, Benito Juarez Road, DhaulaKuan, New Delhi, 110021, India*

7.1 Introduction

To ensure sustainable development, there has been growing concern toward environmental changes over the past decades. Major focus has been on reducing greenhouse gases (CO_2, CH_4, etc.) and exploring other methods for energy production [1]. To monitor the different environmental changes, the production and applications of nanoscale material are increasing remarkably. Among nanomaterials, carbon nanomaterials (CNMs) possess unique and tunable properties to develop new technologies for identifying and addressing environmental challenges [2].

Carbon is one of the most interesting nonmetals, having capability to form wide range of structures and different properties. This property of making different structures with same atom is known as allotropy. The different allotropes of carbon are feasible due to its thermodynamically favorable configurations from sp^3 to sp^2, depending on the heat of formation and pressure conditions. The unique geometry and type of chemical bonds in the structure decide its properties. On the other hand, nanomaterials possess one or more external dimensions in range from 1 to 100 nm. Hence CNMs are substances made from carbon with one or more dimension confined to nanoscale [3].

After the successful synthesis of carbon-based nanomaterials, phenomenal research has been carried out for its applications in environmental devices [4]. CNMs are important as they possess rich pore structure, high surface area, high flexibility, mechanical and electrical strength, and thermal conductivity that make it a potential candidate for various applications [5]. The CNMs possessing porous structure with large surface area includes fullerenes, expanded graphite (EG), carbon nanotubes (CNTs), graphene, graphene oxide (GO), etc. [6]. Despite CNMs exceptional properties, they have limitations in binding with different organic solvents or polymers and hence lack direct practical applications [7]. The introduction of organic functionalities in these materials has opened up new branch materials whose properties can be modified according to the need of required applications [8].

Environmental Applications of Carbon Nanomaterials-Based Devices, First Edition.
Edited by Shadpour Mallakpour and Chaudhery Mustansar Hussain.

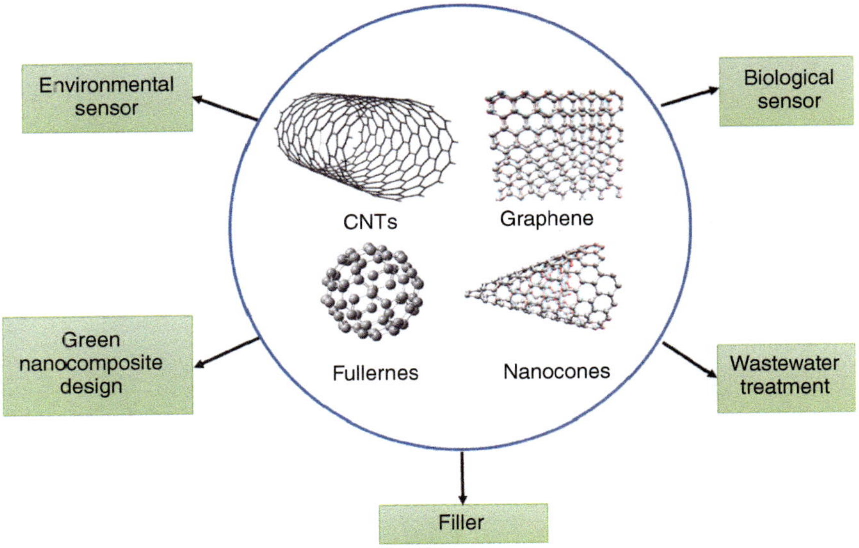

Figure 7.1 Some of carbon nanomaterials and their applications.

Hence the prepared materials possess the properties of both materials. The modification of CNMs with other materials including functionalized hydrocarbons, ionic liquids, etc. makes it more attractive and definitive candidate in applications that include nanodevices, nanosensors, biosensors, specific absorbents, etc. [9]. This chapter briefly discusses the different functionalized CNMs including CNTs, graphene, fullerenes, carbon nanocones (CNCs), etc. and their uses in environmental devices, biological sensors, absorbents for heavy metals, and detection of organic pollutants (see Figure 7.1).

7.2 Synthesis

Different methods are used in the synthesis of CNMs. This section briefly discusses the methods in synthesis of CNTs, graphene, fullerenes, and CNCs.

7.2.1 Carbon Nanotubes

CNTs can be prepared with different methods that includes arc discharge, chemical vapor deposition (CVD), laser ablation, electrolysis, flame synthesis, etc. [10]. The carbon source and parameters like temperature and pressure strongly affect the morphology of the CNTs produced. For example, the use of linear hydrocarbons (like methane, ethylene, acetylene, etc.) as carbon source produces straight hollow CNTs, whereas cyclic hydrocarbons (like benzene, xylene, fullerene, cyclohexene, etc.) produce relatively curved CNTs, having bridged tube wall soften inside. Low-temperature CVD method yields multiwalled carbon nanotubes (MWCNTs); on the other hand high-temperature produces single-walled carbon nanotubes (SWCNTs). CVD is among one of the methods to make high-yield CNTs with

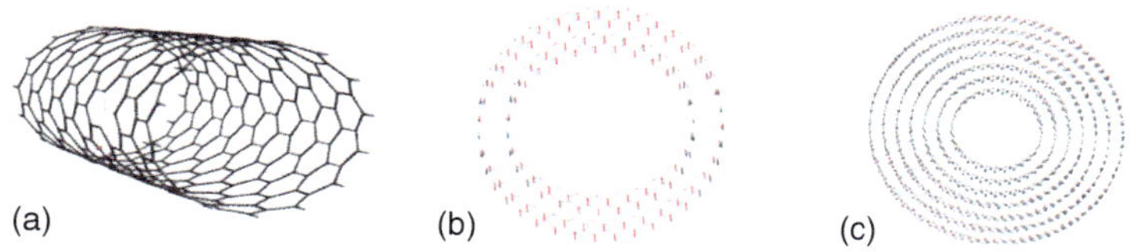

Figure 7.2 Carbon nanotubes: (a) single-walled carbon nanotubes (SWCNTs) (can be zigzag, armchair, and chiral), (b) double-walled carbon nanotubes (made up of two concentric shells), and (c) multiwalled carbon nanotubes (MWCNTs) (made up of several concentric shells).

medium to high purity at low cost. The operating temperature used in CVD is between 500 and 1200 °C. Both the CNTs (single-walled [SWCNT] or double-walled or multi-walled [MWCNT]) can be produced depending upon the precursors used (see Figure 7.2). The diameters of SWCNT and MWCNT range from 0.6 to 4 nm and 10 to 240 nm, respectively [11]. However, in arc discharge method, the temperature used is very high (4000 °C) with low yield of CNTs [10]. Laser ablation method is very expensive that gives low-quality CNTs with low yield [11b].

7.2.2 Graphene

Graphene can be synthesized by a number of methods like liquid-phase and electrochemical exfoliation method, thermal and plasma-enriched CVD techniques, thermal decomposition of SiC, unzipping CNTs, etc. as shown in Figure 7.3 [12]. The CVD methods are costly but produces high-quality graphene that can be used in solar cells, touchscreens, etc. The liquid-phase and electrochemical exfoliation method produces moderate-quality graphene at lower cost that can be used in sensors, field-effect transistors (FETs), transport electrode, etc. The unzipping CNTs produce high-quality graphene at low price that can be used in FETs [13]. One of the methods that is usually used in synthesis of graphene is Hummer's method. It is cost-effective, however the yield of graphene is low. The improved Hummer's method yields high quantity of GO at low cost [14]. Figure 7.4 shows the different steps involved in conversion of graphite to reduced graphene oxide.

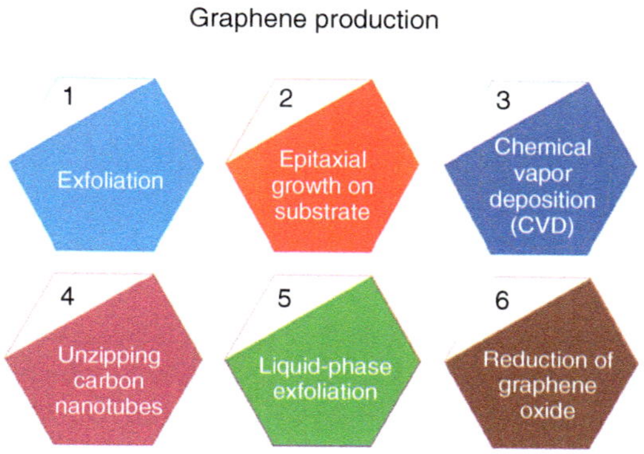

Figure 7.3 Different methods for production of graphene. Source: Based on Choi et al. [12].

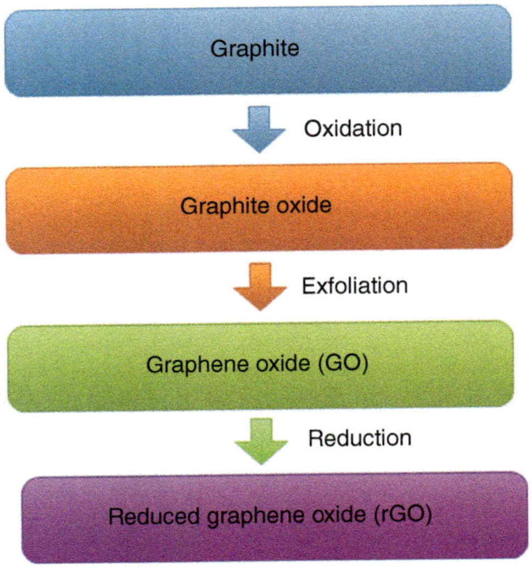

Figure 7.4 Pathways from graphite to reduced graphene oxide. Source: Based on Muzyka et al. [14].

7.2.3 Fullerenes

Fullerene can be synthesized with different methods like bending flat molecules with heat using flash vacuum pyrolysis (FVP), intramolecular C(aryl)–C(aryl) coupling reactions, direct cyclodehydrogenations, etc. [15]. The unfunctionalized hydrocarbons will not work well for the preparative scale synthesis of C_{60} by FVP as it lacks controlled structure growth. The incorporation of halogen atoms or other functionality in precursors could generate radical intermediates to get curvature at high temperatures in the gas phase [16].

7.2.4 Carbon Nanocones

CNCs were first synthesized in 1994 on a graphite substrate using vapor condensation of carbon atoms. Different methods like pyrolysis of hydrocarbons [17], plasma synthesis [18], graphene to nanocones [19], etc. can be used for its synthesis.

7.2.5 Functionalization of Nanomaterials

The CNMs can be functionalized by covalent and non-covalent functionalization [20]. In covalent functionalization, the desired functional group is covalently bonded to the CNMs. The covalent functionalization in nanomaterials is usually achieved by free radical functionalization, fluorination, amidation, carboxylation, etc. [21]. In non-covalent functionalization, adsorption or intercalation of functional ligand molecule on CNMs takes place. The non-covalent functionalization is achieved by ligand non-covalent functionalization, π–π interaction, π–anion or cation interaction, polymer composites, etc. [22].

7.3 Applications

7.3.1 Nanowires

Energy production from the fossil fuels has been known as the main cause of the degradation of environmental quality and the environmental pollution. To address environmental issues, nanotechnology acts as an alternative source for energy conversion and storage, sustainable energy, green technology perspective, and environmental monitoring. Iijima has discovered the CNT, and it resulted in the revolution in the field of nanotechnology. CNTs have unique geometry and fascinating properties [23]. It exhibits large current-carrying capability [24], high thermal conductivity [25], high surface-to-volume ratio, hollow structure, and mechanical strength [26]. CNTs show various applications in the field of material sciences [27], electronics devices [28], optics [29], and biological and environmental fields [30]. Despite the unique properties of the CNTs, some of reports propose CNT hazardous effects on ecosystem and also on organism's health [31]. The reports suggests that inhalation of 0.0003–0.005 μg/ml of MWCNTs could affect immune function [32] and SWCNTs could damage pulmonary system on exposure to 25 μg/ml [33]. Therefore, quantitative information of the CNTs for the environmental use is of great challenge, and there is a need for its assessment. The applications of CNTs in different fields are listed below.

7.3.1.1 Carbon Nanotube as Environmental Sensor

Nowadays, air pollution is a global challenge for scientists [34] because of the air quality degradation that leads to negative effects on the surrounding environment and human health [35]. Before the discovery of CNT, lots of conventional gas sensors are being used for the quantification and protection from air pollution. However, they suffer from many disadvantages, for example, elevated working temperatures and reproducibility problems [36]. CNTs have solved this problem due to their fabulous properties and large surface area. They have been used in adsorption, membrane filtration, disinfection, and photocatalysis for pollution treatment and environmental remediation [37]. Pullithadathil's group [38] have reported excellent gas sensor from polyaniline-supported CNTs (PANI/MWCNTs) for the detection of ammonia (NH_3) gas at trace level of 2–10 ppm concentration. PANI/MWCNTs were synthesized with carboxylated MWCNTs by *in situ* chemical oxidative polymerization with aniline monomer. Transmission electron microscopic analysis confirms that the thickness of PANI was around 7 nm on the outer walls of the CNT. The response and recovery time of the PANI/MWCNT nanocomposite sensor was reported at six seconds for NH_3 gas.

Valentini et al. [39] designed an electrochemical model for the detection of ammonia using CNT/copper material. These CNT or Cu paste electrodes detect the ammonia in the concentration (3–100 μM) in eight seconds [39]. Chen et al. [40] have used pristine CNTs with ultraviolet light during nitric oxide gas sensing for ultralow-level detections. Mishra's [41] group have established MWCNT-based carbon dioxide gas sensor. They have synthesized by varying concentrations of MWCNT with various (0.6–3) wt% in alumina Al_2O_3 sol using low-cost gel-cast method. Uniform self-standing MWCNT film was used for sensing different CO_2 concentrations (from 50 to 450 ppm). The sensor characteristics of MWCNT were examined by the change in resistance on exposure to various concentrations of CO_2 gas. The sensor response (S) is calculated as

$$S\% = \frac{R_{\text{gas}} - R_O}{R_O} * 100$$

where R_{gas} and R_O represent final and initial values of resistances in one cycle of gas exposure.

The charge transfer between carbon material and CO_2 gas leads to change in resistance. It was assumed by the researchers that on exposing to the heat and UV light, CNT requires less recovery time and favors stability. Further, the sensing ability of CNTs was improved by decoration with gold nanoparticles (NPs). This gold-decorated CNT have shown better sensing ability for the detection of NO_2 and CO gases and for benzene [42].

7.3.1.2 Carbon Nanotubes in Wastewater Treatment

CNTs offer a widespread attention for the water treatment and may be the outstanding materials for the elimination of various inorganic and organic pollutants from the water [43] because of their unique and well-defined geometry. In 2008, report suggests a wide use of CNTs in different applications [44], hence causing contamination in the environment at a toxic level. CNTs are hollow and porous nanomaterial that consist of graphene layers and are better than activated charcoal. Functionalization of CNTs improves their properties to interact with the pollutants and impurities. There are lots of reports suggesting strong interactions of the CNTs with different compounds (inorganic or organic), due to the presence of functional group [45]. Therefore, CNTs could be an efficient adsorbent to manage the contamination of environment by various pollutants. A number of mechanisms are proposed to understand the interaction of pollutant with CNTs by variation in different parameters such as pH, concentration, different ions, etc. CNTs can be modified to improve its selectivity and removal of metals or ions from aqueous streams [46]. For wider applications, the growth of CNTs should be economic, facile, greener, and recyclable that can leads to real/dynamic application in polluted water treatment.

Recently, a scientist group designed modified CNTs using ionic liquid for the removal of dyes (sunset yellow FCF dye, orange II, and amaranth dyes) from the water [47] in acidic pH range. The use of acidic pH value favors the adsorption of these dyes. This modified magnetic polymer CNTs reported fast adsorption rate, high adsorption capacity, and easy separation due to magnetism. Peng et al. have reported the adsorption of ofloxacin (OFL) on CNTs by investigating the pH, solubility, and cosolvent effects. They have performed sorption activities of OFL on MWCNTs at various pH ranges. They found that hydrophobic interaction is not prominent factor for sorption mechanism [48]. Xue et al. [49] have reported the different nanomaterials for water treatment such as photocatalytic nanomaterials, nanofiltration membranes, reducing nanomaterials, and adsorption nanomaterial. Many of the reports are for the removal of organics, pathogenic bacteria and virus, and heavy metals using CNTs (single wall or multiwall) [48]. Su and Lu reported that the CNT shows greater activity and less average weight loss than granular activated carbon. CNTs interact with organic compounds and show better adsorption in aqueous environment. These interactions can be due to π–π interactions, hydrophobic effect, electrostatic interactions, hydrogen bonding, van der Waals forces, etc. as shown in Figure 7.5 [50].

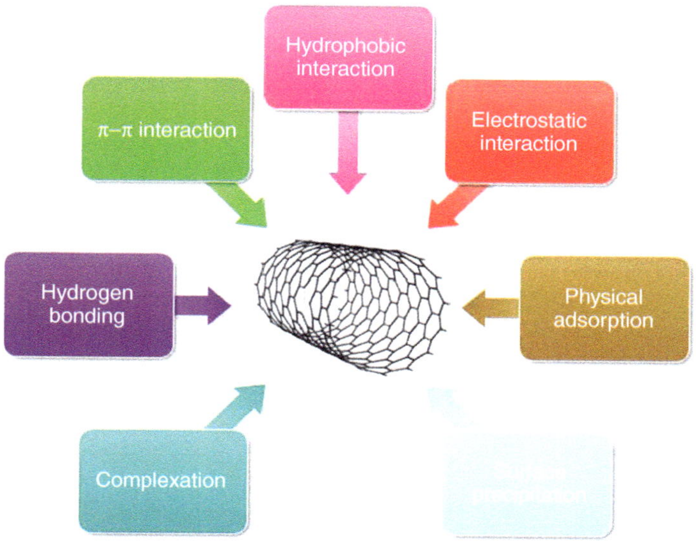

Figure 7.5 Mechanism of interaction of organic pollutants and heavy metal over CNTs. Source: Based on Ahmad et al. [50].

Ahmad et al. [50] reported the use of CNTs for the elimination of organic micropollutants (OMPs) containing the organic moieties, pesticides, antibodies, etc. that on release can cause serious health risks. He et al. [51] developed calibration approaches for predicting quantities of SWCNTs, MWCNTs, and carboxylate MWCNTs in environmental matrices (agricultural soil and anaerobic sludge).

7.3.1.3 Carbon Nanotubes in Green Nanocomposite Design

Soil microbes have direct relation with the soil quality and the plant growth [52]. Therefore, there is a big task to protect soil microbes. Bucheli group [53] have reported that 40% of carbon-based materials have been used in agriculture as additive, as growth stimulator, or as other active components. In this regard, lots of the reports have focused on the use of CNTs in agriculture field. Detailed information about CNTs and soil microbe interaction is not available, so direct exposure of CNTs may be toxic to the soil microbes and plant [54]. Kasel's group have investigated the transport and retention of functionalized CNT using column and lysimeter experiment [55]. They found CNTs retain in soil and did not enter in plant. CNTs had been used in different plants for plant rooting and growth and as a carrier [53–55]. Details are given in Table 7.1.

7.3.1.4 CNT as Biological Sensor

CNT functionalizations by biologically recognized molecules constitute a material that could be used for sensor devices [59]. CNTs can be used with immobilizing redox enzyme and enhance the electron transfer reactions [60]. This property makes them ideal biosensors with enhanced analytical performances. CNT-based biosensors show various application in different fields including biochemistry, gas sensors for pollutant monitoring, clinical analytical chemistry, and agrifood quality [61]. A brief timeline for the development of

Table 7.1 CNTs material used in plant growth.

Carbon-based material	Function	Plant name	Year	References
CNT	Plant rooting, growth, and histology	Blackberry (*Rubusa adenotrichos*)	2014	[56]
CNT	Good delivery carrier for biomolecules	Plant cells (canola and carrot cells)	2012	[57]
CNT	To penetrate plant seed coat and dramatically affect seed germination and plant growth	Tomato plant	2009	[58]

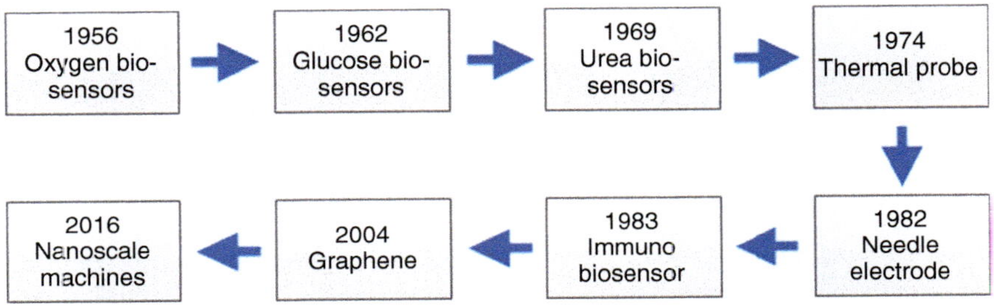

Figure 7.6 Development of sensors timeline. Source: Based on Salajegheh et al. [62].

biosensors is shown in Figure 7.6. Salajegheh et al. [62] had studied the molecularly imprinted composite film for the determination of theophylline by theoretical model. The sodium arginate–MWCNTs have been demonstrated by cyclic voltammetry (CV) and found high conductivity and electrocatalytic activity. Kim's group have synthesized bioactivated MWCNTs deposited on micropore filter paper for the detection of prostate cancer [63]. Comba et al. reported a biosensor for the detection of glucose quantitatively using enzymatic matrix and CNT–mucin composite [64]. Sobhan et al. have designed a reusable biosensor from SWCNT for electrochemical detection of peanut allergen [65].

7.3.1.5 CNT as Filler

CNTs can act as nanofillers. They have been broadly used as functional filler for enhancing the mechanical properties of epoxy composites [66], polymer composites [67], etc. Gong et al. have reported the hybrid material with the assembly of graphene and CNTs [68]. However, there is a weak interaction between the CNT and the polymer due to the less dispersion of carbon material because CNT is unstable in organic solvents due to its inertness [69]. A lot of other carbon-based materials also has been used as filler due to the former drawback of the CNTs. However, all these materials cause pollution due to the toxic chemicals. So now scientists are using the biopolymers from plants, for example, soy protein isolate (SPI). These are alternative source of fossil fuel too. Jin et al. [70] have improved the water resistance and mechanical properties of SPI for designing its nanocomposite with on carbon material by using casting method.

Yu et al. designed a nanoscale hybrid filler composed of SWCNT and graphite nanoplatelets. This hybrid filler enhances the thermal conductivity of epoxy composites due to synergistic effect and also reduces the thermal interface resistance due to the significant enhancement in percolating network [71]. Hu et al. has studied the CNT nanocomposite electrical properties using statistical model. In this study, numerical analysis was used to explore the electrical conductivity, the aspect ratio effects, and the CNT geometry effects on the electrical properties of the nanocomposites [72].

7.3.2 Graphene

Graphene is a carbon sheet with sp^2 bounded carbon atoms extended in 2D with one-atom-carbon thickness. It possesses high surface area of 2630 m^2/g, high electron mobility, high electrical conductivity, and good reflecting index. Pure graphene sheets possess excellent characteristics; however, it has many limitations, for example, acting as switch, absorption of heavy metals, etc. The introduction of controlled functionalities in graphene and its derivatives (see Figure 7.7) will make them suitable for required application [73]. Hence, an increased attention and efforts for real applications in energy and environmental for functional graphene and its derivatives are seen [74].

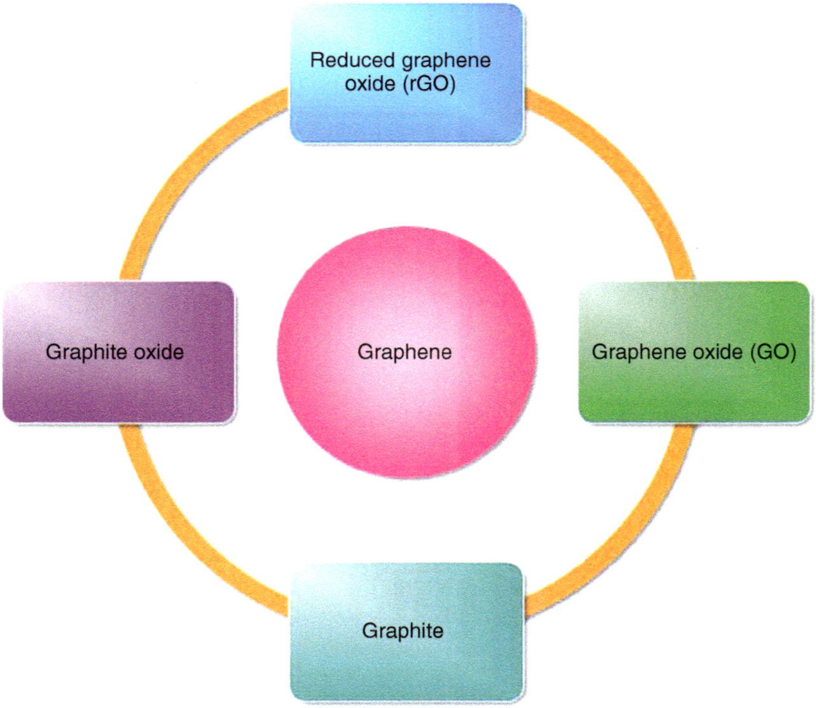

Figure 7.7 Derivatives of graphene.

7.3.2.1 Graphene as Environmental Sensors

Functionalized graphene has been applied in environmental applications, for example, in gas sensing, monitoring, and removal [75]. Carbon dioxide is one of the greenhouse gases, and its emission needs to be monitor to check its effect on the global climate change. Ali et al. [76] performed spin-polarized density functional theory (DFT) over Fe scattered graphene for the detection of CO_2 molecules. The studies revealed that the Fe scattered as adatom and clusters are both active toward CO_2 detection. The incorporation of Fe on graphene makes it ferromagnetic; the magnetic moment of different catalyst was studied to check its activity. The magnetic moment of Fe adatoms on graphene remains same for all the concentration of CO_2, however it decreases from 10.03 to 4 μB for Fe cluster by increasing the concentration of CO_2 from one to four molecules. Fe adatom showed best adsorption capacity, and Fe clusters showed best detection. Wu et al. [77] synthesized hydroquinone (HQ)-modified reduced graphene oxide hydrogel for CO_2 and NO_2 detection. The sensor achieved detection of 200 ppb NO_2 and 20 ppm CO_2 as lower limit of detection (LOD). Shao et al. [78] reported the on-chip assembly of three-dimensional (3D) interconnected porous polypyrrole (PPy)-coupled graphene/$W_{18}O_{49}$ nanowire aerogels and studied it for NO_2 detection. The process involves *in situ* synthesis involving redox reaction between graphene and pyrrole followed by integration of $W_{18}O_{49}$ nanowire during the assembly of chip to increase the sensitivity toward NO_2. The sensor showed nonlinear relationship between the sensor response ($\Delta R/R_0$) and gas concentration for NO_2 above 200 ppb that was reported due to chemiresistive in sensor.

Omidvar and Mohajeri [79] performed theoretical studies on graphene nanoflakes (GNFs) as gas sensor. Different armchair GNFs including pristine, doped (boron and nitrogen), and functionalized (with COOH, CN, and NO_2) were explored for O_2, N_2, CO, and NO sensors. Among all GNFs, functionalized with COOH showed the best interaction with NO molecules. Later, You et al. [80] theoretically found that Fe-doped graphene structure showed strong interaction with NO_2 molecules and is more suited for gas capturing applications.

Jiang et al. [81] prepared sulfonic acid (using 3-mercaptopropyl trimethoxysilane [MPTMS])- functionalized graphene oxide (fSGO) electrochemical alcohol fuel cell sensor (AFCS) for the detection of ethanol vapor. The film can effectively respond to an ethanol standard solution equivalent to blood alcohol content (BAC) = 10 mg/dl blood, corresponding the actually detection limit of 25 ppm ethanol in human breath. Kaniyoor et al. [82] reported Pt-based graphene sensor (Pt/f-G) for hydrogen gas detection by drop-casting technique. The parameters evaluated in the performance of these sensors were "sensitivity" and their "response time" as discussed in Section 7.3.1.1. The Pt/f-G showed increased in the sensitivity toward hydrogen at room temperature (4 vol% in air) compared with Pt/F-MWNT (MWNT = multiwalled nanotube). These sensors showed good stable for repeated hydrogenation and dehydrogenation cycles. Johnson et al. [83] studied Pd-functionalized multilayer graphene snanoribbon (MLGN) for hydrogen sensor. The Pd-functionalized MLGN networks showed high sensitivity to hydrogen in 40 ppm H_2 with $\Delta R/R \sim 55\%$ at room temperature. Ghosh et al. [84] studied Pt-functionalized reduced graphene oxide (rGO) composite and applied it for H_2 sensors (see Figure 7.8). The material synthesized with Pt:rGO (3 : 1) wt% showed the best results. It showed a response of 57% (for 5000 ppm) with response time and recovery time of 65 seconds and 230 seconds, respectively, against in air.

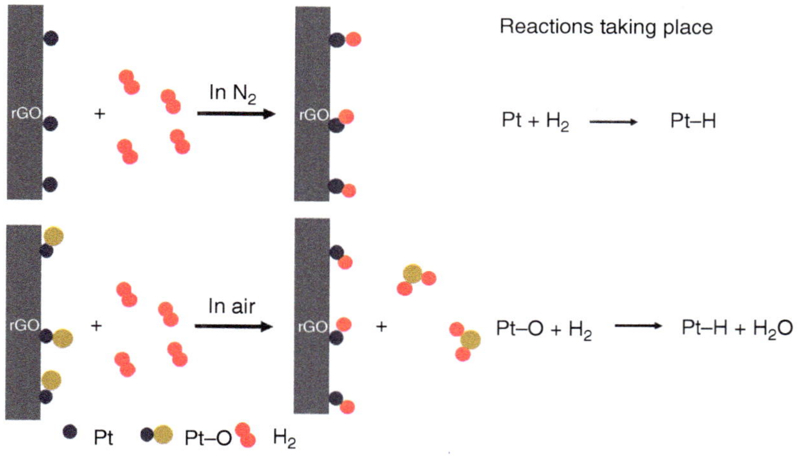

Figure 7.8 Catalytic activity of Pt-functionalized reduced graphene oxide as H_2 sensors in nitrogen and air environment.

Kumar et al. [85] fabricated *meta*-toluic acid fSGO using Langmuir–Blodgett (LB) technique to obtain thin and uniform films on SiO_2/Si wafers. The sensor was tested for ammonia gas detection with concentration varying from 100 to 2000 ppm. However, compared with other sensors, the higher recovery time (~50–80 seconds) in the present study was reported due to the lower desorption rate caused by chemisorptions of NH_3 on the sensing surface.

Zhou et al. [86] first prepared Cu_2O-functionalized graphene sheet nanocomposite using copper acetate ($Cu(Ac)_2$) aqueous solution as Cu source. To mount the functionalized material, patterned microelectrodes using photolithography and sputtering of Au/Cr layer on the Si/SiO_2 substrate were prepared. The functionalized material was then deposited on the sensor, and it was tested for H_2S gas detection. The sensor showed high sensitivity in detecting 5 ppb of H_2S at room temperature.

7.3.2.2 Graphene in Wastewater Treatment

Graphene and GO-based nanomaterials find their application in wastewater remediation for heavy metal removal. Different heavy metals such as cadmium (Cd(II)), lead (Pb(II)), copper (Cu(II)), and chromium(Cr(VI)) [87] are being added to water resources in the form of industrial and other wastes. It is necessary to remove these metals from water resources to preserve the flora and fauna in it [88].

Li et al. [89] chitosan/sulfydryl-functionalized graphene oxide composite (CS/GOSH) using covalent modification and electrostatic self-assembly. The resultant composite material possessed multifunctional groups (–OH, –COOH, –SH, and –NH_2). The adsorbent capability of material was tested for single- and multi-metal ion systems (with 250 mg/l concentration of each ion) for removal of Cu(II), Pb(II), and Cd(II). Metal ion adsorption for mono-component solution was reported as 235, 226, and 117 mg/g for Cu(II), Pb(II), and Cd(II), respectively. In multi-metal ion solution, the absorption was in order Cd(II) > Cu(II) > Pb(II), which might be due to the interaction of metal ions in the mixed solution.

Azamat et al. [90] applied DFT studies to investigate Zn^{2+} ion removal using functionalized graphene nanosheets (GNs). The results showed that functionalized graphene nanosheets can remove Zn^{2+} ions when an electric field of 5 and 2.5 V was applied.

Hu et al. [91] prepared hyperbranched polyamine fSGO for the removal of Pb(II). The synthesis involved preparation of hyperbranched polyamine followed by its modification with GO. For 60 mg/l of Pb(II) at 25 °C, at pH = 6, the materials showed good removal efficiency greater than 95% with a slight decrease (~90%) in removal efficiency for 5th recycle. Madadrang et al. [92] functionalized GO with ethylenediaminetetraacetic acid (EDTA) for Pb(II) removal. For pH 6.8, absorption process was completed within 20 minutes with adsorption capacity of (479 ± 46) mg/g that is four times higher than oxidized CNTs. Hassan et al. [93] reported magnetic sporopollenin (MSp)-functionalized GO using 3-aminopropyltrimethoxysilane for Pb(II) removal. The different functional groups, hydroxyl, carboxyl, and amine, showed different Pd(II) absorption mechanism. The maximum absorption capacity of 323 mg/g was reported; after that equilibrium was reached.

Chen et al. [94] prepared sensors by thermally rGO with thioglycolic acid (TGA)-functionalized gold nanoparticles (Au NPs) (or rGO/TGA-Au NP hybrid structures). The sensor was used for detecting mercury(II) ion inaqueous solutions and it detected lowest concentration of 2.5×10^{-8} M with rapid response of 10 seconds.

Zhang et al. [95] modification graphene surface with 1-octadecanethiol, and the resulted material was tested for Hg^{2+} metal sensors. The sensor was able to detect Hg^{2+} in 10 ppm solution concentration. Chandra and Kim [96] reported polypyrrole–reduced graphene oxide composite with 15 wt% of GO loading for Hg^{2+} removal. Depending upon pH, mercury exists as different species: for pH < 3.0, it exists as Hg^{2+}, for pH > 5, as $Hg(OH)_2$; and for pH ranging 3–5, as Hg^{2+}, $HgOH^+$, and $Hg(OH)_2$. The adsorbent showed high adsorption capacity of 980 mg/g with desorption capacity of 92.3%.

Chandra et al. [97] synthesized Fe nanoparticle–rGO (magnetic in nature) for the removal of arsenic (As(III), As(V) ions). Arsenic ions are removed by electrostatic attraction between the negatively charged arsenic/arsenous acid and positively charged magnetic–rGO surface. The effect of pH and temperature on absorption of As(V) and As(III) was also reported. Fan et al. [98] fabricated magnetic β-cyclodextrin/graphene oxide nanocomposite (MCGN) and applied it for Cr(IV) removal. β-Cyclodextrin (β-CD) is a cyclic oligosaccharide possessing hydrophobic inner cavity and hydrophilic exterior, hence providing cavities for host–guest ion binding. The pH 3 maximum adsorption on MCGN was 103.41 mg/g; however it decreases from 100 to 55 mg/g when the pH value increased from 3 to 12.

Sui et al. [99] prepared CNT–graphene hybrid aerogels using supercritical CO_2 drying and utilize it for desalination of brackish water. Under high concentration of NaCl (up to 35 g/l), high desalination capacity (633.3 mg/g) of hybrid aerogel was reported, which is 15.8 times higher than electrodes used for capacitive deionization (CDI) (maximum capacity of 40.1 mg/g).

Chen et al. [100] reported graphene oxide–chitosan (GO–CS) composite hydrogels for broad-spectrum adsorbents for water purification. It showed good absorption for Cu(II) and Pb(II), with the increase in GO for GO–CS ratio; the high absorption of Pb(II) might be due to the higher electronegativity of Pb(II) ions.

Petrescu et al. [101] synthesized chloroacetic acid-treated GO-functionalized crown ether for metal ion sequestration (Cu, Fe, Na, and K). For Cu and Fe, adsorption capacity increased with increasing initial higher concentration of transition metal ion; however, removal remained almost constant. In the case of Na and K, no result was concluded due to retention of metal ions by graphene.

Luo et al. investigated poly 3-aminopropyltriethoxysilane-grafted graphene oxide (PAS-GO) for metal adsorption in actual wastewater (containing Cr, Ni, Pb, Cd, Cu, Fe, Mn, and Zn with the corresponding concentrations of 0.042, 0.285, 3.866, 5.02, 10.296, 12.65, 15.9, and 112 mg/l, respectively). Before the addition of PAS-GO, the pH of water was adjusted to 3.8. PAS-GO absorbent (1 g/l) showed 33% Pb, 55% Fe, and complete Cr adsorption with no interaction with other ions. With the increase in absorbent (3 g/l), complete removal of Pb and Fe attained with no change in other ion concentrations. With further increase in absorbent amount (5 g/l), about 87.3% Cu and 93.4% Pb were preferentially removed. However, no direct absorbent and metal ion interaction was reported [102].

7.3.2.3 Graphene as Biological Sensors

Graphene can also be functionalized with different biomolecules to form biosensors. The addition of biomolecules improves its biocompatibility, solubility, and selectivity toward different targeted molecules [8b, 103].

Liu et al. [104] reported thymine-rich DNA-fSGO-based magnetic nanoparticles for Hg^{2+} selective detection of via fluorescence changes. 1-Pyrenebutanoic acid succinimidyl ester (PASE) was incorporated on rGO–Fe_3O_4 followed by introduction of amino-Hg-DNA (amino-5′-CTTCTTTCTTCCCCTTGTTTGTTG) to form resulted DNA/PASE–rGO-Fe_3O_4. In the presence of Hg^{2+}, the DNA forms a hairpin structure, which with SYBR Green I, persistent organic pollutants (POP), signal/noise (S/N) (cyanine dye, used as a nucleic acid stain) gives fluorescence. In the absence of Hg^{2+}, the DNA forms a random coil structure, which gives a weak fluorescence with SYBR Green I. The catalyst was highly selective for Hg^{2+} concentration of 0.82 ng/ml with adsorption capacity of 180.8 mg/g. Zhang et al. [105] applied molecular beacon (MB)–GO hybrids to detect small-molecule compounds and heavy metals using fluorescence quenching properties. The sensitivity toward specific DNA, thrombin, Ag^+, Hg^{2+}, and cysteine was reported as 1, 5, 20, 5.7, and 60 nM, respectively.

FET sensors with metallothionein type II (MT-II) protein over rGO were prepared and applied for various heavy metal ion detection. The MT-II-modified rGO FET sensor was reported to detect 1 nM Hg^{2+} (signal-to-noise ratio of 25–30) and Cd^{2+} (signal-to-noise ratio of 15–20) with no response to K^+, Na^+, Ca^{2+}, and Mg^{2+}. Since all the metals studied were positively charged, the study concluded that the sensing mechanism was based on conformational changes in MT-II on heavy metal binding and has no effect of field induced by the binding metal ions [106]. Liu et al. [107] reported covalently grafted (4-ferrocenylethyne) phenylamine (FEPA) on GO for nitrite sensor. The sensor was able to detect 0.1 μmol/l with linearly dependence in the range of 0.3–3100 μmol/l.

1,2-Bis-(2-pyren-1-ylmethylamino-ethoxy) ethane (NPEY) was stacked on graphene nanosheets (GNs) via π–π interaction to form a fluorescence sensor for manganese(II) ions. The sensor showed detection limit of 4.6×10^{-5} M toward Mn^{2+} *in vitro* and in living cells. The blue fluorescence emission in the case of Mn^{2+} is due to binding of NPEY at oxygen donor and imino groups to metal ion that forbids the electron transfer between NPEY and GNs [108].

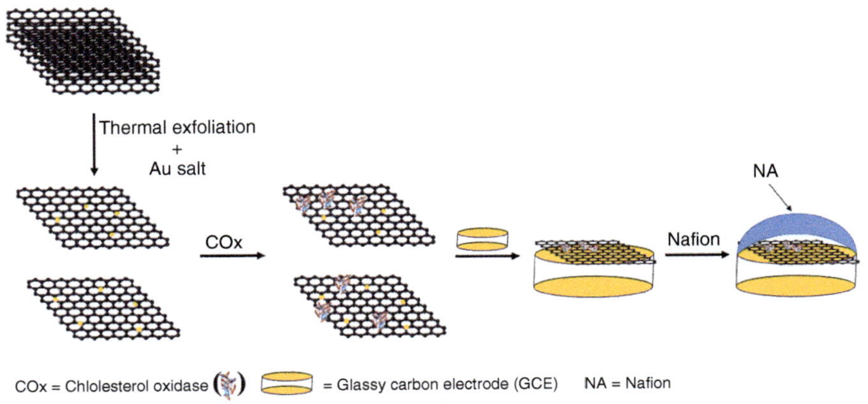

Figure 7.9 Schematic representation for fabrication of cholesterol oxidase biosensor. After this other paragraph (Ionic liquid 1-(3-aminopropyl)-3-methylimidazolium bromide-functionalized graphene was prepared, and the fabricated sensor was applied for detection of pesticides. For detection of carbaryl and monocrotophos in spiked tomato juice, the biosensors were more active in detecting lower concentration of pesticides: 5.3×10^{-15} M (linear range from 1.0×10^{-14} to 1.0×10^{-8} M) and 4.6×10^{-14} M (linear range from 1.0×10^{-13} to 5.0×10^{-8} M), respectively. Source: Based on Zheng et al. [109].

Cholesterol oxidase (COx) biosensor-functionalized gold nanoparticles over GO have been reported for the detection of cholesterol (see Figure 7.9). The calibration curve showed a linear response for the detection of cholesterol up to 135 μM with sensitivity of 314 nA/μM cm^2 [110]. In other study, hemin-functionalized graphene nanosheets (H-GNs) were explored for H_2O_2 and glucose detection. The detection was reported by amperometric method and colorimetric methods. The detection limits for amperometric method for H_2O_2 and glucose reached were 0.2 and 0.3 μM, respectively. For colorimetric methods, glucose oxidase (GOx) is added with 3,3,5,5-tetramethylbenzidine (TMB) (O_2 + glucose $\xrightarrow{\text{Gox}}$ H_2O_2 + gluconic acid) so as to produce H_2O_2. The H-GNs oxidize TMB with H_2O_2 to produce a blue color reaction. The detection limits for H_2O_2 and glucose were as low as 20 and 30 nM, respectively, with colorimetric method [111]. In other study, hemin-functionalized GO sheets were used in detection of ascorbic acid, dopamine, and uric acid. The detection limits of 0.3, 0.17, and 0.17 μmol/l (S/N = 3) were reported for ascorbic acid, dopamine, and uric acid, respectively [112].

7.3.2.4 Graphene for Removing Organic Pollutants

The restoration of water resources contaminated with organic and inorganic impurity is one of the biggest challenges [113]. Erogul et al. prepared (3-aminopropyl)triethoxysilane (APTES) coated with Fe_3O_4 nanoparticles on GO, incorporated on film of glassy carbon electrode (GCE) followed by electrodeposition of gold nanoparticles (Au NPs) for detection of catechol (CC) (1,2-dihydroxybenzene) and hydroquinone (HQ) (1,4-dihydroxybenzene). Both CC and HQ are isomers of phenolic compounds and interfere in the identification of each other. The detection limits for CC and HQ were found to be 0.8 and 1.1 mM, respectively. The amperometric linear response was in the range of 2–145 and 3–137 mM for CC and HQ, respectively [114].

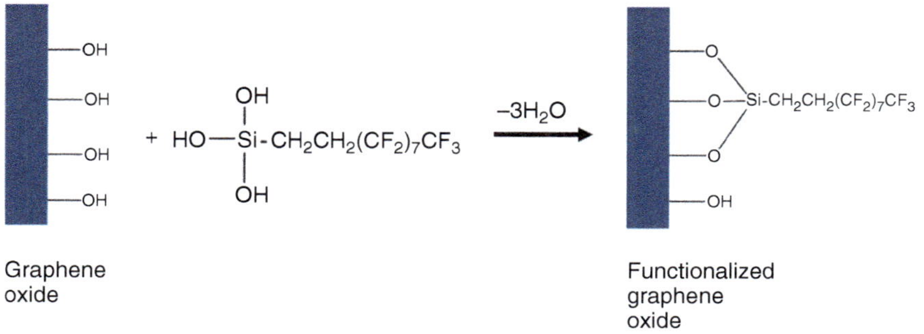

Figure 7.10 Functionalization of graphene oxide. Source: Based on Liu et al. [115].

Fluorinated modified graphene aerogel possessing high porosity and hydrophobicity was applied for absorption of oils and organic solvent (see Figure 7.10). The materials showed absorption capacity up to 11 200% of its weight with no remarkable change in regeneration capability [116].

β-Cyclodextrin-functionalized graphene nanosheet-modified glassy carbon electrode (CD-GNs/GCE) sensor was applied for *o*-nitrophenol (o-NP) detection. The lowest detection limit of *o*-NP was 0.3 μM (signal-to-noise ratio = 3) with linear relationship in the range of 5–400 μM. The CD-GNs/GCE showed no potential change for meta and para NP [115].

Ag-functionalized graphene is applied for nitroaromatic compound (NAC) detection as they are high-toxicity environmental pollutants. NACs with a lower detection limit of nitrobenzene (NB), 2,4-dinitrotoluene (DNT), and 2,4,6-trinitrotoluene (TNT) of 0.60, 0.21, 0.45 ppm, respectively, were reported [117].

Graphene nanosheet-functionalized octadecylamine (ODA) were synthesized for detection of aromatic hydrocarbons like benzene, toluene, ethylbenzene, xylenes, and cyclohexane. For the analysis, low concentration of the aromatic hydrocarbons was dissolved in water. The study reported linear response from 5 to 100 ppm with response time of less than one minute [118].

β-Cyclodextrin-grafted graphene oxide (GO-β-CD) was prepared and applied for absorption of dye pollutants in water (fuchsin acid, methyl orange, and methylene blue). GO-β-CD showed the absorption capacities of 0.88, 0.79, and 0.35 for fuchsin acid, methylene blue, and methyl orange, respectively. The difference in absorption was due to different molecular size and polarity of dyes. When the size of inner cavity of β-CD matches with dye size, the stronger the dye interacted with the β-CD [119].

Magnetic β-cyclodextrin-graphene oxide (MCG) was applied for *p*-phenylenediamine adsorption. Due to its magnetic properties, MCG is easily recovered from reaction mixture. At 45 °C and pH 8, maximum adsorption capacity of 1102.58 mg/g was reported with slight decrease in adsorption capacity (81%) after five cycles [120].

Graphene nanosheets modified with hydroxypropyl-β-cyclodextrin (HP-β-CD) was incorporated on carbon electrode and studied for detection of phenolic organic pollutants. The lowest concentration of *o*-NP detection was 1×10^{-8} M with linearly behavior in the range of 5×10^{-8} to 1×10^{-4} M [121].

Mono(6-ethanediamine-6-deoxy)-β-cyclodextrin (NH_2-β-CD) and graphene nanosheets that were non-covalently bridged with 3,4,9,10-perylene tetracarboxylic acid (CD-PTCA-GNs) are synthesized. The activity of CD-PTCA-GNs on GCE was studied for 1-aminonaphthalene (1-NA) organic pollutants (POPs). The electrode can detect 1-NA with 1.0 nm (S/N = 3), and linear response range was 10–550 nm [122].

7.3.3 Fullerenes

Fullerenes attracted many researchers' attention due to its exceptional physical and chemical properties. It shows various applications in different fields such as environmental monitoring, gas adsorption/storage, water treatment, medicine, semiconducting, pharmaceuticals, etc. Lots of derivatives of fullerenes have been developed by functionalization [123]. When a metal ion is inserted into a fullerene, it forms cage-like structure and results into endo-fullerene and exo-fullerene. This has been used in environmental treatment and monitoring [124].

7.3.3.1 Fullerene as Environmental Sensor

Nasri et al. [125] studied the molecular single-electron transistor properties using a fullerene (C_{60}) molecule. This was the first time researchers have used the fullerene as sensor material for the ammonia detection. Endocrine disruptors (EDCs) (polybrominated biphenyls, plastics, bisphenol A [BPA], plasticizers) are one of the environmental pollutants that disturb the hormonal balance in human beings. Rather and De Wael developed an electrochemical sensor using fullerene for the BPA detection at trace level [126]. Bashiri et al. have used the pristine, boron, gallium, silicon, and many more nonmetal-doped C_{60} fullerenes for the detection of illicit drugs (amphetamine) by using DFT calculations [127]. Kaur et al. also have used the DFT calculation for the detection of toxic sulfur gases (hydrogen sulfide and sulfur dioxide) on the B_{40} fullerene [128]. Arshadi and Anisheh observed the SO_2 and NO_2 molecule adsorption on newly designed Co^{2+}/Cr^{2+} propyrin-induced C_{70} fullerene ($PIC_{70}F$) using DFT calculations [129]. They had investigated chemical and physical properties of SO_2 and NO_2 using $PIC_{70}F$.

7.3.3.2 Fullerene in Wastewater Treatment

Though fullerenes are unstable and less soluble in water, functionalized fullerenes have been used for water purification and other environmental applications. Brunet et al. had designed the fullerene by functionalizing with hydrophilic moieties (such as hydroxyl and amine groups), and it had been used to inactivate pathogenic microorganism through photocatalytic reactions [130]. Fullerene was used in adsorption process for the removal of poisonous organic compounds from water [131]. Some of the fullerene derivative (fullerols and amino fullerene) effectively produced reactive oxygen species (ROS). Therefore, they are being used in water purification and also as disinfectant due to its oxidizing and antibacterial property [132].

7.3.3.3 Fullerene as Biological Sensor

Fullerene size, electronic configuration, hydrophobicity, and 3D cage-like structure are few of the most appealing features of fullerene that have taken active consideration in medical chemistry as an therapeutic agents [133]. Fullerene has been used as the potential drug carriers due to its good biocompatibility, stability, and selective targeted delivery of drugs. If any hydrophilic species attached on the surface of fullerene, it become water soluble, and this property makes them exceptional in drug delivery [134]. Fullerenes also have many conjugated double bonds, so it can react with free radicals. Therefore, it can act as antioxidant (radical scavenger) and antiviral agents [133d, 135]. Szunerit's group had studied the multivalent interactions against virus and bacteria. They have given details about the use of different nanostructural materials such as CNTs, Ag nanoparticles, and fullerenes and their derivatives that possess antiviral activity and had been employed as anti-HIV agents [136].

7.3.3.4 Fullerene in Agriculture

Fullerenes are also used for the removal of various organic pollutants, e.g. lindane, trichloroethylene (agricultural insecticide), etc. [137] and persistent polycyclic aromatic hydrocarbon-polychlorinated biphenyls [138]. Their high surface area, stability, and high chemical affinity make these carbon-based materials unique [139]. Fullerene are being used as fungicides against plant pathogenic fungi such as *Fusarium graminearum* and *Fusarium poae*. It shows less activity than CNT because there should be strong interaction between the fungi and CNM for a good antifungal agent [140]. There are limited numbers of reports for the fullerene uses in seed germination, plant growth, and so on. Fullerene did not respond for seed germination of different species such as *Cucumis sativus*, *Vigna radiate*, *Arabidopsis thaliana*, and many more [141]. They cause negative effects on some plants like *Lemna gibba* that inhibit chlorophyll accumulation and its growth when exposed to fullerene [142]. Similarly phytoplankton stops photosynthesis process in the presence of fullerene [143]. Fullerene and its derivative like fullerols also show some positive effects on the some plants. Cell divisions in *Pseudokirchneriella subcapitata* and hypocotyl growth in *A. thaliana* algae are stimulated by fullerol [144]. Hydroxy fullerenes also improve product quality and crop yields. Fullerol also helps in seed dressings, increase fruit growth, and increase bioactive compounds and numbers in *Momordica charantia* [145]. Still, there are lots of drawbacks in the use of fullerene and its derivative because of its presence in some plants residue and in the food chain.

7.3.4 Carbon Nanocones

The CNCs can be synthesized using direct current plasma-enhanced CVD method [18].

7.3.4.1 Carbon Nanocones as Environmental Sensors

The pyrolytic multiwalled CNC materials were applied for sensing relative humidity (RH). The acid treated films showed 15% higher response to change in RH during 30 minutes' cycle [146].

Density functional calculations were performed surfaces of oxygen-functionalized CNC for its application in the adsorption of sulfur dioxide (SO_2). The results showed that adsorptions of SO_2 on functionalized nanocones are exothermic in nature. The study showed linearity dependencies between absolute adsorption energy (E_{ad}) and orbital energy [147]. In other study, the adsorption of an HCl molecule on the exterior and interior surface of a B and Al atom-modified CNC was studied using DFT calculations. The apex C atom replaced with B or Al showed improved reactivity with no significant change in sensitivity toward HCl [148].

7.4 Conclusion

This chapter briefly summarizes the role of functionalized CNMs in environmental devices and techniques. The synthesis of different CNMs including CNTs, graphene, fullerene, and nanocones with different functional groups is discussed also briefly. CNMs possess high surface area, uniform pore distribution, flexibility, high mechanical and electrical strength, and thermal conductivity. The functionalized CNMs are discussed for environmental sensors, in waste water treatment, as biosensor, etc. In environmental sensors, the detection of different gases including CO_2, NO_2, CH_4, etc. using sensors is also reviewed. In wastewater treatment, detection and removal of heavy metal ions are included. For biosensors, two different applications, i.e. detection of heavy metal and detection of biological molecules, are addressed. The functionalization of CNMs with specific functionalities enhances their potential for specific applications. Modification of CNMs with different functionalities according to the application requirement broadens their scope for various applications.

Important Websites

https://onlinelibrary.wiley.com/doi/10.1002/9783527808854.ch10
https://www.coursera.org/lecture/nanotechnology/what-is-nanotechnology-3R3cX
https://www.understandingnano.com/nanotubes-carbon.html
https://www.coursera.org/lecture/nanotechnology2/sensing-application-of-carbon-nanotubes-941EH
https://www.nanowerk.com/what_is_graphensssse.php

References

1 Bahri, S., Patra, T., and Upadhyayula, S. (2019). *Microporous Mesoporous Mater.* 275: 1–13.
2 (a) Peng, Z., Liu, X., Zhang, W. et al. (2020). *Environ. Int.* 134: 105298. (b) Zaytseva, O. and Neumann, G. (2016). *Chem. Biol. Technol. Agric.* 3: 17.
3 (a) Manawi, Y.M., Samara, A., Al-Ansari, T., and Atieh, M.A. (2018). *Materials* 11: 822. (b) Wang, Y., Yang, P., Zheng, L. et al. (2020). *Energy Storage Mater.* 26: 349–370.
4 (a) Wang, Z., Yu, J., Gui, R. et al. (2016). *Biosens. Bioelectron.* 79: 136–149. (b) Hussain, C.M. (2018). *Handbook of Nanomaterials for Industrial Applications*. Elsevier. (c) Hussain, C.M. (2019). *Handbook of Environmental Materials Management*. Springer. (d) Hussain, C.M. (2020). *Handbook of Functionalized Nanomaterials for Industrial Applications*. Elsevier. (e) Hussain, C.M. and Kecili, R. (2019). *Modern Environmental Analysis Techniques for Pollutants*. Elsevier. (f) Hussain, C.M. and Kharisov, B. (2016). *Advanced Environmental Analysis: Applications of Nanomaterials, Volume 1*. Royal Society of Chemistry. (g) Hussain, C.M. and Mishra, A.K. (2018). *Nanotechnology in Environmental Science, 2 Volumes*. Wiley.

5 Yi, H., Huang, D., Qin, L. et al. (2018). *Appl. Catal., B* 239: 408–424.

6 (a) Mallakpour, S. and Khodadadzadeh, L. (2018). Chapter 7: Biocompatible and biodegradable Chitosan nanocomposites loaded with carbon nanotubes. *Biodegradable and Biocompatible Polymer Composites: Processing, Properties and Applications.* In: (ed. N.G. Shimpi), 187–221. 1e. Woodhead Publishing. (b) Navinchandra, G.S. (2018). *Biodegradable and Biocompatible Polymer Composites Processing, Properties and Applications*. Elsevier Ltd. (c) Palit, S. and Hussain, C.M. (2020). Chapter 1: Modern manufacturing and nanomaterial perspective. *Handbook of Nanomaterials for Manufacturing Applications Micro and Nano Technologies*. In: (ed. C.M. Hussain), 3–20. Elsevier.

7 Chen, W., Duan, L., and Zhu, D. (2007). *Environ. Sci. Technol.* 41: 8295–8300.

8 (a) Chang, H. and Wu, H. (2013). *Energy Environ. Sci.* 6: 3483–3507. (b) Wang, Y., Li, Z., Wang, J. et al. (2011). *Trends Biotechnol.* 29: 205–212.

9 Mukherjee, A., Majumdar, S., Servin, A.D. et al. (2016). *Front. Plant Sci.* 7: 172.

10 Shah, K.A. and Tali, B.A. (2016). *Mater. Sci. Semicond. Process.* 41: 67–82.

11 (a) Nessim, G.D. (2010). *Nanoscale* 2: 1306–1323. (b) Saifuddin, N., Raziah, A., and Junizah, A. (2013). *J. Chem.* 2013: 18. (c) Takagi, D., Homma, Y., Hibino, H. et al. (2006). *Nano Lett.* 6: 2642–2645.

12 Choi, W., Lahiri, I., Seelaboyina, R., and Kang, Y.S. (2010). *Crit. Rev. Solid State Mater. Sci.* 35: 52–71.

13 Bhoria, R.S. (2015). *Indian J. Pure Appl. Phys.* 53: 501–513.

14 Muzyka, R., Kwoka, M., Smędowski, Ł. et al. (2017). *New Carbon Mater.* 32: 15–20.

15 Scott, L.T. (2004). *Angew. Chem. Int. Ed.* 43: 4994–5007.

16 Geckeler, K.E. and Samal, S. (1999). *Polym. Int.* 48: 743–757.

17 Naess, S.N., Elgsaeter, A., Helgesen, G., and Knudsen, K.D. (2009). *Sci. Technol. Adv. Mater.* 10: 065002.

18 Tan, C.K., Loh, K.P., Thong, J.T. et al. (2005). *Diamond Relat. Mater.* 14: 902–906.

19 Lin, C.-T., Lee, C.-Y., Chiu, H.-T., and Chin, T.-S. (2007). *Langmuir* 23: 12806–12810.

20 Zhang, B.-T., Zheng, X., Li, H.-F., and Lin, J.-M. (2013). *Anal. Chim. Acta* 784: 1–17.

21 Liu, J., Ye, Y., Xue, Y. et al. (2017). *J. Polym. Sci., Part A: Polym. Chem.* 55: 622–631.

22 Hu, C.Y., Xu, Y.J., Duo, S.W. et al. (2009). *J. Chin. Chem. Soc.* 56: 234–239.

23 Li, Y., Hori, N., Arai, M. et al. (2009). *Composites Part A* 40: 2004–2012.

24 Wei, B., Vajtai, R., and Ajayan, P. (2001). *Appl. Phys. Lett.* 79: 1172–1174.

25 Collins, P.G., Hersam, M., Arnold, M. et al. (2001). *Phys. Rev. Lett.* 86: 3128.

26 Berber, S., Kwon, Y.-K., and Tománek, D. (2000). *Phys. Rev. Lett.* 84: 4613.

27 Mallakpour, S. and Khadem, E. (2016). *Chem. Eng. J.* 302: 344–367.

28 Avouris, P., Chen, Z., and Perebeinos, V. (2010). Carbon-based electronics. *Nanoscience and Technology: A Collection of Reviews from Nature Journals.* In: (ed. P. Rodgers), 174–184. World Scientific.

29 Misewich, J., Martel, R., Avouris, P. et al. (2003). *Science* 300: 783–786.

30 (a) Yu, M.-F., Lourie, O., Dyer, M.J. et al. (2000). *Science* 287: 637–640. (b) Lu, F., Gu, L., Meziani, M.J. et al. (2009). *Adv. Mater.* 21: 139–152.

31 van der Zande, M., Junker, R., Walboomers, X.F., and Jansen, J.A. (2011). *Tissue Eng. Part B* 17: 57–69.

32 Mitchell, L.A., Gao, J., Wal, R.V. et al. (2007). *Toxicol. Sci.* 100: 203–214.

33 Cui, D., Tian, F., Ozkan, C.S. et al. (2005). *Toxicol. Lett.* 155: 73–85.

34 (a) Wan, J., Zeng, G., Huang, D. et al. (2018). *J. Hazard. Mater.* 343: 332–339. (b) Wu, H., Shi, H., Zhang, H. et al. (2014). *Biomaterials* 35: 5369–5380.

35 (a) Chen, M., Xu, P., Zeng, G. et al. (2015). *Biotechnol. Adv.* 33: 745–755. (b) Cheng, M., Zeng, G., Huang, D. et al. (2016). *Chem. Eng. J.* 284: 582–598. (c) Liang, J., Yang, Z., Tang, L. et al. (2017). *Chemosphere* 181: 281–288. (d) Ren, X., Zeng, G., Tang, L. et al. (2018). *Sci. Total Environ.* 610–611: 1154–1163.

36 (a) Bai, H. and Shi, G. (2007). *Sensors* 7: 267–307. (b) Barsan, N., Koziej, D., and Weimar, U. (2007). *Sens. Actuators, B* 121: 18–35.

37 (a) Wei, L., Shizhen, H., and Wenzhe, C. (2010). *J. Semicond.* 31: 024006. (b) Penza, M., Rossi, R., Alvisi, M. et al. (2009). *Thin Solid Films* 517: 6211–6216.

38 Abdulla, S., Mathew, T.L., and Pullithadathil, B. (2015). *Sens. Actuators, B* 221: 1523–1534.

39 Valentini, F., Biagiotti, V., Lete, C. et al. (2007). *Sens. Actuators, B* 128: 326–333.

40 Chen, G., Paronyan, T.M., Pigos, E.M., and Harutyunyan, A.R. (2012). *Sci. Rep.* 2: 1–7.

41 Ahmad, Z., Manzoor, S., Talib, M. et al. (2020). *Mater. Sci. Eng., B* 255: 114528.

42 Zanolli, Z., Leghrib, R., Felten, A. et al. (2011). *ACS Nano* 5: 4592–4599.

43 (a) Mallakpour, S. and Behranvand, V. (2018). *J. Cleaner Prod.* 190: 525–537. (b) Mallakpour, S., Behranvand, V., and Mallakpour, F. (2019). *Csarbohydr. Polym.* 224: 115138. (c) Mallakpour, S. and Khadem, E. (2019). Chapter 8: Carbon nanotubes for heavy metals removal. *Composite Nanoadsorbents.* In: (eds. G.Z. Kyzas and A.C. Mitropoulos), 181–210. Elsevier. (d) Mallakpour, S. and Rashidimoghadam, S. (2019). Carbon nanotubes for dyes removal. *Composite Nanoadsorbents.* In: (ed. G.Z. Kyzas and A.C. Mitropoulos), 211–243. Elsevier.

44 Lines, M. (2008). *J. Alloys Compd.* 449: 242–245.

45 Ren, X., Chen, C., Nagatsu, M., and Wang, X. (2011). *Chem. Eng. J.* 170: 395–410.

46 Balcajin, O., Noy, A., Fornasiero, F. et al. (2009). Chapter 6: Nanofluidic carbon nanotube membranes: applications for water purification and desalination. *Nanotechnology Applications for Clean Water.* In: (ed. N. Savage, M. Diallo, J. Duncan, et al.), 77–93. William Andrew Publishing.

47 Gao, H., Zhao, S., Cheng, X. et al. (2013). *Chem. Eng. J.* 223: 84–90.

48 Peng, H.B., Pan, M., Wu, R. et al. (2012). Adsorption of ofloxacin on carbon nanotubes: solubility, pH and cosolvent effects. *J. Hazard. Mater.* 211–212: 342–348.

49 Xue, X.-y., Cheng, R., Shi, L. et al. (2017). *Environ. Chem. Lett.* 15: 23–27.

50 Ahmad, J., Naeem, S., Ahmad, M. et al. (2019). *J. Environ. Manage.* 246: 214–228.

51 He, Y., Al-Abed, S.R., and Dionysiou, D.D. (2019). *Environ. Nanotechnol. Monit. Manage.* 11: 100204.

52 Kennedy, A.C. and Smith, K. (1995). *Plant Soil* 170: 75–86.

53 Gogos, A., Knauer, K., and Bucheli, T.D. (2012). *J. Agric. Food. Chem.* 60: 9781–9792.

54 (a) Simonet, B.M. and Valcárcel, M. (2009). *Anal. Bioanal. Chem.* 393: 17. (b) Dinesh, R., Anandaraj, M., Srinivasan, V., and Hamza, S. (2012). *Geoderma* 173: 19–27.

55 Kasel, D., Bradford, S.A., Šimůnek, J. et al. (2013). *Environ. Pollut.* 180: 152–158.

56 Flores, D., Chacón, R., Alvarado, L. et al. (2014). *Am. J. Plant Sci.* 5: 3510.

57 Hao, Y., Yang, X., Shi, Y. et al. (2012). *J. Nanosci. Nanotechnol.* 12: 6287–6293.

58 Khodakovskaya, M., Dervishi, E., Mahmood, M. et al. (2009). *ACS Nano* 3: 3221–3227.

59 (a) Boujtita, M. (2014). Chemical and biological sensing with carbon nanotubes (CNTs). *Nanosensors for Chemical and Biological Applications*. In: (ed. K.C. Honeychurch), 3–27. Woodhead Publishing. (b) Vashist, S.K., Zheng, D., Al-Rubeaan, K. et al. (2011). *Biotechnol. Adv.* 29: 169–188.

60 Tîlmaciu, C.-M. and Morris, M.C. (2015). *Front. Chem.* 3: 59 (1–21).

61 De Micheli, G., Boero, C., Baj-Rossi, C. et al. (2012). *Proceedings of the 49th Annual Design Automation Conference*, DAC 2012, San Francisco, California, USA (3–7 June 2012). DOI: 10.1145/2228360.2228363. 6–11.

62 Salajegheh, M., Ansari, M., Foroghi, M.M., and Kazemipour, M. (2019). *J. Pharm. Biomed. Anal.* 162: 215–224.

63 Ji, S., Lee, M., and Kim, D. (2018). *Biosens. Bioelectron.* 102: 345–350.

64 Comba, F.N., Romero, M.R., Garay, F.S., and Baruzzi, A.M. (2018). *Anal. Biochem.* 550: 34–40.

65 Sobhan, A., Oh, J.-H., Park, M.-K., and Lee, J. (2020). *Microelectron. Eng.* 225: 111281.

66 Shen, X.-J., Pei, X.-Q., Liu, Y., and Fu, S.-Y. (2014). *Composites Part B* 57: 120–125.

67 Li, K., Jin, S., Han, Y. et al. (2017). *Polymers* 9: 321.

68 Gong, S., Cui, W., Zhang, Q. et al. (2015). *ACS Nano* 9: 11568–11573.

69 (a) Wang, C., Yang, S., Ma, Q. et al. (2017). *Carbon* 118: 765–771. (b) González, A. and Alvarez Igarzabal, C.I. (2015). *Food Hydrocolloids* 43: 777–784.

70 Jin, S., Li, K., and Li, J. (2018). *Polymers* 10: 270.

71 Yu, A., Ramesh, P., Sun, X. et al. (2008). *Adv. Mater.* 20: 4740–4744.

72 Hu, N., Masuda, Z., Yan, C. et al. (2008). *Nanotechnology* 19: 215701.

73 (a) Allen, M.J., Tung, V.C., and Kaner, R.B. (2010). *Chem. Rev.* 110: 132–145. (b) Mallakpour, S., Abdolmaleki, A., and Borandeh, S. (2017). *J. Thermoplast. Compos. Mater.* 30: 358–380. (c) Mallakpour, S., Abdolmaleki, A., and Karshenas, A. (2017). *Catal. Commun.* 92: 109–113. (d) Mallakpour, S., Abdolmaleki, A., Mahmoudian, M. et al. (2017). *J. Mater. Sci.* 52: 9683–9695. (e) Mallakpour, S. and Khadem, E. (2018). *Int. J. Biol. Macromol.* 120: 1451–1460.

74 (a) Chabot, V., Higgins, D., Yu, A. et al. (2014). *Energy Environ. Sci.* 7: 1564–1596. (b) Lü, K., Zhao, G., and Wang, X. (2012). *Chin. Sci. Bull.* 57: 1223–1234. (c) Wang, T., Huang, D., Yang, Z. et al. (2016). *Nano-Micro Lett.* 8: 95–119. (d) Abdolmaleki, A., Mallakpour, S., and Borandeh, S. (2016). *Polym. Compos.* 37: 1924–1935. (e) Abdolmaleki, A., Mallakpour, S., and Karshenas, A. (2017). *J. Solid State Chem.* 253: 398–405. (f) Abdolmaleki, A., Mallakpour, S., and Karshenas, A. (2017). *Mater. Sci. Eng., B* 217: 26–35. (g) Abdolmaleki, A., Mallakpour, S., and Mahmoudian, M. (2017). *ChemistrySelect* 2: 11211–11217.

75 Ratinac, K.R., Yang, W., Ringer, S.P., and Braet, F. (2010). *Environ. Sci. Technol.* 44: 1167–1176.

76 Ali, M., Tit, N., and Yamani, Z.H. (2020). *Appl. Surf. Sci.* 502: 144153.

77 Wu, J., Tao, K., Zhang, J. et al. (2016). *J. Mater. Chem. A* 4: 8130–8140.

78 Shao, G., Ovsianytskyi, O., Bekheet, M.F., and Gurlo, A. (2020). *Chem. Commun.* 56: 450–453.

79 Omidvar, A. and Mohajeri, A. (2014). *Sens. Actuators, B* 202: 622–630.

80 You, Y., Deng, J., Tan, X. et al. (2017). *Phys. Chem. Chem. Phys.* 19: 6051–6056.

81 Jiang, G., Goledzinowski, M., Comeau, F.J. et al. (2016). *Adv. Funct. Mater.* 26: 1729–1736.

82 Kaniyoor, A., Jafri, R.I., Arockiadoss, T., and Ramaprabhu, S. (2009). *Nanoscale* 1: 382–386.

83 Johnson, J.L., Behnam, A., Pearton, S., and Ural, A. (2010). *Adv. Mater.* 22: 4877–4880.
84 Ghosh, R., Santra, S., Ray, S.K., and Guha, P.K. (2015). *Appl. Phys. Lett.* 107: 153102.
85 Kumar, R., Kumar, A., Singh, R. et al. (2020). *Mater. Chem. Phys.* 240: 121922.
86 Zhou, L., Shen, F., Tian, X. et al. (2013). *Nanoscale* 5: 1564–1569.
87 Mallakpour, S. and Khadem, E. (2019). *Sci. Total Environ.* 690: 1245–1253.
88 (a) Lim, J.Y., Mubarak, N., Abdullah, E. et al. (2018). *J. Ind. Eng. Chem.* 66: 29–44. (b) Sherlala, A., Raman, A., Bello, M., and Asghar, A. (2018). *Chemosphere* 193: 1004–1017. (c) Ahmad, S.Z.N., Salleh, W.N.W., Ismail, A.F. et al. (2020). *Chemosphere* 248: 126008.
89 Li, X., Zhou, H., Wu, W. et al. (2015). *J. Colloid Interface Sci.* 448: 389–397.
90 Azamat, J., Sattary, B.S., Khataee, A., and Joo, S.W. (2015). *J. Mol. Graphics Modell.* 61: 13–20.
91 Hu, L., Yang, Z., Cui, L. et al. (2016). *Chem. Eng. J.* 287: 545–556.
92 Madadrang, C.J., Kim, H.Y., Gao, G. et al. (2012). *ACS Appl. Mater. Interfaces* 4: 1186–1193.
93 Hassan, A.M., Ibrahim, W.A.W., Bakar, M.B. et al. (2020). *J. Environ. Manage.* 253: 109658.
94 Chen, K., Lu, G., Chang, J. et al. (2012). *Anal. Chem.* 84: 4057–4062.
95 Zhang, T., Cheng, Z., Wang, Y. et al. (2010). *Nano Lett.* 10: 4738–4741.
96 Chandra, V. and Kim, K.S. (2011). *Chem. Commun.* 47: 3942–3944.
97 Chandra, V., Park, J., Chun, Y. et al. (2010). *ACS Nano* 4: 3979–3986.
98 Fan, L., Luo, C., Sun, M., and Qiu, H. (2012). *J. Mater. Chem.* 22: 24577–24583.
99 Sui, Z., Meng, Q., Zhang, X. et al. (2012). *J. Mater. Chem.* 22: 8767–8771.
100 Chen, Y., Chen, L., Bai, H., and Li, L. (2013). *J. Mater. Chem. A* 1: 1992–2001.
101 Petrescu, S., Avramescu, S., Musuc, A. et al. (2020). *Mater. Res. Bull.* 122: 110643.
102 Luo, S., Xu, X., Zhou, G. et al. (2014). *J. Hazard. Mater.* 274: 145–155.
103 (a) Pavlidis, I.V., Patila, M., Bornscheuer, U.T. et al. (2014). *Trends Biotechnol.* 32: 312–320. (b) Rabti, A., Raouafi, N., and Merkoçi, A. (2016). *Carbon* 108: 481–514.
104 Liu, F., Ha, H.D., Han, D.J., and Seo, T.S. (2013). *Small* 9: 3410–3414.
105 Zhang, M., Yin, B.-C., Tan, W., and Ye, B.-C. (2011). *Biosens. Bioelectron.* 26: 3260–3265.
106 Sudibya, H.G., He, Q., Zhang, H., and Chen, P. (2011). *ACS Nano* 5: 1990–1994.
107 Liu, M., Wang, L., Meng, Y. et al. (2014). *Electrochim. Acta* 116: 504–511.
108 Mao, X., Su, H., Tian, D. et al. (2013). *ACS Appl. Mater. Interfaces* 5: 592–597.
109 Zheng, Y., Liu, Z., Jing, Y. et al. (2015). *Sens. Actuators, B* 210: 389–397.
110 Aravind, S.S.J., Baby, T.T., Arockiadoss, T. et al. (2011). *Thin Solid Films* 519: 5667–5672.
111 Guo, Y., Li, J., and Dong, S. (2011). *Sens. Actuators, B* 160: 295–300.
112 Zou, H.L., Li, B.L., Luo, H.Q., and Li, N.B. (2015). *Sens. Actuators, B* 207: 535–541.
113 Sajid, M., Nazal, M.K., Baig, N., and Osman, A.M. (2018). *Sep. Purif. Technol.* 191: 400–423.
114 Erogul, S., Bas, S.Z., Ozmen, M., and Yildiz, S. (2015). *Electrochim. Acta* 186: 302–313.
115 Liu, J., Chen, Y., Guo, Y. et al. (2013). *J. Nanomater.* 2013.
116 Hong, J.-Y., Sohn, E.-H., Park, S., and Park, H.S. (2015). *Chem. Eng. J.* 269: 229–235.
117 Lu, X., Qi, H., Zhang, X. et al. (2011). *Chem. Commun.* 47: 12494–12496.
118 Myers, M., Cooper, J., Pejcic, B. et al. (2011). *Sens. Actuators, B* 155: 154–158.
119 Wang, S., Li, Y., Fan, X. et al. (2015). *Front. Chem. Sci. Eng.* 9: 77–83.
120 Wang, D., Liu, L., Jiang, X. et al. (2015). *Appl. Surf. Sci.* 329: 197–205.

121 Xu, C., Wang, J., Wan, L. et al. (2011). *J. Mater. Chem.* 21: 10463–10471.
122 Zhu, G., Zhang, X., Gai, P., and Chen, J. (2012). *ChemPlusChem* 77: 844–849.
123 Nimibofa, A., Newton, E.A., Cyprain, A.Y., and Donbebe, W. (2018). *J. Mater. Sci. Res.* 7: 22.
124 Dorn, H.C. and Fatouros, P.P. (2010). *Nanosci. Nanotechnol. Lett.* 2: 65–72.
125 Nasri, A., Boubaker, A., Hafsi, B. et al. (2018). *IEEE Sens. J.* 18: 248–254.
126 Rather, J.A. and De Wael, K. (2013). *Sens. Actuators, B* 176: 110–117.
127 Bashiri, S., Vessally, E., Bekhradnia, A. et al. (2017). *Vacuum* 136: 156–162.
128 Kaur, J., Kumar, R., Vohra, R., and Sawhney, R.S. (2019). *J. Mol. Model.* 26: 17.
129 Arshadi, S. and Anisheh, F. (2017). *J. Sulfur Chem.* 38: 357–371.
130 Brunet, L., Lyon, D.Y., Hotze, E.M. et al. (2009). *Environ. Sci. Technol.* 43: 4355–4360.
131 Snow, S.D., Lee, J., and Kim, J.-H. (2012). *Environ. Sci. Technol.* 46: 13227–13234.
132 (a) Kang, S., Mauter, M.S., and Elimelech, M. (2009). *Environ. Sci. Technol.* 43: 2648–2653. (b) Lyon, D.Y., Adams, L.K., Falkner, J.C., and Alvarez, P.J. (2006). *Environ. Sci. Technol.* 40: 4360–4366. (c) Fortner, J., Lyon, D., Sayes, C. et al. (2005). *Environ. Sci. Technol.* 39: 4307–4316. (d) Burakov, A.E., Galunin, E.V., Burakova, I.V. et al. (2018). *Ecotoxicol. Environ. Saf.* 148: 702–712.
133 (a) Fowler, P.W. and Ceulemans, A. (1995). *J. Phys. Chem.* 99: 508–510. (b) Iwamoto, Y. and Yamakoshi, Y. (2006). *Chem. Commun.* 46: 4805–4807. (c) Hetzer, M., Bayerl, S., Camps, X. et al. (1997). *Adv. Mater.* 9: 913–917. (d) Friedman, S.H., DeCamp, D.L., Sijbesma, R.P. et al. (1993). *J. Am. Chem. Soc.* 115: 6506–6509.
134 (a) Fowler, P.W., Heine, T., Mitchell, D. et al. (1996). *J. Chem. Soc., Faraday Trans.* 92: 2203–2210. (b) Foley, S., Crowley, C., Smaihi, M. et al. (2002). *Biochem. Biophys. Res. Commun.* 294: 116–119.
135 Gharbi, N., Pressac, M., Hadchouel, M. et al. (2005). *Nano Lett.* 5: 2578–2585.
136 Szunerits, S., Barras, A., Khanal, M. et al. (2015). *Molecules* 20: 14051–14081.
137 Srivastava, M., Abhilash, P., and Singh, N. (2011). *J. Biomed. Nanotechnol.* 7: 172–174.
138 (a) Wang, L., Fortner, J.D., Hou, L. et al. (2013). *Environ. Toxicol. Chem.* 32: 329–336. (b) Zhang, L., Wang, L., Zhang, P. et al. (2011). *Environ. Sci. Technol.* 45: 1341–1348.
139 Yoo, J., Ozawa, H., Fujigaya, T., and Nakashima, N. (2011). *Nanoscale* 3: 2517–2522.
140 Wang, X., Liu, X., Chen, J. et al. (2014). *Carbon* 68: 798–806.
141 (a) Liu, Q., Zhao, Y., Wan, Y. et al. (2010). *ACS Nano* 4: 5743–5748. (b) Kumar, S., Patra, A., Datta, S. et al. (2015). *Int. J. Adv. Res.* 3: 854–865.
142 Santos, S.M., Dinis, A.M., Rodrigues, D.M. et al. (2013). *Aquat. Toxicol.* 142: 347–354.
143 Tao, X., Yu, Y., Fortner, J.D. et al. (2015). *Chemosphere* 122: 162–167.
144 Gao, J., Wang, Y., Folta, K.M. et al. (2011). *PLoS One* 6 (5): 1–8.
145 Kole, C., Kole, P., Randunu, K.M. et al. (2013). *BMC Biotech.* 13: 37.
146 Svåsand, E. and Häberle, P. (2014). *J. Nanosci. Nanotechnol.* 14: 3575–3580.
147 Najafi, M. (2017). *J. Cluster Sci.* 28: 1345–1353.
148 Vessally, E., Behmagham, F., Massoumi, B. et al. (2016). *Vacuum* 134: 40–47.

8

Fundamental of Functionalized Carbon Nanomaterials for Environmental Devices and Techniques

Baskaran Ganesh Kumar[1,2], *P. PonSathieshkumar*[3], *and K.S. Prakash*[4]

[1] *P.S.R. Arts and College (Affiliated to Madurai Kamaraj University, Madurai), Department of Chemistry, Sivakasi, 626124, Tamilnadu, India*
[2] *P.S.R. Engineering College (Affiliated to Anna University, Chennai), Department of Science and Humanities, Sivakasi, 626124, Tamilnadu, India*
[3] *University of Hyderabad, School of Chemistry, Hyderabad, 500 046, India*
[4] *Bharathidasan Government College for Women (Autonomous) (Affiliated to Pondicherry University, Pondicherry), Department of Chemistry, Muthialpet, Puducherry (UT), 605003, India*

8.1 Introduction

Current development of modern electronic devices created the large havoc on the Earth that includes electronic waste, dyes in water, and toxic industrial waste [1]. The negative impact of the technologies affects land, water, human, animals, and plants. The same modern technologies can also be solution for the problem. However, novel and modern materials should emerge to meet the demand of negative impact. One of such novel materials is carbon nanomaterials (CNMs). The CNMs (fullerenes, carbon nanotubes [CNTs], graphene, etc.) are versatile and exploited to making many environmental devices such as sensors, solar cells, dye degradation, and many more [2]. The abundance of carbon and rich chemistry made the CNMs very attractive. Moreover, new class of carbon nanodots and carbon fibers also attracted enormous interest on CNMs [3]. It is widely believed that the functionalization pave the way to next-generation advanced nanomaterials. However, the scope of CNMs can be extended, if we functionalize the CNM with conventional active functional groups using covalent and non-covalent interaction [4]. The functionalization provides handle for application and additional usability. Further, the carbon chemistry is being well established through years, and it can be applied easily for functionalization [5]. Therefore, the functionalization possesses crucial importance, and it is mandatory to realize the immense potential of the CNM; the functionalization of the CNM are inevitable.

The Industrial Revolution created large industries releasing large amount of toxic wastes to earth, water, and atmospheric air. Hence, the health of human, animal, and plants is in huge risk. The branch of organic optoelectronics evolved over time by

Environmental Applications of Carbon Nanomaterials-Based Devices, First Edition.
Edited by Shadpour Mallakpour and Chaudhery Mustansar Hussain.

CNMs to address the public health and safety [6]. The devices are called environmental devices and gained much importance due to its processability, mechanical ability, and low cost. The CNMs has provided advanced properties and can be used to unprecedented environmental devices [7]. Noticeably, due to the low-weight CNMs, the environmental devices provided the ultraportability after fabrication. Moreover, the CNM can be processed in any substrate due to its processability and mechanical ability. Hence, any medical diagnostic devices (blood testing) can be made in paper and can be disposed after testing by burning [8]. The paper-based CNM devices revolutionized the medical diagnosis and shifted the benchtop diagnosis to the point-of-care diagnosis. Due to the versatility, CNMs are integrated into energy storage devices, sensors for environmental monitoring, energy efficient lighting, membrane filters, and environmental remediation [9]. The advanced CNM-based environmental devices proved to provide high sensitivity and low detection limit in the environmental monitoring (detection of pesticides, toxic ions, bio-organisms in water, and pH detection). The present work briefly highlighted the important and representative examples of the CNM-based environmental devices.

The functionalization means the carbon atoms is connected with active chemical functional groups such as alcohol, acid, and polymers [10]. Hence the functionalization has the use of both CNM and functional groups. Consequently, using the functional group, the optical, mechanical, magnetic, electronical, and thermal properties can be modified by the functionalization. The functionalization also tunes hydrophobicity (water insoluble) of carbon to hydrophilicity (water soluble). The tuning of solubility revolutionized optoelectronic industry by solution-state processing of the carbonaceous functional materials. Importantly, the functionalization is also able to tune band gap and make the energy harvesting efficient [11]. Since the tunability is very high, the CNMs can be specifically tuned to intended applications. It is widely believed that functionalization of CNMs expected to break scientific and technological limits.

The present work exclusively provides up-to-date review of functionalization and its advantages in the applications. We initially described the fundamental properties and advantages of the CNMs. We also elaborately discuss the structure, bonding, electron transport, and morphology of CNMs. Then, we critically explored the different types of CNMs based on the dimensions. The functionalization are classified into two types: covalent and non-covalent functionalizations, and these are also discussed in this chapter. Then the various functionalization methods are explained with the examples. For each example, we discussed the synthetic methods and interaction between the functional groups. The overview includes the basic (–COOH addition to CNMs) to advanced functionalization (π–π stacking). The recent understanding of novel carbon dots is also discussed. The CNMs has huge potential in many industrial applications. We noted the important and representative examples of CNMs in the field of environmental optoelectronic devices. Our discussion on environmental devices includes toxic gas sensing, renewable energy, metal detection, humidity sensing, water purification, and energy efficient lighting. To facilitate the CNM-based technologies, we assessed the scalability/mass production of advanced CNMs and practicality in the synthesis. Finally, the summary, wide scope, challenges, and future directions are given in the later part of this chapter.

8.2 Results and Discussion

8.2.1 What Are Carbon Nanomaterials?

Carbon is the one of the most unique and an indispensable element in the world. Relative abundance of carbon is between 180 and 270 ppm, and it plays a vital role in forming strong bonds in various forms. Carbon has the ability to form long chains at the atomic levels. The most important property is that it has four electrons in the outer electron layer and hence forms single or double or triple bonds with itself and also with other elements. By arranging, same types of atoms in the substances in a different manner with different properties are called allotropes. Diamond and graphite are the allotropes of carbon present in nature. The sp^3 hydridization of C—C bonds in diamond makes the hardest material in nature. Rather, graphite is in the form of lubricant solid because of its weak interlamellar coupling among sheets. These carbon materials would have unique properties, with the size of "nano." In general, the materials containing particle size in between 1 and 100 nm at least in one dimension is defined as nanomaterials. All nanomaterials composed of carbon atoms are termed as "carbon nanomaterials" (CNMs). In recent years, plenty of new CNMs are being reported in the literature in the form of CNT, fullerene, graphene, mesoporous carbon, etc. Figures 8.1 and 8.2 clearly explain the formation of CNMs from graphite to fullerenes, CNTs, and graphene sheets. CNMs are classified based on their structure,.

8.2.1.1 Fullerene

Buckminsterfullerene (C_{60}) was the first discover as fullerene molecule. The discovery expanded the number of allotropes in carbon, and it inspired many researchers to predict and analyze their unique properties. This molecule had enormous applications in many fields, for example, photoelectric device making [12], renewable energy [13], etc. The structure of fullerene is cage-like (truncated icosahedrons) and also a close resemblance of a soccer ball. C_{60} contains 20 carbon hexagons and 12 carbon pentagons, which is connected with a bond along each polygon edge (Figure 8.2). The bond between two hexagons has shorter distance than the bond between a hexagon and pentagon. The nucleus-to-nucleus diameter is around 0.71 nm, and van der Waals diameter of the molecule is around 1.01 nm [14].

8.2.1.2 Carbon Nanotubes

Since the discovery in 1991, CNTs are considered as important allotrope of carbon. It has found to have remarkable properties and enormous application in various fields. Many researchers are earnestly involved in this field, because of its promising anomalous properties and its applications. The structure of CNT is reported as hallow cylinder of carbon by scrolling single- or multilayered graphene (Figure 8.2). CNT has mainly two types based on their structure, single-walled carbon nanotubes (SWCNTs) and multiwalled carbon nanotubes (MWCNTs). In MWCNTs, the distance between two layers is 0.34 nm, and the chirality of each layer is different. It has been found that CNTs exhibit high Young's modulus, high tensile strength, and high electrical properties. Yu et al. found the Young's modulus is around 320–1470 GPa by atomic force microscopy [15]. It has high catalytic properties in many reactions such as high catalytic activity in the oxygen reduction reactions [16].

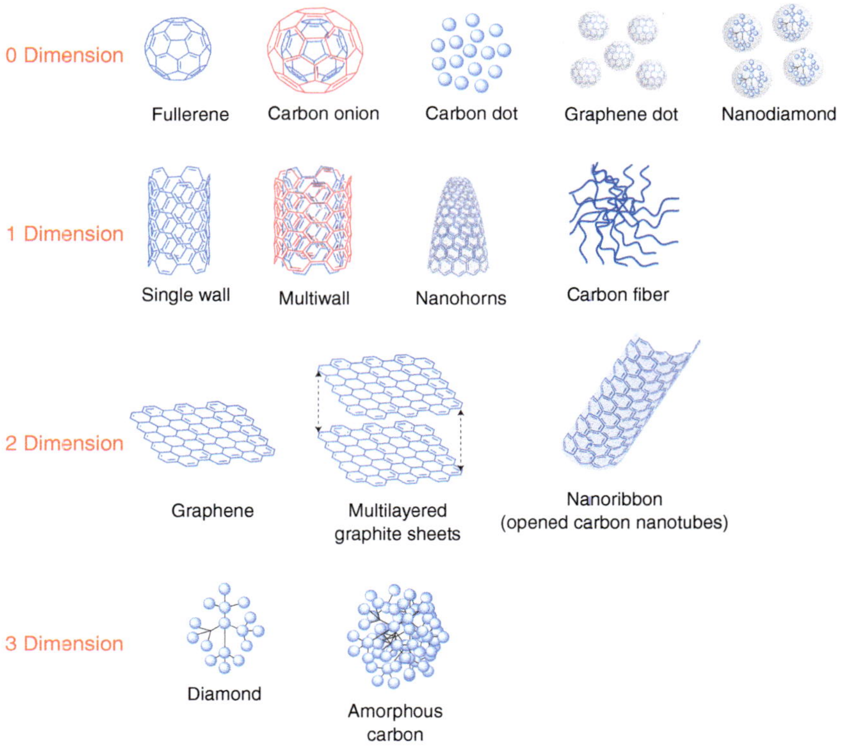

Figure 8.1 Schematic representation of carbon nanomaterials. The classification based on the dimensions of carbon nanomaterials.

8.2.1.3 Graphene

Novoselov, Geim, et al. discovered a single atomic layer of carbon (graphene) for the first time in 2004. Graphene is two-dimensional atomic crystals made up of carbon atoms arranged in a hexagonal lattice [17]. It is considered to be the fundamental structure in the carbon universe [18]. As depicted in Figure 8.2, graphene is the mother of all graphitic forms and also is a thinnest compound known to the mankind (million times thinner than human hair) and also the strongest and the lightest compound with extremely flexibility. It has good electrical, mechanical, and thermal properties, so they can be widely used in various electronic devices [19]. The tensile strength and Young's modulus enable this material to have potential application in the field of ultrastrong and composite materials [20]. This material has potential applications in energy conversion and storage material because of its high surface area and high thermal conductivity [21].

8.2.2 Functionalization of CNMs

8.2.2.1 Need for Functionalization

Functionalization is the introduction of functional groups in the parent material. The main purpose of the functionalization is creating better contact of filler to polymer matrix. When

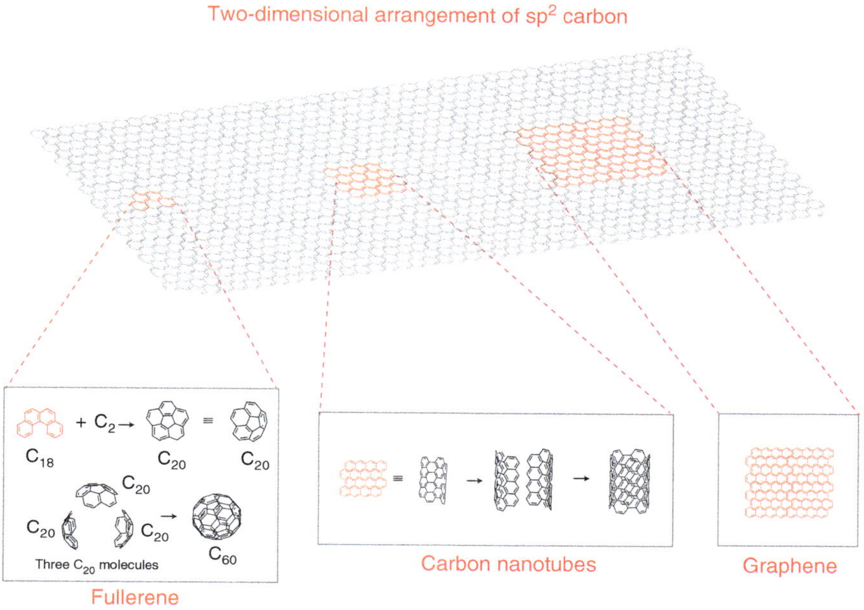

Figure 8.2 Formation of CNMs from graphite to fullerenes, CNTs, and graphene sheets. The graphical representation of folding is shown.

chemical bonds increase between nanofillers and matrix, it leads to increasing in interfacial energy and toughness. The property of material is completely modified. More specifically, there are mainly in two ways: (i) homogeneously dispersion of the nanofillers throughout the matrix and (ii) creating effective interfacial bonding between the nanofillers and the matrix materials.

To resolve many problems in CNMs, it has been directed toward developing methods to modify surface properties of CNMs. Smooth surfaces without any hanging bonds makes pristine CNTs incompatible with most of the organic and inorganic solvents. For example, without functionalization of CNTs, solution-based process is difficult to achieve. This issue creates major drawback in the field of any device manufacturing. These kinds of problems can easily be sorted out by improving functional groups using covalent or non-covalent modifications (*vide infra*). Functionalization of CNMs can increase the water miscibility and improve biocompatibility. Hence, it is most essential for producing novel materials, and those are highly applicable in various fields. For examples, CNTs are functionalized by various methods, and it may be via either covalent or non-covalent (Figure 8.3). Functionalization are mainly classified into two categories, which are described in Sections 8.2.2.2 and 8.2.2.3 in detail. Important functionalization of CNMs, precursor used, and their applications are tabulated (Table 8.1).

8.2.2.2 Covalent Functionalization

Covalent functionalization is the attachment of functional group/moiety via the formation of new covalent bond, which is equivalent sharing of one pair of electrons by CNMs and

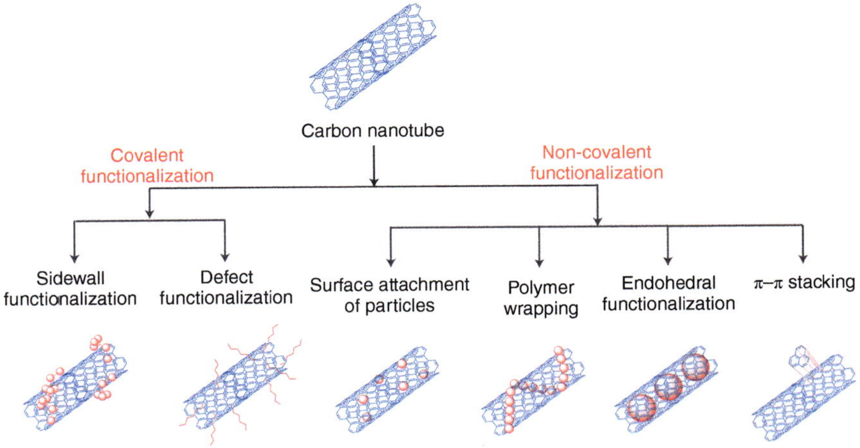

Figure 8.3 Types of functionalization in carbon nanotubes (CNTs). The class action is based on the nature of bonding between CNMs and guest material.

the introduced chemical moiety (Figure 8.4). The covalent functionalization can otherwise be call as defect functionalization because only defective carbon atoms present in the sidewall or at the end of the materials can be functionalized easily using suitable reagents. Initially, defective carbon atoms oxidized with strong oxidizing agents are subsequently converted into carboxylic acid groups, which can be further modified to various functional groups like amide or ester. [57] and some of the addition reaction with CNMs [58]. In this type, the introduction of new functional group/moiety is comparatively easy. After that, it is possible to modify with other groups as per need basis, since carbonyl group has both nucleophilic and electrophilic sites. As already discussed, oxidation is one of the most important techniques in the aspect of functionalization of CNMs, because carbon atoms can easily be oxidized using oxidizing agents.

Fullerenes oxidized with suitable oxidizing agents, for example, polyhydroxylated derivatives (fullerenols or fullerols), were synthesized with dil. H_2SO_4 and HNO_3 or with dil. NaOH in the presence of H_2O_2 or other peroxides. Graphene is somewhat chemically inert; introducing functionalization in graphene enhances its properties including change in band gap, conductivity, improving solubility, and becoming more stable. It is made with exclusive sp^2 carbon atoms with its π electrons delocalized in the entire 2D network (Figure 8.2). Hence, it needs reactive reagent to make covalent adducts – the formation of new covalent bonds with graphene via the formation of reactive species like radicals, nitrenes, carbenes, and arenes. These reactive species covalently modify the materials through free radical addition, CH insertion, or cycloadditions. Among these, free radical addition is the most common, involving electron transfer from graphene to aryl diazonium salts and peroxides. Nitrenes are electron-deficient species that can be generated by thermal or photochemical reactions of organic azides. For example, perfluorophenylnitrenes show enhanced reactivity than alkyl or aryl nitrenes. Nitrenes are preferentially used than carbenes, because carbenes can undergo CH insertion and C=C cycloaddition with graphene.

Table 8.1 Functionalization of carbon nanomaterials.

S. No.	Materials	Methods	Precursors	Key properties	References
1	Water-soluble C_{60}-PDA-GSH nanoparticles	Michael addition/ polymerization	PDA-modified C_{60}	Cytoprotective effect/human cells at a low concentration of 2 μg/ml	[22]
2	[60]Fullerene-polyglycidyl nitrate	Modified Bingel reaction	C_{60}/bromomelonic acid	Energetic binder	[23]
3	Fullerene phenylalanine lead salt	Acidic mediated reaction	Fullerene phenylalanine/ $Pb(NO_3)_2$	Catalyst	[24]
4	Fullerene ethylenediamine nitrate	Ion exchange reaction	Ethylenediamine/dil. nitric acid	Catalyst	[25]
5	Trinitrophenyl C_{60}	Nitration	Trinitrochlorobenzene/sodium azide	Energetic materials	[26]
6	GA/Fe_2O_3/AP	Facile sol–gel method	Graphene aerogel/ammonium perchlorate	Propellants	[27]
7	Graphene nanoplatelets by aryne	Aryne cycloaddition	2-(Trimethylsilyl)aryl triate	Dispersible in DMF, *o*-DCB, ethanol, chloroform, and water	[28]
8	Poly(3-hexylthiophene)-GO (P3HT-GO)/C_{60}	Esterification	GO with $-CH_2OH$ terminated regioregular P3HT	200% increase in the power conversion efficiency in comparison with a pure P3HT/ C_{60}	[29]
9	GO/porphyrins	Amidation	GO and porphyrins	Better optical limiting properties compared with C_{60}/ dispersibility in organic solvents	[30]

(Continued)

Table 8.1 (Continued)

S. No.	Materials	Methods	Precursors	Key properties	References
10	PEGylated GO	Amidation	GO and polyethylene glycol	Important candidate for the delivery of hydrophobic drugs in biological systems	[31]
11	GO/polystyrene (GO/PS)	Diazonium addition	GO and polystyrene	Enhancing the glass transition temperature of the polymer by 15 °C	[32]
12	Imidazolium-modified GO	Amidation	1-(3-Aminopropyl)-imidazole	Used in polymer solar cells	[33]
13	G with hydrogen plasma	Adsorption	Cold hydrogen plasma	Changes their hybridization state to sp^3 from sp^2	[34]
14	Fluorographene	Fluorination	XeF_2	Changes its structure and electronic and optical properties	[35]
15	Amine group-modified CNT	Supercritical water oxidation	MWCNT, O_2, 30% ammonia solution	High solubility in water	[36]
16	Amine group-modified CNT	Radical polymerization	MWCNT	High-capacity dye removal	[37]
17	Chlorinated CNT	Chlorination	DWCNT, CCl_4	Humidity sensor	[38]
18	Thiol-functionalized CNT	Amidation	COOH-functionalized MWCNT and cysteamine	Adsorbent for dye	[39]
19	Boron and chlorine dual-doped CNT	Chemical tailoring technique	Boron carbide nanoflakes	Efficient oxygen reduction reaction (ORR) catalyst	[40]
20	Nitrogen-doped CNT	Pyrolysis	Iron(II) phthalocyanine	ORR catalyst	[41]
21	Nitrogen-doped CNT composite	Plasma-enhanced chemical vapor deposition	N_2, CH_4, Fe, Ni	Improves performance of Li-ion batteries	[42]
22	MnO_2@MWCNT fiber	Fabrication and twisting	MWCNT, $KMnO_4$	Solid-state capacitors	[43]

23	Pd-decorated SWCNT	Electron beam evaporation	SWCNT	H_2 gas sensor	[44]
24	Pt- and Au-functionalized MWCNT	PECVD/sputtering	MWCNT	NO_2 and NH_3 gas detectors	[45]
25	Chlorosulfonated PE, hydroxypropyl cellulose, Pd-decorated SWCNT	Coating and metal sputtering	SWCNT	NO_2, HCN, HCl, Cl_2, acetone, and benzene gas detection and discrimination	[46]
26	Pd-loaded SWCNT	Metal sputtering	SWCNT	Methane gas sensor	[47]
27	Thiol-functionalized magnetic CNT	Mechanical stirring	MCNT	Extraction of sulfonamide antibiotics	[48]
28	Surface-oxidized CNT	Oxidation	CNT	Removal of uranium(IV)	[49]
29	Fe(III) oxide-coated ethylenediamine-functionalized CNT	Oxidation and amidation	MWCNT	Adsorption of arsenate	[50]
30	Dopamine, PP-functionalized CNT	Mechanical	MWCNT	Adsorption of Cu^{2+}	[51]
31	Ceria-adhered CNT	Mechanical	CNT	Adsorption chromium ions	[52]
32	ZnS/CNT	Nanoparticle growth	MWCNT	Removal of dye methylene blue organic dye	[53]
33	CdS/CNT	Polyol method	MWCNT	Degradation of brilliant red X-3B dye	[54]
34	Ceria/CNT	Oxidation and deposition	CNT	Removal of arsenate from water	[55]

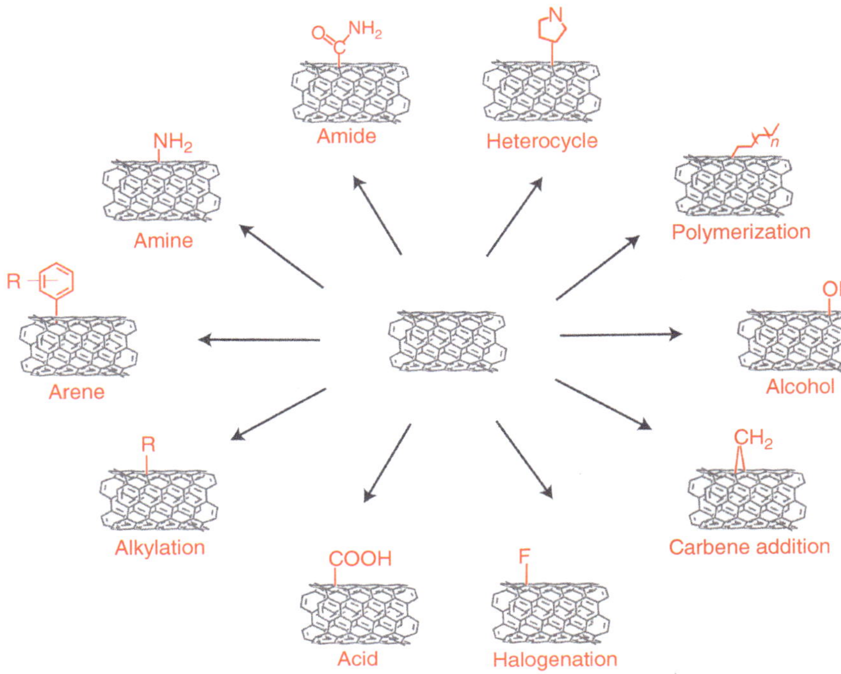

Figure 8.4 Functionalization of CNTs with various functional groups. Source: Modified from Moradi et al. [56].

Another important oxidation in graphite plays a vital role; graphene oxide (GO) can be prepared by the reaction of oxidizing agents with graphite. With the oxidation of graphite, it is oxidized to intersperse the carbon layers with oxygen; finally the carbon layers separated completely. The interplanar spacing between layers of graphite is increased. It should be noted that the presence of oxygen atom in GO increases its ionic and non-ionic interaction with wide range of molecules. This material is a perfect platform for molecular engineering. Modifying the GO with other functional groups like epoxy and hydroxyl groups on either side of the GO has bifunctional behavior. This bifunctional property can be acted as a structural node within metal–organic frameworks (MOFs). Generally, functionalization of GO can highly influence the change in electrical properties, compared with its precursors. GO or chemically modified graphenes (CMGs) would become highly adaptable in many applications. For example, diamide-functionalized polyethylene glycol (PEG)-based material has superstability in biological solutions. *N*-Hydroxysuccinimde (NHS) is added as co-catalyst during amidation of GO functionalizations. Carbodiimide-activated esterification on GO and further functionalization with polymer and its solubilization are reported by Vaca et al. [59]. Initially, carboxylic group is converted into corresponding acylchloride and followed by esterification with alcohols or polymers containing alcohol groups (polyvinyl alcohols). This leads to the formation of polyvinylalcohol (PVA)-functionalized GO [60]. Similarly, PVC-modified GO is also prepared in the literature [61]. Epoxy functional group is also one of the important methods in GO. Shan group successfully functionalized poly-L-lysine covalently with graphene by modifying epoxy groups. These materials are found to have potential application in the field of biomedical [62].

In general, the material has no oxygen atoms; radical reaction and 1,3-dipolar cycloaddition are good choices for the functionalization. Diazotization proceeds via the formation of diazonium salt (combination of amine and nitride in a low-temperature reaction condition), and XPS analysis clearly stated the formation of single electron transfer between reagent and graphene [63]. Swager and coworker reported an efficient method for the functionalization of CNMs (CNTs and graphite) using the reduction of CNMs with sodium naphthalide, followed by the addition reaction of diaryliodonium salts. They extended the same protocol to single-walled, double-walled, and multiwalled carbon nanotubes as well as graphite [64]. 1,3-Dipolar addition in GO [65] was also reported. Similarly, there are various reports found in the literature for the functionalization of GO and its applications. The chemical functionalization may be either direct or defect functionalization.

Similarly, one of the milestones in nanotube chemistry is the oxidative purification of CNTs by liquid- or gas-phase oxidation. Oxidation of CNTs leads to the formation of defects in the walls, introduction of oxygen functional groups, opening of end caps of closed CNTs, cutting of CNT, and also removing impurities present in the CNTs (metal catalyst as impurities in metal-mediated synthesis). Usually, oxidation reaction of CNTs is performed in the presence of boiling HNO_3, H_2SO_4, and piranha solutions (sulfuric acid with hydrogen peroxide). Other reagents like O_2, air, ozone, HF, OsO_4, H_2O_2, $K_2Cr_2O_7$, $KMnO_4$, etc. can also be used as oxidizing agent with different reaction conditions.

Direct functionalization is usually adopted for the oxygenation of CNMs while other functionalizations such as halogenations, amidation, esterification, cycloaddition, and nucleophilic and electrophilic addition can be done through direct or defect functionalization. Generally, most of the defect modifications of CNMs having oxidized CNMs are used as starting materials. If carboxylic group (–COOH) is introduced successfully in a CNM, then it is ready for further functionalization. For example, –COOH group can easily form a zwitterion with amines (R-NH_2). This reaction is used to solubilize CNTs using octadecylamine as a solvent [66, 67]. Amidation of CNMs from the carboxylic group can easily be achieved using corresponding of amines. The new bond is formed between –CO– and –NH. During amide formation reaction, carboxylic group needs to be activated by the formation of acyl chlorides using reagent thionyl chloride. Another easy modification from carboxylic functional group is esterification, which is usually achieved using corresponding alcohols (ROH).

Another important CNM is fullerenes; the major problem in the field of biological applications of fullerene is solubility issue in water and low solubility in many organic solvents. However, this problem can easily sort out by functionalization of fullerenes [68]. The presence of double bonds in the fullerenes opens many reactions to modify the surface of the fullerenes. Because of double bonds, easily achievable addition reactions are considered as key reaction for functionalization of fullerenes. In addition reactions, functional groups are introduced into the surface of fullerenes to modify its physical or chemical properties. If the polar groups such as –OH and –NH_2 are introduced, certain polarity to the material thus rendering molecule becomes soluble in water. For example, surface hydroxylkation and amidation reactions are used for introducing –OH and –NH_2 group, respectively, and hydrogen bonding is used to improve the solubility. Major drawback of the reaction is that the density of functional group is not controlled; hence, there is a lack of reproducibility specifically when materialize. However, this functionalization can successfully done while using precise chemical composition. In other CNMs (graphene, CNTs, etc.), the oxygen-based functional groups is introduced in the presence of strong acids at high-temperature conditions. Similar reaction in fullerenes leads to the formation of polyhydroxylated products. An alternative route for introducing oxygen-based functional group is the use of a basic condition. This reaction was carried out in the presence of catalytic amount of tetrabutylammonium hydroxide with benzene [69]. Amidation was done using neat reaction conditions; fullerenes are mixed with primary amines for 16–24 hours. The mechanism for the formation of products is the formation of C_{60} anion through single electron transfer process. This anion is green in color, and the color of the reaction is changed into brownish that clearly indicates the formation of product [70]. Another important reaction in fullerenes is cycloaddition. It behaves as 2π component; the 1,3-dipolar cycloaddition of azomethineylide to fullerenes produces a stable pyrrolidinofullerene complex. There is a possibility of introducing a wide range of substituents in this reaction [71]. The introduction of large number of complex molecules to fullerenes in a one-pot approach is also possible.

Other important CNM is CNTs, and those are considered an excellent vehicle for drug delivery fields. As depicted in Figure 8.3, covalent functionalization in CNTs can be further divided into two categories: (i) oxidation and end/defect functionalization and (ii) sidewall covalent functionalization. The main purpose of the oxidation is to purify the material because this process shortens the length of CNTs, producing holes with oxygenated functional groups and removing other impurities in the materials. Most importantly, this process introduces many oxygen functional groups at the end of CNTs [72, 73]. Sidewall functionalization involves the breaking of sp^2 bonding networks to form sp^3-hybridized binding in sidewalls of CNTs. Hence, it requires harsh reaction conditions, and the properties of CNTs changed very much. Many reactions are possible in CNTs, such as fluorination, carbine addition, arylation, 1,3-dipolar cycloaddition, nitrene addition, radical addition, nucleophilic addition, hydrogenation, and polymerizations [56]. Functionalization of CNTs with various groups is depicted in Figure 8.4.

8.2.2.3 Non-covalent Functionalization

Non-covalent functionalization is the interaction of core molecule with other functional materials without involving covalent bond. In this type, binding of new group/moiety includes hydrophobic interactions, π–π stacking, and electrostatic binding with parent material. The stability of the interaction is explained by energies of non-covalent/supramolecular interactions. Compared with non-covalent/supramolecular interaction, this is lower than those in covalent bonds, but when combined over large surfaces, they can have relative strength like covalent bonds. The energy ranges of different interactions were summarized in the literature [74–76]. Some of energy ranges is given here. Dissociation energies of ionic bonds range from 100 to 350 kJ/mol, organometallic metal–carbon bond ranges from 150 to 400 kJ^{-1}, and coordinate covalent bond ranges from 50 to 200 kJ^{-1} (coordination complexes or metal bonds in crown ethers). However, in noncovalent systems, the dissociation energies are different. It allows the systems to be reversible or kinetically labile and also self-correcting.

The hydrophobic nature of CNMs can interact with aliphatic parts of surfactants via hydrophobic interactions. Hence, other hydrophilic part of surfactant is ready to interact with the environment, making the material with potential change in its properties. Another important interaction is used for effective functionalization for the materials that are possessing charge sites, that is, electrostatic interactions. For example, graphene-based material (GO) are negatively charged; some positive functional groups can effectively functionalized with the GO surface. In CNTs, there are four approaches that have been taken for non-covalent functionalization such as (i) surface attachment of particles, (ii) wrapping of polymers around the CNT sidewall, (iii) endohedral functionalization, and (iv) π–π stacking [77]. Normally, conjugated polymers are used to wrap CNTs. The aromatic rings in the conjugated polymer and surface of CNTs are mainly responsible for this kind of functionalization through π–π stacking and van der Waals interactions [78]. For example, pyrene ring-based polymer was prepared via atom transfer radical polymerization, and the same was utilized for the dispersion of CNT by Petrov group [79]. This type of functionalization was also achieved through polymer wrapping method by utilizing polyaniline [80], poly-L-lysine [81], and oligothiophene-based PEG [82] and π–π interaction between the aromatic ring of functional material and π electrons of graphite sheets on the surface of CNTs [83, 84]. This type of coating will disrupt the van der Waals interactions and subsequently SWCNTs to form aggregate as bundles [85]. There are two prevalent requirements for π–π interactions: (i) the existence of π systems and (ii) geometry of interacting species.

Another important application in biomedical field is the formation of electrostatic complex with DNA and RNA. This functionalization improves the solubility of the parent material and also delivers genes into the cells [86, 87]. In graphene, there are two types of π–π interactions (face-to-face and edge-to-face arrangements) occurring between electron-rich and electron-poor regions; those mainly influence the interaction with other molecules or nanomaterials (Table 8.2).

8.2.3 CNMs for Environment Devices

Due to versatile properties of CNMs, these can be used in multiple optoelectronic devices. The most exploited property of the CNMs is mechanical flexibility with optical transparency, and it revolutionized wearable electronics industry. Additionally, CNMs have more surface area and obviously increase performance of the optoelectronic devices. These can be made as suspensions. The suspension of CNMs simplified the device making by spin coating, and it encouraged many researchers to make devices at the tabletop conditions. The suspensions also greatly reduced the cost, provide rapid deposition time, and are free to use any substrates (glass, plastic, paper, and rubber). The colloidal processing methods expand the scope of the CNMs and provided the extraordinary progress in the field of CNMs, hence not only CNM devices for environment but also the manufacturing of devices in green condition (zero energy, water processing, etc.). The tunable properties (band gap) also influenced many researchers to consider CNMs as active materials for the devices. Within the scope of the present work, we cover the overview of few devices in the field of the environmental applications (Figure 8.5). We also noted the limitations wherever possible. The discussion includes solar cell, gas sensors, humidity sensors, light-emitting

Table 8.2 Functionalized carbon nanomaterials based devices.

S. No.	Materials	Device	Purpose	Key properties	References
1	CNT–azobenzene	Solar cell	High-density batteries	Efficiency = 7.1, high energy density than Li batteries	[88]
2	MWCNTs–lysozyme	Solar cell	Counter electrode, improve the dispersion	Large surface area, efficiency 11.6%, I_{sc} = 9.23 mA/cm^2	[89]
3	CNT-COOH/TiO_2	Solar cell	Linking dye and TiO_2	I_{sc} increased 25 times, V_{oc} increased up to 100 mV	[90]
4	Polymer–CNT	Solar cell	Band gap shift and high charge mobility	Efficiency increased by 26%, I_{sc} = 0.88 mA/cm^2	[91]
5	Graphene oxide–CNT	Solar cell	Counter electrode, colloidal processing	Efficiency up to 5.29%, V_{oc} = 0.68 V, FF = 51.05%, I_{sc} = 15.25 mA/cm^2	[92]
6	CNT-COOH	Gas sensor	Inkjet processing, ethanol sensing	13 ppm detection, recovery time = 60 s, sensitivity = 24.4×10^{-4}	[93]
7	Pd-CNT	Gas sensor	H_2 sensing	400 ppm detection, response time = 5 s, scalable synthesis	[94]
8	SWNT-PABS	Gas sensor	NH_3	Selectivity, 5 ppm detection, variability 10–15%	[95]
9	PANI/acid–CNT	Gas sensor	Chloroform	50 ppm detection, reusability, response time = 2 min	[96]
10	Nano ITO–CNT	Gas sensor	NH_3	Plastic substrate, 13 ppb detection limit, threefold sensitivity	[97]
11	CNT-COOH	Humidity sensor	Humidity analysis	Paper substrate, low hysteresis, response time = 30 s, recovery time = 25 s	[98]

12	FeO-Au-CNT	Humidity sensor	H_2O and D_2O analysis	High stability, response time = 70 s, high selectivity	[99]
13	CNT/silicone composite	Humidity sensor	Humidity analysis	Response time = 30 s, recovery time = 27 s, low hysteresis, high sensitivity	[100]
14	Graphene oxide–diamine	Humidity sensor	Humidity analysis	Response time = 52 s, recovery time = 72 s, high sensitivity	[101]
15	Graphene–polyethylenimine	Humidity sensor	Humidity analysis	Graphene increases the performance 10 times, response time < 20 s, RH = 50%	[102]
16	QD/graphene	CNT LEDs	Hybrid LEDs	FWHM < 80 nm, high color purity, efficiency = 3.47 cd/A, EQE = 1.28%, L_{max} = 390 cd/m^2	[103]
17	NH_2–graphene/QD	CNT LEDs	White LEDs	Pure white CIE coordinate (0.33, 0.37), tunable emission, efficiency = 35 lm/W	[104]
18	Silane carbon dots	CNT LEDs	White LEDs	CRI = 88, pure white CIE coordinate (0.33, 0.33), efficiency = 27 lm/W	[105]
19	Octadecylamine-graphene QD	CNT LEDs	White LEDs	L_{max} = 464 cd/m^2, pure white CIE coordinate (0.27, 0.29), FRET efficiency = 11.6%	[106]
20	Propynyloxybenzyl-graphene quantum dots	CNT LEDs	White LEDs	Coordinate (0.336, 0.394), tunable white LEDs, few step functionalization	[107]
21	CNT–oxide groups	Metal absorption	CrO_4^{2-} adsorption	Adsorption capacity = 249 μmol/g, surface area adsorption capacity = 0.83 μmol/m^2	[108]

(Continued)

Table 8.2 (Continued)

S. No.	Materials	Device	Purpose	Key properties	References
22	CNT-SH	Metal absorption	Hg^{2+} and Pb^{2+}	Adsorption capacity = 65.52 mg/g	[109]
23	CNT–ethylene diamine	Metal absorption	Cd^{2+}	Adsorption capacity = 25.7 mg/g	[110]
24	CNT-PAMAM	Metal absorption	Zn^{2+} and CO^{2+}	High adsorption, adsorption capacity = 494 mg/g	[111]
25	CNT–iminodiacetic acid	Metal absorption	Pb^{2+}, As^{3+}, V^{5+}	Adsorption capacity = 1.3 ng/l	[112]
26	CNT-OH	Water purification	Sodium ion filter	Adsorption capacity = 122 mg/g	[113]
27	CNT–cellulose acetate	Water purification	Membrane filter	Pore size < 6 nm with concentration = 0.0005 wt%	[114]
28	CNT–phenylenediamine	Water purification	Membrane filter	Tuning of membrane hydrophilicity, concentration = 0.5 wt%	[115]
29	CNT–organosilane	Water purification	Membrane filter	Facile preparation process, superhydrophobic, lowest flux = 37 530 l/m^2 h bar	[116]
30	CNT–cyclodextrin	Water purification	Membrane filter	Removing organic pollutant up to 0.01 ppb, absorption efficiency >80%	[117]

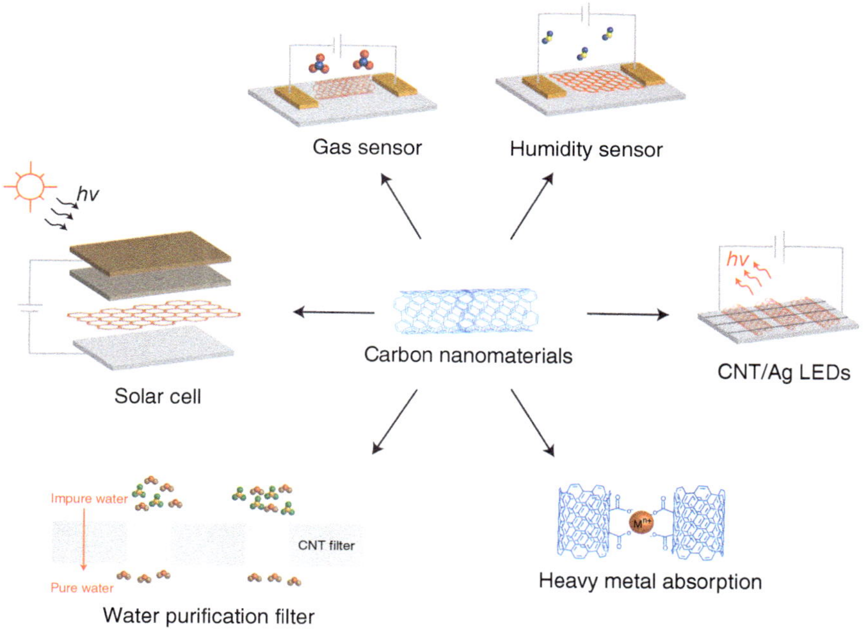

Figure 8.5 Applications of carbon nanomaterials (CNMs) in the field of environmental devices.

devices (LEDs), heat metal absorption, and water purification. The CNM-based devices are expected to provide sustainable future for human societies.

8.2.3.1 Solar Cell

The solar cell is energy-harvesting device, and it converts the solar energy into electrical energy. The solar cells are expected to play huge impact in the environment because conventional resources produce electricity by burning the large amount of coal. The burning of the coal is less efficient and releases large amount of the toxic gases. However, the solar cell generates the energy from free energy source with green energy generation. Since the electricity is used in every household, the foremost impact of CNMs might be raised from CNM-based solar cell. The CNMs can be used in many categories in the solar cell such as co-active layers, transparent electrodes, electrolyte boosters, dye sensitization, electron transport layers for electron collection, heterojunction materials, and quantum dot–CNM active layers. The CNMs provide flexibility, transparency, colloidal processing, and low-temperature processing. These functions of CNMs directly and indirectly improvise the efficiency and adoptability of the CNM solar cell over others. In this section, we discuss the few solar cell application of the functionalized CNMs and important results.

Kim and coworkers [90] reported solar cell using integration of acid-functionalized carbon nanotubes (f-CNTs) with the TiO_2 film for anchoring the dye molecules. The functionalization provided anchoring and reduces the distance between the TiO_2 and dye molecule. The integration renders the active materials water soluble and made the processing and device making simple. The anchoring enhanced the performance of up to 25% when compared with pristine CNT-based solar cell. The short circuit value also increased up to 100 mV. The charge transport properties of the f-CNTs channelized the charge flow and enhance the overall performance. Similarly, Chen and coworkers [92] demonstrated the novel idea of solar cell by integrating the GO with the CNTs. The integration was possible by the using the GO as the solvent for CNT nanoparticle preparation. The novel combination gives the efficiency of up to 5.29%, and the value was higher than the pristine CNT-based solar cell. Even though the idea was novel, the preparation was simple and can be expanded to industrial scale. The fabricated solar cell exhibited the fill factor of 51.05%, open circuit voltage of 0.68 V, and short circuit current of 15.25 mA/cm^2. The impressive results are due to the enhanced conductivity of GO that decreases the sheet resistance of the device.

8.2.3.2 Gas Sensors by Functionalized CNMs

Industrial revolution generates the toxic gases to the environments and created the huge amount air pollution. Additionally modern institutions like hospitals, manufacturing sectors, and food contaminations constantly demand the detection of the NH_3, H_2S, chloroform, hydrogen, and many more. Hence, gas sensing paramount for the environment safety and management. The CNMs can be used for gas sensing with complimented properties. The CNMs can provide the flexibility and colloidal processability with high selectivity, low limit of the detection, power efficient, and fast recovery/response time. Additionally, when compared with conventional toxic metal oxide sensors, CNTs are made up of carbon and environmentally compatible. Since the whole sensor industry moved to automation, electronically rich CNMs expected to play major role in the revolution. The basic principle of detection of gases is change in the electronic properties (e.g. resistance) by the contact of the toxic gases. Based on the signal quality, the detection and concentration of the analyte will be identified. Most of the cases, CNMs are used as active materials and have excellent sensing properties. Hence, we provide special attention to the CNM-based sensor for environment contamination. Based on the requirements, the sensors can be used for detection, screening, and isolation. In this section, we discuss the few representative sensor applications of the functionalized CNMs and important results.

Wang and coworkers [94] demonstrated that Pd nanoparticle-functionalized CNTs for the H_2 gas detection. The hydrogen gas monitoring is crucial for leakage detection and used to control H_2 fuel distribution. The device is made with colloidal solution of the Pd/CNT mixture, spin coated for device. The integration provides 400 ppm level detection of H_2 gas, response time of five seconds, and scalable synthesis. The palladium nanoparticle integration shifted the band gap by lowering the work function. The lowering of band gap enhances the charge transfer and hence increases the sensitivity. Haddon and coworkers [95] explored ammonia gas sensors based on *m*-aminobenzene sulfonic acid-functionalized SWCNTs. The ammonia is a toxic gas and is dangerous to humans and plants. The sensor has the excellent sensitivity with the detection limit of as low as 5 ppm. The f-CNT integration increased the conduction to manyfold from 5.4×10^{-7} to 5.6×10^{-3} S/cm. The key factor for the impressive performance is functionalization of CNTs with *m*-aminobenzene sulfonic acid. The NH_3 gas molecules interact with sensor and modify the band alignment by its basicity. The whole device was designed with the idea of exploitation of the functionalization of CNTs.

8.2.3.3 Humidity Sensors by Functionalized CNMs

Humidity sensor is unique class of gas sensor and is discussed in a separate section. The word humidity refers to water in gaseous form. The humidity sensor is used in many industries like in medicine, agriculture, fuel cell water detection, and electronics instrument protection from water. The water is polar molecule due to its electronegative oxygen atom. Hence, it can deposit and interact with any interacting system. Even trace amount of water can vary the chemical, mechanical, electrical, and physical properties of both nature and manmade materials. So, the high humidity spoils any system and renders to malfunction. Both in household and industries, by knowing the safety, quality and performance of the devices can be varied. Consequently, in view of safety and security, humidity sensors gained much importance. The humid sensors measure both air and temperature of the environment. The humidity detection can be resistive, capacitive, or thermal-dependent conductivity. Since the CNMs possess rich electronic properties, they are ideal for the humidity sensing application. The CNMs can act as active material (material that senses water molecule) in humidity sensor and has the excellent response. The CNM-based humidity sensors exhibited good sensitivity, low response time, longevity, and high selectivity. In this section, we discuss the few representative humidity sensor of functionalized CNMs and important results.

Zhang and coworkers [100] fabricated the humidity sensor from silicone f-CNTs. They found the humidity using water concentration vs. capacitance value changes. They have achieved the RH value of 11% with low hysteresis. The low RH value is due to the condensation of vapors on the porous CNT surfaces due to capillary condensation effects. The devices exhibited high sensitivity with the response time of 30 seconds and recovery time of 27 seconds. Response time higher than the recovery time indicates the water condensation is more and water takes more time to evaporate. Similarly Lu and coworker [101] demonstrated the diamine-functionalized GO films. To identify the humidity, they have used the electrical properties vs. concentration of water molecules. The functionalized GO synthesis was by simple room-temperature stirring and can be optimized to the industrial scale. They have used screen printing method to integrate the functionalized CNTs with the electrodes. The CNTs provide the flexibility and portability to the device. The devices showed a response time of 52 seconds and recovery time of 72 seconds with high sensitivity. Due to the f-CNT integration, they have achieved the sensitivity of up to 90% with less hysteresis. They found that the mechanism of the humid sensing is due to ion transport mechanism in f-CNTs.

8.2.3.4 LEDs by Functionalized CNMs

LEDs are modern light sources and indirectly contribute to the environment by the low energy consumption. The conventional light bulbs used 3% efficiency, and the remaining was lost as unusable heat. However, the light-emitting diodes are reaching almost 60% and dominating the industry and academia. Hence the LEDs are energy efficient and consume the less amount of electricity generated from burning large amount of coal. So, the LEDs prevent the release of large amount of greenhouse gases and protect the atmosphere from the global warming. The same energy efficiency also provides long operation time for the attached battery, giving huge impact in underdeveloped countries like Africa. The LEDs are also made from the environmental friendly materials such as indium, but the conventional bulbs uses toxic mercury. Modern materials are contributing toward the LED technology, and the CNMs are one of them. Since there is no moving components like tungsten in LEDs, it is compactly packed, and the lifetime of device will be more. The CNMs can be used to reinforce active layer, improvise the electron transport, or generate the flexible substrates. The CNMs provide better electron injection and ultraportability to the LEDs. The band gap tuning ability of CNMs further increases the efficiency of the LEDs.

Rhee and coworkers [103] reported quantum dot-functionalized graphene with amines for color-pure light-emitting diodes. By the integration of quantum dots, they could achieve the FWHM < 80 nm and give an exceptional color purity. The functionalization of the amines with the graphene generated extra energy states and created the smaller line width of output spectra. Hence, new heteroatoms N and O generate band gap tuning on graphene quantum dot architecture, controlling the degree of functionalization through the different reaction time. Through the functionalization, they have achieved the efficient of 3.47 cd/A, EQE of 1.28%, and L_{max} of 390 cd/m^2. Xie and coworkers [105] reported that silane-functionalized carbon dots for white LEDs. They have successfully tuned the properties of carbon dot and achieve green and red quantum dots by the same procedure. The silane dots were synthesized by the solvothermal methods, and LEDs are fabricated using the down-conversion method. The carbon dots generate 3D cross-links during the functionalization by simple synthetic procedure. The white light was generated by combining blue LED with dry deposition green and red quantum dots on it. The white light is very close to very pure white light with CIE coordinate of (0.33, 0.33) and with the efficiency of 27 lm/W.

8.2.3.5 Metal Absorption by Functionalized CNMs

Among the various application of the CNMs, metal absorption is an unusual application that utilizes the porosity and physical structure of CNMs. The metal distribution in the environment is nowadays everywhere due to the huge electronic waste and used light bulbs. The electronic wastes generate the cancerous chromium and lead, and the light bulbs release the large amount of the mercury. Through the poor waste management, metals entered into the water, land, and air. When compared with other pollutants, metal pollution is very dangerous. Because metals are not biodegradable, it can penetrate into aquatic life. Those metal ions can actively replace or coordinate with biological important molecules and can impair them. The metals have both interfering effect and cumulative effect in the biological system. Hence, these metals might be removed from the environment for the of animals, plants, and humans. Additionally, manufacturing sectors like metallurgy, battery, and automobile require to remove the toxic metals from the wastewater stream. Since CNMs filter the metal ion from the water, it is usually called as the molecular straws. The CNMs absorb the metals by its polarity and porosity. In this section, we discuss the few representative metal-absorbing functionalized CNMs and important results.

Su and coworkers [108] demonstrated that chromium adsorption using oxygen-functionalized CNTs. They have successfully removed the anionic CrO_4^{2-} from the wastewater. For the functionalization, CNTs were treated with acid and have COOH and oxygen functionalized. The oxygen content was found to be 22%. The adsorption capability f-CNTs 249 μmol/g was increased 15-fold than pristine CNTs. The enhanced adsorption capability is ascribed to anchoring ability of oxygen present in the surface of the functionalized CNTs. The CNTs has the nanoscale and has more surface area for the effective adsorption. The surface adsorption capacity of CNTs is 0.83 μmol/m^2. Wang et al. [112] demonstrated that iminodiacetic acid-functionalized CNTs for the metal ion adsorption. The functionalization was carried out at low-temperature stirring. Through the successful functionalization, they have removed the Pb^{2+}, As^{3+}, and V^{5+} from the water stream with the detection limit of 1.3 ng/l. The desorption is an important parameter for the reuse of the f-CNTs, and it was demonstrated using HNO_3 washing. The f-CNTs have successful selection over other similarly sized or charged cations such as Fe^{3+}, Mn^{2+}, and Zn^{2+}. The synthesized f-CNTs are completely biofriendly, and they demonstrated successfully on the food products such as shrimp and fish.

8.2.3.6 Water Purification by Functionalized CNMs

Water is an integral part of the environment, and every living being is benefited from it. Hence, improving water quality alone impacts all life in underdeveloped countries such as Somalia. At the same time providing pure water is also a major issue, and many technologies are being developed to solve the water contamination problems. The contaminations include biological, chemicals, metals, and dyes, and it is originated from various industries. Hence, water contaminants have higher risk value than any known weapon to the humanity. The contaminants can be filtered out by the various types of membrane filters. The membranes are known best to scientific and industrial community due to their low cost, easy operation, high efficiency, and reusability. Due to the huge demand, all sorts of the water filters are evolved over time, and advanced solution is evolved from functionalized CNMs. The CNMs has tunable pore size is expected to dominate the membrane purification. Additionally, CNMs has exhibited excellent selectivity and permeability for the water purification. The permeability is important because the membrane should allow the necessary biofriendly ions through the filter that impact taste and quality of water. If CNMs can meet industrial level mass production, then the technology will be indispensable among the others.

Chen and coworkers [116] explored silane-functionalized CNTs for the separation of oil from the water, for example, in the recent times, oil contamination in sea due to wreck-based leakage. The integration of silane with the CNTs provides the superhydrophobicity and has excellent water-repellent properties for the oil-repellent behavior. The functionalization carried out using the chemical method, and it generates the cross-link to entirely cover the CNTs. The functionalization not only improves the hydrophobicity but also improves the film-forming ability of the CNTs, and they achieve the films of up to 4 cm. When combined with f-CNT, the oil flammability also suppressed, and hence the fire risk is also reduced. Similarly, Durbach and coworkers [117] demonstrated cyclodextrin-functionalized CNTs for water purifications. The functionalization carried out chemically by using linker with CNT and cyclodextrin. The carboxylic group in the CNTs is reacted with OH in the cyclodextrin and provided functionalization. The CNTs showed the excellent adsorption facility of up to 80% of organochlorines. Organochlorine is an important pollutant since it is being widely present in the Earth's crust. Among the various organochlorines, trichloroethylene exhibited 99% adsorption with CNTs. The performance of functionalized CNTs is better than pristine CNTs. More related useful references are listed at the end of reference part [118–139].

8.3 Conclusion, Challenges, and Future Prospects

Since its first invention, in less than 20 years, CNMs gained attention both in academia and industry. Within the time of 20 years, many forms of CNMs were demonstrated such as fullerenes, carbon dots, graphene, nanotubes, nanoribbons, carbon nanofibers, nanohorns, and many more. We believe functionalization offers important advantage in the CNMs. Hence, we discussed and summarized the efforts of the various functionalization methods. Our approach is indicating the possibility in synthesis and highlighted its selectivity. The tutorial describes various CNM formations by hybridization, bending, and folding. We summarized and consolidated various efforts and hinted the recent advances. We suggested main and alternative approaches. The functionalization includes covalent functionalization such as side wall and defect functionalization. We also discussed non-covalent functionalization such as surface attachment, polymer wrapping, endohedral functionalization, and π–π stacking. Particularly, the physical properties and their importance in the preparation of advanced hybrid materials for structures, solar cells, fuel cells, water treatment/remediation, molecular detection and quantification in atmosphere, heavy metal absorption, etc. have been discussed. We highlighted the standard and running horse materials in the each criterion. We streamlined the synthetic procedures for making CNMs and suggest to avoid serendipitous approach. Overall, our observations indicated that functionalization could potentially revolutionize the CNM applications.

The future of CNMs is very promising, and few issues must be addressed. The issues are described below. (1) The CNMs are near to industrial acceptance and should be optimized for industrial needs. The industrial application needs simple preparation, high reproducibility, and large-scale preparation. Hence new automated methods to synthesize the CNMs has to be evolved. (2) The synthesis and functionalization is being task of specialist/expert. However, the CNMs are used in the biology, physics, and materials science. Hence the simplicity in the synthesis should be established. (3) Various electron transport mechanisms are not well established in the various CNMs. The mechanism should be explored to fully utilized potential. (4) The starting point of synthesis of CNMs is still hit-and-trial approach and serendipity. Hence, guided theoretical calculations are required to establish the foundation of synthesis of the targeted nanomaterials. (5) The CNM for environmental devices is attractive in recent times with many airborne diseases. Consequently, the cost-efficient mass production, cost-effective device manufacturing, and ultraportable monitors should be identified (as a handheld kit). (6) The scope of the CNMs is mainly limited to materials science. Through functionalization, the carbon is water soluble and can be utilized to biology effectively. The CNMs has carbon-based architecture, and biology also has carbon-based architecture. Therefore, biocompatibility should be high, and demonstrative examples must be pursued to widen the scope.

Acknowledgments

Dr. Baskaran Ganesh Kumar acknowledges P.S.R. Arts and Science College, Tamil Nadu, India, for the infrastructure and lab facilities. Dr. K. S. Prakash thanks Bharathidasan Government College for Women (Autonomous), Puducherry, for the infrastructure.

Important Websites

https://onlinelibrary.wiley.com/journal/19390041
https://www.beilstein-journals.org/bjoc/series/42
https://link.springer.com/referenceworkentry/10.1007%2F978-90-481-9751-4_48

References

1 Freeman, H., Harten, T., Springer, J. et al. (1992). Industrial pollution prevention! A critical review. *J. Air Waste Manage. Assoc.* 42 (5): 618–656.

2 Jariwala, D., Sangwan, V.K., Lauhon, L.J. et al. (2013). Carbon nanomaterials for electronics, optoelectronics, photovoltaics, and sensing. *Chem. Soc. Rev.* 42 (7): 2824–2860.

3 Xu, Q., Pu, P., Zhao, J. et al. (2015). Preparation of highly photoluminescent sulfur-doped carbon dots for Fe(III) detection. *J. Mater. Chem. A* 3 (2): 542–546.

4 Hirsch, A. (2002). Functionalization of single-walled carbon nanotubes. *Angew. Chem. Int. Ed.* 41 (11): 1853–1859.

5 Holzinger, M., Vostrowsky, O., Hirsch, A. et al. (2001). Sidewall functionalization of carbon nanotubes. *Angew. Chem. Int. Ed.* 40 (21): 4002–4005.

6 Arduini, F., Cinti, S., Scognamiglio, V. et al. (2017). How cutting-edge technologies impact the design of electrochemical (bio)sensors for environmental analysis. A review. *Anal. Chim. Acta* 959: 15–42.

7 Gao, C., Guo, Z., Liu, J.H., and Huang, X.J. (2012). The new age of carbon nanotubes: an updated review of functionalized carbon nanotubes in electrochemical sensors. *Nanoscale* 4 (6): 1948–1963.

8 Huang, J., Liu, Y., and You, T. (2010). Carbon nanofiber based electrochemical biosensors: a review. *Anal. Methods* 2 (3): 202–211.

9 Hussain, C.M. and Mishra, A.K. (2019). *Nanotechnology in Environmental Science, 2 Volumes*. Wiley.

10 Katz, E. and Willner, I. (2004). Biomolecule-functionalized carbon nanotubes: applications in nanobioelectronics. *ChemPhysChem* 5 (8): 1084–1104.

11 Minot, E.D., Yaish, Y., Sazonova, V. et al. (2003). Tuning carbon nanotube band gaps with strain. *Phys. Rev. Lett.* 90 (15): 156401.

12 Li, C.-Z., Yip, H.-L., and Jen, A.K.Y. (2012). Functional fullerenes for organic photovoltaics. *J. Mater. Chem.* 22 (10): 4161–4177.

13 Chen, L.-M., Hong, Z., Li, G., and Yang, Y. (2009). Recent progress in polymer solar cells: manipulation of polymer: fullerene morphology and the formation of efficient inverted polymer solar cells. *Adv. Mater.* 21 (14–15): 1434–1449.

14 He, Y., Chen, H.-Y., Hou, J., and Li, Y. (2010). Indene–C_{60} bisadduct: a new acceptor for high-performance polymer solar cells. *J. Am. Chem. Soc.* 132 (4): 1377–1382.

15 Yu, M.F., Files, B.S., Arepalli, S., and Ruoff, R.S. (2000). Tensile loading of ropes of single wall carbon nanotubes and their mechanical properties. *Phys. Rev. Lett.* 84 (24): 5552–5555.

16 Chen, T., Cai, Z., Yang, Z. et al. (2011). Nitrogen-doped carbon nanotube composite fiber with a core–sheath structure for novel electrodes. *Adv. Mater.* 23 (40): 4620–4625.

17 Novoselov, K.S., Geim, A.K., Morozov, S. et al. (2004). Electric field effect in atomically thin carbon films. *Science* 306: 666–669.

18 Geim, A.K. and Novoselov, K.S. (2007). The rise of graphene. *Nat. Mater.* 6: 183–191.

19 Bolotin, K.I., Sikes, K., Jiang, Z. et al. (2008). Ultrahigh electron mobility in suspended graphene. *Solid State Commun.* 146 (9–10): 351–355.

20 Lee, C., Wei, X., Kysar, J.W., and Hone, J. (2008). Measurement of the elastic properties and intrinsic strength of monolayer graphene. *Science* 321: 385–388.

21 Balandin, A.A., Ghosh, S., Bao, W. et al. (2008). Superior thermal conductivity of single-layer graphene. *Nano Lett.* 8 (3): 902–907.

22 Zhang, X., Ma, Y., Fu, S., and Zhang, A. (2019). Facile synthesis of water-soluble fullerene (C_{60}) nanoparticles via mussel-inspired chemistry as efficient antioxidants. *Nanomaterials* 9: 1647.

23 Gong, W., Jin, B., Peng, R. et al. (2015). Poly(glycidyl nitrate) and its thermal decomposition synthesis and characterization of [60]fullerene-poly(glycidyl nitrate) and its thermal decomposition. *Ind. Eng. Chem. Res.* 54: 2613–2618.

24 Guan, H.J., Peng, R.F., Jin, B. et al. (2014). Preparation and thermal performance of fullerene-based lead salt. *Bull. Korean Chem. Soc.* 35 (8): 2257–2262.

25 Chen, B.-L., Jin, B., Peng, R.-F. et al. (2014). Synthesis, characterization and thermal decomposition of fullerene-ethylenediamine dinitramide. *Chin. J. Energetic Mater.* 22 (4): 467–472.

26 Wang, N., Li, J., and Ji, G. (1996). Synthesis of trinitrophenyl C_{60} derivative. *Propellants Explos. Pyrotech.* 4 (1): 317–318.

27 Lan, Y., Jin, M., and Luo, Y. (2015). Preparation and characterization of graphene aerogel/Fe_2O_3/ammonium perchlorate nanostructured energetic composite. *J. Sol-Gel Sci. Technol.* 74 (1): 161–167.

28 Zhong, X., Jin, J., Li, S. et al. (2010). Aryne cycloaddition: highly efficient chemical modification of graphene. *Chem. Commun.* 46 (39): 7340–7342.

29 Yu, D., Yang, Y., Durstock, M. et al. (2010). Soluble P3HT-grafted graphene for efficient bilayer-heterojunction photovoltaic devices. *ACS Nano* 4 (10): 5633–5640.

30 Xu, Y., Liu, Z., Zhang, X. et al. (2009). A graphene hybrid material covalently functionalized with porphyrin: synthesis and optical limiting property. *Adv. Mater.* 21 (12): 1275–1279.

31 Liu, Z., Robinson, J.T., Sun, X., and Dai, H. (2008). PEGylated nanographene oxide for delivery of water-insoluble cancer drugs. *J. Am. Chem. Soc.* 130 (33): 10876–10877.

32 Fang, M., Wang, K., Lu, H. et al. (2009). Covalent polymer functionalization of graphene nanosheets and mechanical properties of composites. *J. Mater. Chem.* 19 (38): 7098–7105.

33 Valentini, L., Cardinali, M., Bittolo Bon, S. et al. (2010). Use of butylamine modified graphene sheets in polymer solar cells. *J. Mater. Chem.* 20 (5): 995–1000.

34 Savchenko, A. (2009). Transforming graphene. *Science* 323: 589–591.

35 Robinson, J.T., Burgess, J.S., Junkermeier, C.E. et al. (2010). Properties of fluorinated graphene films. *Nano Lett.* 10: 3001–3005.

36 Chun, K.Y., Moon, I.K., Han, J.H. et al. (2013). Highly water-soluble multi-walled carbon nanotubes amine-functionalized by supercritical water oxidation. *Nanoscale* 5: 10171.

37 Maleki, A., Hamesadeghi, U., Daraei, H. et al. (2017). Amine functionalized multi-walled carbon nanotubes: single and binary systems for high capacity dye removal. *Chem. Eng. J.* 313: 826–835.

38 Bulusheva, L.G., Sysoev, V.I., Lobiak, E.V. et al. (2019). Chlorinated holey double-walled carbon nanotubes for relative humidity sensor. *Carbon (NY)* 148: 413–420.

39 Robati, D., Mirza, B., Ghazisaeidi, R. et al. (2016). Adsorption behavior of methylene blue dye on nanocomposite multi-walled carbon nanotube functionalized thiol (MWCNT-SH) as new adsorbent. *J. Mol. Liq.* 216: 830–835.

40 Kou, Z., Guo, B., He, D. et al. (2018). Transforming two-dimensional boron carbide into boron and chlorine dual-doped carbon nanotubes by chlorination for efficient oxygen reduction. *ACS Energy Lett.* 3 (1): 184–190.

41 Paul, D.R., Koros, W.J., Liu, R.Y.F. et al. (2009). Nitrogen-doped carbon nanotube. *Science* 323: 760–764.

42 Shin, W.H., Jeong, H.M., Kim, B.G. et al. (2012). Nitrogen-doped multiwall carbon nanotubes for lithium storage with extremely high capacity. *Nano Lett.* 12 (5): 2283–2288.

43 Shi, P., Li, L., Hua, L. et al. (2017). Design of amorphous manganese oxide@multiwalled carbon nanotube fiber for robust solid-state supercapacitor. *ACS Nano* 11 (1): 444–452.

44 Mayers, B., Gates, B., Yin, Y., and Xia, Y. (2001). Functionalized carbon nanotubes for molecular hydrogen sensors. *Adv. Mater.* 13 (18): 1384–1386.

45 Penza, M., Cassano, G., Rossi, R. et al. (2007). Enhancement of sensitivity in gas chemiresistors based on carbon nanotube surface functionalized with noble metal (Au, Pt) nanoclusters. *Appl. Phys. Lett.* 90 (17): 1–4.

46 Lu, Y., Partridge, C., Meyyappan, M., and Li, J. (2006). A carbon nanotube sensor array for sensitive gas discrimination using principal component analysis. *J. Electroanal. Chem.* 593 (1–2): 105–110.

47 Lu, Y., Li, J., Han, J. et al. (2004). Room temperature methane detection using palladium loaded single-walled carbon nanotube sensors. *Chem. Phys. Lett.* 391 (4–6): 344–348.

48 Nasir, A.N.M., Yahaya, N., Zain, N.N.M. et al. (2019). Thiol-functionalized magnetic carbon nanotubes for magnetic micro-solid phase extraction of sulfonamide antibiotics from milks and commercial chicken meat products. *Food Chem.* 276: 458–466.

49 Schierz, A. and Zänker, H. (2009). Aqueous suspensions of carbon nanotubes: surface oxidation, colloidal stability and uranium sorption. *Environ. Pollut.* 157 (4): 1088–1094.

50 Zhang, X., Huang, Q., Liu, M. et al. (2015). Preparation of amine functionalized carbon nanotubes via a bioinspired strategy and their application in Cu^{2+} removal. *Appl. Surf. Sci.* 343: 19–27.

51 Di, Z.C., Ding, J., Peng, X.J. et al. (2006). Chromium adsorption by aligned carbon nanotubes supported ceria nanoparticles. *Chemosphere* 62 (5): 861–865.

52 Feng, S.A., Zhao, J.H., and Zhu, Z.P. (2008). The manufacture of carbon nanotubes decorated with ZnS to enhance the ZnS photocatalytic activity. *New Carbon Mater.* 23 (3): 228–234.

53 Ma, L.L., Sun, H.Z., Zhang, Y.G. et al. (2008). Preparation, characterization and photocatalytic properties of CdS nanoparticles dotted on the surface of carbon nanotubes. *Nanotechnology* 19 (11): 115709.

54 Peng, X., Luan, Z., Ding, J. et al. (2005). Ceria nanoparticles supported on carbon nanotubes for the removal of arsenate from water. *Mater. Lett.* 59 (4): 399–403.

55 Veličković, Z., Vuković, G.D., Marinković, A.D. et al. (2012). Adsorption of arsenate on iron(III) oxide coated ethylenediamine functionalized multiwall carbon nanotubes. *Chem. Eng. J.* 181–182: 174–181.

56 Moradi, O., Yari, M., Zare, K. et al. (2012). Carbon nanotubes: a review of chemistry principles and reactions. *Fullerenes Nanotubes Carbon Nanostruct.* 20 (2): 138–151.

57 Sun, Y.P., Fu, K., Lin, Y., and Huang, W. (2002). Functionalized carbon nanotubes: properties and applications. *Acc. Chem. Res.* 35 (12): 1096–1104.

58 Dyke, C.A. and Tour, J.M. (2004). Overcoming the insolubility of carbon nanotubes through high degrees of sidewall functionalization. *Chem. Eur. J.* 10 (4): 812–817.

59 Veca, L.M., Lu, F., Meziani, M.J. et al. (2009). Polymer functionalization and solubilization of carbon nanosheets. *Chem. Commun.* 18: 2565–2567.

60 Salavagione, H.J., Gómez, M.A., and Martínez, G. (2009). Polymeric modification of graphene through esterification of graphite oxide and poly(vinyl alcohol). *Macromolecules* 42 (17): 6331–6334.

61 Salavagione, H.J. and Martínez, G. (2011). Importance of covalent linkages in the preparation of effective reduced graphene oxide-poly(vinyl chloride) nanocomposites. *Macromolecules* 44 (8): 2685–2692.

62 Shan, C., Yang, H., Han, D. et al. (2009). Water-soluble graphene covalently functionalized by biocompatible poly-l-lysine. *Langmuir* 25 (20): 12030–12033.

63 Bardhan, N.M. (2017). 30 years of advances in functionalization of carbon nanomaterials for biomedical applications: a practical review. *J. Mater. Res.* 32 (1): 107–127.

64 He, M. and Swager, T.M. (2016). Covalent functionalization of carbon nanomaterials with iodonium salts. *Chem. Mater.* 28 (23): 8542–8549.

65 Hudson, J.L. et al. (2004). Water-soluble, exfoliated, nonroping single-wall carbon nanotubes. *J. Am. Chem. Soc.* 126 (36): 11158–11159.

66 Chattopadhyay, D., Lastella, S., Kim, S., and Papadimitrakopoulos, F. (2002). Length separation of zwitterion-functionalized single wall carbon nanotubes by GPC. *J. Am. Chem. Soc.* 124 (5): 728–729.

67 Li, J. and Zhang, Y. (2006). Large-scale aligned carbon nanotubes films. *Physica E* 33 (1): 235–239.

68 Nakamura, E. and Isobe, H. (2003). Functionalized fullerenes in water. The first 10 years of their chemistry, biology, and nanoscience. *Acc. Chem. Res.* 36 (11): 807–815.

69 Li, J., Takeuchi, A., Ozawa, M. et al. (1993). C_{60} fullerol formation catalysed by quaternary ammonium hydroxides. *J. Chem. Soc., Chem. Commun.* 23: 1784–1785.

70 Rašović, I. (2017). Water-soluble fullerenes for medical applications. *Mater. Sci. Technol.* 33: 777–794.

71 Maggini, M., Scorrano, G., and Prato, M. (1993). Addition of azomethine ylides to C_{60}: synthesis, characterization, and functionalization of fullerene pyrrolidines. *J. Am. Chem. Soc.* 115 (24): 9798–9799.

72 Barinov, A., Gregoratti, L., Dudin, P. et al. (2009). Imaging and spectroscopy of multiwalled carbon nanotubes during oxidation: defects and oxygen bonding. *Adv. Mater.* 21 (19): 1916–1920.

73 Hiura, H., Ebbesen, T.W., and Tanigaki, K. (1995). Opening and purification of carbon nanotubes in high yields. *Adv. Mater.* 7 (3): 275–276.

74 Atkins, P. and dePaula, J. (2002). *Physical Chemistry*. New York: W. H. Freeman.

75 Steed, J.W. and Atwood, J.L. (2013). *Supramolecular Chemistry*. New York: Wiley.

76 Skinner, H.A. and Connor, J.A. (1985). Metal-ligand bond-energies in organometallic compounds. *Pure Appl. Chem.* 57 (1): 79–88.

77 Star, A. et al. (2001). Preparation and properties of polymer-wrapped single-walled carbon nanotubes. *Angew. Chem. Int. Ed.* 40 (9): 1721–1725.

78 Jeon, I.-Y., Chang, D.W., Nanjundan, A.K., and Baek, J.-B. (2011). *Functionalization of Carbon Nanotubes: In Functional Molecular Nanostructures*. Germany: Springer.

79 Petrov, P., Stassin, F., Pagnoulle, C., and Jérôme, R. (2003). Noncovalent functionalization of multi-walled carbon nanotubes by pyrene containing polymers. *Chem. Commun.* 23: 2904–2905.

80 Zhang, H., Li, H.X., and Cheng, H.M. (2006). Water-soluble multiwalled carbon nanotubes functionalized with sulfonated polyaniline. *J. Phys. Chem. B* 110 (18): 9095–9099.

81 Ling, X., Wei, Y., Zou, L., and Xu, S. (2014). Functionalization and dispersion of multiwalled carbon nanotubes modified with poly-l-lysine. *Colloids Surf., A* 443: 19–26.

82 Lee, J.U., Huh, J., Kim, K.H. et al. (2007). Aqueous suspension of carbon nanotubes via non-covalent functionalization with oligothiophene-terminated poly(ethylene glycol). *Carbon* 45 (5): 1051–1057.

83 Chen, J., Liu, H., Weimer, W.A. et al. (2002). Noncovalent engineering of carbon nanotube surfaces by rigid, functional conjugated polymers. *J. Am. Chem. Soc.* 124 (31): 9034–9035.

84 Gómez, F.J, Chen, R.J., Wang, D. et al. (2003). Ring opening metathesis polymerization on non-covalently functionalized single-walled carbon nanotubes. *Chem. Commun.* 2: 190–191.

85 Curran, S.A., Ajayan, P.M., Blau, W.J. et al. (1998). A composite from poly(*m*-phenylenevinylene-*co*-2,5-dioctoxy-*p*-phenylenevinylene) and carbon nanotubes: a novel material for molecular optoelectronics. *Adv. Mater.* 10 (14): 1091–1093.

86 Singh, R., Pantarotto, D., Mccarthy, D. et al. (2005). Binding and condensation of plasmid DNA onto functionalized carbon nanotubes: toward the construction of nanotube-based gene delivery vectors. *J. Am. Chem. Soc.* 127 (12): 4388–4396.

87 He, C., Chung, T.S., and Goh, S.H. (2005). Polyethylensimine-grafted multiwalled carbon nanotubes for secure noncovalent immobilization and efficient delivery of DNA. *Angew. Chem. Int. Ed.* 44 (30): 4782–4785.

88 Kolpak, A.M. and Grossman, J.C. (2011). Azobenzene-functionalized carbon nanotubes as high-energy density solar thermal fuels. *Nano Lett.* 11 (8): 3156–3162.

89 Zhao, J., Ma, J., Nan, X., and Tang, B. (2016). Application of non-covalent functionalized carbon nanotubes for the counter electrode of dye-sensitized solar cells. *Org. Electron.* 30: 52–59.

90 Jang, S.R., Vittal, R., and Kim, K.J. (2004). Incorporation of functionalized single-wall carbon nanotubes in dye-sensitized TiO_2 solar cells. *Langmuir* 20 (22): 9807–9810.

91 Lee, R.H., Lee, L.Y., Huang, J.L. et al. (2011). Conjugated polymer-functionalized carbon nanotubes enhance the photovoltaic properties of polymer solar cells. *Colloid. Polym. Sci.* 289 (15–16): 1633–1641.

92 Ma, J., Zhou, L., Li, C. et al. (2014). Surfactant-free synthesis of graphene-functionalized carbon nanotube film as a catalytic counter electrode in dye-sensitized solar cells. *J. Power Sources* 247: 999–1004.

93 Alshammari, A.S., Alenezi, M.R., Lai, K.T., and Silva, S.R.P. (2017). Inkjet printing of polymer functionalized CNT gas sensor with enhanced sensing properties. *Mater. Lett.* 189: 299–302.

94 Yu, S., Welp, U., Hua, L.Z. et al. (2005). Fabrication of palladium nanotubes and their application in hydrogen sensing. *Chem. Mater.* 17 (13): 3445–3450.

95 Bekyarova, E., Davis, M., Burch, T. et al. (2004). Chemically functionalized single-walled carbon nanotubes as ammonia sensors. *J. Phys. Chem. B* 108 (51): 19717–19720.

96 Kar, P. and Choudhury, A. (2013). Carboxylic acid functionalized multi-walled carbon nanotube doped polyaniline for chloroform sensors. *Sens. Actuators, B* 183: 25–33.

97 Rigoni, F., Drera, G., Pagliara, S. et al. (2014). High sensitivity, moisture selective, ammonia gas sensors based on single-walled carbon nanotubes functionalized with indium tin oxide nanoparticles. *Carbon* 80: 356–363.

98 Han, J.W., Kim, B., Li, J., and Meyyappan, M. (2012). Carbon nanotube based humidity sensor on cellulose paper. *J. Phys. Chem. C* 116 (41): 22094–22097.

99 Lee, J., Mulmi, S., Thangadurai, V., and Park, S.S. (2015). Magnetically aligned iron oxide/gold nanoparticle-decorated carbon nanotube hybrid structure as a humidity sensor. *ACS Appl. Mater. Interfaces* 7 (28): 15506–15513.

100 Wang, D., Huang, Y., Cai, W. et al. (2014). Functionalized multi-wall carbon nanotubes/silicone rubber composite as capacitive humidity sensor. *J. Appl. Polym. Sci.* 131 (11).

101 Su, P.G. and Lu, Z.M. (2015). Flexibility and electrical and humidity-sensing properties of diamine-functionalized graphene oxide films. *Sens. Actuators, B* 211: 157–163.

102 BenAziza, Z., Zhang, K., Baillargeat, D., and Zhang, Q. (2015). Enhancement of humidity sensitivity of graphene through functionalization with polyethylenimine. *Appl. Phys. Lett.* 107 (13): 134102.

103 Kwon, W., Kim, Y.H., Kim, J.H. et al. (2016). High color-purity green, orange, and red light-emitting diodes based on chemically functionalized graphene quantum dots. *Sci. Rep.* 6 (1): 1–10.

104 Tetsuka, H., Nagoya, A., and Asahi, R. (2015). Highly luminescent flexible amino-functionalized graphene quantum dots@cellulose nanofiber–clay hybrids for white-light emitting diodes. *J. Mater. Chem. C* 3 (15): 3536–3541.

105 Wang, Y., Wang, K., Han, Z. et al. (2017). High color rendering index trichromatic white and red LEDs prepared from silane-functionalized carbon dots. *J. Mater. Chem. C* 5 (37): 9629–9637.

106 Kim, D.H. and Kim, T.W. (2018). Ultrahigh-luminosity white-light-emitting devices based on edge functionalized graphene quantum dots. *Nano Energy* 51: 199–205.

107 Sekiya, R., Uemura, Y., Murakami, H., and Haino, T. (2014). White-light-emitting edge-functionalized graphene quantum dots. *Angew. Chem. Int. Ed.* 53 (22): 5619–5623.

108 Xu, Y.J., Rosa, A., Xi, L.I.U., and Su, D.S. (2011). Characterization and use of functionalized carbon nanotubes for the adsorption of heavy metal anions. *New Carbon Mater.* 26 (1): 57–62.

109 Zhang, C., Sui, J., Li, J. et al. (2012). Efficient removal of heavy metal ions by thiol-functionalized superparamagnetic carbon nanotubes. *Chem. Eng. J.* 210: 45–52.

110 Vuković, G.D., Marinković, A.D., Čolić, M. et al. (2010). Removal of cadmium from aqueous solutions by oxidized and ethylenediamine-functionalized multi-walled carbon nanotubes. *Chem. Eng. J.* 157 (1): 238–248.

111 Hayati, B., Maleki, A., Najafi, F. et al. (2018). Heavy metal adsorption using PAMAM/CNT nanocomposite from aqueous solution in batch and continuous fixed bed systems. *Chem. Eng. J.* 346: 258–270.

112 Wang, J., Ma, X., Fang, G. et al. (2011). Preparation of iminodiacetic acid functionalized multi-walled carbon nanotubes and its application as sorbent for separation and preconcentration of heavy metal ions. *J. Hazard. Mater.* 186 (2–3): 1985–1992.

113 Mishra, A.K. and Ramaprabhu, S. (2011). Functionalized graphene sheets for arsenic removal and desalination of sea water. *Desalination* 282: 39–45.
114 El Badawi, N., Ramadan, A.R., Esawi, A.M., and El-Morsi, M. (2014). Novel carbon nanotube–cellulose acetate nanocomposite membranes for water filtration applications. *Desalination* 344: 79–85.
115 Rahimpour, A., Jahanshahi, M., Khalili, S. et al. (2012). Novel functionalized carbon nanotubes for improving the surface properties and performance of polyethersulfone (PES) membrane. *Desalination* 286: 99–107.
116 Gu, J., Xiao, P., Huang, Y. et al. (2015). Controlled functionalization of carbon nanotubes as superhydrophobic material for adjustable oil/water separation. *J. Mater. Chem. A* 3 (8): 4124–4128.
117 Salipira, K.L., Mamba, B.B., Krause, R.W. et al. (2007). Carbon nanotubes and cyclodextrin polymers for removing organic pollutants from water. *Environ. Chem. Lett.* 5 (1): 13–17.
118 Hussain, C.M. and Kharisov, B. (2017). *Advanced Environmental Analysis: Application of Nanomaterials*. The Royal Society of Chemistry.
119 Hussain, C.M. (2018). *Handbook of Nanomaterials for Industrial Applications*. Elsevier.
120 Hussain, C.M. (2019). *Handbook of Environmental Materials Management*, 1e. Springer International Publishing.
121 Hussain, C.M. and Keçili, R. (2019). *Modern Environmental Analysis Techniques for Pollutants*, 1e. Elsevier.
122 Hussain, C.M. (2020). *Handbook of Functionalized Nanomaterials for Industrial Applications*. Elsevier.
123 Hussain, C.M. (2020). *Handbook of Manufacturing Applications of Nanomaterials*. Elsevier.
124 Mallakpour, S. and Soltanian, S. (2016). Surface functionalization of carbon nanotubes: fabrication and applications. *RSC Adv.* 6: 109916–109935.
125 Mallakpour, S. and Behranvand, V. (2018). Synthesis of mesoporous recycled poly(ethylene terephthalate)/MWNT/carbon quantum dot nanocomposite from sustainable materials using ultrasonic waves: application for methylene blue removal. *J. Cleaner Prod.* 190: 525–537.
126 Mallakpour, S., Behranvand, V., and Mallakpour, F. (2019). Synthesis of alginate/carbon nanotube/carbon dot/fluoroapatite/TiO_2 beads for dye photocatalytic degradation under ultraviolet light. *Carbohydr. Polym.* 224: 115138.
127 Mallakpour, S. and Khadem, E. (2019). Linear and nonlinear behavior of crosslinked chitosan/N-doped graphene quantum dot nanocomposite films in cadmium cation uptake. *Sci. Total Environ.* 690: 1245–1253.
128 Mallakpour, S. and Khadem, E. (2016). Carbon nanotube–metal oxide nanocomposites: fabrication, properties and applications. *Chem. Eng. J.* 302: 344–367.
129 Mallakpour, S. and Rashidimoghadam, S. (2021). Utilization of starch and starch/carbonaceous nanocomposites for removal of pollutants from wastewater. In: *Handbook of Polymer Nanocomposites for Industrial Applications* (ed. C.M. Hussain), 477–502. Elsevier.
130 Mallakpour, S., Abdolmaleki, A., and Borandeh, S. (2017). Fabrication of amino acid-based graphene-zinc oxide (ZnO) hybrid and its application for poly(ester–amide)/graphene-ZnO nanocomposite synthesis. *J. Thermoplast. Compos. Mater.* 30: 358–380.

131 Mallakpour, S., Abdolmaleki, A., and Karshenas, A. (2017). Graphene oxide supported copper coordinated amino acids as novel heterogeneous catalysts for epoxidation of norbornene. *Catal. Commun.* 92: 109–113.

132 Mallakpour, S., Abdolmaleki, A., Mahmoudian, M. et al. (2017). Synergetic effect of synthesized sulfonated polyaniline/quaternized graphene and its application as a high-performance supercapacitor electrode. *J. Mater. Sci.* 52: 9683–9695.

133 Mallakpour, S. and Khadem, E. (2018). Construction of crosslinked chitosan/nitrogen-doped graphene quantum dot nanocomposite for hydroxyapatite biomimetic mineralization. *Int. J. Biol. Macromol.* 120: 1451–1460.

134 Abdolmaleki, A., Mallakpour, S., and Karshenas, A. (2017). Synthesis and characterization of new nanocomposites films using alanine-Cu-functionalized graphene oxide as nanofiller and PVA as polymeric matrix for improving of their properties. *J. Solid State Chem.* 253: 398–405.

135 Abdolmaleki, A., Mallakpour, S., and Mahmoudian, M. (2017). Preparation and evaluation of edge selective sulfonated graphene by chlorosulfuric acid as an active metal-free electrocatalyst for oxygen reduction reaction in alkaline media. *ChemistrySelect* 34: 11211–11217.

136 Abdolmaleki, A., Mallakpour, S., and Karshenas, A. (2017). Facile synthesis of glucose-functionalized reduced graphene oxide (GFRGO)/poly(vinyl alcohol) nanocomposites for improving thermal and mechanical properties. *Mater. Sci. Eng., B* 217: 26–35.

137 Mallakpour, S. and Khodadadzadeh, L. (2018). Chapter 7: Biocompatible and biodegradable chitosan nanocomposites loaded with carbon nanotubes. In: *Biodegradable and Biocompatible Polymer Composites Processing, Properties and Applications* (ed. N.G. Shimpi), 187–221. Sawston, Cambridge, United Kingdom: Elsevier, Woodhead Publishing. https://doi.org/10.1016/B978-0-08-100970-3.00007-9.

138 Mallakpour, S. and Khadem, E. (2019). Chapter 8: Carbon nanotubes for heavy metals removal. In: *Composite Nanoadsorbents* (eds. G. Kyzas and A.C. Mitrpoulos), 181–210. Amsterdam, Netherlands: Elsevier. https://doi.org/10.1016/B978-0-12-814132-8.00009-5, eBook ISBN: 9780128141335.

139 Mallakpour, S. and Rashidimoghadam, S. (2019). Chapter 9: Carbon nanotubes for dyes removal. In: *Composite Nanoadsorbents* (eds. G. Kyzas and A.C. Mitrpoulos), 211–244. Amsterdam, Netherlands: Elsevier. https://doi.org/10.1016/B978-0-12-814132-8.00010-1. eBook ISBN: 9780128141335.

9

Functionalized Magnetic Carbon Nanomaterials for Environmental Remediation

Ambika[1] *and Pradeep Pratap Singh*[2]

[1] *University of Delhi, Hansraj College, Department of Chemistry, Delhi, 110007, India*
[2] *University of Delhi, Swami Shraddhanand College, Department of Chemistry, Delhi, 110036, India*

9.1 Introduction

Rapid industrialization and exponential population growth have led to an increased discharge of wastewater containing various pollutants such as heavy metals, dyes, bacteria/viruses, other organic pollutants that pose harmful effects on the environment [1–10]. These harmful and toxic contaminants may lead to various life-threatening diseases [11]. Therefore, there is an urgent need to develop a technology that can effectively control water pollution. Nanotechnology can provide opportunities for environmental remediation [12–44]. In the past few decades, nanocomposites (NCs) have attracted the attention of the scientific community due to their high surface area to volume ratio of the reinforcing nanoparticles compared to conventional composites. The reinforcing material can be prepared using different particles, sheets, or fibers [45]. Carbon nanomaterials (CNMs), such as fullerenes, carbon nanotubes (CNTs), and graphene, have been utilized as ideal reinforcing fillers for the preparation of different carbon nanocomposites (CNCs) for their unique physical and chemical properties, which offer potential advances in various fields [46, 47]. Recently, magnetic carbon nanocomposites (MCNCs) have attracted scientists' attention due to their porous structure, easy recovery and recyclability and benign biocompatibility which may be attributed to their property to respond to an external stimulus such as a magnetic or electrical field. The agglomeration of iron oxide can be reduced by functionalization with carbon nanoparticles. Magnetic separation could replace filtration, reducing the loss of materials. These nanomaterials have been employed for various applications such as catalysis, energy storage, sensors, electronics, biomedical field, and environmental remediation (Scheme 9.1) [48, 49]. This chapter aims to discuss the preparation of functionalized carbon based magnetic nanocomposites (CMNCs) and their application in environmental remediation.

Environmental Applications of Carbon Nanomaterials-Based Devices, First Edition.
Edited by Shadpour Mallakpour and Chaudhery Mustansar Hussain.

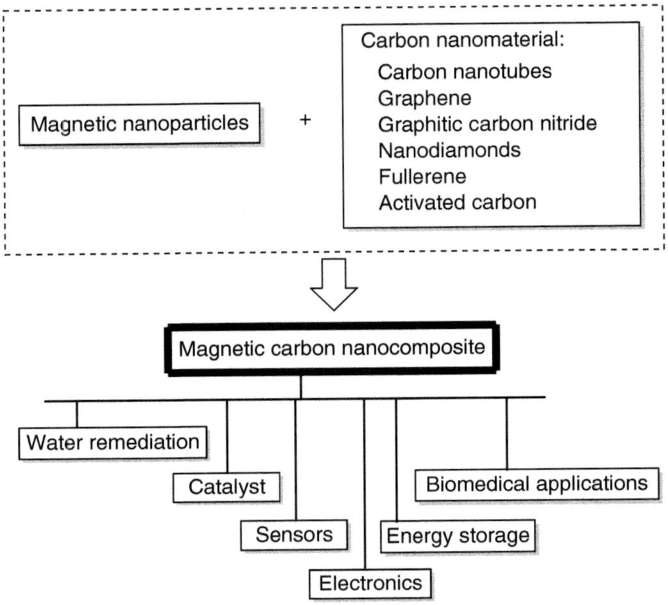

Scheme 9.1 Applications of magnetic carbon nanocomposites. Source: Based on Siddiqui et al. [48], Janusz et al. [49].

9.2 Types of Carbon-Based Magnetic Nanocomposites Used in Pollutants Removal from Environment

The carbon nanomaterial functionalization can modify their physicochemical properties suitable for chemical and biological applications. Various MCNCs such as multi-walled carbon nanotubes (MWCNTs)/magnetite hybrids, carbon-coated magnetic nanoparticles, iron oxide/graphene NC, iron-encapsulated carbon sphere can be formed by different techniques [50–53]. Few of these NCs have been discussed below.

9.2.1 Carbon Nanotubes Based Magnetic Nanocomposites

CNTs are carbon nanomaterials having diameters in nanometers. They possess extraordinary mechanical, thermal, and electrical properties [54, 55]. The surface functionalization could be enhanced by the decoration of CNTs with various compounds such as inorganic nanoparticles to either the external/internal surface of the CNTs [56–58]. These NCs have been utilized in electronic, aerospace, bio-imaging applications, and environmental remediation [59, 60].

9.2.2 Graphene and Its Derivative Based Magnetic Nanocomposites

Graphene is an allotrope of carbon with a honeycomb-like structure and a zero bandgap. It possesses a higher surface area as compared to other carbonaceous materials [17]. Graphene possesses unique morphology, chemical structure, and electronic properties [61–65]. Graphene-based multifunctional magnetic NCs can be prepared by the surface functionalization of graphene and its derivative using magnetic nanoparticles by different methods. These NCs have been employed for various applications such as data storage and security/sensors to biomedical applications and environmental remediation [66–69].

9.2.3 Fullerenes Based Magnetic Nanocomposites

A fullerene consists of hollow molecular cages, balls, or tubes of strongly bonded carbon atoms connected by single and double bonds [70]. Unique properties of fullerenes and iron oxide magnetic nanoparticles could be combined to increase their activities. These NCs possess potential applications in different fields such as nano-electronics, energy storage, energy conversion, catalysis, sensors, medical biology, photoelectrochemistry, environmental remediation [71–75].

9.2.4 Nanodiamond-Filled Magnetic Nanocomposites

Nanodiamond particles (NDs) are carbon nanomaterials with spherical particles of various sizes and surface structure, high surface area, high thermal, electrochemical, and optical properties [76, 77]. These NCs have been employed for biomedical and environmental remediation [78–81].

9.2.5 Graphitic Carbon Nitride Based Magnetic Nanocomposites

Graphitic carbon nitride (g-C3N4) is a graphite analog and stable allotrope of carbon nitride at ambient conditions. It possesses a tunable bandgap (1.8–2.7 eV), efficient intercalation of salts, high semi-conductive properties, and good biocompatibility. Recently, g-C_3N_4 has been employed as nanomaterial support for the preparation of NCs [82]. The different properties of the NCs and their separation process can be enhanced by incorporating magnetic compounds over the g-C_3N_4 sheets. These NCs have been employed as heterogeneous catalysts in environmental remediation [83, 84].

9.3 Different Processing Methods for Magnetic Carbon-Based Nanocomposites

The magnetic carbon-based NCs can be processed by employing the following methods.

9.3.1 Melt Blending

In this method, a viscous liquid is prepared by melting the polymer pellets followed by the high shear force for the dispersal of the nanofillers such as CNTs and graphene. [47, 85]. This method is simple, eco-friendly, cost-effective, but high shear force and high temperature can deteriorate NC property. Therefore the mixing process should be fine-tuned to achieve optimum conditions.

9.3.2 Hydrothermal Method

This method is a solution reaction-based approach employed to prepare a broad range of nanomaterials such as magnetic nanomaterials and CNCs. The particle size, morphology, and properties of the product depend on the type and concentration of the precursors, solvent, capping agents, temperature, and reaction time [86]. It offers several advantages, relatively mild operating conditions, one-step synthetic procedure, eco-friendly, and good dispersion in solution.

9.3.3 Co-Precipitation Method

This method is used to prepare different nanomaterials such as metal oxides nanoparticles, mixed metal, or metal ceramics NCs. The precursors such as inorganic salts are dissolved in water or any other solvents to form a homogenous solution of ions (Scheme 9.2). These salts undergo nucleation and growth phases and precipitates as hydroxides or oxalates after achieving the critical concentration of species [87]. The particles' size and shape depend on the pH, temperature, and concentration of the solution [88]. After precipitation, filtration and washing are carried out, followed by calcination to convert hydroxide into oxides with a definite crystalline structure. The agglomeration which tampers the particle size of the composites can be avoided by adding surfactants [89].

9.3.4 *In Situ* Polymerization

In this method, before polymerization, the preformed magnetic nanostructures are dispersed directly into the monomer or monomer solution. It allows for inorganic particles to be either physically trapped within the matrix or covalently bound to the polymers [90]. The controlled synthesis of the polymeric matrix in the presence of the inorganic phase allows the adjustment of composite characteristics such as functionality, stability, or biocompatibility required, especially in biomedical applications (Scheme 9.3).

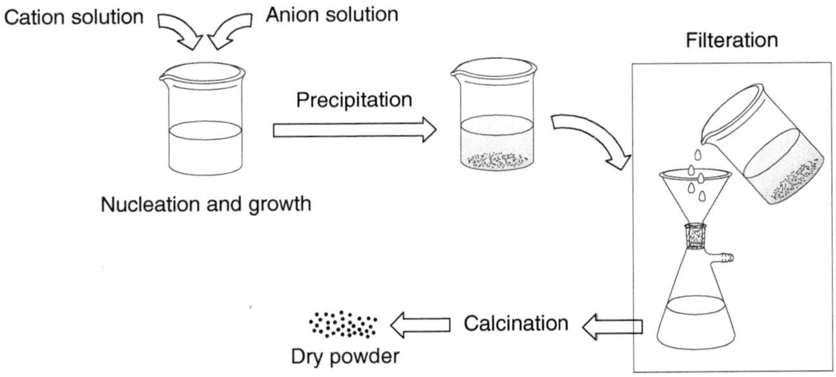

Scheme 9.2 Schematic representation of co-precipitation method.

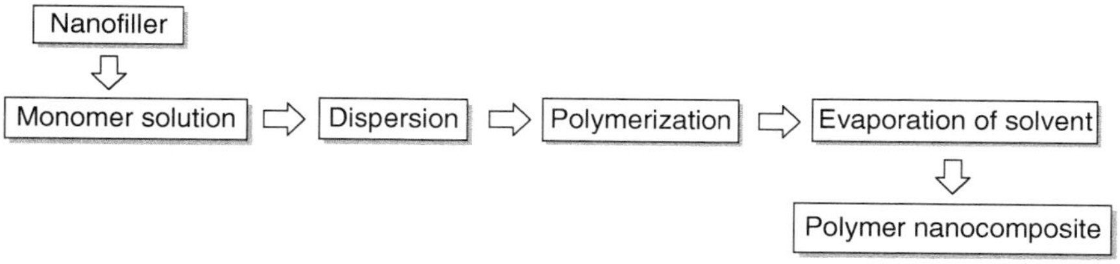

Scheme 9.3 Different steps involved in *in situ* polymerization.

9.3.5 Sol–Gel Method

Sol–gel method can be employed to produce micro and nanostructures, either film of metal oxides or colloidal powder. The sol can be prepared by using inorganic salts or metal-organic compounds [91]. The hydrolysis and condensation reactions lead to form an integrated network as gel [91]. Various properties like shape, structure, and size depends on pH, solvent, temperature, and concentration. (Scheme 9.4). The main advantages of sol–gel process are low-temperature operation, which prevents oxidation of precursors, and low cost.

9.4 Applications of Magnetic Carbon-Based Nanocomposites

Due to the exponential population growth and rapid industrialization, the environment is getting contaminated with hazardous pollutants like heavy metals, dyes, pesticides, and antibiotics. Various new materials such as MCNCs and treatment approaches have been developed for the remediation of the environment. Few examples have been discussed below.

9.4.1 Adsorption of Heavy Metals

Heavy metals are among the most common contaminants released into the environment through various natural and anthropogenic activities like mining and other industries. These heavy metals are highly toxic and hazardous to human beings and animals. They tend to bioaccumulate into the environment and consequently enter the food chains,

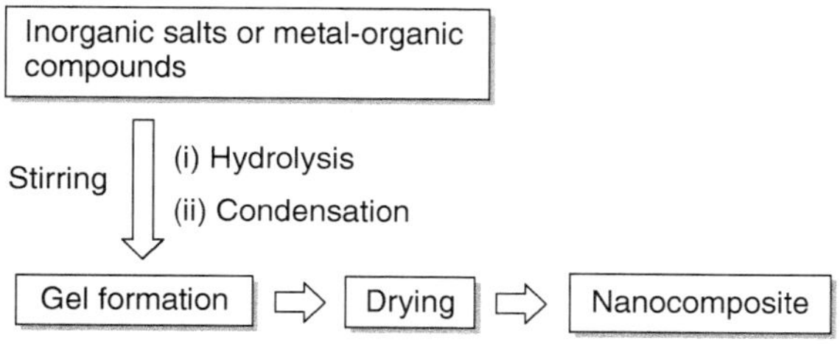

Scheme 9.4 Sol–gel method for the production of different types of nanostructures.

resulting in several life-threatening diseases. Various MCNCs have been developed and employed for the removal of these contaminants. For example, magnetic bio-nanocomposite have been fabricated using eggshell membrane-zeolitic imidazolate framework (ZIF-67@Fe_3O_4@ESM) and was utilized for the removal of Cu(II) and basic red 18 (BR 18). The above NC displayed a higher adsorption rate and removal percentage than pure ZIF-67 [92]. Sulfonated magnetic graphene oxide (MGO) composite has been used for the removal of Cu(II) from water with high sorption capacity at pH 4.68 [93]. Magnetite/porous graphene NCs displayed high adsorption capacities of dyes methyl violet (MV) and heavy metal such as Pb(II) and Cu(II) ions from wastewater [94]. Copper, chromium, iron, lead, and cyanate can be removed from wastewater by employing immobilized cyanate hydratase on magnetic-multiwall carbon nanotubes (m-MWCNTs-rTl-Cyn) (Scheme 9.5) [95].

Graphene oxide (GO) anchored on magnetite was utilized for Pb(II) uptake with a sorbtion capacity of 588.24 mg/g [96]. Magnetic graphene oxide functionalized tryptophan (MGO-Trp) NCs have been employed for the adsorption of Pb(II), Cd(II), and Cr(VI) ions with high adsorption capacity under microwave irradiation. Pb(II) and Cd(II) can be efficiently adsorbed at pH 7, while Cr(VI) can be extracted at pH 2 [97]. K_2CO_3 activated porous biochar (KBC) loaded with nano-zero-valent iron (nZVI)/nano-α-hydroxy-iron oxide (nHIO) has been employed for the adsorption of cadmium (Cd) in aqueous solution. The strong synergistic effects between biochar and nZVI/nHIO enhances the removal efficiency of cadmium (Cd) in an aqueous solution. The loading of nZVI/nHIO reduces the specific surface area, whereas the adsorption capacity of Cd increased more than threefold compared to KBC [98]. Magnetic CS/GO NCs can be used for the pH-dependent removal of contaminants such as heavy metal ions [Cr(VI), Pb(II)] and dyes [Orange (O 7) and methylene blue (MB)] from wastewater. The above NC could be recycled for several cycles without appreciable loss in activity [99–102]. Similarly, Cr(VI) can be adsorbed by applying magnetically modified GO/CS/ferrite NCs at pH 2.0 with high adsorption capacity [103]. Chitin/magnetite/MWCNTs demonstrated higher surface area and significant magnetic properties than the natural chitin. The above NC has been employed to remove Cr(VI) [104]. Magnetic chitosan (Chi@Fe_3O_4) NC modified with graphene oxide (Chi@Fe_3O_4GO) has been employed as an adsorbent for the

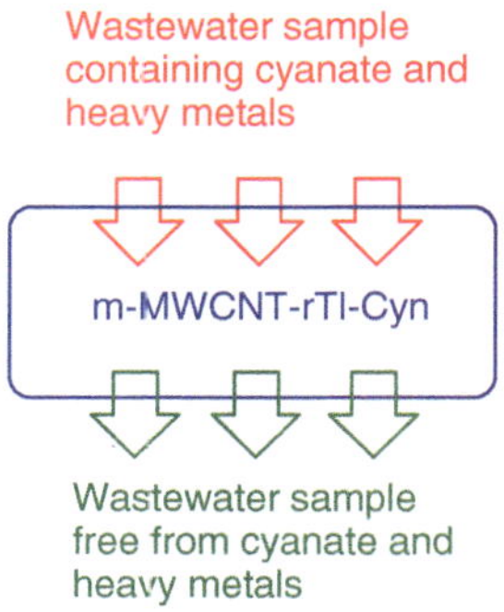

Scheme 9.5 Removal of heavy metals from wastewater by employing immobilized cyanate hydratase on magnetic-multiwalled carbon nanotubes. Source: Based on Ranjan et al. [95].

removal of Cr(VI) from water. The adsorbents exhibited varied Cr(VI) removal efficiency at solution pH 2. The maximum adsorption capacity was found to be 142.32 and 100.51 mg/g for Chi@Fe_3O_4 and Chi@Fe_3O_4GO, respectively. The above NC can be reused up to four cycles [105]. Graphene sheets decorated with zero-valent iron nanoparticles were utilized for pH-dependent Cr(VI) removal [106]. Magnetic graphene NC displayed good sorption ability for Cr(VI) removal from water treatment compared to pristine graphene [107]. Graphene nanohybrid (magnetic) anchored via core@double shell nanoparticles of crystalline iron oxide and inside a shell of amorphous Si–S–O has been employed for the nearly 100% Cr(VI) uptake within five minutes at 3.0 g/l dose and under acidic pH (1–3) [108, 109]. MGO, magnetic chemically-reduced graphene (MCRG), and magnetic annealing-reduced graphene (MARG) have been employed for the adsorption of tetracycline (TC), Cd(II), and As(V). The adsorption depends on the pH of the solution. The above NCs displayed the following order of adsorption capability MGO > MCRG > MARG. The superior adsorption capability of MGO could be attributed to its high dispersibility, thin nanosheets, various O-containing functional groups, and electrostatic attraction. The presence of TC can significantly suppress the adsorption of As(V), whereas As(V) does not have any significant impact on the TC adsorption [110]. Chitosan-magnetic-graphene oxide (CMGO) NC demonstrated efficient arsenic adsorption. The removal efficiency increases from acidic to neutral pH, whereas it starts decreasing at alkaline conditions. The NC could be easily separated and recovered using an external magnetic field [111]. A macroporous 3D GO hydrogel (MGOH) has been employed for the adsorption of As(III) and As(V) with a maximum sorption capacity of 25.1 and 74.2 mg/g, respectively [112]. $Ni_{0.5}Zn_{0.5}Fe_2O_4$ Ferrite (NZF) and $Ni_{0.5}Zn_{0.5}Fe_2O_4$/CNTsNC (CNZF) were utilized as magnetic adsorbents for the removal of arsenic(V) contaminated anions from simulated industrial wastewater with maximum adsorption capacities of 56 and 66 mg/g, respectively [113]. A novel magnetic NCs of iron oxide nanoparticles encapsulated by carbon layer (Fe_3O_4@C) has been used to remove arsenate. The removal efficiency of the Fe_3O_4@C NC increases with increasing carbon concentration [114]. GO/ferric hydroxide material was utilized for >95% sorption of As(V) in polluted drinking water [115–117]. GO-magnetite has been employed to remove As(III) and As(V) in water. The sorption capacity reported was 99.9% at 1.0 μg/l concentration [118]. Mn-magnetite-graphene hybrid magnetic material was employed for 99.9% arsenite removal at pH = 7. The used sorbent could be separated by a magnetic field [119]. Manganese ferrite/graphene oxide ($MnFe_2O_4$/GO) NCs have been used to adsorb nickel ions (Ni^{2+}), MB from water with a high adsorption capacity of 152.67 (Ni^{2+}) and 89.29 mg/g, (MB) respectively and could be easily separated by an external magnetic field [120]. Cobalt oxide nanoparticles have been adsorbed from the water by utilizing magnetized α-cellulose-fibers from wastepaper biomass (M-WP-αCFs) with high adsorption capacity [121]. Magnetic polyethyleneimine-functionalized carbon quantum dots/$MnFe_2O_4$ (PECQDs/$MnFe_2O_4$) NCs have been employed to adsorb heavy metal ions such as U(VI) from wastewater. The above NC displayed superior construction stability and adsorption capacity of 194 mg/g (Scheme 9.6) [122].

PANI/reduced graphene oxide (RGO)/Fe_3O_4 adsorbent has been utilized to remove Hg(II) in an aqueous solution. The above NC displayed enhanced adsorption capacity as compared to RGO/Fe_3O_4. The NC could be easily separated and regenerated at pH 2 [123].

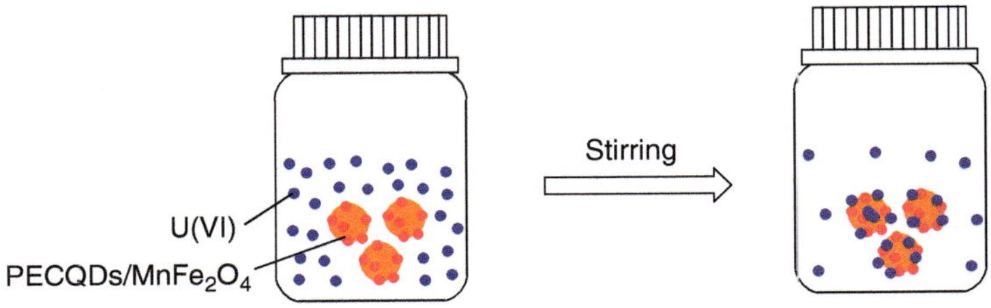

Scheme 9.6 Adsorption of U(VI) from wastewater using PECQDs/$MnFe_2O_4$ NCs. Source: Modified from Huang et al. [122].

9.4.2 Adsorption of Organic Dye

Dyes are one of the toxic and harmful pollutants which contribute to the pollution of the environment. They have been widely employed in various industries like cosmetic, petroleum, printing, and textile tannery. These dye molecules are difficult to biodegrade and may pose several ill effects on living organisms [124]. Recently, various MCNCs have been employed for the removal of toxic dyes from wastewater. For example, fuchsine dye (FS) can be adsorbed effectively on magnetic CS/GO NC at pH 5.5. This may be attributed to its amido groups become protonated in acidic conditions and increase its solubility in water [125]. Fuchsine dye can also be efficiently removed from water by employing graphene-magnetite nanohybrid (G/Fe_3O_4) NC. The above NC can be reused for five cycles without any loss in sorption capacity [126]. G/Fe_3O_4 can also be utilized to remove various dyes like congo red (CR), MB, and pararosaniline (PR) [127, 128]. Different dyes like victoria blue (VB), methyl orange (MO), brilliant yellow (BY), neutral red (NR), alizarin red (AR), and safranine T (ST) can be adsorbed from water using graphene sulfonic-magnetite with excellent sorption capacity. It may be due to the electron-donating effect of the amino group of cationic dyes. The sorbent could be recycled for six cycles without any loss in sorption capacities [129]. Magnetic Co-Fe_2O_4 graphene sheets displayed efficient removal of MO in water with 71.54 mg/g sorption capacity [130]. Fe^0 nano-crystals intercalated within graphene lakes have been applied for the adsorption of both cationic and anionic dyes such as MO, MB, crystal violet (CV), and Rhodamine B (RhB) dye from wastewater [131]. Similarly, MO and MB can be removed from water by employing a simple and cost-effective industrial waste-based novel magnetic nanocomposite (Fe@GPHC). The above NC displayed a maximum adsorption capacity of 11 and 8.9 mg/g for MB and MO, respectively [132]. Mesoporous carbon/cobalt ferrite NC has been applied for the removal of both cationic dye (MB) and anionic dye (CR) within a wide solution pH range (from 3 to 11). The adsorption of the dyes could be attributed to the ion-π interaction between microcrystal graphite in the NC and dye molecules. The above NC could be easily recovered by applying a magnetic field without significant efficiency loss [133]. Methyl blue (MTB) can be efficiently adsorbed using a reusable magnetic CS-GO NC [102, 134]. MB can be removed from water using a cost-effective magnetic β-cyclodextrin-chitosan-GO NC with high sorption capacity [135]. Similarly, MB can also be removed from wastewater by utilizing magnetic cellulose/GO NC under alkaline conditions [136]. A highly recyclable, magnetic TiO_2/graphene NC (TiO_2/Fe_3O_4/GN-SO_3H) was employed for the efficient photocatalytic removal of MB from water. The above NC demonstrated good water-dispersibility and convenient magnetic separation

[137]. MB and *p-tert*-butylphenol (PTBP) could be efficiently removed using magnetic iron oxide/graphene oxide (MGO) NC. The above NC could be reused for at least five cycles without appreciable removal efficiency loss [138]. Magnetic MOF@GO and MOF@CNT hybrid NCs have been utilized for the removal of MB from water. The above NCs displayed enhanced adsorption capacity toward MB compared to the parent materials, which may be attributed to the synergetic effect of covalent bonding between the parent materials and to the unique features of the nanoscale metal-organic framework (MOF) [139]. GO/g-C_3N_4-Fe_3O_4 NC was employed to adsorb TC antibiotic and MB dye. The above NC displayed pH-dependent adsorption with a maximum adsorption capacity of 120 mg/g at pH = 3 for TC and MB 220 mg/g at pH = 9. The high adsorption efficiency for TC and MB could be attributed to π–π and hydrogen bonding interaction. The adsorbed TC and MB on the GO/g-C_3N_4-Fe_3O_4 surface were recovered and reused up to five cycles (Scheme 9.7) [140].

Magnetic multi-walled carbon nanotubes/cerium dioxide (MMWCNTs-CeO2) NC has been employed for the enhanced photocatalytic degradation of MB in the presence of H_2O_2. The introduction of magnetic may lead to the easy separation of photocatalyst from the solution [141]. A magnetic CNT-reduced graphene oxide-silver NC was employed for the adsorption and simultaneous catalytic reduction of MB and 4-nitrophenol (4-NP). The above NC could be easily recovered using a magnet and recycled for more than 15 cycles without any significant loss in activity [142]. $CoFe_2O_4$/mpg-C_3N_4 NC was utilized for the photocatalytic degradation of malachite green (MG), MB, acid orange 7 (AO 7), and RhB under UV-light irradiation. The above NC could be magnetically recovered and reused for five runs without an appreciable loss in activity [143]. Graphene quantum dots integrated with magnetic graphene oxide (GO/Fe_3O_4/GQD) has been utilized as a filler and was incorporated *in situ* within PEG-1500 induced macroporous polyvinyl alcohol/carboxymethyl cellulose hydrogel beads. The above hydrogel NC can be employed to remove cationic pollutants such as MB, RhB [144]. Magnetically recoverable g-C_3N_4/Fe_3O_4/Ag_2WO_4/AgBr (gCN/M/AgW/AgBr) NCs, as efficient visible-light-active photocatalysts, have been utilized for the degradations of RhB, MB, MO, and fuchsine pollutants [145]. Activated carbon/α-Fe_2O_3 NC has been applied for the adsorption of acid yellow 17 (AY 17) dye [146]. $ZnFe_2O_4$@g-C_3N_4 NCs have been employed for the photocatalytic degradation of RhB in the presence of H_2O_2 under visible light irradiation. The above NC displayed a high degradation rate as compared to the pure $ZnFe_2O_4$. The NCs could also be recycled for at

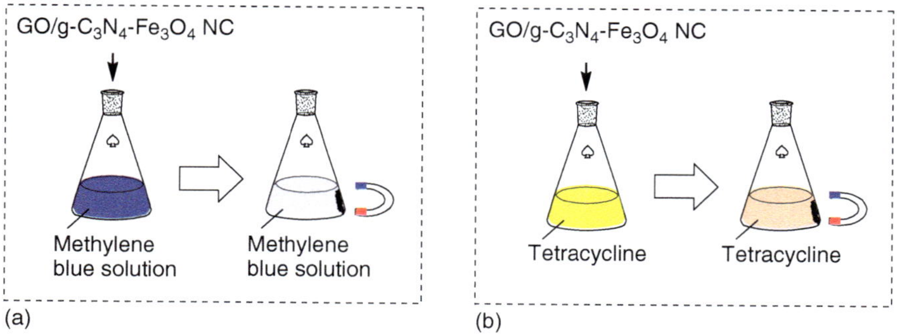

Scheme 9.7 Removal of a) methylene blue; (b) tetracycline using GO/g-C_3N_4-Fe_3O_4 NC. Source: Based on Sahoo et al. [140].

least five cycles without appreciable loss of photocatalytic activity [147]. A magnetic amine-functionalized carbon nanotube (NH_2-CNT/Fe_2O_3)-zeolitic imidazolate framework-8 (ZIF-8) NCs (NH_2-CNT/Fe_2O_3/ZIF-8: NCFZ) with different amounts of NH_2-CNT/Fe_2O_3 (5, 10, and 15 wt% denoted as NCFZ-5, NCFZ-10, and NCFZ-15) were applied for selective removing cationic dyes (MG and RhB) from a binary system [148]. Magnetic zeolitic imidazolate framework-8/cobalt ferrite/graphene oxide (ZIF-8/$CoFe_2O_4$/GO: ZCG) NCs have been applied for the removal ability of MG [149]. A reusable magnetite-RGO (MRGO) hybrid material displayed excellent MG and RhB dyes removal. The above NC can be recycled for several cycles without any appreciable loss in activity [150]. (ZnO/CuO)/RGO ternary NCs have been used for the photocatalytic degradation of RhB dye and 4-chlorophenol under visible light irradiation [151]. Highly porous N/S doped magnetic carbon aerogel (N/S-MCA) displayed 98–99% removal of bisphenol-A (BPA) from aqueous solution at pH 7 [152]. A magnetic titanium nanotube/carbon nanotube NC (magnetite TNT@CNT) demonstrated oxidative degradation of BPA from high saline polycarbonate plant wastewater (PCW) using catalytic wet peroxide oxidation (CWPO). The NC could be reused for four consecutive cycles with a very slight loss in its catalytic activity [153]. *N,N′*-carbonyldiimidazole functionalized β-cyclodextrin ligand (β-CD-CDI) was assembled on the surface of Fe_3O_4 by Zn-N interaction to prepare magnetic bifunctional β-cyclodextrin NC (Fe_3O_4@β-CD-CDI). The above NC can be employed for the adsorption and degradation of persistent organic pollutants in an aqueous solution. Further, the NC can be employed as a heterogeneous catalyst to remove BPA [154]. Graphitic carbon nitride modified MCNCs (Fe_3O_4@C/g-C_3N_4) has been used prepared for the activation of peroxymonosulfate to degrade AO 7. The above NCs displayed enhanced AO 7 degradation efficiency with easy magnetic separation, good stability, and reusability [155]. The magnetite/carbon NC has been employed for the adsorption of Nylosan blue (NB), Chromazurol S (Ch S), and basic red 2 (BR 2) dyes from aqueous solutions. The above NC displayed enhanced removal efficiency with maximum adsorption capacities of 223.82, 114.68, and 286.91 mg/g for NB, Ch S, BR 2, respectively. The NC could be regenerated and reused for seven cycles [156]. Anionic azo dye Reactive black 5 (RB 5) can be effectively adsorbed using CS/RGO mesoporous NC. The enhanced adsorption may be attributed to the strong electrostatic interactions, hydrogen bonding, and Van der Waals forces [157].

9.4.3 Other Organic Pollutants

Toxic pollutants other than heavy metals and dyes are disposed of in huge volumes into the water bodies from various industrial activities such as agricultural runoffs, oil refining, and medicines. These include herbicides, pesticides, aromatic compounds, phenols, amines, polyhalogenated compounds, and antibiotics[158]. Recently, MCNCs have been employed for the removal of various pollutants. For example, cationic, anionic, and organic pollutants such as copper, phosphate, and toluene in highly acidic water could be removed by employing magnetic NC alginate beads [159]. The carbon-dots/fatty acid-coated magnetic nanoparticles (MNPs) (CDs/C_{11}-Fe_3O_4) displayed higher adsorption performance for benzo[*a*]pyrene as compared to C_{11}-Fe_3O_4 NCs [160]. Phenol-formaldehyde/polydopamine/modified halloysite nanotubes (PF/PDA/HNTs) hybrid NCs were utilized to adsorb phenolic compounds. The enhanced adsorption capacity for phenol by above NC could be attributed to the

H-bonding and hydrophobic interactions of PF matrix with phenols [161]. Fe_3O_4@Chitosan-AgNP NC has been used as an effective solid-phase catalyst for the reduction of *p*-nitrophenol. The above NC could be magnetically separated and reused for more than 15 cycles. Also, it displayed an excellent antifungal activity against agricultural pathogens such as *Colletotrichum coccodes*, *Aspergillus niger*, and *Pyricularia* sp. [162]. The superparamagnetic carbon/$ZnFe_2O_4$ (C/$ZnFe_2O_4$) NCs have been employed to remove naphthalene and 2-naphthol from the aquatic system. The NC could be used for four cycles without appreciable loss of activity [163]. Oxidized multi-walled carbon nanotube (OMWCNT) with non-magnetic property and OMWCNT-Fe_3O_4, and OMWCNT-κ-carrageenan-Fe_3O_4 NCs with magnetic property have been used for the adsorption of diquat dibromide (DQ) herbicide from water. OMWCNT displayed faster adsorption and higher maximum adsorption capacity value than magnetic adsorbents. The maximum adsorption capacity value of OMWCNT was approximately 2.8-fold higher than magnetic OMWCNT-Fe_3O_4 and 5.4-fold higher than magnetic OMWCNT-κ-carrageenan-Fe_3O_4 at 25 °C. Furthermore, OMWCNT's highest herbicide adsorption capacity could be regenerated and reused at least five times [164]. 2,4-Dichlorophenoxyacetic acid can be efficiently removed from aqueous solution using magnetically separable Fe_3O_4/polyaniline/polypyrrole NCs [165]. Ciprofloxacin (CIP) can be effectively degraded by employing a reusable CuS/Fe_2O_3/Mn_2O_3 magnetic NC. The above NC displayed higher catalytic performance as compared to bare CuS and Fe_2O_3/Mn_2O_3 composite. The degradation efficiency of CIP depends on the common anions in order of $Cl^- > SO_4^{2-} > NO_3^- > HCO_3^- >$ Humic acid [166]. Sulfamethoxazole (SMX) can be adsorbed using mesoporous silica-magnetic graphene oxide NC material (MGO-Si). The adsorption of SMX decreases with increases in pH, which may be due to the electrostatic repulsion between the dissociated SMX and MGO-Si at basic pH [167]. Biowaste-derived substances derived carbon-coated magnet-sensitive nanoparticles (bio-based substances-green compost [BBS-GC]/carbon shell) has been employed to adsorb polycyclic aromatic hydrocarbons [168]. MOFs-derived Fe/C nanomaterial (DMOFs) has been used to catalyze gamma irradiation-induced degradation of antibiotics, cephalosporin C (CEP-C), and sulfamethazine (SMT) in an aqueous solution [169]. Amino-functionalized MIL-88B(Fe)-based porous carbon has been applied for CIP's adsorption [170]. Fe/C/Pd NPs have been used for the removal. Fe/C/Pd NPs displayed enhanced adsorption of the halogenated organic pollutants (eosin Y) in comparison to Fe/C NPs [171]. Humic acid (HA) could be adsorbed by employing magnetic polyaluminium chloride (M-PAC)-MWCNT NC from an aqueous solution. The above NC displayed enhanced adsorption compared to MWCNTs, which may be attributed to the interaction between PAC and HA through hydrogen-bond and electrostatic attraction [172]. Organic pollutants (e.g. aniline) can be removed efficiently from an aqueous solution by applying magnetic MWCNT/ferrite ($NiFe_2O_4$) NC. Moreover, the above NC displayed could be easily separated from the aqueous solution by employing an ordinary magnet [173].

9.5 Future Prospects

MCNCs possess high specific surface areas, excellent electrical, optical, thermal, and chemical activity. These NCs offer huge environmental remediation opportunities due to their high adsorption capacity and easy and convenient recycling using a permanent magnet. However, the development of new and simple synthetic, cost-effective, and environmentally benign methodologies is needed to prepare various MCNCs with largely saturated magnetization and high specific surface area. The research should be focused on the large-scale commercial production of these NCs. Further, research should be emphasized to develop anti-corrosive coating materials on the surface of magnetic NCs with many adsorption sites for different pollutants, which can prevent proton permeation to the magnetic core. However, the coating layer should. Research should emphasize the development of high-value and safer products, having optimized properties with more sustainable and environmentally friendly materials. Also, various research studies should be based on treating real natural and wastewater from industries to evaluate the efficiency and applicability of diverse MCNCs in wastewater treatment. The cytotoxicity and risk assessment for various MCNCs should also be studied to understand their effect on the health of human beings, living organisms, and ecosystems.

9.6 Conclusions

Magnetic carbon-NCs offer opportunities to develop better environmental remediation materials due to their unique physical and chemical properties. These NCs could be prepared using melt blending, hydrothermal, co-precipitation, *in situ* polymerization, and sol–gel method. Also, these NCs possess a significant ability to be utilized as a photocatalyst for water treatment. These NCs possess huge potential to be used in environmental remediation due to their large adsorption capacity and convenient recycling using a permanent magnet. These NCs can remove various types of toxic contaminants such as heavy metals, dyes, pesticides, and antibiotics. Hence, they can be used as a substitute in industries for the purification of wastewater.

Important Websites

https://www.google.com/search?q=Functionalized+Magnetic+Carbon+Nanomaterials+For+Environmental+Remediation&rlz=1C1CHBF_enUS858US858&oq=Functionalized+Magnetic+Carbon+Nanomaterials+For+Environmental+Remediation&aqs=chrome..69i57.1243j0j7&sourceid=chrome&ie=UTF-8.

https://www.google.com/search?q=Functionalized+Magnetic+Carbon+ Nanomaterials+For+Environmental+Remediation&newwindow=1& rlz=1C1CHBF_enUS858US858&sxsrf=ALeKk01yBAOloXSZyBra- SZGZGzGeaBSmQ:1603724326537&source=lnms&tbm=isch&sa=X&ved=2ahUKEwiAzrPWwtLsAhXqQxUIHc-UCpYQ_AUoA3oECAYQBQ&biw=1368&bih=802.

References

1 Reddy, K.R. and Cameselle, C. (eds.) (2009). *Electrochemical Remediation Technologies for Polluted Soils, Sediments and Groundwater*. New Jersey: Wiley.

2 Vikrant, K., Kumar, V., Vellingiri, K., and Kim, K.H. (2019). Nanomaterials for the abatement of cadmium(II) ions from water/wastewater. *Nano Res.* 12: 1489–1507.

3 Vikrant, K. and Kim, K.H. (2019). Nanomaterials for the adsorptive treatment of Hg(II) ions from water. *Chem. Eng. J.* 358: 264–282.

4 Maitlo, H.A., Kim, J.H., Kim, K.H. et al. (2019). Metal-air fuel cell electrocoagulation techniques for the treatment of arsenic in water. *J. Cleaner Prod.* 207: 67–84.

5 Zhao, B., Shao, Q., Hao, L. et al. (2018). Yeast template synthesized Fe-doped cerium oxide hollow microspheres for visible photodegradation of acid orange 7. *J. Colloid Interface Sci.* 511: 39–47.

6 Zhao, J., Ge, S., Liu, L. et al. (2018). Microwave solvothermal fabrication of zirconia hollow microspheres with different morphologies using pollen templates and their dye adsorption removal. *Ind. Eng. Chem. Res.* 57: 231–241.

7 Zhao, J., Ge, S., Pan, D. et al. (2018). Solvothermal synthesis, characterization and photocatalytic property of zirconium dioxide doped titanium dioxide spinous hollow microspheres with sunflower pollen as bio-templates. *J. Colloid Interface Sci.* 529: 111–121.

8 Zhao, J., Ge, S., Pan, D. et al. (2019). Microwave hydrothermal synthesis of In_2O_3-ZnO nanocomposites and their enhanced photoelectrochemical properties. *J. Electrochem. Soc.* 166: H3074–H3083.

9 Wang, W., Hao, X., Chen, S. et al. (2018). pH-responsive Capsaicin@chitosan nanocapsules for antibiofouling in marine applications. *Polymer* 158: 223–230.

10 Zhang, P., Ge, T., Yang, H. et al. (2018). Antifouling of titania nanostructures in real maritime conditions. *Sci. Adv. Mater.* 10: 1216–1223.

11 Chansuvarn, W. and Jainae, K. (2018). Determination of contaminated heavy metals in lacquer thinner. *Appl. Mech. Mater.* 879: 144–148.

12 Hussain, C.M. and Kharisov, B. (eds.) (2017). *Advanced Environmental Analysis-Application of Nanomaterials*. The Royal Society of Chemistry.

13 Hussain, C.M. (ed.) (2018). *Handbook of Nanomaterials for Industrial Applications*. Elsevier.

14 Hussain, C.M. and Mishra, A.K. (2019). *Nanotechnology in Environmental Science*, vol. 2. Wiley.

15 Hussain, C.M. (ed.) (2019). *Handbook of Environmental Materials Management*, 1e. Springer International Publishing.

16 Hussain, C.M. and Kecili, R. (eds.) (2019). *Modern Environmental Analysis Techniques for Pollutants*, 1e. Elsevier.

17 Hussain, C.M. (ed.) (2020). *Handbook of Functionalized Nanomaterials for Industrial Applications*. Elsevier.

18 Hussain, C.M. (ed.) (2020). *Handbook of Manufacturing Applications of Nanomaterials*. Elsevier.

19 Mallakpour, S. and Soltanian, S. (2016). Surface functionalization of carbon nanotubes: fabrication and applications. *RSC Adv.* 6: 109916–109935.

20 Mallakpour, S. and Behranvand, V. (2018). Synthesis of mesoporous recycled poly(ethylene terephthalate)/MWNT/carbon quantum dot nanocomposite from sustainable materials using ultrasonic waves: application for methylene blue removal. *J. Cleaner Prod.* 190: 525–537.

21 Mallakpour, S., Behranvand, V., and Mallakpour, F. (2019). Synthesis of alginate/carbon nanotube/carbon dot/fluoroapatite/TiO_2 beads for dye photocatalytic degradation under ultraviolet light. *Carbohydr. Polym.* 224: 115138.

22 Mallakpour, S. and Khadem, E. (2019). Linear and nonlinear behavior of crosslinked chitosan/N-doped graphene quantum dot nanocomposite films in cadmium cation uptake. *Sci. Total Environ.* 690: 1245–1253.

23 Mallakpour, S. and Khadem, E. (2016). Carbon nanotube-metal oxide nanocomposites: fabrication, properties and applications. *Chem. Eng. J.* 302: 344–367.

24 Mallakpour, S. and Rashidimoghadam, S. (2020). Utilization of starch and starch/carbonaceous nanocomposites for removal of pollutants from wastewater. In: *Handbook of Polymer Nanocomposites for Industrial Applications* (ed. C.M. Hussain), 477–502. Amsterdam: Elsevier.

25 Mallakpour, S., Abdolmaleki, A., and Borandeh, S. (2017). Fabrication of amino acid-based graphene-zinc oxide (ZnO) hybrid and its application for poly(ester–amide)/graphene-ZnO nanocomposite synthesis. *J. Thermoplast. Compos. Mater.* 30: 358–380.

26 Mallakpour, S., Abdolmaleki, A., and Karshenas, A. (2017). Graphene oxide supported copper coordinated amino acids as novel heterogeneous catalysts for epoxidation of norbornene. *Catal. Commun.* 92: 109–113.

27 Mallakpour, S., Abdolmaleki, A., Mahmoudian, M. et al. (2017). Synergetic effect of synthesized sulfonated polyaniline/quaternized graphene and its application as a high-performance supercapacitor electrode. *J. Mater. Sci.* 52: 9683–9695.

28 Mallakpour, S. and Khadem, E. (2018). Construction of crosslinked chitosan/nitrogen-doped graphene quantum dot nanocomposite for hydroxyapatite biomimetic mineralization. *Int. J. Biol. Macromol.* 120: 1451–1460.

29 Abdolmaleki, A., Mallakpour, S., and Karshenas, A. (2017). Synthesis and characterization of new nanocomposites films using alanine-Cu-functionalized graphene oxide as nanofiller and PVA as polymeric matrix for improving of their properties. *J. Solid State Chem.* 253: 398–405.

30 Abdolmaleki, A., Mallakpour, S., and Borandeh, S. (2016). Improving interfacial interaction of l-phenylalanine-functionalized graphene nanofiller and poly(vinyl alcohol) nanocomposites for obtaining significant membrane properties: morphology, thermal, and mechanical studies. *Polym. Compos.* 37: 1924–1935.

31 Abdolmaleki, A., Mallakpour, S., and Mahmoudian, M. (2017). Preparation and evaluation of edge selective sulfonated graphene by chlorosulfuric acid as an active metal-free electro-catalyst for oxygen reduction reaction in alkaline media. *ChemistrySelect* 34: 11211–11217.

32 Abdolmaleki, A., Mallakpour, S., and Karshenas, A. (2017). Facile synthesis of glucose-functionalized reduced graphene oxide (GFRGO)/poly(vinyl alcohol) nanocomposites for improving thermal and mechanical properties. *Mater. Sci. Eng., B* 217: 26–35.

33 Mallakpour, S. and Khodadadzadeh, L. (2018). Biocompatible and biodegradable chitosan nanocomposites loaded with carbon nanotubes. In: *Biodegradable and Biocompatible Polymer Composites Processing, Properties and Applications* (ed. N.G. Shimpi), 187–221. Sawston, Cambridge: Elsevier, Woodhead Publishing.

34 Mallakpour, S. and Khadem, E. (2019). Carbon nanotubes for heavy metals removal. In: *Composite Nanoadsorbents* (eds. G. Kyzas and A.C. Mitrpoulos), 181–210. Amsterdam: Elsevier.

35 Mallakpour, S. and Rashidimoghadam, S. (2019). Carbon nanotubes for dyes removal. In: *Composite Nanoadsorbents* (eds. G. Kyzas and A.C. Mitrpoulos), 211–244. Amsterdam: Elsevier.

36 Ambika and Singh, P.P. (2020). Advances in carbon nanomaterial-based green nanocomposites. In: *Emerging Carbon-Based Nanocomposites for Environmental Applications* (eds. C.M. Hussain, A.K. Mishra and S.B. Mishra), 175–202. Wiley.

37 Ambika and Singh, P.P. (2020). Natural polymers-based hydrogels for adsorption applications. In: *Natural Polymers-Based Green Adsorbents for Water Treatment* (ed. S. Kalia). Amsterdam: Elsevier (In press).

38 Singh, P.P. and Ambika (2018). Ruthenium compounds: a new approach in nanochemistry. In: *Ruthenium Chemistry* (eds. A.K. Mishra and L. Mishra), 91–110. Singapore: Pan Stanford.

39 Singh, P.P. and Ambika (2018). Nanotechnology: an emerging field for sustainable water resources. In: *Nanotechnology for Sustainable Water Resources* (eds. C.M. Hussain and A.K. Mishra), 73–101. Wiley.

40 Singh, P.P. and Ambika (2017). Biopolymers: recent trends and their applications. In: *Biopolymers: Structure, Performance and Applications* (eds. A.K. Mishra, C.M. Hussain and S.B. Mishra), 271–286. New York: Nova.

41 Singh, P.P. and Ambika (2016). Polymeric nanospheres in organic waste removal. In: *Smart Materials for Waste Water Applications* (ed. A.K. Mishra), 237–256. New Jersey: Wiley.

42 Singh, P.P. and Ambika (2018). Nanotechnology: greener approach for sustainable environment. In: *Nanotechnology in Environmental Science* (eds. C.M. Hussain and A.K. Mishra), 805–824. Wiley-VCH.

43 Singh, P.P. and Ambika (2018). Environmental remediation by nano adsorbents based polymer nanocomposite. In: *New Polymer Nanocomposites for Environmental Remediation* (eds. C.M. Hussain and A.K. Mishra), 223–241. Amsterdam: Elsevier.

44 Singh, P.P. and Ambika (2018). Dimensions of nanocomposites in pollution control. In: *Nanocomposites for Pollution Control* (eds. C.M. Hussain and A.K. Mishra), 107–126. Singapore: Pan Stanford.

45 Khan, W.S., Hamadneh, N.N., and Khan, W.A. (2016). Polymer nanocomposites – synthesis techniques, classification and properties. In: *Science and Applications of Tailored Nanostructures*, 1e (ed. P.D. Sia), 50–66. Cheshire, UK: One Central Press.

46 Singh, P.P. and Ambika (2021). Carbon based nanocomposites: preparation and application in environmental pollutants removal. In: *Environmental Remediation Through Carbon Based Nano Composites*, 1e (eds. M. Jawaid, A. Ahmad, N. Ismail and M. Rafatullah), 203–229. Singapore: Springer.

47 Ambika and Singh, P.P. (2020). Carbon nanocomposites: the potential heterogeneous catalysts for organic transformations. *Curr. Org. Chem.* 24: 1–20.

48 Siddiqui, M.T.H., Nizamuddin, S., Baloch, H.A. et al. (2018). Synthesis of magnetic carbon nanocomposites by hydrothermal carbonization and pyrolysis. *Environ. Chem. Lett.* 16: 821–844.

49 Janusz, J.T., Nikos, N.G., Zolnierkiewicz, G. et al. (2019). Magnetic properties of TiO_2/graphitic carbon nanocomposites. *Rev. Adv. Mater. Sci.* 58: 107–122.

50 Jia, B., Gao, L., and Sun, J. (2007). Self-assembly of magnetite beads along multiwalled carbon nanotubes via a simple hydrothermal process. *Carbon* 45: 1476–1481.

51 Zhang, L., Ni, Q.Q., Natsuki, T., and Fu, Y. (2009). Carbon nanotubes/magnetite hybrids prepared by a facile synthesis process and their magnetic properties. *Appl. Surf. Sci.* 255: 8676–8681.

52 Zhang, S., Niu, H., Hu, Z. et al. (2010). Preparation of carbon coated Fe_3O_4 nanoparticles and their application for solid-phase extraction of polycyclic aromatic hydrocarbons from environmental water samples. *J. Chromatogr. A* 1217: 4757–4764.

53 Wang, M.Y., Shen, T., Wang, M. et al. (2014). One-pot synthesis of α-Fe_2O_3 nanoparticles-decorated reduced graphene oxide for efficient nonenzymatic H_2O_2 biosensor. *Sens. Actuators, B* 190: 645–650.

54 Lourie, O., Cox, D.M., and Wagner, H.D. (1998). Buckling and collapse of embedded carbon nanotubes. *Phys. Rev. Lett.* 81: 1638–1641.

55 Lukic, B., Seo, J.W., Bacsa, R.R. et al. (2005). Catalytically grown carbon nanotubes of small diameter have a high Young's modulus. *Nano Lett.* 5: 2074–2077.

56 Kuang, Q., Li, S.F., Xie, Z.X. et al. (2006). Controllable fabrication of SnO_2-coated nanotubes by chemical vapor multiwalled carbon deposition. *Carbon* 44: 1166–1172.

57 Han, L., Wu, W., Kirk, F.L. et al. (2004). A direct route toward assembly of nanoparticle-carbon nanotube composite materials. *Langmuir* 20: 6019–6025.

58 Qu, L.T., Dai, L., and Osawa, E. (2006). Shape/size-controlled syntheses of metal nanoparticles for site-selective modification of carbon nanotubes. *J. Am. Chem. Soc.* 128: 5523–5532.

59 Zhou, L., Pan, S., Chen, X. et al. (2014). Kinetics and thermodynamics studies of pentachlorophenol adsorption on covalently functionalized Fe_3O_4@SiO_2-MWCNTs core-shell magnetic microspheres. *Chem. Eng. J.* 257: 10–19.

60 Choi, J.H., Nguyen, F.T., Barone, P.W. et al. (2007). Multimodal biomedical imaging with asymmetric single-walled carbon nanotube/iron oxide nanoparticle complexes. *Nano Lett.* 7: 861–867.

61 Mayorov, A.S., Gorbachev, R.V., Morozov, S.V. et al. (2011). Micrometer-scale ballistic transport in encapsulated graphene at room temperature. *Nano Lett.* 11: 2396–2399.

62 Stoller, M.D., Park, S., Zhu, Y. et al. (2008). Graphene-based ultra-capacitors. *Nano Lett.* 8: 3498–3502.

63 Li, X., Zhu, Y., Cai, W. et al. (2009). Transfer of large-area graphene films for high-performance transparent conductive electrodes. *Nano Lett.* 9: 4359–4363.

64 Lee, E., Hong, J.Y., Kang, H., and Jang, J. (2012). Synthesis of TiO_2 nanorod-decorated graphene sheets and their highly efficient photocatalytic activities under visible-light irradiation. *J. Hazard. Mater.* 219: 13–18.

65 Moser, J., Barreiro, A., and Bachtold, A. (2007). Current-induced cleaning of graphene. *Appl. Phys. Lett.* 91: 163513.

66 Yang, M.Q., Zhang, N., Pagliaro, M., and Xu, Y.J. (2014). Artificial photosynthesis over graphene-semiconductor composites. Are we getting better. *Chem. Soc. Rev.* 43: 8240–8254.

67 Georgakilas, V., Otyepka, M., Bourlinos, A.B. et al. (2012). Functionalization of graphene: covalent and non-covalent approaches, derivatives and applications. *Chem. Rev.* 112: 6156–6214.

68 Kubacka, A., Fernandez-Garcia, M., and Colon, G. (2011). Advanced nano-architectures for solar photocatalytic applications. *Chem. Rev.* 112: 1555–1614.

69 Tong, H., Ouyang, S., Bi, Y. et al. (2012). Nano-photocatalytic materials: possibilities and challenges. *Adv. Mater.* 24: 229–251.

70 Nasir, S., Hussein, M.Z., Zainal, Z., and Yusof, N.A. (2018). Carbon-based nanomaterials/allotropes: a glimpse of their synthesis, properties and some applications. *Materials* 11: 295.

71 Innocenzi, P. and Brusatin, G. (2001). Fullerene-based organic–inorganic nanocomposites and their applications. *Chem. Mater.* 13: 3126–3139.

72 Djordjevic, A., Merkulov, D.S., Lazarevic, M. et al. (2018). Enhancement of nano titanium dioxide coatings by fullerene and polyhydroxy fullerene in the photocatalytic degradation of the herbicide mesotrione. *Chemosphere* 196: 145–152.

73 Li, J., Zhou, S.L., Hong, G.B., and Chang, C.T. (2013). Hydrothermal preparation of P25-graphene composite with enhanced adsorption and photocatalytic degradation of dyes. *Chem. Eng. J.* 219: 486–491.

74 Dai, K., Yao, Y., Liu, H. et al. (2013). Enhancing the photocatalytic activity of lead molybdate by modifying with fullerene. *Mol. Catal.* 374–375: 111–117.

75 Shan, S.J., Zhao, Y., Tang, H., and Cui, F.Y. (2017). A mini-review of carbonaceous nanomaterials for removal of contaminants from wastewater. *IOP Conf. Ser.: Earth Environ. Sci.* 68: 012003.

76 Mochalin, V.N., Shenderova, O., Ho, D., and Gogotsi, Y. (2012). The properties and applications of nanodiamonds. *Nat. Nanotechnol.* 7: 11–23.

77 Mochalin, V.N. and Gogotsi, Y. (2015). Nanodiamond-polymer composites. *Diamond Relat. Mater.* 58: 161–171.

78 Zhang, Q., Mochalin, V.N., Neitzel, I. et al. (2011). Fluorescent PLLA-nanodiamond composites for bone tissue engineering. *Biomaterials* 32: 87–94.

79 Zhang, Q., Mochalin, V.N., Neitzel, I. et al. (2012). Mechanical properties and biomineralization of multifunctional nanodiamond-PLLA composites for bone tissue engineering. *Biomaterials* 33: 5067–5075.

80 Brady, M.A., Renzing, A., Douglas, T.E.L. et al. (2015). Development of composite poly(lactide-*co*-glycolide)-nanodiamond scaffolds for bone cell growth. *J. Nanosci. Nanotechnol.* 15: 1060–1069.

81 Parizek, M., Douglas, T.E., Novotna, K. et al. (2012). Nanofibrous poly(lactide-*co*-glycolide) membranes loaded with diamond nanoparticles as promising substrates for bone tissue engineering. *Int. J. Nanomed.* 7: 1931–1951.

82 Zhao, Z., Sun, Y., and Dong, F. (2015). Graphitic carbon nitride based nanocomposites: a review. *Nanoscale* 7: 15–37.

83 Zhu, J., Xiao, P., Li, H., and Carabineiro, S.A.C. (2014). Graphitic carbon nitride: synthesis, properties, and applications in catalysis. *ACS Appl. Mater. Interfaces* 6 (19): 16449–16465.

84 Wang, A., Wang, C., Fu, L. et al. (2017). Recent advances of graphitic carbon nitride-based structures and applications in catalyst, sensing, imaging, and LEDs. *Nano-Micro Lett.* 9: 47.

85 Al Sheheri, S.Z., Al-Amshany, Z.M., Al Sulami, Q.A. et al. (2019). The preparation of carbon nanofillers and their role on the performance of variable polymer nanocomposites. *Des. Monomers Polym.* 22: 8–53.

86 Chiu, C.T. and Chen, D.H. (2018). One-step hydrothermal synthesis of three-dimensional porous Ni–Co sulfide/reduced graphene oxide composite with optimal incorporation of carbon nanotubes for high performance supercapacitors. *Nanotechnology* 29: 175602.

87 Stankic, S., Suman, S., Haque, F., and Vidic, J. (2016). Pure and multi metal oxide nanoparticles: synthesis, antibacterial and cytotoxic properties. *J. Nanobiotechnol.* 14: 73.

88 Rangari, V. (2011). Polymer nanocomposite materials for structural applications. In: *Advances in Nanocomposites: Synthesis, Characterization and Industrial Applications* (ed. B. Reddy), 61–84. London: IntechOpen.

89 Jadhav, A.P., Kim, C.W., Cha, H.G. et al. (2009). Effect of different surfactants on the size control and optical properties of $Y_2O_3:Eu^{3+}$ nanoparticles prepared by coprecipitation method. *J. Phys. Chem. C* 113: 16652–16657.

90 Pavlidou, S. and Papaspyrides, C.D. (2008). A review on polymer-layered silicate nanocomposites. *Prog. Polym. Sci.* 33: 1119–1198.

91 Singh, P.P. and Ambika (2018). Ceramic materials: preparation, properties and applications. In: *Sol-Gel Based Nanoceramic Materials: Preparation, Properties and Applications* (ed. A.K. Mishra), 1–31. Singapore: Pan Stanford.

92 Mahmoodi, N.M., Taghizadeh, M., Taghizadeh, A. et al. (2019). Bio-based magnetic metal-organic framework nanocomposite: ultrasound-assisted synthesis and pollutant (heavy metal and dye) removal from aqueous media. *Appl. Surf. Sci.* 480: 288–299.

93 Hu, X.J., Liu, Y.G., Wang, H. et al. (2013). Removal of Cu(II) ions from aqueous solution using sulfonated magnetic graphene oxide composite. *Sep. Purif. Technol.* 108: 189–195.

94 Bharath, G., Alhseinat, E., Ponpandian, N. et al. (2017). Development of adsorption and electrosorption techniques for removal of organic and inorganic pollutants from wastewater using novel magnetite/porous graphene-based nanocomposites. *Sep. Purif. Technol.* 88: 206–218.

95 Ranjan, B., Pillai, S., Permaul, K., and Singh, S. (2019). Simultaneous removal of heavy metals and cyanate in a wastewater sample using immobilized cyanate hydratase on magnetic-multiwall carbon nanotubes. *J. Hazard. Mater.* 363: 3–80.

96 Yang, X., Chen, C., Li, J. et al. (2012). Graphene oxide-iron oxide and reduced graphene oxide-iron oxide hybrid materials for the removal of organic and inorganic pollutants. *RSC Adv.* 2: 8821–8826.

97 Mahmoud, M.E., Osman, M.M., Gehan, H.A.A., and Nabil, M. (2020). Microwave-assisted adsorption of Cr(VI), Cd(II) and Pb(II) in presence of magnetic graphene oxide-covalently functionalized-tryptophan nanocomposite. *J. Alloys Compd.* 823: 153855.

98 Zhu, L., Tong, L., Zhao, N. et al. (2019). Coupling interaction between porous biochar and nano zero valent iron/nano α-hydroxyl iron oxide improves the remediation efficiency of cadmium in aqueous solution. *Chemosphere* 219: 493–503.

99 Debnath, S., Maity, A., and Pillay, K. (2014). Magnetic chitosan-GO nanocomposite: synthesis, characterization and batch adsorber design for Cr(VI) removal. *J. Environ. Chem. Eng.* 2: 963–973.

100 Fan, L., Luo, C., Sun, M. et al. (2013). Highly selective adsorption of lead ions by water-dispersible magnetic chitosan/graphene oxide composites. *Colloids Surf., B* 103: 523–529.

101 Sheshmani, S., Ashori, A., and Hasanzadeh, S. (2014). Removal of acid orange 7 from aqueous solution using magnetic graphene/chitosan: a promising nano-adsorbent. *Int. J. Biol. Macromol.* 68: 218–224.

102 Fan, L., Luo, C., Sun, M. et al. (2012). Preparation of novel magnetic chitosan/graphene oxide composite as effective adsorbents toward methylene blue. *Bioresour. Technol.* 114: 703–706.

103 Phiri, J., Johansson, L.S., Gane, P., and Maloneya, T. (2018). A comparative study of mechanical, thermal and electrical properties of graphene, graphene oxide- and reduced graphene oxide-doped microfibrillated cellulose nanocomposites. *Composites Part B* 147: 104–113.

104 Salam, M.A. (2017). Preparation and characterization of chitin/magnetite/multiwalled carbon nanotubes magnetic nanocomposite for toxic hexavalent chromium removal from solution. *J. Mol. Liq.* 233: 197–232.

105 Subedi, N., Lahde, A., Abu-Danso, E. et al. (2019). A comparative study of magnetic chitosan (Chi@Fe_3O_4) and graphene oxide modified magnetic chitosan (Chi@Fe_3O_4GO) nanocomposites for efficient removal of Cr(VI) from water. *Int. J. Biol. Macromol.* 137: 948–959.

106 Jabeen, H., Chandra, V., Jung, S. et al. (2011). Enhanced Cr(VI) removal using iron nanoparticle decorated graphene. *Nanoscale* 3: 3583–3585.

107 Zhu, J., Wei, S., Chen, M. et al. (2013). Magnetic nanocomposites for environmental remediation. *Adv. Powder Technol.* 24: 459–467.

108 Zhu, C., Han, L., Hu, P., and Dong, S. (2012). *In situ* loading of well-dispersed gold nanoparticles on two-dimensional graphene oxide/SiO_2 composite nanosheets and their catalytic properties. *Nanoscale* 4: 1641–1646.

109 Zhu, J., Wei, S., Gu, H. et al. (2012). One-pot synthesis of magnetic graphene nanocomposites decorated with core@double-shell nanoparticles for fast chromium removal. *Environ. Sci. Technol.* 46: 977–985.

110 Huang, D., Wu, J., Wang, L. et al. (2019). Novel insight into adsorption and co-adsorption of heavy metal ions and an organic pollutant by magnetic graphene nanomaterials in water. *Chem. Eng. J.* 358: 1399–1409.

111 Sherlala, A.I.A., Raman, A.A.A., Bello, M.M., and Buthiyappan, A. (2019). Adsorption of arsenic using chitosan magnetic graphene oxide nanocomposite. *J. Environ. Manage.* 246: 547–556.

112 Liang, J., He, B., Li, P. et al. (2019). Facile construction of 3D magnetic graphene oxide hydrogel via incorporating assembly and chemical bubble and its application in arsenic remediation. *Chem. Eng. J.* 358: 552–563.

113 Ahangari, A., Raygan, S., and Ataie, A. (2019). Capabilities of nickel zinc ferrite and its nanocomposite with CNT for adsorption of arsenic(V) ions from wastewater. *Engineering* 7: 103493.

114 Huong, P.T.L., Huy, L.T., Lan, H. et al. (2018). Magnetic iron oxide-carbon nanocomposites: impacts of carbon coating on the As(V) adsorption and inductive heating responses. *J. Alloys Compd.* 739: 139–148.

115 Zhang, H.B., Zheng, W.G., Yan, Q. et al. (2010). Electrically conductive polyethylene terephthalate/graphene nanocomposites prepared by melt compounding. *Polymers* 51: 1191–1196.

116 Zhang, K., Dwivedi, V., Chi, C., and Wu, J. (2010). Graphene oxide/ferric hydroxide composites for efficient arsenate removal from drinking water. *J. Hazard. Mater.* 182: 162–168.

117 Zhang, M.H., Zhao, Q.I., Bai, X., and Ye, Z.F. (2010). Adsorption of organic pollutants from coking wastewater by activated coke. *Colloids Surf., A* 362: 140–146.

118 Chandra, V., Park, J., Chun, Y. et al. (2010). Water-dispersible magnetite-reduced graphene oxide composites for arsenic removal. *ACS Nano* 4: 3979–3986.

119 Nandi, D., Gupta, K., Ghosh, A.K. et al. (2012). Manganese incorporated iron(III) oxide-graphene magnetic nanocomposite, synthesis, characterization, and application for the arsenic(III)-sorption from aqueous solution. *J. Nanopart. Res.* 14: 1272.

120 Thy, L.T.M., Kiem, N.H., Tu, T.H. et al. (2020). Fabrication of manganese ferrite/graphene oxide nanocomposites for removal of nickel ions, methylene blue from water. *Chem. Phys.* 533: 110700.

121 Kadam, A., Saratale, R.G., Shinde, S. et al. (2019). Adsorptive remediation of cobalt oxide nanoparticles by magnetized α-cellulose fibers from waste paper biomass. *Bioresour. Technol.* 273: 386–393.

122 Huang, S., Jiang, S., Pang, H. et al. (2019). Dual functional nanocomposites of magnetic $MnFe_2O_4$ and fluorescent carbon dots for efficient U(VI) removal. *Chem. Eng. J.* 368: 941–950.

123 Li, R., Liu, L., and Yang, F. (2015). Polyaniline/reduced graphene oxide/Fe_3O_4 nanocomposite for aqueous Hg(II) removal. *Water Sci. Technol.* 72: 2062–2070.

124 Lellis, B., Polonio, C.Z.F., Pamphile, J.A., and Polonio, J.C. (2019). Effects of textile dyes on health and the environment and bioremediation potential of living organisms. *Biotechnol. Res. Innovations* 3: 275–290.

125 Li, L., Fan, L., Luo, C. et al. (2014). Study of fuchsine adsorption on magnetic chitosan/graphene oxide. *RSC Adv.* 4: 24679–24685.

126 Wang, C., Feng, C., Gao, Y. et al. (2011). Preparation of a graphene-based magnetic nanocomposite for the removal of an organic dye from aqueous solution. *Chem. Eng. J.* 173: 92–97.

127 Yao, Y.J., Miao, S.D., Liu, S.Z. et al. (2012). Synthesis, characterization, and adsorption properties of magnetic Fe_3O_4@graphene nanocomposite. *Chem. Eng. J.* 184: 326–332.

128 Ai, L., Zhang, C., and Chen, Z. (2011). Removal of methylene blue from aqueous solution by a solvothermal-synthesized graphene/magnetite composite. *J. Hazard. Mater.* 192: 1515–1524.

129 Wang, S., Wei, J., Lv, S. et al. (2013). Removal of organic dyes in environmental water onto magnetic-sulfonic graphene nanocomposite. *Clean-Soil, Air, Water* 41: 992–1001.

130 Li, B.J., Cao, H.Q., Yin, G. et al. (2011). Cu_2O@reduced graphene oxide composite for removal of contaminants from water and super capacitor. *J. Mater. Chem.* 21: 10645–10648.

131 Ansari, N., Feghhi, F., Montazeri, M. et al. (2019). Facile synthesis of water-stable iron intercalated multi layered graphene nanocomposite with large magnetic moments as superior water pollutant remediators. *Synth. Met.* 255: 116105.

132 Saygili, H. (2019). Hydrothermal synthesis of magnetic nanocomposite from biowaste matrix by a green and one-step route: characterization and pollutant removal ability. *Bioresour. Technol.* 278: 242–247.

133 Wu, C., Dong, D., Yu, X. et al. (2020). Mesoporous carbon/cobalt ferrite nanocomposite: a charge and pH independent magnetic adsorbent for dye pollutant treatment. *Diamond Relat. Mater.* 105: 107796.

134 Fan, L., Luo, C., Li, X. et al. (2012). Fabrication of novel magnetic chitosan grafted with graphene oxide to enhance adsorption properties for methyl blue. *J. Hazard. Mater.* 215–216: 272–279.

135 Fan, L., Luo, C., Sun, M. et al. (2013). Synthesis of magnetic β-cyclodextrinchitosan/graphene oxide as nanoadsorbent and its application in dye adsorption and removal. *Colloids Surf., B* 103: 601–607.

136 Shi, H., Li, W., Zhong, L., and Xu, C. (2014). Methylene blue adsorption from aqueous solution by magnetic cellulose/graphene oxide composite: equilibrium, kinetics, and thermodynamics. *Ind. Eng. Chem. Res.* 53: 1108–1118.

137 Zhang, L., Hu, X., Wang, C., and Tai, Y. (2018). Water-dispersible and recyclable magnetic TiO_2/graphene nanocomposites in wastewater treatment. *Mater. Lett.* 231: 80–83.

138 Zhang, B., Li, Y., Wu, T. et al. (2018). Magnetic iron oxide/graphene oxide nanocomposites: formation and interaction mechanism for efficient removal of methylene blue and *p-tert*-butylphenol from aqueous solution. *Mater. Chem. Phys.* 205: 240–252.

139 Jabbari, V., Veleta, J.M., Chaleshtori, M.Z. et al. (2016). Green synthesis of magnetic MOF@GO and MOF@CNT hybrid nanocomposites with high adsorption capacity towards organic pollutants. *Chem. Eng. J.* 304: 774–783.

140 Sahoo, S.K., Padhiari, S., Biswal, S.K. et al. (2020). Fe_3O_4 nanoparticles functionalized GO/g-C_3N_4 nanocomposite: an efficient magnetic nanoadsorbent for adsorptive removal of organic pollutants. *Mater. Chem. Phys.* 244: 122710.

141 Feng, K., Song, B., Li, X. et al. (2019). Enhanced photocatalytic performance of magnetic multi-walled carbon nanotubes/cerium dioxide nanocomposite. *Ecotoxicol. Environ. Saf.* 171: 587–593.

142 Islam, M.R., Ferdous, M., Sujan, M.I. et al. (2020). Recyclable Ag-decorated highly carbonaceous magnetic nanocomposites for the removal of organic pollutants. *J. Colloid Interface Sci.* 562: 52–62.

143 Hassani, A., Eghbali, P., Ekicibil, A., and Metin, O. (2018). Monodisperse cobalt ferrite nanoparticles assembled on mesoporous graphitic carbon nitride ($CoFe_2O_4$/mpg-C_3N_4): a magnetically recoverable nanocomposite for the photocatalytic degradation of organic dyes. *J. Magn. Magn. Mater.* 456: 400–412.

144 Sarkar, N., Sahoo, G., and Swain, S.K. (2020). Graphene quantum dot decorated magnetic graphene oxide filled polyvinyl alcohol hybrid hydrogel for removal of dye pollutants. *J. Mol. Liq.* 302: 112591.

145 Mousavi, M. and Yangjeh, A.H. (2018). Magnetically recoverable highly efficient visible-light-active g-C_3N_4/Fe_3O_4/Ag_2WO_4/AgBr nanocomposites for photocatalytic degradations of environmental pollutants. *Adv. Powder Technol.* 29: 94–105.

146 Ranjithkumar, V., Sangeetha, S., and Vairam, S. (2014). Synthesis of magnetic activated carbon/α-Fe_2O_3 nanocomposite and its application in the removal of acid yellow 17 dye from water. *J. Hazard. Mater.* 273: 127–135.

147 Borthakur, S. and Saikia, L. (2019). $ZnFe_2O_4$@g-C_3N_4 nanocomposites: an efficient catalyst for Fenton-like photodegradation of environmentally pollutant Rhodamine B. *J. Environ. Chem. Eng.* 7: 103035.

148 Mahmoodi, N.M., Oveisi, M., Taghizadeh, A., and Taghizadeh, M. (2019). Novel magnetic amine functionalized carbon nanotube/metal-organic framework nanocomposites: from green ultrasound-assisted synthesis to detailed selective pollutant removal modelling from binary systems. *J. Hazard. Mater.* 368: 746–759.

149 Huang, L., Miao, J., and Shuai, Q. (2020). Carboxyl-functionalized magnetic porous organic polymers as efficient adsorbent for wastewater remediation. *J. Taiwan Inst. Chem. Eng.* 109: 97–102.

150 Sun, H.M., Cao, L.Y., and Lu, L.H. (2011). Magnetite/reduced graphene oxide nanocomposites, one step solvothermal synthesis and use as a novel platform for removal of dye pollutants. *Nano Res.* 4: 550–562.

151 Kumaresan, N., Sinthiya, M.M.A., Ramamurthi, K. et al. (2020). Visible light driven photocatalytic activity of ZnO/CuO nanocomposites coupled with rGO heterostructures synthesized by solid-state method for RhB dye degradation. *Arabian J. Chem.* 13: 3910–3928.

152 Ahamad, T., Ruksana, M.N., Alhabarah, A.N., and Alshehri, S.M. (2019). N/S doped highly porous magnetic carbon aerogel derived from sugarcane bagasse cellulose for the removal of bisphenol-A. *Int. J. Biol. Macromol.* 132: 1031–1038.

153 Mirzaee, S.A., Jaafarzadeh, N., Gomes, H.T. et al. (2019). Magnetic titanium/carbon nanotube nanocomposite catalyst for oxidative degradation of Bisphenol A from high saline polycarbonate plant effluent using catalytic wet peroxide oxidation. *Chem. Eng. J.* 370: 372–386.

154 Liu, D., Huang, Z., Li, M. et al. (2020). Construction of magnetic bifunctional β-cyclodextrin nanocomposites for adsorption and degradation of persistent organic pollutants. *Carbohydr. Polym.* 230: 115564.

155 Guo, F., Lu, J., Liu, Q. et al. (2018). Degradation of acid orange 7 by peroxymonosulfate activated with the recyclable nanocomposites of g-C_3N_4 modified magnetic carbon. *Chemosphere* 205: 297–307.

156 Muntean, S.G., Nistor, M.A., Muntean, E. et al. (2018). Removal of colored organic pollutants from wastewaters by magnetite/carbon nanocomposites: single and binary systems. *J. Chem.* 2018: 6249821.

157 Cheng, J.S., Du, J., and Zhu, W. (2012). Facile synthesis of three-dimensional chitosan-graphene mesostructures for reactive black 5 removal. *Carbohydr. Polym.* 88: 61–67.

158 Abdel-Shafy, H.I. and Mansour, M.S.M. (2016). A review on polycyclic aromatic hydrocarbons: source, environmental impact, effect on human health and remediation. *Egypt. J. Pet.* 25: 107–123.

159 Phiri, I., Ko, J.M., Mushonga, P. et al. (2019). Simultaneous removal of cationic, anionic and organic pollutants in highly acidic water using magnetic nanocomposite alginate beads. *J. Water Process Eng.* 31: 100884.

160 Yang, D., Tammina, S.K., Li, X., and Yang, Y. (2019). Enhanced removal and detection of benzo[a]pyrene in environmental water samples using carbon dots-modified magnetic nanocomposites. *Ecotoxicol. Environ. Saf.* 170: 383–390.

161 Hatami, M., Yazdan, M., and Mahmoudian, P.M. (2020). Facile production of HNTs\PDA\PF nanocomposites by unique and environment-friendly method for the removal of phenolic pollutants in water as an environmental adsorbent. *J. Taiwan Inst. Chem. Eng.* 108: 1–5.

162 Tomke, P.D. and Rathod, V.K. (2020). Facile fabrication of silver on magnetic nanocomposite (Fe_3O_4@Chitosan-AgNP nanocomposite) for catalytic reduction of anthropogenic pollutant and agricultural pathogens. *Int. J. Biol. Macromol.* 149: 989–999.

163 Sharma, A., Siddiqi, Z.M., and Pathania, D. (2017). Adsorption of polyaromatic pollutants from water system using carbon/$ZnFe_2O_4$ nanocomposite: equilibrium, kinetic and thermodynamic mechanism. *J. Mol. Liq.* 240: 361–371.

164 Duman, O., Ozcan, C., Polat, T.G., and Tunc, S. (2019). Carbon nanotube-based magnetic and non-magnetic adsorbents for the high-efficiency removal of diquat dibromide herbicide from water: OMWCNT, OMWCNT-Fe_3O_4 and OMWCNT-κ-carrageenan-Fe_3O_4 nanocomposites. *Environ. Pollut.* 244: 723–732.

165 Goswami, B. and Mahanta, D. (2020). Polyaniline-Fe_3O_4 and polypyrrole-Fe_3O_4 magnetic nanocomposites for removal of 2,4-dichlorophenoxyacetic acid from aqueous medium. *J. Environ. Chem. Eng.* 8: 103919.

166 Huang, Y., Nengzi, L., Zhang, X. et al. (2020). Catalytic degradation of ciprofloxacin by magnetic $CuS/Fe_2O_3/Mn_2O_3$ nanocomposite activated peroxymonosulfate: influence factors, degradation pathways and reaction mechanism. *Chem. Eng. J.* 388: 124274.

167 Ninwiwek, N., Hongsawat, P., Punyapalakul, P., and Prarat, P. (2019). Removal of the antibiotic sulfamethoxazole from environmental water by mesoporous silica-magnetic graphene oxide nanocomposite technology: adsorption characteristics, coadsorption and uptake mechanism. *Colloids Surf., A* 580: 123716.

168 Nistico, R., Cesano, F., Franzoso, F. et al. (2018). From biowaste to magnet-responsive materials for water remediation from polycyclic aromatic hydrocarbons. *Chemosphere* 202: 686–693.

169 Yang, Q., Chen, D., Chu, L., and Wang, J. (2020). Enhancement of ionizing radiation-induced catalytic degradation of antibiotics using Fe/C nanomaterials derived from Fe-based MOFs. *J. Hazard. Mater.* 389: 122148.

170 Tran, T.V., Nguyen, D.T.C., Le, H.T.N. et al. (2019). Amino-functionalized MIL-88B(Fe)-based porous carbon for enhanced adsorption toward ciprofloxacin pharmaceutical from aquatic solutions. *C.R. Chim.* 22: 804–812.

171 Haham, H., Grinblat, J., Sougrati, M.T. et al. (2015). Engineering of air-stable Fe/C/Pd composite nanoparticles for environmental remediation applications. *J. Magn. Magn. Mater.* 389: 82–89.

172 Li, S., Li, Z., Ke, B. et al. (2019). Magnetic multi-walled carbon nanotubes modified with polyaluminium chloride for removal of humic acid from aqueous solution. *J. Mol. Liq.* 279: 241–250.

173 Salam, M.A., Gabal, M.A., and Obaid, A.Y. (2012). Preparation and characterization of magnetic multi-walled carbon nanotubes/ferrite nanocomposite and its application for the removal of aniline from aqueous solution. *Synth. Met.* 161: 2651–2658.

10

Functionalized Carbon Nanotubes for Ammonia Sensors

Rakshith K. Srinivasreddy and Ravi-Kumar Kadeppagari

Jyothy Institute of Technology, Centre for Incubation, Innovation, Research and Consultancy (CIIRC), Department of Food Technology, Tataguni, Off Kanakapura Main Road, Bengaluru, 560082, India

10.1 Introduction

Ammonia detecting sensors are useful for environmental applications [1]. A sensor that can monitor environmental ammonia levels will be helpful for taking precautionary measures since higher levels of ammonium ions are hazardous to mammals and aquatic organisms and can lead to adverse effects [2, 3]. Consumption of food or water contaminated with ammonium can elevate the ammonium levels in the blood (hyperammonemia), and this can severely affect the function of brain; cause hyperexcitability, coma, and convulsions; and finally lead to death [4]. Inhalation of air containing ammonia will cause a burning sensation to the eyes, skin, throat, nose, gastrointestinal tract, or lungs in humans. These ill effects might be serious enough to cause permanent blindness, lung disease, or death. Ammonia is highly soluble in water and can disrupt the pH of biological systems [5]. It can also lead to direct toxicity to plants [6, 7] and loss of plant diversity [8]. Toxicity of ammonium is believed to be the source of losses in fish hatcheries [9].

Hence, sensors are necessary for rapid determination of ammonia at number of areas like healthcare, food, resin production using urea, fertilizer plants, explosives/munitions plants, semiconductor production, nylon production, water and wastewater facilities, and clandestine drug labs in addition to environmental monitoring.

10.2 Ammonia Sensors

Ammonia sensors based on metal oxide (SnO_2) [10] operate based on the change of conductance due to the chemisorption of ammonia gas molecules on the sensing layer. The major drawback with these sensors is that they are not selective to the particular gas. Another kind of ammonia sensors are based on the change of charge carrier concentration in the catalytic metal in response to the concentration of the ammonia gas [11, 12]. Change

Environmental Applications of Carbon Nanomaterials-Based Devices, First Edition.
Edited by Shadpour Mallakpour and Chaudhery Mustansar Hussain.

in charge carriers was quantified by using the field effect devices like capacitor or transistor [12]. However, this kind of sensor is not selective, and the response depends on the different parameters like catalytic metal, morphology of the metal layer, and operating temperature.

Potentiometric ammonia sensor based on zirconium titanium phosphate ion exchanger was developed. This sensor showed potentiometric response to other cations, even though it has long life [13]. A fluorosensor for ammonia detection was constructed by using rhodamine-incorporated polyvinyl chloride (PVC) [14]. An ammonium electrode based on PVC membrane containing palmitic acid and nonactin as an ammonium ionophore for the determination of ammonia was constructed [15]. A conductometric sensor for ammonium detection was constructed by using PVC membrane containing nonactin [16]. However, this sensor is not specific for the ammonium.

Colorimetric and optical approaches were extensively used for the detection of ammonium [17, 18]. Classical colorimetric methods were based on Nessler's reagent [19] and Berthelot's indophenol reaction [18]. However, in the colorimetric methods, the phenolic reagent and its by-product generated during the reaction were highly toxic [19]. High-performance liquid chromatogram coupled with fluorescence detector was used for the ammonium detection [20, 21]. However, in the chromatographic methods, the samples should be subjected to pre-column derivatization for detecting the fluorescence [22]. Ammonium was detected by a system with flow injection analysis, and it contains immobilized enzyme reactor and chemiluminescence detector [23]. Here, ammonium is converted to glutamic acid by glutamate dehydrogenase (GlDH) in the presence of 2-oxoglutarate and NADH. Then glutamic acid is oxidized by glutamate oxidase to produce hydrogen peroxide that is detected by a chemiluminometric method using luminol and hexacyanoferrrate(III). An optical and dual enzyme-based biosensor was developed by using GlDH and diaphorase for the detection of ammonia [24]. However, this sensor did not work well at higher concentrations of ammonia.

Enzyme-based amperometric biosensors gained a lot of importance in the last decade due to the high selectivity of the bioreceptor and the sensitivity of electrochemical signal transduction. This led to the specific detection and quantification of target analyte in the complex matrices. In this kind of sensors, the enzyme acts as a bioreceptor, and electrons are generated when the target analyte comes in contact with the enzyme. These electrons are transduced to the working electrode (WE) faster since the enzyme was coated on the WE and electrons are generated in close vicinity of the WE. As a result of early transduction of electrons, the sensitivity of the sensor gets improved. The selectivity of the sensor improves since the enzyme is specific to the targeted analyte. A Gwent Sensor Ltd. has developed an amperometric biosensor for the detection of ammonium, and it was based on the GlDH. However, its working range is 1–10 ppm, which is limited. Another sensor was constructed for the detection of ammonium by immobilizing the two enzymes, GlDH and glutamate oxidase, on a Clark-type oxygen electrode [22]. However, this sensor lost around 60% of its response after 18 days. Yet another amperometric ammonia sensor based on the two enzymes, GlDH and diaphorase, was developed [25]. Nevertheless, it shows linear response only at lower concentrations of ammonium (2.5–500 μM). Moreover, the sensor based on two enzymes is difficult to manage since each enzyme needs different optimum conditions and it is difficult to maintain both the conditions simultaneously, and it affects the stability of the sensor. Hence, single-enzyme system is advantageous.

By using the enzyme AlaDH, a reagent less amperometric ammonia biosensor was developed [1]. This enzyme converts pyruvate to alanine and generates two electrons in the presence of ammonium and NADH. However, this sensor response is linear only at higher levels of ammonium (10–100 mM), and the signal level is very low in response to the different concentrations of ammonium. In the earlier decades, a considerable amount of analytical research has been done for the electrochemistry of the NAD^+/NADH redox reactions coupled to various electrodes [26]. However, only a limited number of electrochemical sensors based on dehydrogenases were reported due to large overvoltage encountered for NADH oxidation at electrodes [27] and surface fouling associated with the accumulation of reaction products [26]. This problem was addressed by applying functionalized carbon nanotubes (CNTs) to the electrodes since they offered an accelerated electron transfer along with minimization of surface fouling [28–30].

10.3 Types and Synthesis of Carbon Nanotubes

CNTs are made up of carbon, and they are tubelike materials with a diameter in the nanometer range. They are synthesized from graphite sheet. Graphite layers are rolled up in to unbreakable hexagonal mesh-like structure with carbon molecules appearing at the apices of hexagonal structures. The CNTs can be single-walled carbon nanotubes (SWCNTs), double-walled carbon nanotubes (DWCNTs), or multi-walled carbon nanotubes (MWCNTs), if single, double, or more than two carbon layers are present, respectively [31]. The SWCNTs were compared with MWCNTs in Table 10.1. They can be manufactured by three important methods, i.e. chemical vapor deposition (CVD), electric arc discharge, or laser ablation method.

Fabrication methods at high temperatures (laser ablation or electric arc discharge) were initially used for synthesizing CNTs. However, currently these methods have been replaced by CVD method where low temperatures (<800 °C) were used. In CVD method, nanotube diameter, length, purity, alignment, orientation, and density can be accurately controlled due to operations at relatively lower temperatures [32]. Electric arc discharge method uses higher temperatures (>1700 °C) that will typically lead to the expansion of CNTs with

Table 10.1 Comparison between single-walled carbon nanotubes (SWCNTs) and multiwalled carbon nanotubes (MWCNTs).

SWCNTs	MWCNTs
Graphene single layer will be present	Graphene multiple layers will be present
Synthesized with catalyst	Catalyst is not essential for the synthesis
Not easy for mass production	Easy for mass production
Less pure	More pure
During functionalization there are more chances for defects	Less chances for defects during functionalization, but difficult to repair
Easy to characterize and evaluate	It is complex
It can be twisted easily	It cannot be twisted easily

lesser number of structural defects compared with other methods [33]. Most common methods use arc discharge between highly pure graphite electrodes with diameters in the range of 6–12 mm and with a separation of 1–2 mm in a helium-filled chamber at subatmospheric pressure (500 Torr). Helium can be replaced with hydrogen or methane [34]. In this method, CNTs can be synthesized with or without the use of catalyst precursors. Usually, MWCNTs were synthesized without the use of catalyst precursors, whereas SWCNTs utilize catalyst precursors [35]. The SWCNTs utilize complex anode (made up of graphite and metal) for the expansion in arc discharge. The Ni–Y-graphite mixtures were shown to produce high yields (<90%) of SWCNTs with an average diameter of 1.4 nm.

In the laser ablation method, a quartz tube containing a pure graphite block and Ar atmosphere will be heated in a furnace at 1200 °C by using high-power laser [36]. The laser will vaporize the graphite within the quartz. The diameter of the tubes decreases with the increase in the laser pulse power [37]. However, this method is not economically advantageous. Standard method for CNT synthesis is CVD. There are different kinds of CVD like catalytic chemical vapor deposition (CCVD) that is either thermal [38] or plasma enhanced, water-assisted CVD, oxygen-assisted CVD, microwave plasma-enhanced chemical vapor deposition (MPECVD), hot filament chemical vapor deposition (HFCVD), and radiofrequency chemical vapor deposition (RF-CVD) [39]. However, CCVD is the standard method for the synthesis of CNTs at present. The process of growing CNTs in this method is same as electric arc discharge method. Exciting carbon atoms that are in contact with metallic catalysts were also similar to the arc discharge method. Chemical breakdown of hydrocarbon on the substrate happens during CCVD, and this method allows CNTs to expand on different materials. During CVD process, metal catalyst layer will be prepared, and substrate will be processed at 700 °C. Generally used metal catalysts are nickel, cobalt, iron, or a combination of them [40]. The purpose of using the metal nanoparticles along with catalyst supports like MgO or Al_2O_3 is to provide higher surface area for the by-product of the catalytic reaction between the pure carbon and the metal particles. In the initial step of CNT expansion, two kinds of gases will fuel the fluidized bed reactor, a commonly used reactor. Those gases are carbon-containing gas like methane, ethylene, or acetylene and process gas like ammonia, nitrogen, or hydrogen. The carbon-containing gas will be broken on the surface of the catalyst, and the carbon becomes visible at the edges of the nanoparticle where the CNTs can be produced. The CCVD method is economically suitable for large-scale CNT production compared with laser ablation method. Three methods for the synthesis of CNTs were compared in Table 10.2.

The CNTs show different characteristic properties like high thermal conductivity, elasticity, and low density. They are chemically inert and show high surface-to-volume ratio. Functionalization of the surface of CNTs will lead to highly soluble and compatible materials, which in turn can be derivatized with active molecules for different applications. Both pristine and functionalized CNTs were used for environmental applications like sensing ammonia.

10.4 Carbon Nanotube-Based Ammonia Sensors

SWCNTs were used for the development of ammonia sensors. Chemiresistor ammonia sensor based on SWCNTs was developed, which could detect the ammonia levels up to

Table 10.2 Comparison among CNT synthesis methods.

Method	CVD	Electric arc discharge	Laser ablation
Rate of yield	More than 75%	More than 75%	More than 75%
Type of CNTs produced	Single walled and multiwalled	Single walled and multiwalled	Single walled and multiwalled
Advantages	Low temperature High purity Aligned growth Production at large scale	Inexpensive High-quality CNTs	Relatively pure
Disadvantages	Defective CNTs	High temperature CNTs to be purified Tangled CNTs	Limited to lab-scale Production CNTs to be purified

3 ppb in the urban environment [41]. Ammonia gas sensing properties were enhanced after compositing CNTs with tungsten oxide nanobricks [42]. The synergistic sensory enhancement of composite material was attributed to the hole transfer between the inversion layer on tungsten nanobricks and the CNT surface. Electrical properties of sensors based on poly{(9,9-dioctylfluorene)-2,7-diyl-alt-[3,6-dithiene-2-yl]-2,5-di(2-ethylhexyl)pyrrolo[3,4-*c*]pyrrole-1,4-dione-5,5″-diyl} (PF8-DPP)-sorted SWCNTs were investigated, and they selectively detected ammonia at room temperature [43]. They showed high sensitivity, good stability, fast response (30 seconds), and better reversibility within one minute at 1.5 V [43]. The SWCNT-coated polymethyl methacrylate optical fiber was used as fiber optic sensor for the detection of ammonia in the range of 0–500 ppm at room temperature [44]. However, this sensor was also responding to methanol and ethanol vapors, suggesting its less selectivity to the ammonia.

Vertically gown and aligned MWCNTs directly on the surfaces of Ni electrodes and SiO_2 surfaces showed sensibility to the pollutant, ammonia at ambient temperature [45]. MWCNT/copper composite paste electrode was developed for the detection of ammonium in drinking water [46]. These composite electrodes exhibited significantly better performance compared with the electrodes obtained from CNT or Cu paste. The CNT/copper composite paste electrode showed lower detection limit (3.47 μM), response time of eight seconds, higher sensitivity (16.45 $\mu A/\mu mol\, l\, cm^2$), and acceptable reproducibility [46].

Polyaniline (PANI)-coated MWCNTs showed a linear response to ammonia in the 0.2–15 ppm range [47]. Ternary nanocomposite with polyaniline/multiwalled carbon nanotubes/molybdenum disulfide (PANI/MWCNTs/MoS_2) was used for developing ammonia gas sensor [48]. This nanocomposite was synthesized by *in situ* chemical oxidative polymerization method and characterized by Fourier transform infrared (FT-IR) spectrometry, X-ray diffraction (XRD), scanning electron microscopy (SEM), energy-dispersive spectrometry (EDS), transmission electron microscopy (TEM), and X-ray photoelectron spectroscopy (XPS). Ammonia sensor based on this nanocomposite showed fast response and recovery time and better repeatability that were better than PANI, PANI/MWCNT, or PANI/MoS_2 films [48]. Better ammonia sensing of the PANI/MWCNT/MoS_2 nanocomposite film was attributed to the special interactions at p–n heterojunction and synergistic effect of the ternary nanomaterials. The PANI/CNT composite prepared by doping with camphor sulfonic acid showed better response to ammonia gas [49].

The cellulose–TiO_2–MWCNT nanocomposite was synthesized by hydrothermal process [50]. This hybrid nanocomposite showed higher response to ammonia gas and also faster recovery in the range of 50–500 ppm [51]. This hybrid ammonia gas sensor is cheap and flexible [52]. The $Co_{1-x}Ni_xFe_2O_4$/MWCNT ($x = 0.2, 0.4, 0.5, 0.6, 0.8$) nanocomposites were synthesized by *in situ* solvothermal method by using ethylene glycol, and these nanocomposites showed high selectivity to ammonia compared with either MWCNTs or $Co_{1-x}Ni_xFe_2O_4$ nanoparticles [52].

An array of CNTs on the template of anodic aluminum oxide was used for the development of ammonia sensor [53]. Here, open-ended MWCNTs were vertically deposited on the inner wall of aluminum oxide. Hence, ammonia will be diffused in to nanochannels for the adsorption. This sensor showed response time of less than a minute, high sensitivity, and good reproducibility. The MWCNTs embedded in porous alumina templates were used for sensing ammonia [54]. In this study, alumina templates that were highly ordered were fabricated by an anodization process that has two steps, and vertically aligned MWCNT arrays were grown in these templates by xylene pyrolysis without using a catalyst. These were integrated into a design of resistive sensor. In another study [55], a novel CNT integration method was developed for manufacturing conductance-based ammonia gas sensors, and this method showed higher yield (>90%) at the wafer scale. Here, dense CNT arrays were grown locally in a horizontal manner *in situ* at wafer scale. Selective and directional catalytic growth of CNTs along with required design and fabrication of the catalyst support will enable the direct integration of CNT arrays into heterogeneous devices. Ammonia sensors based on these horizontal CNT arrays showed good reproducibility, long-term stability, and detection limit of 1 ppm [55]. Cross sensitivity by relative humidity was found to have no adverse effect on the response to ammonia.

One study showed improved ammonia sensing performance for SWCNTs of (6,5) chirality; those were ultrasonically treated at high power and bundled into nanotube network by thermal annealing [56]. Two pathways of chemiresistive sensing were observed for CNT networks that lead to increased and decreased resistance in response to concentrations of ammonia from 10^2 to 10^4 ppm. This kind of multidirectional chemiresistive response can be explained by mechanisms like neutralization of p-type conductivity of the CNTs either partially or completely, Schottky barrier change, and interface dipoles between CNTs and electrodes. In another study, drop-deposited and semiconductor-enriched SWCNT networks were used as ammonia sensor that was operated based on impedance [56]. At all concentrations, the effect of ammonia on high-frequency arc resistance is linear, which suggests the use of impedance spectra as signal for main sensing. Impedance metric mode of operation showed the 2.1 times improvement in the sensitivity of ammonia compared with the traditional direct current (DC) measurement [57]. In addition, impedance metric sensing is less susceptible to noise problems compared with DC when there is an ionic contamination or dissociation of ammonia molecules.

10.5 Functionalization of Carbon Nanotubes

In 1991, Ijima brought CNTs to the attention of scientific community. Outstanding properties of CNTs like higher mechanical strength, flexibility, and better electrical and thermal properties make them potential and attractive candidates for different applications. However, there are few limitations for their usage. Surface energy of CNTs is highly different from that of common organic solvents and polymers, and due to this, CNTs will not have chemical affinity to the organic solvents. Hence, the dispersion of CNTs into the solvents is the major obstacle for various applications [58]. Also, the surface of CNTs will not provide physical interaction with matrix or solvent [59]. Hence, the modification of CNTs makes them more attractive and useful for different applications like nanodevices, organic electronics, etc. The CNTs are modified with different types of polymers that are important functionally to improve the characteristics of end product. Covalent (chemical) and non-covalent (physical) methods are frequently used to functionalize CNTs.

Covalent or chemical functionalization is based on the covalent bonding of functional groups with carbon of CNTs that can be done at the end caps of CNTs or at sidewalls that have many defects. Sidewall functionalization by direct covalent bonding will change the hybridization from sp^2 to sp^3 and simultaneously lead to the loss of p-conjugation system on the graphene layer. This functionalization can be done by using highly reactive molecules like fluorine. Fluorination of CNTs is the popular method of covalent functionalization, and fluorinated CNTs have weaker C—F bonds than those in alkyl fluorides [60]. Hence, C—F bonds become substitution sites for additional functionalization on CNTs [61]. Thus, fluorine atoms have been successfully replaced with amino, alkyl, and hydroxyl groups [62]. Other functionalizations like cycloaddition, chlorination, bromination, hydrogenation, and carbene and nitrene additions have been successfully carried out.

Another way to covalent functionalization is through defects of CNTs. Intrinsic defects of CNTs can be enriched by inducing oxidative damage to the CNTs by using strong acids, which leads to the functionalization of holes with oxygenated groups [63]. Specifically, treatment of nanotubes with strong acids like nitric acid, sulfuric acid, and their mixture [64], strong oxidizing agents like potassium permanganate [65], reactive plasma [66], or ozone [67] will open CNTs and subsequently lead to the generation of oxygenated functional groups like carboxylic acid, ketone, alcohol, or ester groups. These groups will help to tether different kinds of molecules onto the defects and ends of CNTs through the chemical reactions like silanation, esterification, and polymer or biomolecule linking or thiolation. Covalent functionalized CNTs are advantageous since they are soluble in organic solvents due to their polar and nonpolar functional groups.

However, covalent methods have major drawbacks like inevitable creation of excessive defects on the CNTs, and sometimes they are fragmented in to small pieces.

In the non-covalent functionalization, conjugated system of the side walls of CNT was not affected. Thereby, the final structure of the CNT was not disturbed, and this is the advantage of non-covalent method that can be used for tuning the interfacial properties of CNTs. In this method, CNTs were functionalized non-covalently by employing π–π stacking or hydrophobic interactions with the help of aromatic compounds, polymers, or surfactants. In the non-covalent approaches, the modifications of CNTs will not only preserve their properties but also improve their solubilities significantly. Non-covalent methods include aromatic small-molecule absorption, surfactants, biopolymers, polymer wrapping, and endohedral method.

Functionalization of CNTs will make them compatible with different systems due to their easy solubility, and especially it will be helpful for using CNTs in the biological systems and in the development of biosensors.

10.6 Functionalized Carbon Nanotubes for Ammonia Sensors

Gas sensors based on SWCNT layers deposited on plastic substrates displayed a high sensitivity to ammonia, and the sensitivity reached to as low as 3 ppb after combining with the low noise [41]. In addition, functionalization of SWCNT bundle layers with indium tin oxide nanoparticles led to the threefold increase in the sensitivity compared with pristine CNTs for ammonia concentrations above 200 ppb [41]. These functionalized SWCNTs responded to water vapor that allows to tailor the selectivity of the sensor with respect to humidity interference during outdoor environmental monitoring [41].

Zinc oxide layer overlapped with COOH-functionalized MWCNTs that were screen-printed and Pd-doped (Pd-COOH-MWCNTs) was used for sensing ammonia [68] This sensor was insensitive to humidity but could detect ammonia up to 50 ppm that is not better compared with those sensors reported in the literature. Acid-functionalized MWCNTs were bonded with tetra-α-isopentyloxyphthalocyanine copper or nickel or lead for developing ammonia sensors [69]. Here π–π stacking interaction-based solution self-assembly method was used to obtain metal–MWCNT hybrids. Among these, copper–MWCNT hybrids showed detection limit of 75 ppb for ammonia, recovery time of 180 seconds, and excellent reproducibility, selectivity, and stability. The improved ammonia sensing performance is attributed to the synergistic effect between metallic pentyloxyphthalocyanine and MWCNTs. In another study, MWCNTs deposited with titanium dioxide by atomic layer deposition were used for sensing ammonia [70]. Functionalization of MWCNTs by carboxyl plasma polymer before the deposition of titanium dioxide led to enhanced sensitivity toward ammonia [70]. The HCl-doped MWCNTs/PANI nanocomposites were used for the sensing of ammonia, and they showed high sensing response and reproducibility [71]. These nanocomposites were prepared by *in situ* oxidative polymerization of aniline in the presence of cetyltrimethylammonium bromide and different concentrations of MWCNTs [72].

An ammonium biosensor-based on amperometry, alanine dehydrogenase of *Bacillus subtilis*, and functionalized MWCNTs was constructed by using screen-printed electrodes. The biosensor output current was increased 2.4 times after the addition of functionalized MWCNTs [30]. The biosensor exhibited better detection limit (0.001 mM NH_4^-), broader linearity range, and lesser response time (30 seconds) due to the incorporation of the functionalized MWCNTs [30]. Output current of enzyme–CNTs/carbon working electrode at the oxidation peak was more than two times higher compared with that of enzyme/carbon working electrode in the cyclic voltammogram. This result indicated a role for functionalized MWCNTs in the promotion of electron transfer from the enzyme to the working electrode, thereby increasing the signal, sensitivity, and response time.

However, studies on the use of functionalized CNTs for ammonia sensing are less, and still there is a more scope for using functionalized CNTs for sensing ammonia.

10.7 Conclusions and Future Perspectives

The better performance of the ammonia sensors based on CNTs could be due to the following advantages of CNTs. Higher surface-to-volume ratio of CNTs will increase the surface area of the working electrode or probe responsible for the sensing of ammonia. Due to their better efficiency with respect to electron transfer, CNTs will enhance the ammonia sensor's sensitivity and response time. However, more studies are required on the use of functionalized CNTs for the sensing of ammonia so that applications of functionalized CNTs can be not only readily explored for environmental applications but also extended to the food and health sectors for the monitoring of ammonia.

Acknowledgments

Authors are thankful to Sri SharadaPeetham, Sringeri, and Jyothy Charitable Trust, Bengaluru, for their support and facilities.

Important Websites

https://pubs.acs.org/doi/abs/10.1021/am301229w.
https://pubs.acs.org/doi/abs/10.1021/jp0471857.
https://www.sciencedirect.com/science/article/pii/S0925400510009263?casa_token=yQx_garotLcAAAAA:na06WSKoMGua3zMfUFHwHiJwv8upiU12scKpYxJQ7u4luVdJpy22FDXLnCVR21ihBqNs2Rdu6CU.
https://link.springer.com/content/pdf/10.1007/s10854-007-9144-5.pdf.
https://www.sciencedirect.com/science/article/abs/pii/S0008622314008148.

References

1 Tan, L.L., Musa, A., and Lee, Y.H. (2011). Determination of ammonium ion using a reagent less amperometric biosensor based on immobilized alanine dehydrogenase. *Sensors* 11 (10): 9344–9360.
2 Nandra, K., George, I.M., Mushotzky, R.F. et al. (1997). ASCA observations of Seyfert 1 galaxies. II. Relativistic iron Kα emission. *Astrophys. J.* 477: 602.
3 Dobler, L.J., Gibbons, J.M., Evtodienko, V.Y., and Hach Company (2006). Quick acting toxic ammonia test for aqueous samples. US Patent 7, 033, 839.
4 Wilkinson, D.J., Smeeton, N.J., and Watt, P.W. (2010). Ammonia metabolism, the brain and fatigue; revisiting the link. *Prog. Neurobiol.* 91 (3): 200–219.

5 Laukkanen, S., Kangas, A., and Kangas, J. (2002). Applying voting theory in natural resource management: a case of multiple-criteria group decision support. *J. Environ. Manage.* 64 (2): 127–137.
6 Sheppard, P., Comrie, A., Packin, G.D. et al. (2002). The climate of the US Southwest. *Clim. Res.* 21 (3): 219–238.
7 Krupa, S.V. (2003). Effects of atmospheric ammonia (NH_3) on terrestrial vegetation: a review. *Environ. Pollut.* 124 (2): 179–221.
8 Ellis, J.D., Evans, J.D., and Pettis, J. (2010). Colony losses, managed colony population decline, and colony collapse disorder in the United States. *J. Apic. Res.* 49 (1): 134–136.
9 Hussain, C.M. and Mishra, A.K. (2019). *Nanotechnology in Environmental Science*, vol. 2. Wiley.
10 Zakrzewska, K. (2001). Mixed oxides as gas sensors. *Thin Solid Films* 391 (2): 229–238.
11 Winquist, F., Spetz, A., Lundström, I., and Danielsson, B. (1984). Determination of ammonia in air and aqueous samples with a gas-sensitive semiconductor capacitor. *Anal. Chim. Acta* 164: 127–138.
12 Lundström, I., Spetz, A., Winquist, F. et al. (1990). Catalytic metals and field-effect devices—a useful combination. *Sens. Actuators, B* 1 (1–6): 15–20.
13 Hassan, S.S., Marei, S.A., Badr, I.H., and Arida, H.A. (2001). Novel solid-state ammonium ion potentiometric sensor based on zirconium titanium phosphate ion exchanger. *Anal. Chim. Acta* 427 (1): 21–28.
14 Preininger, C. and Mohr, G.J. (1997). Fluorosensors for ammonia using rhodamines immobilized in plasticized poly(vinyl chloride) and in sol-gel; a comparative study. *Anal. Chim. Acta* 342 (2): 207–213.
15 Karakuş, E., Pekyardımcı, Ş., and Kılıç, E. (2006). A new potentiometric ammonium electrode for biosensor construction. *Artif. Cells, Blood Substitutes, Immobilization Biotechnol.* 34 (5): 523–534.
16 Khadro, B., Namour, P., Bessueille, F. et al. (2009). A novel conductometric sensor based on a PVC membrane containing nonactin for ammonium determination. *Int. J. Environ. Anal. Chem.* 89 (1): 11–19.
17 Dubas, S.T. and Pimpan, V. (2008). Green synthesis of silver nanoparticles for ammonia sensing. *Talanta* 76 (1): 29–33.
18 Lau, K.T., Edwards, S., and Diamond, D. (2004). Solid-state ammonia sensor based on Berthelot's reaction. *Sens. Actuators, B* 98 (1): 12–17.
19 Zdráhal, Z. and Večeřa, Z. (1994). Photometric detection inside the stream of sample sheath fluid. *Anal. Lett.* 27 (1): 195–213.
20 Mopper, K. and Lindroth, P. (1982). Diel and depth variations in dissolved free amino acids and ammonium in the Baltic Sea determined by shipboard HPLC analysis. *Limnol. Oceanogr.* 27 (2): 336–347.
21 Nishikawa, Y. and Kuwata, K. (1984). Liquid chromatographic determination of low molecular weight aliphatic amines in air via derivatization with 7-chloro-4-nitro-2,1,3-benzoxadiazole. *Anal. Chem.* 56 (11): 1790–1793.
22 Kwan, R.C., Hon, P.Y., Mak, W.C. et al. (2006). Biosensor for rapid determination of 3-hydroxybutyrate using bienzyme system. *Biosens. Bioelectron.* 21 (7): 1101–1106.
23 Tabata, M., Murachi, T., Endo, J., and Totani, M. (1992). Use of various types of column reactors for flow-injection analysis. *J. Chromatogr. A* 597 (1): 435–442.

24 Azmi, N.E., Abdullah, J., Ahmad, M. et al. (2012). An optical based biosensor for the determination of ammonium in aqueous environment. *Am. J. Anal. Chem.* 3 (5): 364.

25 Rahman, S.A., Abdullah, J., Sidek, H., and Azmi, N.E. (2012). Mediated amperometric biosensor for the determination of ammonium. *Anal. Bioanal. Electrochem.* 4: 262–276.

26 Wang, Z., Etienne, M., Quilès, F. et al. (2012). Durable cofactor immobilization in sol-gel bio-composite thin films for reagentless biosensors and bioreactors using dehydrogenases. *Biosens. Bioelectron.* 32 (1): 111–117.

27 Wang, J. (2005). Carbon-nanotube based electrochemical biosensors: a review. *Electroanalysis* 17 (1): 7–14.

28 Musameh, M., Wang, J., Merkoci, A., and Lin, Y. (2002). Low-potential stable NADH detection at carbon-nanotube-modified glassy carbon electrodes. *Electrochem. Commun.* 4 (10): 743–746.

29 Park, J.K., Yee, H.J., Lee, K.S. et al. (1999). Determination of breath alcohol using a differential-type amperometric biosensor based on alcohol dehydrogenase. *Anal. Chim. Acta* 390 (1): 83–91.

30 Dave, U.C., Ingale, D.V., Venkatesh, K. et al. (2017). Multi walled carbon nanotubes enhance the response and sensitivity of the ammonium biosensor based on alanine dehydrogenase. *J. Electroanal. Chem.* 784: 102–108.

31 Hussain, C.M. (2020). *Handbook of Functionalized Nanomaterials for Industrial Applications*. Elsevier.

32 Mallakpour, S. and Khadem, E. (2016). Carbon nanotube–metal oxide nanocomposites: fabrication, properties and applications. *Chem. Eng. J.* 302: 344–367.

33 Mallakpour, S. and Soltanian, S. (2016). Surface functionalization of carbon nanotubes: fabrication and applications. *RSC Adv.* 6: 109916–109935.

34 Grobert, N. (2007). Carbon nanotubes—becoming clean. *Mater. Today* 10 (1): 28–35.

35 Mallakpour, S. and Behranvand, V. (2018). Synthesis of mesoporous recycled poly(ethylene terephthalate)/MWNT/carbon quantum dot nanocomposite from sustainable materials using ultrasonic waves: application for methylene blue removal. *J. Cleaner Prod.* 190: 525–537.

36 Abbasi, E., Sedigheh, F.A., Abolfazl, A. et al. (2014). Dendrimers: synthesis, applications, and properties. *Nanoscale Res. Lett.* 9 (1): 247–255.

37 Jose-Yacaman, M., Miki-Yoshida, M., Rendon, L., and Santiesteban, J.G. (1993). Catalytic growth of carbon microtubules with fullerene structure. *Appl. Phys. Lett.* 62 (2): 202–204.

38 Vander Wal, R.L., Berger, G.M., and Ticich, T.M. (2003). Carbon nanotube synthesis in aflame using laser ablation for in situ catalyst generation. *Appl. Phys. A* 77 (7): 885–889.

39 Eatemadi, A., Daraee, H., Karimkhanloo, H. et al. (2014). Carbon nanotubes: properties, synthesis, purification, and medical applications. *Nanoscale Res. Lett.* 9: 393–405.

40 Landi, B.J., Raffaelle, R.P., Castro, S.L., and Bailey, S.G. (2005). Single-wall carbon nanotube—polymer solar cells. *Prog. Photovoltaics Res. Appl.* 13 (2): 165–172.

41 Rigoni, F., Tognolini, S., Borghetti, P. et al. (2014). Environmental monitoring of low-ppb ammonia concentrations based on single-wall carbon nanotube chemiresistor gas sensors: detection limits, response dynamics, and moisture effects. *Proc. Eng.* 87: 716–719.

42 Vuong Le, X., AnhLuu, T.L., HuuLamNguyen, H., and TuNguyen, C. (2019). Synergistic enhancement of ammonia gas-sensing properties at low temperature by compositing carbon nanotubes with tungsten oxide nanobricks. *Vacuum* 168: 108861.

43 Zhang, X., Zhao, J., Tange, M. et al. (2015). Sorting semiconducting single walled carbon nanotubes by poly(9,9-dioctylfluorene) derivatives and application for ammonia gas sensing. *Carbon* 94: 903–910.

44 Manivannan, S., Saranya, A.M., Renganathan, B. et al. (2012). Single-walled carbon nanotubes wrapped poly-methyl methacrylate fiber optic sensor for ammonia, ethanol and methanol vapors at room temperature. *Sens. Actuators, B* 171–172: 634–638.

45 Arab, M., Berger, F., Picaud, F. et al. (2006). Direct growth of the multi-walled carbon nanotubes as a tool to detect ammonia at room temperature. *Chem. Phys. Lett.* 433: 175–181.

46 Valentini, F., Biagiotti, V., Lete, C. et al. (2007). The electrochemical detection of ammonia in drinking water based on multi-walled carbon nanotube/copper nanoparticle composite paste electrodes. *Sens. Actuators, B* 128 (1): 326–333.

47 He, L., Jia, Y., Meng, F. et al. (2009). Gas sensors for ammonia detection based on polyaniline-coated multi-wall carbon nanotubes. *Mater. Sci. Eng., B* 163 (2): 76–81.

48 Zhang, D., Wu, Z., Li, P. et al. (2018). Facile fabrication of polyaniline/multi-walled carbon nanotubes/molybdenum disulfide ternary nanocomposite and its high-performance ammonia-sensing at room temperature. *Sens. Actuators, B* 258: 895–905.

49 Eising, M., Eduardo Cava, C., Villegas Salvatierra, R. et al. (2017). Doping effect on self-assembled films of polyaniline and carbon nanotube applied as ammonia gas sensor. *Sens. Actuators, B* 245: 25–33.

50 Mun, S., Chen, Y., and Kim, J. (2012). Cellulose–titanium dioxide–multiwalled carbon nanotube hybrid nanocomposite and its ammonia gas sensing properties at room temperature. *Sens. Actuators, B* 171–172: 1186–1191.

51 Mallakpour, S., Behranvand, V., and Mallakpour, F. (2019). Synthesis of alginate/carbon nanotube/carbon dot/fluoroapatite/TiO_2 beads for dye photocatalytic degradation under ultraviolet light. *Carbohydr. Polym.* 224: 115–138.

52 Tang, Y., Zhang, Q., Li, Y., and Wang, H. (2012). Highly selective ammonia sensors based on $Co_{1-x}Ni_xFe_2O_4$/multi-walled carbon nanotubes nanocomposites. *Sens. Actuators, B* 169: 229–234.

53 Hoa, N.D., Quy, N.V., Cho, Y., and Kim, D. (2007). An ammonia gas sensor based on non-catalytically synthesized carbon nanotubes on an anodic aluminum oxide template. *Sens. Actuators, B* 127 (2): 447–454.

54 Mangu, R., Rajaputra, S., Clore, P. et al. (2010). Ammonia sensing properties of multiwalled carbon nanotubes embedded in porous alumina templates. *Mater. Sci. Eng., B* 174: 2–8.

55 Guerin, H., Poche, H., Pohle, R. et al. (2014). High-yield, in-situ fabrication and integration of horizontal carbon nanotube arrays at the wafer scale for robust ammonia sensors. *Carbon* 78: 326–338.

56 Lutsyk, P.M., Shankar, P., Rozhin, A.G., and Kulinich, S.A. (2019). Surface sensitivity of ultrasonically treated carbon nanotube network towards ammonia. *Surf. Interfaces* 17: 100363.

57 Pyo Hong, H., Hyub Kim, J., Jin Lee, C., and KiMin, N. (2015). In-plane impedancemetric ammonia sensing of solution-deposited, highly semiconductor-enriched single-wall carbon nanotube submonolayer network gas sensors. *Sens. Actuators, B* 220: 27–32.

58 Song, Y.S. and Youn, J.R. (2005). Influence of dispersion states of carbon nanotubes on physical properties of epoxy nanocomposites. *Carbon* 43: 1378–1385.

59 Ajayan, P.M., Schadler, L.S., Giannaris, C., and Rubio, A. (2000). Single-walled carbon nanotube–polymer composites: strength and weakness. *Adv. Mater.* 12: 750–753.

60 Kelly, K.F., Chiang, I.W., and Mickelson, E.T. (1993). Insight into the mechanism of sidewall functionalization of single-walled nanotubes: an STM study. *Chem. Phys. Lett.* 313: 445–450.

61 Touhara, H., Inahara, J., and Mizuno, T. (2002). Fluorination of cup-stacked carbon nanotubes, structure and properties. *Fluorine Chem.* 114: 181–188.

62 Stevens, J.L., Huang, A.Y., and Peng, H. (2003). Sidewall amino-functionalization of SWNTs through fluorination and subsequent reactions with terminal diamines. *Nano Lett.* 3: 331–336.

63 Chen, J., Hamon, M.A., Hu, H. et al. (1998). Solution properties of single-walled carbon nanotubes. *Science* 282 (5386): 95–98.

64 Liu, J., Rinzler, A.G., Dai, H. et al. (1998). Fullerene pipes. *Science* 280: 1253–1256.

65 Yu, R., Chen, L., Liu, Q. et al. (1998). Platinum deposition on carbon nanotubes via chemical modification. *Chem. Mater.* 10: 718–722.

66 Wang, S.C., Chang, K.S., and Yuan, C.J. (2009). Enhancement of electrochemical properties of screen-printed carbon electrodes by oxygen plasma treatment. *Electrochim. Acta* 54 (21): 4937–4943.

67 Sham, M.-L. and Kim, J.-K. (2006). Surface functionalities of multi-wall carbon nanotubes after UV/ozone and TETA treatments. *Carbon* 44 (4): 768–777.

68 Tulliani, J.M., Cavalieri, A., Musso, S. et al. (2011). Room temperature ammonia sensors based on zinc oxide and functionalized graphite and multi-walled carbon nanotubes. *Sens. Actuators, B* 152 (2): 144–154.

69 Kang, D., Wang, B., Wang, X. et al. (2017). Stably dispersed metallophthalocyanine noncovalently bonded to multiwalled carbon nanotubes for ammonia sensing at room temperature. *Sens. Actuators, B* 246: 262–270.

70 Kaushik, P., Eliáš, M., Michalička, J. et al. (2019). Atomic layer deposition of titanium dioxide on multi-walled carbon nanotubes for ammonia gas sensing. *Surf. Coat. Technol.* 370: 235–243.

71 Ansari, M.O., Ansari, S.P., Yadav, S.K. et al. (2014). Ammonia vapor sensing and electrical properties of fibrous multi-walled carbon nanotube/polyaniline nanocomposites prepared in presence of cetyl-trimethylammonium bromide. *J. Ind. Eng. Chem.* 20 (4): 2010–2017.

72 Mallakpour, S. and Rashidimoghadam, S. (2019). Carbon nanotubes for dyes removal. In: *Composite Nanoadsorbents* (eds. G. Kyzas and A.C. Mitrpoulos), 211–244. Amsterdam, Netherlands: Elsevier.

11

Functionalized Carbon Nano Lab-on-a-Chip Devices for Environment

RaviPrakash Magisetty[1,2], *Naga Srilatha Cheekuramelli*[1,2], *and Radhamanohar Aepuru*[1,2]

[1]*Defence Institute of Advanced Technology, Materials Engineering Division, Pune, India*
[2]*Universidad Tecnológica Metropolitana, Departamento de Ingeniería Mecánica, Facultad de Ingeniería, Santiago, Chile*

11.1 Introduction

Miniaturization is an important aspect of electronic technology, ranging from microelectronics to nanoelectronics, developed to fabricate devices for various applications in wide-field areas. This technology began in the late 1050s with the advent of planar process aimed to integrate a number of circuit components within a single small area [1]. This miniaturization was described as lab on a chip (LoC), which was probably first coined by Moser et al. for miniaturized thin-film glutamate and glutamine biosensor [1]. Later on, researchers described the LoC as a miniaturized device capable of scale-down and can integrate and automate the conventional multiple laboratory operations into a system that fit on a chip with size ranging from mm to few square cm [1].

In the field of science and technology, utilization of carbon-based materials and its allotropic forms such as amorphous carbon, graphite, and diamonds to newly introduced auspicious carbon nanotubes (CNTs), grapheme oxides, graphene quantum dots, and fullerene is increasing rapidly due to their functional characteristics and potential applications and has become a part of everyday human life [2, 3]. To achieve the carbon material demand for various applications, various synthetic methods have been developed, which will provide a clean and green sustainable pollution-free environment and reduce fossil fuel utilization. Recently LoCs has been successfully employed in various applications, including renewable energy, agriculture applications, biomedical applications, ocean, and atmospheric and space applications [4–11]. However, the environment and the ecosystem's sustainable developments are demanding the use of carbon-based materials and its allotropic family members for LoC devices and lab-on-box (LoB) systems [2, 3] because carbon materials possess significant properties, which can be altered by flexibly by controlling its molecular chemistry and its chemical structure, leading to a great diversity and versatility of chemical and physical functionalities. Even these carbon-based materials and its allotropic family members are useful in the field of biomedical engineering with the additional functionality

Environmental Applications of Carbon Nanomaterials-Based Devices, First Edition.
Edited by Shadpour Mallakpour and Chaudhery Mustansar Hussain.

of biocompatibility facilitates for the possible biomedication with the compatible biological system [12, 13]. Therefore, researchers are demanding carbon-based materials for LoC or lab-on-box (LoB) systems [4–11].

In this context, the advancement of carbon-based programmable/digital coding metamaterials as LoC devices has been explored for renewable energy, atmosphere, and space applications. Moreover, carbon-based metamaterials as LoC devices are described for detection, monitoring, and diagnosing in the field of agriculture and biomedical applications. Also, carbon-based LoC sensor devices are presented for understanding of biogeochemical processes from the marine environment. In addition, the need and advancement of carbon-based LoC or LoB devices for environmental applications are highlighted.

11.2 Need for Carbon Nano Lab-on-a-Chip Devices for Environment, and Its Advancement

LoC or LoB science and technology alludes operations on a very miniaturized scale within a portable or handheld device, where analysis of samples can take place *in situ*, exactly where the samples regenerated, rather than being transported around to large laboratory facility. Though LoC or LoB device fabrication drawbacks were reported since last decade, the advancement in materials and micro/nanofabrication techniques could facilitate a novel path for producing a number of efficient LoC or LoB devices for the environment. Additive fabrication/3D printing/rapid prototyping conserves natural resources compared to other manufacturing technologies, demonstrating its far less impact on our environment. Thus advancement in fabrication technologies such as the additive manufacturing technology with nanoscale probability could allow us an easy way to fabricate efficient eco-friendly devices within the advisable economical boundaries.

Since the last decade, carbon and its allotropic configurations such as carbon nanotubes and graphene and its functionalized forms produce electronic, chemical, and physical functionalities that can extend its network toward electronic device scaling and enhance performance. Despite the interaction between adjacent layers of these allotropic forms, renderers compromise properties if advanced fabrication techniques enabling production-level viable growth lead to efficiency with wide functionality for various applications. Complementary metal-oxide-semiconductor (CMOS) compatible process that can enable carbon-based materials and its functional forms has long-term potential to further enhance carbon-based devices' applicability for wide applications by providing other channel technology options for the fabrication of high-mobility planar devices. Most interestingly in addition to these unique properties, carbon allotropes and other carbon-based hybrids have the potential to decouple electronic devices from silicon substrate that would lead to true 3D on chip integration. Such 3D chip will advance chip scaling and power scaling while increasing performance. Carbon-based electronics have the potential for other device applications that require circuit enhancements such as high radiation resistance and high-temperature operating conditions that make them attractive for harsh environments such as space automotive, nuclear environments, and many other needs.

Though the carbon and its derivatives were widely explored for various applications, for any new material technology, industrial environmental, safety, and health (ESH), acceptance is a critical requirement and becomes a part of infrastructure requirements at both suppliers and end user. Although initial studies have been conducted on ESH aspects of carbon materials, its allotropic forms, and its functionalized hybrids, comprehensive understanding and regulations are still lacking, and they must be addressed the issues to accelerate the industry acceptance of carbon technology for electronics.

Despite the hurdles always associated with the commercialization of new technology, viable examples of these devices are becoming to appear on the market. It seems that LoC technology will become increasingly important in the coming years in the medical world, as well as in the chemical, agricultural, and energy fields.

11.3 Carbon Nano Lab-on-a-Chip Devices for Environment

11.3.1 Renewable Energy Applications

The growing demand of energy across the globe's attention is toward sustainable energy generation as an alternative to the increasing environmental issues attained by nonconventional energy generation. Commonly adopted renewable energy resources include solar energy that is sustainable, totally inexhaustible, and nonpolluting. However, solar energy extraction devices such as solar panels demonstrate low efficiency. The absorbed light is converted into electricity, but this absorption cannot be done efficiently using conventional materials. The absorption coefficient is an inherent characteristic of the material that cannot be improved using synergistic methods. However, this drawback can be overcome using light-trapping devices. In the context of microsystems, integrated electronics and LoC or LoB devices would be preeminent for reliable power generation. Metamaterial-based LoC devices using solar cells can offer unique advantages compared with the conventional solar cells such as ease of integration, portability with low cost, and high surface-area-to-volume ratio.

Metamaterials are a class of engineered materials with exotic properties that cannot be observed using conventional materials. Metamaterial subwavelength structures can efficiently couple light with a semiconductor layer in the solar cell. In other means, metamaterials as LoC devices provide easy integration with the solar cell, so that the light can be efficiently coupled with the solar semiconductor layer. However, near-perfect trapping of light of broad spectral range into a semiconducting layer could be attained via programmable/reconfigurable metamaterial-based LoC devices. Further, these devices facilitate the realization of low-cost and highly efficient photovoltaic devices.

The requirements of modern and scientific-technological developments have stretched the metamaterial-based LoC device research to programmable/digital coding metamaterials based on LoC devices for light-trapping applications. These programmable/digital coding metamaterial-based LoC devices have long been (from 2014) a wide-ranging research topic of interest for scientists and engineers due to the controllability of EM wave behavior and its functionalities from microwave to the visible range for various applications [14]. For programmable metamaterial-based LoC devices, engwwineered structural design materials that control light (functionalities with respect to incident angle and irradiation in wideband frequency) would be an alternative solution to enhance light-trapping capability and generate electricity efficiently [14–16].

On the other hand, carbon-based materials such as graphene have been explored as tunable/reconfigurable metamaterials for manipulation of electromagnetic (EM) wave. Particularly graphene is widely explored materials for EM wave manipulation, which is attributable due to the charge density-enabled active surfaces, which can be controlled via pulsating static electricity as DC voltage. More importantly, 2D graphene sheets facilitate as meta-atoms of metasurfaces, enabling versatile LoC or LoB devices [17].

In summary, combined technology of carbon-based material and programmable metamaterial, i.e. carbon-based programmable/tunable/reconfigurable metamaterial, renders prominent LoC or LoB devices for near-perfect trapping of light over a broad spectral range into a semiconducting layer of the solar cell. Further, this significant functionality could facilitate the realization of eco-friendly, highly efficient, and stable photovoltaic cells for a low cost.

11.3.2 Agriculture Applications

Nanotechnology application in the field of agriculture is mainly for the production of food, feed, and fiber. This demand is tremendously increasing day by day with the increasing population while natural resources are being depleted continuously. Consequently, the crop production cost is increasing at an alarming rate. To overcome these limitations, advanced technology-instigated precision farming is required, which is a better solution to maximize agricultural production with a reduction in price [18]. Recently, advancement in nanotechnology continues, and various technologies of nanotechnology such as smart delivery of fertilizers and pesticides, detection of contaminants and pathogens have become available to improve the practices of precision farming [18]. Carbon-based nanomaterials like fullerene, graphene, carbon nanotubes, nanodiamonds, and carbon nanodots have received significant consideration among the scientific community due to their distinctive physiochemical properties [2]. The best example is allotropic forms of carbon-posed unique functional properties that facilitate extensive use in the agricultural industry. The use of carbon nanotubes as controllable release fertilizers has an excellent impact on seed germination. Researchers revealed that up to a twofold increase in germination rate is attributable to the nanotube penetration into the seed that ameliorating effects on germination and growth [19]. This is further confirmed by the report of carbon nanotubes suggesting that it can be used as an excellent source of nutrient supply to the plant cell [20]. Moreover, improved germination of tomato seeds by using carbon nanotube has been reported. They suggested that carbon nanotube acts as pore/channels, allowing better penetration of water into the seeds [19]. Carbon-based nanoparticles are utilized for the improvement in the growth and development of crops and better yield. These are also suggested for assimilation purposes, translocation of nutrients, and storage of compounds [21]. Carbon-coated gold nanoparticles were utilized as a gene gun for transmitting DNA into the *Oryza sativa*, *Leucaena leucocephala*, and *Nicotiana tabacum*. Plant injuries are almost significant and recovery of plant tissue is easier and valuable [22]. Carbon nanotubes have been studied and suggested when biomolecule is bound to CNTs, which prevents degradation and provides stabilized cellular biometabolism as compared with the free molecules [23].

Since the last decade, extensive research on carbon-based materials for agriculture suggests that these materials could be considered as a prominent material among the other material community due to their distinctive physiochemical properties. Further, these reports suggested that the carbon-based nanomaterials standout as unique materials for the development of devices for detection, monitoring, and diagnosing in the agriculture field environment. However, a few reports on carbon-based LoC or LoB devices investigated the mechanism of cell propagation and differentiation, development, and formation of tissues, the design of highly organized biomaterial scaffolds to provide controlled cell adhesion, the spatial organization of cell modulation in the scaffolds, local environment biomolecule regulation capability, surface biological molecule surface modification at the level of nanoscale precision capabilities for enhanced biological compatibility and activity, single-molecule manipulation, and drug delivery in a controlled and targetable way. Although there is a lot of unknown developments available today, agricultural and nanotissue engineering is not just a dream but, maybe with the advancement in nanotechnology-empowered LoC or LoB technologies, could facilitate a path for quick advancement in Agri field. For example, recently researchers have developed a nondiagnostic kit also called as LoB used for measuring biological aspects of plants in a convenient way [24]. The developed small LoB-based devices can help detect pathogens and can help the farmers in the prevention of plant-based diseases [25, 26]. This device consists of four sensors that are able to detect ZEA, T-2/HT-2, deoxynivalenol (DON), and fumonisins (FBI/FB2) toxins, and they suggested the device for detection of fungal to be attached on crops like wheat, barley, and corn [27]. This device is also useful for other purposes, such as detection of antigen and antibodies and nucleotide sequence, gene target, isolation, and purification of specific genes and could be prominently used in the agriculture sector for the required crop production and its improved multifunctionalities have been tested under field trials. Among the developments respective to the agricultural field, sensors are becoming more important because of their wide-ranging sensing capability, various functionalities, and potential to guide the improvement of its operational capability. Currently, nanobiosensors are attractive prospects in comparison with the traditional sensors in the agricultural sector due to their distinct capabilities for bioprocess control, biodefense, quality control of food, and the most important medical applications shown in Figure 11.1. Recently, researchers reported that carbon nanotubes can penetrate into the seeds of tomatoes [28]. This report proposes that the new LoC or LoB system based on carbon nanotubes could facilitate the release and delivery of nutrients that take advantage of nanoscale porous media on the surface of crop plants [28]. Ma et al. presented a nanobiosensor that can detect deep DNA damage using a multiwalled nanotube layer in chitosan placed on the screen-printed carbon electrode [29]. Nanosensors combined with immobilized bioreceptor probes are demanding recent days for selective/targeted molecules [30]. These are new horizons for basic research and for bioanalytical applications. Such nanosensors integrated with LoC or LoB technologies not only facilitate molecular analysis but also help in monitoring many metabolites [31, 32]. These kinds of advanced technologies can detect a range of microorganisms and analytes like insecticides, urea, and glucose.

In summary, carbon-based materials and its functionalized hybrids with wide functionality were widely explored, suggesting that these materials enabled LoC or LoB devices would be an advancement in the field of agriculture for detection, monitoring, and diagnosing of quality crop production with reduced economical burden.

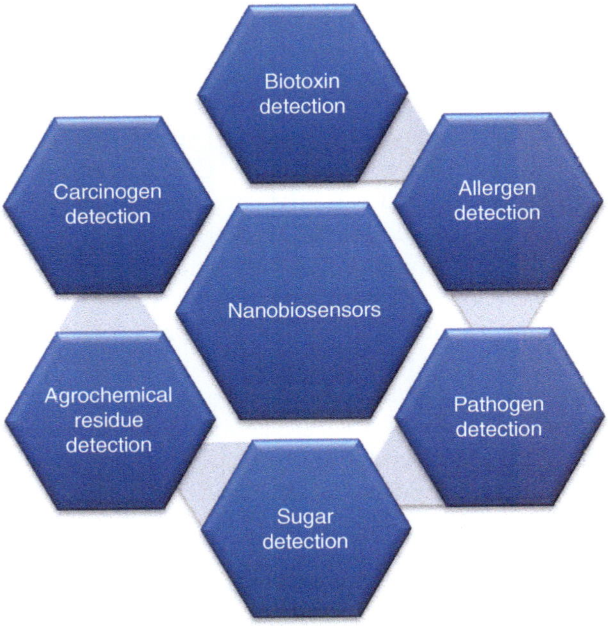

Figure 11.1 Applications of nanobiosensor in agriculture. Source: Based on Tian et al. [28].

11.3.3 Biomedical Applications

Magnetodielectric material-inspired smart antenna sensors are used to measure magnetic and electric fields near a biological system. Electromagnetic field exposure is not a new phenomenon. Recently, electromagnetic field exposure in the environment has been increasing due to the growing demand for electricity and technological advancement in electronic systems [1]. The human body is exposed to either high strength or weak electric and magnetic fields, both at home and at work, from electricity transmission and generation, domestic and industrial systems, telecommunication systems, and broadcasting [1–3]. Evidence of health effects from electromotive force (EMF) including their influence on the brain is inconclusive, and the probability that EMF exposure is a genuine health hazard is currently small [2]. The weaker electric currents induced in a human body in the absence of external electromagnetic fields are the stimulative sources for biological activities [2]. For example, most biological reactions from digestion to brain activities go along with the rearrangement of charged particles; also, the heart is an electrically active system. However, exposure to a high level of nonionizing energy such as radio waves can damage the structure and function of biological/nervous systems, and those are restricted by national and international guidelines. For example, microwave frequencies below 3 GHz can penetrate the outer layers of the skin, by being absorbed in the underlying tissues, and result in biological effects such as heating, including burns, cataracts, and possibly death [3].

Over the source of the past decade, numerous electromagnetic field sources have become the focus of health concerns, including power lines, microwave ovens, computers, TV screens, security devices, radars, and, most recently, mobile phones and their base stations

[1–3]. Researchers have introduced the threshold of a magnetic field with respect to physiological response to a magnetic field. It was suggested that the smallest magnetic field triggers a response in humans [1–3]. Nevertheless, the field-induced effects are due to the change in field directions over time (oscillating nature). When these strong direction shifting magnetic fields are directed on to a human, small electric currents begin to pulse through the body [2]. At above the threshold, the currents can stimulate biological cells, for example, currents to stimulate supersensitive cells in the retina giving the illusion of white light flickering. The heightened risk of childhood leukemia is the most commonly cited potential health consequence obtained by the living near high-voltage power lines [2]. This significant phenomenon of time-varying magnetic fields can be understood by MRI, for example, during a rapid movement of the head under a large magnet that produces ~3 T magnetic field. Moving quickly induces varying magnetic fields, leads to currents in different regions of the brain, causing nausea, loss of brain balance, metallic taste in the mouth, and in some cases magnetophosphenes [4].

To overcome all these abnormalities/damages in human tissue/nervous system, it is required to sense/measure how much (electric or magnetic field) electromagnetic field is influenced/concentrated near the biological system so that the induced effects could be minimized by implementing an alternative protective scheme or avoiding those field sources [5, 6]. The proposed LoC-based smart antenna sensor as an LoC device for the measurement of electric and magnetic field confinement near a biological system has been illustrated in Figure 11.2. Substrate integrated waveguide (SIW) along with the LoC-based smart antennae sensor could lead to reduced losses; consequently, transmitted signal clarity would be obtained [7]. Further, in addition to the SIW technology, negative resistance circuit within an LoC-based sensor antenna system could provide drastic reduction of losses attained during EM field sensing mechanism, or desirable improvement in the quality factor of the LoC-based sensing system could be achieved [7]. Moreover, printed antenna on a magnetodielectric substrate helps in the miniaturization of the LoC-based sensor device producing low-power and efficient LoC device for biomedical applications [8]. Magnetodielectric medium exhibits properties similar to those of free space medium, hence impedance matching over a wide bandwidth (Table 11.1) [1–5].

To explain the sensing mechanism in Figure 11.2, two sensing areas with different dimensions are developed to realize split ring resonate (SRR) structure, wherein meandered slot is employed for retrieving the permittivity, while parallel slots are used to reveal permeability characteristics. The working principle is explained as follows: the meandered line yields large electrical inductance, while the parallel lines can be viewed as a capacitive

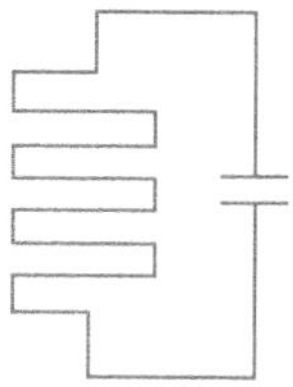

Figure 11.2 LoC-based power smart antenna sensor to measure electric and magnetic fields near to a biological system.

Table 11.1 Material-dependent comparison of properties for smart antenna.

Property	Magnetodielectric materials	Dielectric materials
Impedance match	$\sqrt{\mu/\varepsilon}$	$\sqrt{1/\varepsilon}$
Miniaturization	$\sqrt{\varepsilon\mu}$	$\sqrt{\varepsilon}$
Antenna bandwidth	$1/\varepsilon$	$1/\varepsilon$

Source: Based on Moser [1]; Hu et al. [2]; Vitale et al. [3]; Pol et al. [4]; de Figueiredo and coworkers [5].

electrical element. These two electrical passive components in LoC-based smart antenna sensor device can confine the electrical and magnetic fields, respectively. The magnetic field and electric field magnitude distributions concentrated on the ground plane of the proposed LoC-based smart antenna sensor at the resonance frequency would be improved by introducing a magnetodielectric substrate [8]. Moreover, The introduction of the compact design of sensing areas could facilitate field confinement. These separated magnetic and electric fields, i.e. opposite EM field distributions employing magnetic and electric field sensing parameters obtained by collecting field measurements, have effects on biological tissues or the nervous system that could be minimized by implementing an alternative protective scheme or avoiding those field sources [5, 6].

Carbon-based materials and their allotropic forms such as amorphous carbon, graphite, and diamonds added to newly introduced auspicious CNTs, grapheme oxides, graphene quantum dots, and fullerene have become important in the field of science and technology of biomedical applications. Each and every form of carbon family members exhibits inimitable features that were widely exploited in diverse biomedical applications including drug delivery, biosensing, imaging, and diagnosis [33]. Tunable carbon-based materials such as graphene have extensively explored in biomedical fields and as a metamaterial-based LoC device. 2D graphene sheets facilitate meta-atoms of metasurfaces enabling it to achieve versatile LoC or LoB devices [17], demonstrating that the carbon-based nanomaterials stand out as unique materials for the development of devices for detection, monitoring, and diagnosing in biomedical field areas.

The combined technology of carbon-based material and the antenna sensor takes further on the path toward low power and fast, efficient, and secure LoC- or LoB-based sensing. It also brings us to an important step closer to the realization of protection schemes against abnormal/damages in a human biological system. This technology creates a tremendous opportunity for researchers to better understand the brain and how it is affected by disorders that disrupt neural connections, like Alzheimer's disease and dementia, and to find new ways to treat brain diseases. It may even allow the routine exchange of data between computers and the human biological systems [5, 6].

Smart bioelectromagnetic electronic chip antenna as LoC device to tune and sense brain regions like radios for better memory: Our brains are electrochemical systems, using electricity to manipulate their neurological functions is a well-known proven technique [1]. Electric current-instigated stimulations applied to the brain hold a lot of promise, for example, deep brain stimulations that control the symptoms of depression to zapping our gray matter to improve our vision [1]. In the memory, the brain depends on its many regions working together to perform tasks such as memorizing the person's names or memorizing the past and present tasks. However, many of these are located far apart and are unable to coordinate efficiently during memorizing via communication process, particularly, those who are having damage in the brain or brain cells slows the communication between its distinct regions. Stimulating those patient's brain could help bypass the internal delays and help them to complete everyday tasks more efficiently [1]. Researchers have suggested that, in the brain, if we introduce rhythmic electrical wave activity, which will stimulate neurons to communicate with each other (like in-between neurotransmitter transporter and receptor) and is considered as an essential way to maintain the flow of information. For complex tasks, the accurate electric rhythmic wave signal without hostage of interference is an important prerequisite. In recent years, researchers have performed many experiments on brain neurons by introducing stimulations to mimic brain functions, for example, a microcircuit/silicon chip that could help fight paralysis [3], building brain-inspired computing [2], and electronic chip that could mimic the brain to make memories in a flash [4]. Neural stimulations involve a carbon nanotube electrode array [34] and recording bidirectional with a soft carbon nanotube fiber electrode array [3]. All of these findings suggest that the beat of brain waves is important for task performance and that the regions of the brain should be stimulated externally to communicate.

The electric rhythmic wave signal was obtained from the smart bioelectromagnetic electronic chip antenna as an LoC device to tune brain regions like radios for better memory. A similar proposed chip antenna is shown in Figure 11.2. This LoC antenna helps neural connections happen in the brain through electrical impulses. When tiny energy spikes reach a certain threshold of voltage, the neurons bind together as a result the flow of information from an assigned task. On the LoC antenna, the magnetic field is used to generate a magneto current (if it is an optoelectric antenna, it induces photocurrent) for switching the neurons and is equivalent to the binding and breaking of neural connections, enabling neurons to connect, i.e. induce to learn, or inhibit, i.e. to forget. This is akin to magnetogenetics, where magnetic field-induced stimulation of neurons causes them to either on or off, enabling or inhibiting connections to the next neuron in the chain.

Most biochemical reactions from digestion to brain activities go along with the rearrangement of charged particles, i.e. the influence of human brain just as they influence any other material made up of charged particles. When the magnetic field acts on a magnetoelectric material, they influence the distribution of electric charges on their surface. They cause current to flow through the neurons to stimulate. Material with magnetoelectric effect, i.e. the magnetic field-induced electric field, if such a material enabled antenna chip (LoC) placed near the neurons, allows communication and hence efficient performance of tasks. If magnetoelectric material is exposed to even extremely low magnetic field, they generate their own local electric field at the same frequency, where in turn the electric field can directly couple to the electric circuitry of the neural network. Carbon and its family of allotropic forms as magnetoelectric materials have been widely explored for neurological disorders due to their distinctive physiochemical functionalities. Particularly graphene and its functionalized forms are widely explored materials for neuron stimulations, which is attributable due to the charge density-enabled active surfaces that can be controlled via external stimulative sources such as electricity-induced external stimulations [3, 33]. More importantly, 2D graphene sheets facilitate as meta-atoms of metasurfaces would probably lead to achieving versatile LoC or LoB devices for neuron stimulations [3, 17].

Moreover, the combined (carbon/its allotropic forms with LoC device) technology may offer a new way to interact with computers; if it in reverse, the LoC produces a measurable field in response to the brain's own electrical fields, and it acts as an electronic smart antenna sensor. Our brain states would then become input parameters for computers, which would be able to directly stimulate regions of the brain in return. This combined (carbon/its allotropic forms with LoC device) technology takes further on the path toward fast efficient and secure magnetic field-based computing. It also brings us an important step closer to the realization of a bionic brain, a brain on a chip that can learn from the environment just like humans do. This technology creates a tremendous opportunity for researchers to better understand the brain and how it is affected by disorders that disrupt neural connections, like Alzheimer's disease and dementia, and opening up new ways to treat brain diseases. It may even allow the routine exchange of data between computers and the biological systems.

11.3.4 Ocean and Atmospheric Applications

The transfer of mass and energy between many subcategories of the marine ecosystem are regulated by many biogeochemical processes. Over the past decade, advancements in electroanalytical chemistry and electronic technology have been progressively achieved to probe marine environments and discover how marine chemistry and life interact with each other. In ocean engineering, the major objective is the development of sensors and systems that will provide a significant measurement of complementary processes such as physical and biological processes and variables over a large time intervals and large volumes. Also, molecular processes in the marine ecosystem affected by climate change are evaluated in submillimeter-scale investigations with a large-scale volume of data and a wide variety of interconnected sensors. Above all developments can assess ranging of ocean research field areas from natural biogeochemical cycles to marine pollution monitoring. Traditionally these requirements were met by manual sampling procedures for analysis either on board of research vessels or in the research laboratory. Remote sensing, autonomous underwater vehicles (AUVs), and buoys for automated *in situ* measurements are the other possible measurement technologies. However, traditional developments can assess only local environment variables at a single point in space and time. Traditional fixed interval measurements may not be sufficient for dynamic measurement of a wide variety of parameters and have limitations in terms of energy consumption and nonsteady responses. The deficiency of sensor development is primarily due to the demand of the working environment. Long-term reliability challenges often render even reliable and fully developed sensors unsuitable for ocean applications. Moreover, new challenges include biofouling, electrolysis, corrosion, calibration errors, sensor drift, need for miniaturization, power requirements, development issues, etc. Apart from chemical and biological measurements, certain physical measurements such as pressure or depth, temperature, and conductivity or salinity are also important in ocean metrology. These are primary data for tsunami detection, wave and tide gages, platform leveling applications, pressure sensors also find use in maritime or homeland port security. Moreover, in recent days detecting of surface vessel movements such as ships, boats or buoys, and subsurface systems unmanned submerged vehicles, submarines, mining systems, or surveillance systems have become more important.

In order to overcome all difficulties based on traditional technologies and for understanding of biogeochemical processes from the marine environment, high resolution and advanced measurement technologies such as carbon-based multiparameter assessment LoC devices are necessary. LoCs, i.e. nanotechnology-based chemical and biological parameter analyzers for example; wet chemical analyzers have shown reliable accuracy. Some other promising examples of LoC ocean sensors include a multiparameter detection system using electrodes, LoC microfluidic chemical detection system, *in-situ*-spectrometers. To satisfy reduced power consumption, low production cost and size, multi-environmental stability, used as disposal device, LoCs, and nanotechnology carbon-based materials are considered as the technologies of the future. However, challenges such as packaging for operation in ocean environment and biofouling, etc., need to be overcome to achieve reliable and economical device development.

Metamaterials are artificial periodic structures made of small elements (the size of the element is smaller than the wavelength of interest) and are designed to achieve specific electromagnetic functionality, i.e. an artificial structure can be described by an effective permittivity "ε" and permeability "μ." In 1968 Veselago introduced the concept of left-handed medium with negative permittivity and permeability rendered negative refractive index, he suggested that this significant functionality is attributable to the electric field "E" and magnetic field "H," and the wave vector from a left-handed medium [35]. Since the first experimental demonstration in 2000, these meta mediums have become attractive and essential to research not only because they exhibit extraordinary physical properties but also due to their broad applicability in various research environments such as in electromagnetic fields and physics. Therefore, since then lot of efforts have been attempted by the researchers to study metamaterial's functional properties and its applicability, particularly, in the field of atmospheric and space sciences [14–16, 36, 37]. Some of the researchers suggested that due to these extraordinary functional properties (not observed in the constituent materials), recently, the power of utilization of metamaterials (LH) in the design of electromagnetic wave radiation absorption systems has been increasing to avoid electromagnetic radiation pollution for the eco-friendly and sustainable development. Metamaterial concept with LoC platforms owing to its compactness could be a thought of futuristic radiation pollution control LoC devices [17, 38]. More recently functionalized carbon-based metamaterials are increasing in various fields and are proved in biomedical engineering and as sensors in other applications due to their extraordinary properties and functionalities [17, 38]. Such kind of functionalized carbon-based devices with LoC platforms could be a thought of futuristic multifunctional eco-friend and sustainable devices for atmospheric and space sciences [17, 38].

In atmospheric and space sciences, particle physics has a great impact on every field, though initially, particle accelerators are used to study how this universe is formed and what happens when two particles collide, and the physics behind particles, allowing us to implement these concepts in various fields. In particle physics, particle detection and its implementation is predominant, metamaterial concepts with LoC platforms are utilizing in the design of detectors that identify unknown particles in high energy experiments has been increasing [14–16, 36]. Moreover, there has been significant interest on LoC-based metamaterials (left-handed materials) for the understanding of Cherenkov radiation inside or in the vicinity of electromagnetic structured media within the particle detectors [14–16, 36]. Cherenkov radiation was experimentally discovered by Pavel Cherenkov and theoretically formulated by Ilya Frank and Igor Tamn et al. It is a peculiar form of electromagnetic radiation that produced when charged particles move through a medium with a speed/velocity greater than the phase velocity of light in that medium. This significant effect in the right-handed (RH) medium has proven useful in experimental and applied physics, for the detection of cosmic rays, novel electromagnetic sources, localized sensing in biological systems, spectroscopy of complex nanostructures, and elementary particle detection [36, 37, 39]. The mechanism involves the polarization of the atoms in the traversed medium with subsequent relation to induced dipoles returning to tshe baseline, as an unknown particle with a fixed momentum (faster than the speed of light) travels through a radiator (example of isotropic traditional radiators are one with refractive index $n_1 = 1.05$ (silica aerogel radiator) or one with a refractive index $n_2 = 1.0005$ (CF_4 gas radiator)), consequently, emits Cherenkov radiation, which is in the form of a cone [36, 37, 39]. This emitted Cherenkov cone propagates forward with the particle moment in material with a positive refractive index. Particle momentum and Cherenkov cone together measure the particle velocity, determining particle mass and identification of the particle [36, 37, 39]. However, the difficulty arises during the detection is the saturation of Cherenkov cone opening angle, i.e. the opening angle of Cherenkov cone saturates at higher momenta to a value independent of the particle's mass: $\alpha_{\text{PH Sat}} = \arccos\left(\frac{1}{n}\right)$, where n is the refractive index of the radiator, as a result, it is difficult to detect the particle, predominantly, difficulty in distinguishing the particles at higher momentum [36, 39]. Alternative solution for this problem is the lowering of refractive index of radiators. This indeed enhances the resolving power of detector at higher momenta. Unfortunately, this further leads to a drastic reduction of Cherenkov cone angles as a result number of photons emitted per unit length decreases [39]. In order to overcome these abnormalities during particle detection, medium with optimum refractive index control is preeminent and demanded, and this cannot be achieved with the traditional radiators; metamaterials could be an alternative solution [36, 39].

However, the common restriction in all metamaterials is that each functionality requires a distinct engineered structural design, making each metamaterial suitable only for a specific task and limiting its application. In response to these limitations, the requirements in modern and scientific-technological developments have stretched this metamaterial research to programmable/digital coding metamaterials. These programmable/digital coding metamaterials have been a wide-ranging research topic of interest for scientists and engineers since 2014 due to the controllability of EM wave behavior and its functionalities [14]. Graphene has been proven effective in controlling electromagnetic waves. To date, the graphene-based wavefront control is studied extensively in terahertz or higher frequencies, and it was suggested that the pure resistive nature of graphene and lack of reactance enable phase control. This kind of grapheme-based programmable/reconfigurable metamaterials as LoC devices could be an alternative solution for controlling of refractive index during particle detection in detectors [17, 38].

Programmable metamaterials are engineered structural design materials with controlled EM functionalities (functionalities for example: permeability, permittivity, refractive index, phase, and amplitude) that are programmable by reconfigurable and digital coding approaches, which could be an alternative solution to control radiator refractive index according to the particle momentum during elemental particle detection [14–16]. In programmable metamaterials, reconfigurability is achieved by using components that tune metamaterial response (such as refractive index and/or phase) or that completely change the current flow within the metamaterial medium to tune EM functionalities, including varactors or variable resistors, PIN diodes or switches, microelectromechanical systems [14–16]. On the other hand, these coding metamaterials (single metamaterials can be digitally controlled to obtain distinctly different functionalities) can digitally control EM wave functionalities by applying binary bits (0 and 1) on engineered periodic structural configurations [14–16]. Due to their inherent digital nature, programmable metamaterial associate to the coding metamaterial paradigm, where, the device is encoded using a discrete set of unit cell states. The state of the metamaterial can be thus elegantly described as a state matrix, thus providing a powerful and intuitive design perspective while drawing a clear parallelism with information theory. Carbon-based material technology combined with the programmable/reconfigurable metamaterial technology opens new ways to design and has proven helpful in tackling the challenge of packing multiple EM wave functionalities within a single device and serving multiple purposes concurrently [14–16].

Moreover, significant interest on metamaterials has been providing a new research scope for understanding of the direction of Cherenkov radiation and its effect by analyzing refractive index and its electromagnetic properties such as permeability and permittivity of that media, i.e. metamaterials facilitate to investigate reversed Cherenkov radiation (RCR) and its effect [36]. RCR is an exotic electromagnetic radiation that is emitted opposite to the moving charged particles in left-handed materials (negative refractive index mediums such as metamaterial), where the Cherenkov cone is reversed such that the particle travel in the opposite direction in that medium [36], which we cannot observe in traditional isotropic radiators such as silica aerogel radiator and CF_4 gas radiator. Metamaterial structures contain periodic subdiffraction dimensions, whereby plasma behavior at a particular frequency. Plasmonic behavior allows magnetic and electrostatic properties to be decoupled, giving rise to novel properties such as negative refractive index, as a result, reversed Cherenkov cone. Left-handed materials (LHM) materials are of particular use with RCR as a reversed cone angle yields a wider range to measure compared to the traditional materials, which are limited 0–90°. This wider angle range allows LHM to be used as more accurate detectors since the cone angle is related to the particle velocity [37]. RCR has a distinct advantage that the detectors for particle and the detectors for RCR are naturally separated forward and backward regions, so that the physical interference is minimized and the detection sensitivity can be improved than the traditional detectors. Moreover, oscillators or amplifiers are based on CR effect that employs double negative materials (DNMs) rather than traditional materials. Hence, by tailoring the material properties of DNMs, we can easily obtain slower phase velocity, allowing the oscillator or amplifier can operate at low voltage. In addition, losses in DNMs can be reduced by increasing inductance to capacitance ratio, i.e. L/C ratio [40]. The above delineated significant advantages and unusual properties of metamaterial in addition to the carbon material technology such as recent developments of carbon composites, carbon nanotubes, grapheme hybrids, and composites [41–55] are demanded for the requirements in modern and scientific-technological developments. Carbon materialenabled programmable metamaterials are engineered structural design materials with controlled EM functionalities that are programmable by reconfigurable and digital coding approaches could be an alternative solution for generating RCR to investigate its effect and applicability in atmospheric and ocean science [14–16]. Certainly, there are still challenging concepts that need to be investigated for developing a reliable RCR detector.

In summary, the experimental demonstration of double negative materials and increased control of the properties of metamaterial would allow for new Cherenkov applications beyond particle physics and beyond the relationship described in the Frank–Tamm equation. Here a metamaterial offers a possibility of guiding electromagnetic waves, moreover, programmable metamaterials with controlled EM functionalities via reconfigurable, and digital coding approaches would allow novel and miniaturized detectors. This could be an alternative solution for the limitations by traditional particle detectors. New opportunities for Cherenkov applications appear as nanoscience and materials research advances, with some experimental results long-postulated and others surprising the field.

11.4 Conclusion

Metamaterial-based LoC devices integrated solar cells can offer unique advantages compared to the conventional solar cells. The advantages include ease of integration, portability with low cost, and high surface-area-to-volume ratio. However, carbon-based programmable/digital coding metamaterials as LoC devices facilitate near-perfect trapping of light over the broad spectral range into a semiconducting layer of solar cells for the realization of low cost and highly efficient photovoltaic devices. Further, carbon-based metamaterial concepts with LoC platforms are expanded to the field of atmosphere and space technology for designing of detectors that identify unknown particles in high energy experiments. Also, these carbon-based LoC devices as left-handed mediums could facilitate for understanding of RCR and its effect inside or in the vicinity of electromagnetic structured media via programmable controlling of radiator refractive index according to the particle momentum during elemental particle detection within the particle detector. Moreover, carbon-based metamaterials as LoC devices facilitate smart delivery of fertilizers and pesticides, detection of contaminants, and pathogens for improving the practices of precision farming in the agriculture field. These devices allows detection, monitoring, and diagnosis of the problems in the field of biomedicals for improving the quality of life. Also, carbon-based LoC sensor devices are prescribed for significant measurement of complementary processes such as physical and biological processes and variables over large time intervals and large volumes. Hence, marine ecosystem molecular processes to climate change at the submillimeter scale could be protected by taking prevention strategies against marine pollution. Therefore, carbon and its allotrope family members with wide-ranging functionalities enable LoC devices to be multifunctional, eco-friendly, and sustainable devices and also to facilitate efficient and low-cost eco-friendly and sustainable development.

Important Websites

https://www.google.com/search?client=firefox-b-d&q=Functionalized+Carbon+Nano+Lab+on+Chip+Devices+for+Environment.

https://www.google.com/search?q=Functionalized+Carbon+Nano+Lab+on+Chip+Devices+for+Environment&newwindow=1&client=firefox-b-d&sxsrf=ALeKk002dQXPkbm6NXqX9YcWet2m0IWWcg:1603725527428&source=lnms&tbm=isch&sa=X&ved=2ahUKEwi_mISTx9LsAhUkSBUIHdbvAksQ_AUoAXoECAYQAw&biw=1368&bih=798.

References

1 Moser, I. (1995). Miniaturized thin film glutamate and glutamine biosensors. *Biosens. Bioelectron.* 10 (6–7): 527–532.

2 Hu, Q., Wujcik, E.K., Kelarakis, A. et al. (2017). Carbon-based nanomaterials as novel nanosensors. *J. Nanomater.* 2017: 1–2.

3 Vitale, F., Summerson, S.R., Aazhang, B. et al. (2015). Neural stimulation and recording with bidirectional, soft carbon nanotube fiber microelectrodes. *ACS Nano* 9 (4): 4465–4474.

4 Pol, R., Céspedes, F., Gabriel, D., and Baeza, M. (2017). Microfluidic lab-on-a-chip platforms for environmental monitoring. *TrAC Trends Anal. Chem.* 95: 62–68.

5 Han, A., Hou, H., Li, L. et al. (2013). Microfabricated devices in microbial bioenergy sciences. *Trends Biotechnol.* 31 (4): 225–232.

6 Foudeh, A.M., Fatanat Didar, T., Veres, T., and Tabrizian, M. (2012). Microfluidic designs and techniques using lab-on-a-chip devices for pathogen detection for point-of-care diagnostics. *Lab Chip* 12 (18): 3249.

7 Koev, S.T., Dykstra, P.H., Luo, X. et al. (2010). Chitosan: an integrative biomaterial for lab-on-a-chip devices. *Lab Chip* 10 (22): 3026.

8 Zhou, J., Zheng, J., Shi, X. et al. (2019). Graphene-based fully transparent thin film surface acoustic wave devices for sensing and lab-on-chip applications. *J. Electrochem. Soc.* 166 (6): B432–B440.

9 López-González, B., Jiménez-Valdés, R.J., Moreno-Zuria, A. et al. (2017). Waste-to-energy conversion from a microfluidic device. *J. Power Sources* 360: 80–86.

10 Nascetti, A., Caputo, D., Scipinotti, R., and de Cesare, G. (2016). Technologies for autonomous integrated lab-on-chip systems for space missions. *Acta Astronaut.* 128: 401–408.

11 Takenaga, S., Schneider, B., Erbay, E. et al. (2015). Fabrication of biocompatible lab-on-chip devices for biomedical applications by means of a 3D-printing process. *Phys. Status Solidi* 212 (6): 1347–1352.

12 Mewada, A., Pandey, S., Shinde, S. et al. (2013). Green synthesis of biocompatible carbon dots using aqueous extract of *Trapa bispinosa* peel. *Mater. Sci. Eng. C* 33 (5): 2914–2917.

13 Das, T., Saikia, B.K., Dekaboruah, H.P. et al. (2019). Blue-fluorescent and biocompatible carbon dots derived from abundant low-quality coals. *J. Photochem. Photobiol. B* 195 (1–11): 6–19.

14 Abadal, S., Cui, T.J., Low, T., and Georgiou, J. (2020). Programmable metamaterials for software-defined electromagnetic control: circuits, systems, and architectures. *IEEE J. Emerg. Sel. Top. Circuits Syst.* 10 (1): 6–19.

15 Zhao, J., Yang, X., Dai, J.Y. et al. (2019). Programmable time-domain digital-coding metasurface for non-linear harmonic manipulation and new wireless communication systems. *Natl. Sci. Rev.* 6 (2): 231–238.

16 Bao, L. and Cui, T.J. (2020). Tunable, reconfigurable, and programmable metamaterials. *Microwave Opt. Technol. Lett.* 62 (1): 9–32.

17 Chen, H., Lu, W.-B., Liu, Z.-G., and Geng, M.-Y. (2020). Microwave programmable graphene metasurface. *ACS Photonics* 7 (6): 1425–1435.

18 Ahmed, F., Arshi, N., Kumar, S. et al. (2013). Nanobiotechnology: scope and potential for crop improvement. In: *Crop Improvement Under Adverse Conditions* (ed. N. Tuteja S.S. Gill), 245–269. New York, NY: Springer, Softcover ISBN 978-1-4899-9298-7, eBook ISBN 978-1-4614-4633-0.

19 Khodakovskaya, M., Dervishi, E., Mahmood, M. et al. (2009). Carbon nanotubes are able to penetrate plant seed coat and dramatically affect seed germination and plant growth. *ACS Nano* 3 (10): 3221–3227.

20 González-Melendi, P., Fernández-Pacheco, R., Coronado, M.J. et al. (2008). Nanoparticles as smart treatment-delivery systems in plants: assessment of different techniques of microscopy for their visualization in plant tissues. *Ann. Bot.* 101 (1): 187–195.

21 Nair, R., Varghese, S.H., Nair, B.G. et al. (2010). Nanoparticulate material delivery to plants. *Plant Sci.* 179 (3): 154–163.

22 Kumar, P., Fennell, P., and Robins, A. (2010). Comparison of the behaviour of manufactured and other airborne nanoparticles and the consequences for prioritising research and regulation activities. *J. Nanopart. Res.* 12 (5): 1523–1530.

23 Wu, Y., Phillips, J.A., Liu, H. et al. (2008). Carbon nanotubes protect DNA strands during cellular delivery. *ACS Nano* 2 (10): 2023–2028.

24 Younas, A., Yousaf, Z., Rashid, M. et al. (2020). Nanotechnology and plant disease diagnosis and management. In: *Nanoagronomy* (ed. S. Javad), 101–123. Cham: Springer International Publishing, Hardcover ISBN 978-3-030-41274-6, ISBN 978-3-030-41275-3.

25 Nezhad, A.S. (2014). Future of portable devices for plant pathogen diagnosis. *Lab Chip* 14 (16): 2887–2904.

26 Pimentel, D. Invasive plants: their role in species extinctions and economic losses to agriculture in the USA. In: *Management of Invasive Weeds* (ed. Inderjit), 1–7. Dordrecht: Springer Netherlands, Hardcover ISBN 978-1-4020-9201-5, Softcover ISBN 978-90-481-8089-9.

27 Lattanzio, V.M.T., Nivarlet, N., Lippolis, V. et al. (2012). Multiplex dipstick immunoassay for semi-quantitative determination of Fusarium mycotoxins in cereals. *Anal. Chim. Acta* 718: 99–108.

28 Tian, X., He, W., Cui, J. et al. (2010). Mesoporous zirconium phosphate from yeast biotemplate. *J. Colloid Interface Sci.* 343 (1): 344–349.

29 Ma, Y., Jiao, K., Yang, T., and Sun, D. (2008). Sensitive PAT gene sequence detection by nano-SiO_2/*p*-aminothiophenol self-assembled films DNA electrochemical biosensor based on impedance measurement. *Sens. Actuators, B: Chem.* 131 (2): 565–571.

30 Qu, J.-H., Horta, S., Delport, F. et al. (2020). Expanding a portfolio of (FO$^-$) SPR surface chemistries with the co(III)-NTA oriented immobilization of his 6-tagged bioreceptors for applications in complex matrices. *ACS Sensors* 5 (4): 960–969.

31 Fan, X., White, I.M., Shopova, S.I. et al. (2008). Sensitive optical biosensors for unlabeled targets: a review. *Anal. Chim. Acta* 620 (1–2): 8–26.

32 Yoo, Y.E., Smadja, C., and Ammar, M. (2020). Micro/nano fabrication and packaging technologies for bio systems. In: *Engineering of Micro/Nano Biosystems* (ed. G. Barbillon, A. Bosseboeuf, K. Chun, R. Ferrigno, O. Français). Microtechnology and MEMS, 89–137. Singapore: Springer, Print ISBN: 978-981-13-6548-5, Electronic ISBN: 978-981-13-6549-2.

33 Maiti, D., Tong, X., Mou, X., and Yang, K. (2019). Carbon-based nanomaterials for biomedical applications: a recent study. *Front. Pharmacol.* 9: 1–16.

34 Wang, K., Fishman, H.A., Dai, H., and Harris, J.S. (2006). Neural stimulation with a carbon nanotube microelectrode array. *Nano Lett.* 6 (9): 2043–2048.

35 Veselago, V.G. (1968). The electrodynamics of substances with simultaneously negative values of ε and μ. *Sov. Phys. Uspekhi* 10 (4): 509–514.

36 Duan, Z., Tang, X., Wang, Z. et al. (2017). Observation of the reversed Cherenkov radiation. *Nat. Commun.* 8 (4): 4–10.

37 Shaffer, T.M., Pratt, E.C., and Grimm, J. (2017). Utilizing the power of Cerenkov light with nanotechnology. *Nat. Nanotechnol.* 12 (2): 106–117.

38 Liaskos, C., Tsioliaridou, A., Pitsillides, A. et al. (2015). Design and development of software defined metamaterials for nanonetworks. *IEEE Circuits Syst. Mag.* 15 (4): 12–25.

39 Ginis, V., Danckaert, J., Veretennicoff, I., and Tassin, P. (2014). Controlling Cherenkov radiation with transformation-optical metamaterials. *Phys. Rev. Lett.* 113 (16): 1–5.

40 Duan, Z.Y., Wu, B.I., Xi, S. et al. (2009). Research progress in reversed Cherenkov radiation in double-negative metamaterials. *Prog. Electromagn. Res.* 90: 75–87.

41 Mallakpour, S. and khodadadzadeh, L. (2018). Chapter 7. Biocompatible and biodegradable chitosan nanocomposites loaded with carbon nanotubes. In: *Biodegradable and Biocompatible Polymer Composites Processing, Properties and Applications* (ed. N.G. Shimpi), Woodhead Publishing is an imprint of Elsevier, The Officers' Mess Business Centre, Royston Road, Duxford, CB22 4QH, United Kingdom. 50 Hampshire Street, 5th Floor, Cambridge, MA 02139, United States. The Boulevard, Langford Lane, Kidlington, OX5 1GB, United Kingdom, 187–221.

42 Mallakpour, S. and Khadem, E. (2019). Chapter 8. Carbon nanotubes for heavy metals removal. In: *Composite Nanoadsorbents: A volume in Micro and Nano Technologies* (eds. G. Kyzas and A.C. Mitrpoulos), Imprint Elsevier, ISBN 978-0-12-814132-8, 181–210, https://doi.org/10.1016/C2017-0-00861-3.

43 Mallakpour, S. and Rashidimoghadam, S. (2019). Chapter 9. Carbon nanotubes for dyes removal. In: *Composite Nanoadsorbents: A volume in Micro and Nano Technologie* (eds. G. Kyzas and A.C. Mitrpoulos), Imprint Elsevier, ISBN978-0-12-814132-8, 211–244, https://doi.org/10.1016/C2017-0-00861-3.

44 Mallakpour, S. and Khadem, E. (2016). Carbon nanotube–metal oxide nanocomposites: fabrication, properties and applications. *Chem. Eng. J.* 302: 344–367.

45 Mallakpour, S., Abdolmaleki, A., and Borandeh, S. (2017). Fabrication of amino acid-based graphene-zinc oxide (ZnO) hybrid and its application for poly(ester–amide)/graphene-ZnO nanocomposite synthesis. *J. Thermoplast. Compos. Mater.* 30: 358–380.

46 Mallakpour, S. and Behranvand, V. (2018). Synthesis of mesoporous recycled poly(ethylene terephthalate)/MWNT/carbon quantum dot nanocomposite from sustainable materials using ultrasonic waves: application for methylene blue removal. *J. Cleaner Prod.* 190: 525–537.

47 Mallakpour, S., Behranvand, V., and Mallakpour, F. (2019). Synthesis of alginate/carbon nanotube/carbon dot/fluoroapatite/TiO_2 beads for dye photocatalytic degradation under ultraviolet light. *Carbohydr. Polym.* 224: 115138.

48 Abdolmaleki, A., Mallakpour, S., and Karshenas, A. (2017). Facile synthesis of glucose-functionalized reduced graphene oxide (GFRGO)/poly(vinyl alcohol) nanocomposites for improving thermal and mechanical properties. *Mater. Sci. Eng., B* 217: 26–35.

49 Mallakpour, S., Abdolmaleki, A., and Karshenas, A. (2017). Graphene oxide supported copper coordinated amino acids as novel heterogeneous catalysts for epoxidation of norbornene. *Catal. Commun.* 92: 109–113.

50 Abdolmaleki, A., Mallakpour, S., and Karshenas, A. (2017). Synthesis and characterization of new nanocomposites films using alanine-Cu-functionalized graphene oxide as nanofiller and PVA as polymeric matrix for improving of their properties. *J. Solid State Chem.* 253: 398–405.

51 Abdolmaleki, A., Mallakpour, S., and Borandeh, S. (2016). Improving interfacial interaction of l-phenylalanine-functionalized graphene nanofiller and poly(vinyl alcohol) nanocomposites for obtaining significant membrane properties: morphology, thermal, and mechanical studies. *Polym. Compos.* 37: 1924–1935.

52 Mallakpour, S., Abdolmaleki, A., Mahmoudian, M. et al. (2017). Synergetic effect of synthesized sulfonated polyaniline/quaternized graphene and its application as a high-performance supercapacitor electrode. *J. Mater. Sci.* 52: 9683–9695.

53 Abdolmaleki, A., Mallakpour, S., and Mahmoudian, M. (2017). Preparation and evaluation of edge selective sulfonated graphene by chlorosulfuric acid as an active metal-free electrocatalyst for oxygen reduction reaction in alkaline media. *ChemistrySelect* 34: 11211–11217.

54 Mallakpour, S. and Khadem, E. (2018). Construction of crosslinked chitosan/nitrogen-doped graphene quantum dot nanocomposite for hydroxyapatite biomimetic mineralization. *Int. J. Biol. Macromol.* 120: 1451–1460.

55 Mallakpour, S. and Khadem, E. (2019). Linear and nonlinear behavior of crosslinked chitosan/N-doped graphene quantum dot nanocomposite films in cadmium cation uptake. *Sci. Total Environ.* 690: 1245–1253.

12

Functionalized Carbon Nanotubes (FCNTs) as Novel Drug Delivery Systems: Emergent Perspectives from Applications

Shikha Gulati, Sanjay Kumar, Ayush Mongia, Anchita Diwan, and Parinita Singh

University of Delhi, Sri Venkateswara College, Department of Chemistry, Dhaulan Kuan, Delhi, 110021, India

12.1 About the Chapter

This chapter provides an overview of the advancements in the field of functionalized carbon nanotubes (FCNTs) as novel drug delivery systems. These nanotubes have emerged as a highly efficient drug delivery system, owing to their excellent physical and chemical properties. They are being extensively researched upon with respect to their biomedical applications due to their high drug loading capability and controlled release of the delivered drug. Also, there are certain additional advantages that carbon nanotubes (CNTs) obtain once they are functionalized with suitable agents to modify their physical and biological properties. FCNTs exhibit non-immunogenicity, biocompatibility, and reduced toxicity to cells. There are two widely employed methods to FCNTs: covalent and non-covalent functionalizations, which are discussed briefly. In this chapter, special emphasis is given to those functionalizing agents, which on combining with CNTs improve their functioning as drug delivery systems. This chapter also discusses classification of CNTs, methods for their synthesis, and ways to functionalize them by employing diverse methods. Like many other drug delivery systems, CNTs are also prone to impurities that may hamper their abilities to deliver a drug efficiently, and for this reason, some of the important methods for their purification are delved into. In addition, their potential in drug delivery is discussed followed by an analysis on their drug loading efficiency, which depicts certain studies that are conducted to obtain information about their loading capabilities and how this characteristic influences the desired therapeutic outcome. A very important aspect of any drug delivery system is to achieve targeted delivery of the drug to the affected site. On doing so, the required dose is reduced because of targeted delivery, and moreover, nonspecific accumulation of drugs in healthy cells of the body is prevented. Therefore, this chapter emphasizes this aspect of CNTs for achieving targeted drug

Environmental Applications of Carbon Nanomaterials-Based Devices, First Edition.
Edited by Shadpour Mallakpour and Chaudhery Mustansar Hussain.

delivery by focusing on the role of FCNTs toward the treatment of some chronic diseases and their affected locations within body. The discussions in this area include cancer, brain, lymphatic, and tuberculosis targeting, highlighting the applications of FCNTs and their potential in drug delivery.

12.2 Introduction

The bioavailability and inherent toxicity of various potential drugs affect their pharmacological profile so much that their therapeutic efficacy is strongly compromised. One of the possibilities that can act as a potential source of innovative drug design strategies is the incorporation of nanotechnologies at an untimely step in the drug development process. Several applications of nanotechnology in medicinal field have encouraged the development of nanoparticles, nanostructured surfaces, and nano-analytical techniques for fast and early diagnostics, smart drug delivery, and real-time assessments of therapeutic efficacy. The very aim of generating a nanocarrier drug delivery system is to reduce the toxicity due to high doses that need to be administered and improve the therapeutic effect caused by the agent employed. These systems exhibit vast potential for the mitigation of patient risk and subsequent progression of the disease and the realization of personalized medicine (theranostics). In this regard, CNTs have been extensively explored as multipurpose carriers for achieving enhanced diagnosis and delivery of the required drug. Their multipurpose physicochemical features facilitate the covalent and non-covalent introduction of numerous pharmaceutically applicable entities and permit for coherent design of innovative nanostructures for drug development. CNTs are composed of seamless cylinders (open ended or capped) that are formed by the rolling of graphene sheets. The cylinders have a high aspect ratio with diameters as small as 1 nm and a length of several micrometers. CNTs made from a single graphene sheet results in a single-walled nanotubes (SWNT), while several graphene sheets make up multiwalled carbon nanotubes (MWCNTs) (Figure 12.1). The CNTs were discovered earlier by Bacon in 1960 while the first successful synthesis of MWCNTs was witnessed by the transmission electron microscopy (TEM) microphotograph of a Japanese microscopist, Sumio Iijima. Single-walled carbon nanotubes (SWCNTs) were synthesized in 1993. There has been an elevated interest in these allotropes of carbon, owing to their exceptional physical and chemical properties along with potential uses in various domains ever since their discovery in 1991 by Iijima [1]. However, the poor dispersibility of CNTs has been the utmost obstruction to their use in nanomedicine. Thus, it was only after the development of strategies to functionalize these molecules with organic groups and make them soluble that opened the way to bio-applications of CNTs. Therefore, CNTs can be functionalized due to their high surface area to enhance their solubility in the aqueous phase or to provide the appropriate functional groups that can bind to the desired therapeutic material or the target tissue to elicit a therapeutic effect and improve their biocompatibility. Also, CNTs might aid the penetration of the target cell by the delivered therapeutic molecule to treat several diseases. Our focus is on the advancement in the functionalizations of CNTs, which are

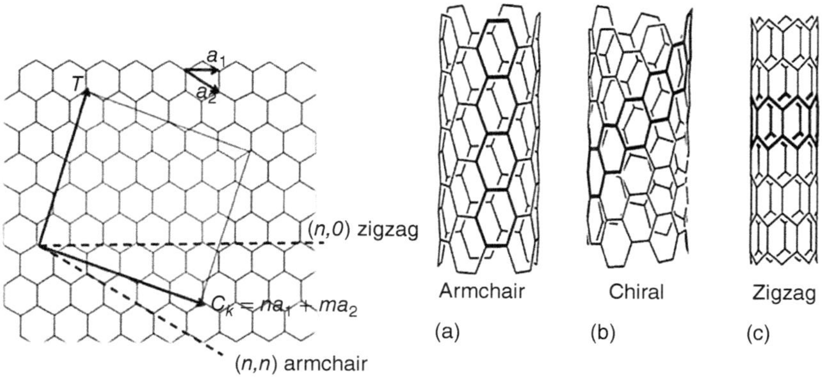

Figure 12.1 Types of SWCNTs: (a) armchair arrangement, (b) chiral arrangement, and (c) zigzag arrangement.

the preconditions for CNT applications in medicine. In general, there are two main strategies based on both covalent and non-covalent linkages between the surface of nanotubes and the different biological moieties employed for the functionalization of CNTs, i.e. (i) covalent and (ii) non-covalent functionalizations. In this chapter, we provide an overview of the medical applications in addition to drug delivery and in treatments of several diseases including the diagnosis and treatment of cancer, lymphatic targeting, infectious diseases, and central nervous system disorders and applications in tissue engineering and the associated potential risks of CNT applications in nanomedicine.

12.3 Carbon Nanotubes (CNTs)

CNTs are unique, sp^2-hybridized, three-dimensional, carbon-based nanomaterials with C–C distance of ~1.4 Å envisioned as seamless tubular cylinder rolled-up graphite planes with open ends and diameter of around tens of nanometers. This carbon-based novel nanomaterial belongs to a subfamily of fullerenes (or bucky balls), i.e. the third allotropic form of carbon along with diamond and graphite. CNTs are hexagonally arranged with honeycombed lattice of carbon atoms with the unique characteristics including high aspect ratio, nanosized diameter, ultralight weight and surface area, rich surface chemistry, neutral electrostatic potential and extremely high drug cargo ability, and excellent material properties, such as ultimate electrical and thermal conductivities and mechanical strength, rendering them attractive valuable nanovehicle for drug delivery [2, 3]. They are recognized as the third allotrope of carbon with length-to-diameter ratio of up to 28 000 000 : 1. Individual CNTs are different from each other depending on the number of carbon atoms, such as C20, C30, C36, C70, and C78, and each of these individual members is called "graphene" [1, 4–6]. These hollow cylindrical tubes collectively form hexagonal structural units because of their arrangement in a specific pattern [7–11].

12.4 Classification of CNTs

CNTs can be broadly classified into two types based on their wall structure, that is, (i) SWCNTs and (ii) MWCNTs:

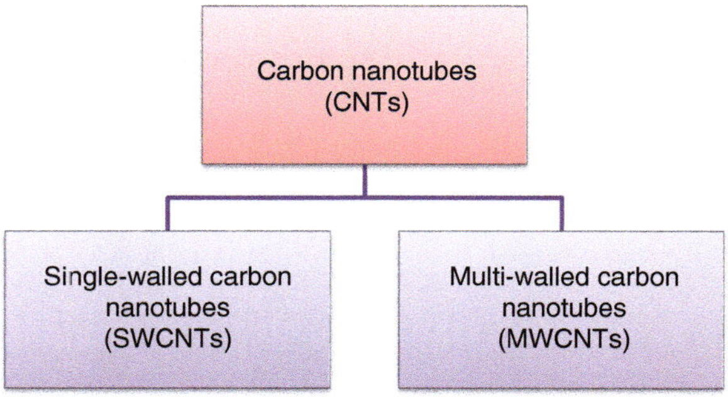

(i) **Single-walled Carbon Nanotubes (SWCNTs)**: A single graphene sheet is folded to form a cylindrical closed structure. SWCNTs are characterized by the hemispherical arrangement of carbon networks represented by the presence of a single graphitic sheet with its both ends capped. These have diameter range from 0.4 to 3.0 nm and length of 20–1000 nm. Also, the SWCNTs may further be divided into categories based on the differences in the carbon atom arrangement (Figure 12.1): (a) the armchair arrangement is typically recognized by the chairs perpendicular to the tube axis, whereas (b) zigzag arrangement consists of a V-shaped perpendicular to the tube axis. The degree of chirality determines the unique electrical, conductive properties and metallic characteristics.

(ii) **Multiwalled Carbon Nanotubes (MWCNTs)**: Around 3–5 sheets of SWNTs are rolled upon each other to form a multilayered structure. The concentric cylindrical layers of graphene sheets have length in the range from 1 to 50 μm and diameter in the range from 1.4 to 100 nm.

Apart from structural differences, dimensional differences also exist between the CNTs. A coaxial structure, containing two concentric graphene cylinders with higher thermal and chemical stability than *SWNTs*, is also found. These are termed *double-walled carbon nanotubes (DWCNTs)* that are another type of CNTs that resemble SWCNTs due to similar morphology. DWCNTs consist of exactly two concentric cylindrical layers, whereas *triple-walled carbon nanotubes (TWCNTs)* are characterized by the presence of three walls. Apart from this, many variations in CNT shapes have been observed including carbon nanohorns (CNHs), nanobuds, and nanotorus, which have huge potential in drug delivery due to the adaptability of their modified structure. Figure 12.2 diagrammatically depicts the various types of CNTs used in drug delivery applications. Further, CNTs are also classified depending upon their synthesizing methods, structural modifications, and solubility, such as conjugated CNTs, functionalized

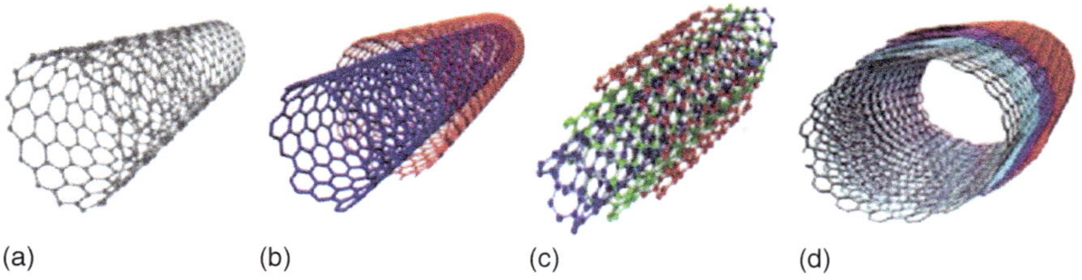

Figure 12.2 Types of carbon nanotubes: (a) single wall, (b) double wall, (c) triple wall, and (d) multiwall.

CNTs, CNTs dispersed in solvent, and surfactant-coated CNTs, which may include drugs, monoclonal antibodies, ligands, etc.

12.4.1 Advantages of Carbon Nanotubes (CNTs)

CNTs may offer following advantages particularly in the context of drug delivery applications:

- Extremely light weight and easy availability.
- Highly ordered structure with uniformity and high aspect ratio.
- Photoluminescence property.
- Highly elastic.
- Biocompatibility and non-immunogenicity.
- Cellular internalization via endocytosis and passive diffusion mechanisms.
- Ferromagnetism and nonlinear property.
- Cell membrane penetration due to its tiny nano-needle tubular structure.
- Open ends on both sides of CNTs that make the inner surface available and further help in incorporation of bioactive molecules within the tubes.
- Longer inner volume for endohedral filling relative to diameter.
- High mechanical strength, thermal conductivity, superconductivity, and metallic or semimetallic behavior.

12.4.2 Disadvantages of Carbon Nanotubes (CNTs)

In addition to the several advantages, there are some disadvantages also associated with CNTs:

- Pristine CNTs exhibit some toxicity; therefore these are not suitable for drug delivery.
- Bundling/aggregation phenomena.
- Low aqueous solubility in organic or nonorganic solvents.
- Accumulation in liver.

However, the above advantages and disadvantages have to be weighed depending upon the case.

12.5 Synthetic Methodologies of CNTs

The most widely used processes for synthesizing CNTs include the following:

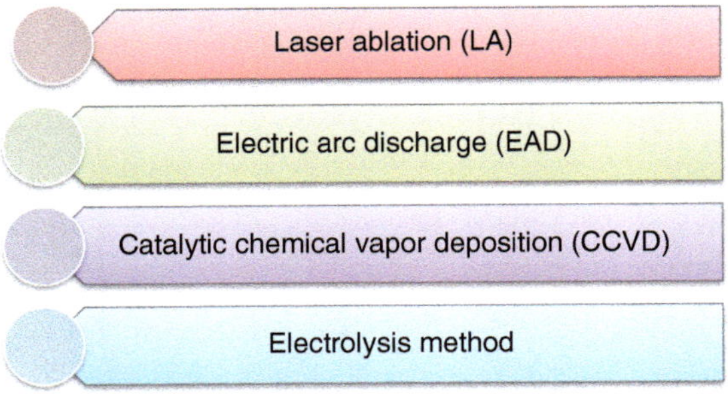

There are certain characteristics on which the selection of a specific synthesis method depends on, which include controllable diameter, purity, cost, quality, minimum impurities, and yield obtained [12, 13]. These methods are discussed below in brief.

12.5.1 Laser Ablation (LA) Method

CNTs can be synthesized through this method inside high-temperature furnace (1200 °C) containing argon (Ar) atmosphere (Figure 12.3). Here, irradiation of a metal-doped solid graphite target is carried out by high-power laser vaporization of pure graphite target. This technique results in the synthesis of CNTs that are uniform, chiral, and highly pure, have fewer structural defects, and have controlled size. The yield obtained would be approximately 70% and can be elevated by porous targets that are opposed to the standard metal targets or by employing rapid laser pulses. The diameter of synthesized CNTs is known to become narrower with elevation in the power of laser pulses [14, 15].

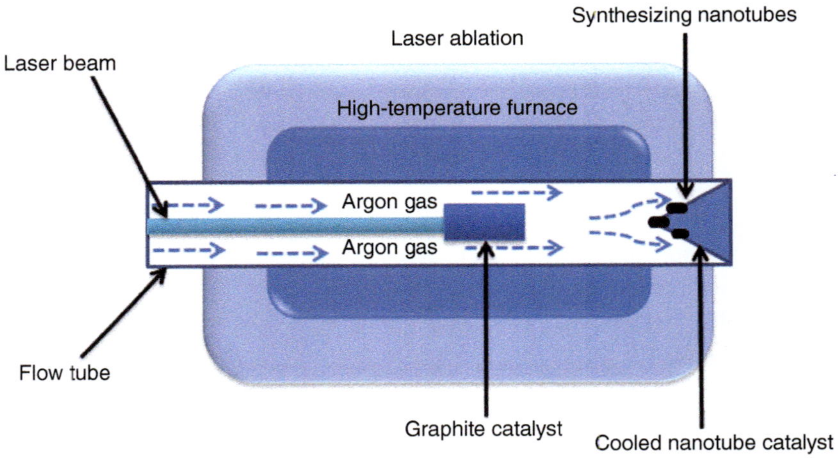

Figure 12.3 Laser ablation (LA) method.

12.5.2 Electric Arc Discharge (EAD) Method

In this technique, an electric current is passed in an inert gas (He or Ar) atmosphere by employing two types of graphite electrodes (anode and cathode) that are pure and different (Figure 12.4). Application of high voltage electron beams (around 100 amp) leads to the synthesis of CNTs at the end of the anode. These beams are generated by the electric arc, which bombards the surface of graphite. The formation of CNTs on a substrate takes place due to the electric arc, which is a plasmon setup made across nanotubes. This technique is economically feasible and generates CNTs with fewer defects; however, inconsistent size and carbonaceous and metallic impurities are observed in SWCNTs that are produced through this technique [15, 16]. About 30–90% yield of SWCNTs (0.5–1.5 nm diameter) can be obtained by employing various metal catalysts like Co, Rh, Pt, Fe, and Ni.

12.5.3 Catalytic Chemical Vapor Deposition (CCVD) Method

In this method, resistive heating or an electron beam is employed to generate energy, and hydrocarbons such as propylene, ethylene, toluene, and acetylene, etc. are employed as raw materials (Figure 12.5). In the presence of an inert gas, pyrolysis of hydrocarbons takes place, leading to the production of reactive carbon species due to the breakdown of molecules at about 500–1000 °C. CNTs are produced due to diffusion and binding of carbon to the heated substrate that is coated with a catalyst [14, 17, 18].

12.5.4 Electrolysis Method

In this method, two electrodes are used that are immersed in molten ionic salts at 600 °C and are subjected to a current, which leads to the production of MWCNTs. This step is followed by dissolving ionic salt in distilled water to obtain carbonaceous material. MWCNTs having a diameter of 10–20 nm along with 10–15 walls and more than 500 nm in length are obtained [19].

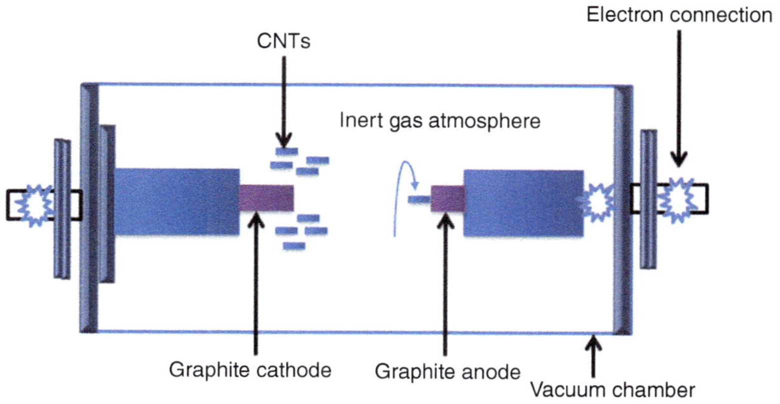

Figure 12.4 Electric arc discharge (EAD) method.

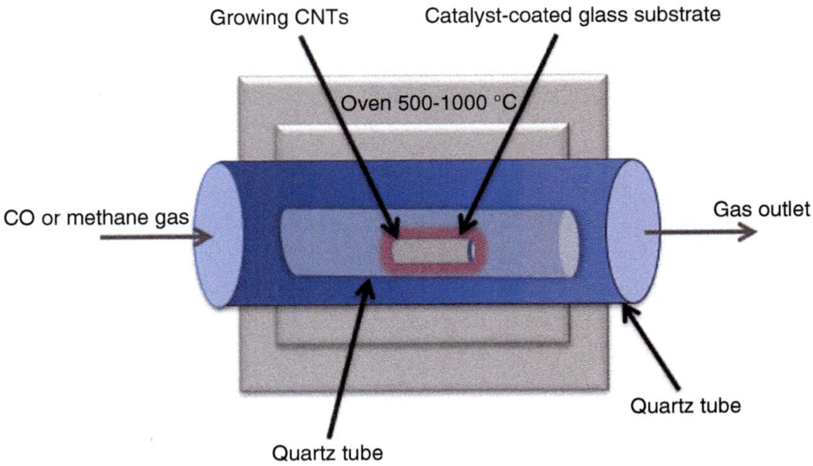

Figure 12.5 Catalytic chemical vapor deposition (CCVD) method.

12.6 Purification Techniques of CNTs

The synthesized CNTs contain a variety of impurities such as amorphous carbon and metal catalysts that are not suitable for drug delivery applications. Hence, several techniques can be used for the purification of CNTs including vacuum oven treatment, microwave treatment, oxidation, surface annealing, microfiltration, ultrasonication, cutting, and chromatography [20, 21] (Figure 12.6).

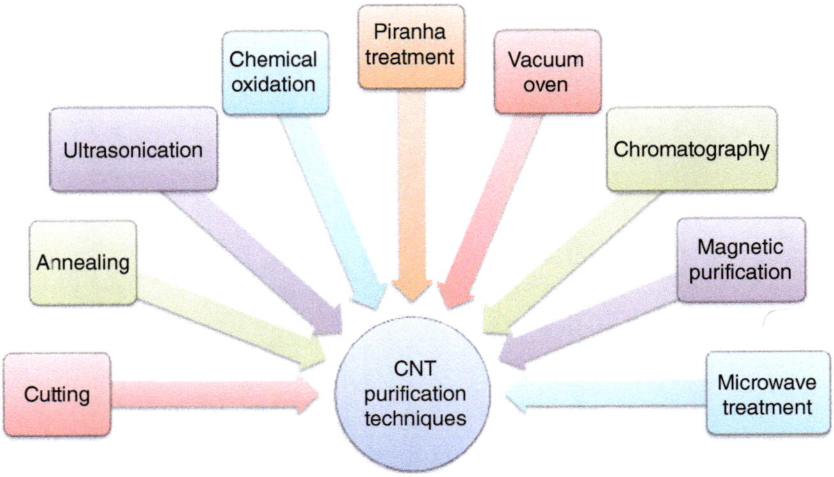

Figure 12.6 Various methods employed for purification of CNTs. Source: Based on Mallakpour et al. [11]; Mallakpour et al. [12].

12.6.1 Vacuum Oven Treatment

The vacuum oven treatment can be employed for the purification of synthesized CNTs. Although this is a simple and easily accessible technique, it is not used much because of the availability of more efficient methods.

12.6.2 Microwave Treatment

The microwave treatment for purification of CNTs is very proficient, faster, and environment friendly since this method requires very fewer amount solvents. High yield, fast reactions, optimum pressure and temperature, less time, and lesser number of steps are the advantages of this procedure.

12.6.3 Chemical Oxidation

The oxidative processes are used to purify CNTs and increase their solubility in water, to conjugate with biomolecules, and to open their ends and fill them with many bioactives. Among the various available procedures for the oxidation of CNTs, wet oxidation process is most proficient, easy, and scalable procedure in which a variety of oxidizing agents such as nitric acid (HNO_3), hydrogen peroxide : ammonium hydroxide ($H_2O_2 : NH_4OH$), nitric acid : sulfuric acid ($HNO_3 : H_2SO_4$), and nitric acid : hydrochloric acid ($HNO_3 : HCl$) are used. Generally, metallic impurities can easily be removed by oxidizing the CNTs in a wet environment, which consists of soluble oxidizing agents (H_2O_2 and H_2SO_4) to clear the metals from the surface of the nanotubes.

In these oxidative processes, sp^2-hybridized carbon atom starts to oxidize and finally generates the carboxylic acid functional group. It has been found that the concentration of total acidic functional groups gets elevated to six hours of reflux time; after that constancy is observed for the rest of the reflux time. Highest defects are found in nanotubes within six hours of refluxing, whereas extended reflux time produces fewer defective nanotubes.

12.6.4 Piranha Treatment

It is the most general method for purification and dispersion of pristine CNTs. This treatment is executed by immersing the nanotubes into piranha solution, where the dispersed form of CNTs is maintained. Generally, $HNO_3 : H_2SO_4$ is used in 1 : 3 ratios. Treatment with piranha solution not only shortens the length of the nanotubes but also results in impurity separation from nanotubes. Generation of carboxylic functional groups occurs at the tips and around the sidewalls, where higher strain is exhibited by the curvature of the nanotubes. It is noted that Piranha solution should be used very cautiously because it is corrosive and explosive. Thus, it should be handed by bearing gloves, avoiding contact with the eyes. Furthermore, it is a combination of strong nitric and sulfuric acid since this is an exothermic process; hence it should precede after proper cooling.

12.6.5 Annealing

This method is carried out at high temperatures approximately 873–1873 K, where defects are consumed and the arrangement of nanotubes takes place. Graphitic carbon and fullerene are pyrolyzed at this temperature.

12.6.6 Ultrasonication

In this approach, ultrasonic vibrations are employed to separate the impurities from nanotubes. The separation of impurities mainly depends on the surfactants, reagents, and solvent used. Ultrasonication is performed in acid solution, and respective purification mainly depends upon the exposure time. At higher exposure time, CNTs are cut into small pieces, whereas the shorter exposure time removes only metal solvates.

12.6.7 Magnetic Purification

This is a simple and nondestructive technique, which allows purification of large quantities of CNTs. Ferromagnetic catalytic impurities are separated from CNTs by suspending them with inorganic nanoparticles (mainly ZrO_2 or $CaCO_3$) in an ultrasonic bath, so that the impurity particles are trapped with permanent magnetic poles. Subsequent chemical treatment produces highly pure CNTs.

12.6.8 Cutting

The cutting is a combination approach based on the chemical cutting, mechanical cutting, or combination of both. It is used to cut nanotubes into smaller pieces by using a fluorinated reagent. Ball mill used for mechanical cutting of nanotubes employs high shear forces that break down the bonds between nanoparticulate impurities and carbon nanotubes. It is a process of combination involving chemical and mechanical cutting of the nanotubes and usually is carried out by ultrasonically induced cutting in an acid solution [22].

12.6.9 Chromatography

Chromatography is also used for purification of nanotubes into different fractions based on their size and diameter. Gel permeation chromatography (GPC) and high-performance liquid chromatography–size exclusion chromatography (HPLC–SEC) columns are employed for the separation of CNTs. The size and size distribution of CNTs are controlled by the pore size of column. Ultrasonication of CNTs functionalized with soluble functional groups helps even further to control the size of CNTs [22].

The selection of a particular method depends upon the efficiency to separate impurities, improving aqueous solubility, generation of functional groups, ease, and scalability. Out of the available purification methods, chemical oxidation is the best and more efficient method for purification utilizing various oxidizing agents, but precautions should be taken.

12.7 Need of Functionalization of Carbon Nanotubes (CNTs)

Even though CNTs possess excellent chemical and physical properties, pristine (non-functionalized) CNTs are intrinsically water insoluble and cannot disperse uniformly in most aqueous media or biological milieu due to their hydrophobic structure and generally contain impurities like catalyst nanoparticles and amorphous carbon. Also, due to their greater flexibility and high surface energy, CNTs have a tendency to aggregate into bundles and superstructures of bundles with a vast extent of entanglement. These bundled CNTs contain a massive number of semiconducting and metallic particles in a ratio of one and two-thirds of the mixture, respectively. This results in a main scientific obstruction and health concern for biocompatibility assessment toward drug delivery applications. Consequently, CNTs have to undergo purification and should be soluble in physiological conditions before they can be employed as drug carriers in nanomedicine. In this regard, the surface modifications are performed to overcome the major limitations associated with CNTs, which are referred to as *functionalization*. Functionalization enhances cellular uptake and solubility and reduces the bundling/aggregation. Various functionalities can be attached to the surface of CNTs to form FCNTs, which facilitate their consumption in various fields [23]. Thus, the functionalization is an imperative facet for the engineering of CNTs, which foresee the very promising nano-candidature of FCNTs in targeted and controlled drug applications.

12.8 Functionalization Strategies of CNTs

To improve their solubility and biocompatibility level, generally, two main strategies (based on both covalent and non-covalent linkages between the surface of nanotubes and the different biological moieties) are employed for the functionalization of CNTs: (i) covalent and (ii) non-covalent chemical functionalizations (Figure 12.7).

12.8.1 Covalent Functionalization

In the case of need of a strong linkage between CNTs and biomolecules, covalent functionalization is a highly effective method that can be used. This covalent bonding depends on the grafting of molecules that are chemically reactive onto their inert sp^2 carbon structure of the π-conjugated skeleton. According to the site of functional groups on the surface, it may be broadly divided into two main functionalization strategies, i.e. (a) "end-defect" functionalization and (b) "sidewall" functionalization:

(a) **End-Defect Covalent Functionalization:** This type of functionalization involves "end" tip oxidation, i.e. the production of various functional groups that contain oxygen such as hydroxyl, alcohol, lactone, carboxylic, phenolic, ketone, ester, etc. by sonication and by employing a mixture of strong oxidizing agents. Several functional

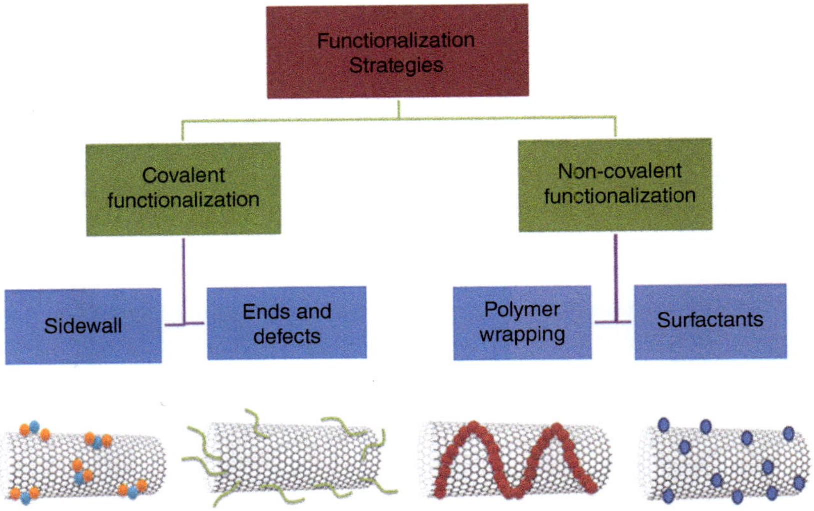

Figure 12.7 Functionalization strategies of carbon nanotubes (CNTs).

groups can be used to enhance solubility of CNTs by varying different factors such as acid, reaction conditions (temperature, sonication, and time), and concentration:

Carboxylation: This is one of the most effective and widely used functionalization methods to link CNTs with several functional groups such as amides and amines. Before covalent functionalization, oxalyl chloride ($C_2O_2Cl_2$), thionyl chloride ($SOCl_2$), active ester, or *n*-hydroxysuccinimide (NHS) is used to activate carboxylic acid (–COOH) to form highly reactive intermediate groups, leading to a stable covalent linkage between nanotubes and biomolecules. Reactivity of "ends and defects" covalent functionalization is higher than the sidewall functionalization [24, 25].

Esterification/amidation reactions: Oxidized CNTs are widely employed for even further modification of CNTs. Generally, CNTs that contain incorporated carboxylic groups are converted to amidated (amidation by the use of ethylene-diamine) and acylated (acylation by the use of oxalyl or thionyl chloride) CNTs. Tetrachlorosilane is used to activate acyl chloride and oxidized CNTs. Consecutive modification on the nanotube sidewall is achieved through the use of obtained chlorosilylesters [26].

(b) **Sidewall Covalent Functionalization**: This is another approach for the production of different functional groups along with avoiding the loss of Van Hove singularities on the sidewall surfaces of the nanotubes. This type of functionalization can be obtained through gas-phase reaction with the nondestructive sidewall [22, 24, 27]. The existence of sidewall defects like pentagon–heptagon pair, i.e. Stone–Wales defects or vacancies, leads to the improvement in local chemical reactivity of the CNTs. Generally, covalent sidewall functionalization produces carbon sites that are sp^3 hybridized, which disturb the band-to-band transitions of ð electrons [26]:

1,3-Dipolar cycloaddition reaction: This is an effective method for covalent sidewall functionalization for enhancing the solubility of nanotubes in aqueous and in several organic solvents. Azomethineylide is an organic 1,3 dipoles that consist of a carbanion situated next to an ammonium ion and forms pyrrolidine intermediates on performing its cycloaddition with dipolarophiles. It can be simply generated via decarboxylation of iminium salts obtained from the condensation of á-amino acids along with aldehydes or ketones. Amine functional groups are introduced on the surface of CNTs through 1,3-dipolar cycloaddition reaction of azomethineylide [26].

Diels–Alder cycloaddition: This reaction can also be employed for the sidewall functionalization of CNTs. The SWCNTs that are fluorinated undergo Diels–Alder cycloaddition with several dienes, such as anthracene, 2-trimethylsiloxyl-1,3,-butadiene, and 2,3-dimethyl-1,3-butadiene, to obtain functionalized SWCNTs. In this reaction, double bonds on the sidewall of SWCNTs were activated by the electron-withdrawing fluorine atoms of the surface of the nanotubes, which enhanced the rate of cycloaddition reaction [26]. The reaction between MWCNTs and 3,6-diamineotetrazine consisted of several possible interactions including coating through ð–ð stacking interactions and cross-linking [26].

12.8.2 Non-covalent Functionalization

Functionalization of CNTs can be carried out with several agents through various non-covalent interactions such as van der Waals, ð–ð stacking, and hydrophobic interactions. A variety of functionalities can be used to functionalize CNTs through non-covalent interactions such as DNA, surfactants, fluorophores, polymers, aromatic organic molecules, lipids, proteins, and endohedral functionalization.

Generally, the molecules that are used for the functionalization of nanotubes belong to three classes: (i) surfactants, (ii) polymers, and (iii) biopolymers (peptides and nucleic acids). The mechanism that is used for this type of functionalization is the self-assembly mechanism, due to the hydrophobic effects [22, 26–28]. Along with suspending the nanotubes in aqueous medium, the surfactants also prevent their aggregation via columbic repulsion. The Pluronic F-127/PEG were employed to adsorb doxorubicin (DOX) onto functionalized CNTs non-covalently [29]. For the dispersion of nanotubes, the cationic, anionic, and non-ionic types of surfactants are available:

π–π stacking interaction: Without affecting the properties and functions, various aromatic molecules can be adsorbed with fused aromatic benzene rings on the surface of CNTs through π–π–π stacking. The supramolecular complex formed as a result of interaction between anthracycline antibiotic, DOX, and nanotubes surface is via π–π stacking interaction because both of them possess π-rich electrons [30, 31]. Through the use of π–π stacking interaction, some complex dendrimers have been immobilized on nanotubes surface. For solubilizing CNTs through π–π stacking between aromatic DNA base units and surface of nanotubes, single-stranded DNA molecules have been extensively used [32].

Non-covalent functionalization by PEGylation: In the case of obtaining long-circulating nanotubes, PEGylation phenomenon is extremely important. The surface of nanotubes is strongly linked to the hydrocarbon chain of the lipids via hydrophilic polyethylene glycol (PEG) chain extending into the aqueous phase, which provides biocompatibility and solubility. PEGylated nanotubes are very stable suspension in serum and in solutions, having high salinity without the need for excess phospholipids–polyethylene glycol (PL–PEG). Specific PEGylated nanotubes for required applications can be obtained through the use of different PEG chain lengths and structure [33]. Non-covalent functionalization-mediated PEGylation of CNTs enhances the solubility, stability, biocompatibility, and circulation period in comparison with the use of covalent functionalization, hence proving to be a better and effective approach for drug delivery purposes.

12.9 Advantages of Functionalized Carbon Nanotubes (FCNTs)

Functionalized CNTs offer the following advantages:

- Reduced bundling effect.
- Improved solubility and dispersibility.
- Increased encapsulation tendency.
- Decreased *in vivo* and *in vitro* toxicity.
- Enhanced biocompatibility, leading to improved cellular internalization.

12.10 Medicinal Applications of Functionalized Carbon Nanotubes (FCNTs)

12.10.1 FCNTs in Drug Delivery

Drug delivery is systemic administration of drugs in a controlled and sustained manner to achieve maximal therapeutic efficacy using transdermal distribution. The primary objectives in developing novel drug delivery systems are to increase the therapeutic capabilities by [34] reducing drug loss and degradation [35], reducing cytotoxicity and harmful side effects by increasing the persistence of a drug [36], enhancing bioavailability [37], and minimizing drug exposure toward nontarget cells [38]. Drug delivery systems are designed with the competency to cross the cell membrane and enhance the therapeutic and pharmacological profile of drug molecules upon administration [39]. CNTs offer an alternative and encouraging platform as drug delivery vehicles or "nanocarriers" due to their unique physicochemical properties. Due to its great potential, these nanocarriers have been employed as efficient drug delivery vehicles as various therapeutic active compounds, like anti-infectives, anti-inflammatory molecules, antineoplastic agents, cardiovascular drugs, and genes. CNTs can be used as drug delivery systems for many biomedical applications that can prolong the release of the drug without causing toxicity to healthy tissue [28] (Figure 12.8). These can easily penetrate the cell membrane and enter the cell nuclei without the cell recognizing it as an intruder due to their needlelike cylindrical structure. Particularly, SWCNTs have attracted interest due to their most important characteristics as nanocarriers, with its ability to penetrate the cell membrane [40] and facilitate intracellular internalization and trafficking within the cell cytoplasm [41]. CNTs offer potential advantages such as intrinsic stability, ability to carry a high cargo loading, and structural flexibility, allowing prolonged circulation time, and hence the bioavailability of the drug molecules carried [42]. Additionally, it was observed that SWCNTs can enter mammalian cells [43–45]. Due to their promising properties, SWCNTs have been employed as potential delivery systems for intracellular transport of proteins [46], nucleic acids [29, 43], and drug molecules [47, 48].

SWCNTs can be functionalized with low-molecular-weight targeting agents and antibodies to provide high efficacy for cellular internalization [49]. These nanocarriers have greater loading competence and can be employed to evade the immune system by the incorporation of targeting agents and stealth molecules [50, 51]. These nanocarriers can be loaded with drug molecules such as DOX via p–p stacking interactions, quercetin, methotrexate, and paclitaxel [52, 53]. Moreover, functionalized CNTs can act as delivery vehicles for antimicrobial molecules like amphotericin B [54] and internalize it within the mammalian cells. CNT-free drug shows higher antifungal toxicity in comparison with CNT formulation (no cell death by CNT formulation as compared with 40% of the cells being killed by CNT-free formulation). Pathogenic bacteria can be captured by surface-engineered CNTs in liquid media [55]. However, the first generation, that is, the SWCNTs, demonstrated to be unfit for pharmacological application as drug delivery vehicles because they tended to accumulate in aqueous solutions. To overcome this drawback, natural polymers such as polysaccharides [56–58] and synthetic polymers such as poly(phenylacetylene) [49] have been put to use to encapsulate SWCNTs via non-covalent interactions, hence enhancing their compatibility with water and physiological environments. For instance, in research done by Sonam Sharma et al., they investigated the effect of functionalization of CNTs as drug delivery systems by targeting MCF-7 breast cancer cell line. Properties of DOX/PEG-FA-MWCNT (PEGylated MWCNTs bearing DOX) and DOX/MWCNT (MWCNTs bearing DOX) formulations were also compared. The tiny nano-needle-shaped structures of the CNTs were clearly revealed by the microscopic image [59]. Due to its nano-needle-specific mechanism and receptor-mediated endocytosis [60], higher fluorescence intensity was observed for DOX/PEG-FA-MWCNTs, clearly suggesting the higher uptake for DOX/PEG-FA-MWCNT nanoformulations. Moreover, prolonged circulation time was observed for PEGylated MWCNTs in comparison with DOX solution [59].

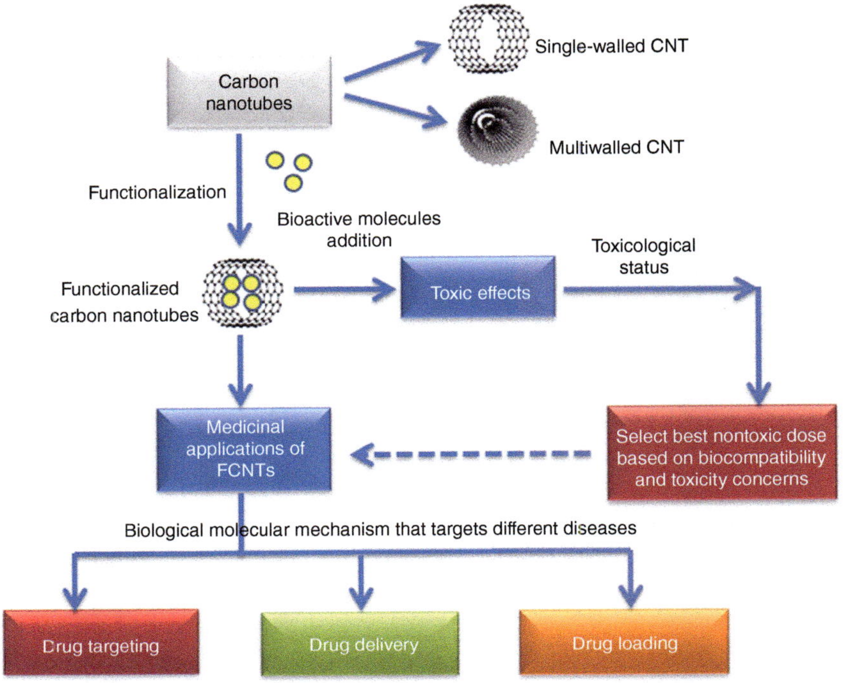

Figure 12.8 Medicinal applications of functionalized carbon nanotubes (FCNTs).

12.10.2 FCNTs in Drug Loading

One of the key measures to be an expedient delivery system is the drug loading process where the active drugs are combined with the carriers to give its terminal form. CNTs are having proved to be more useful because of their spherical shape and high surface-area-to-volume ratio that grants a tremendous potential to accommodate drugs [61]. Besides, drug loading capacity can be enhanced by decorating hydrophilic or amphiphilic polymers on the surface of the nanotubes, which gives rise to some extra space (for instance, see [29]). Another significance of these nanomaterials is that they are capable of carrying pharmaceutical agents via multiple ways, namely, encapsulation within the cavity, tethering on the surface upon functionalization, and adsorption on the wall or among the walls of CNTs. These approaches are primarily applicable to drugs having low surface tension due to the encapsulation that is driven by capillary forces and hydrophobic forces. Besides these, covalent bonds are expected to modulate the molecular structure of the drugs, and hence their specificity and bioactivity could be modified [62]. Efficient functionalization of CNTs makes them more soluble/dispersed in water and enables active sites for the addition of drugs, ligands, and other agents like PEG to attain long blood circulation half-life [48]. The loading of the drugs employing CNTs as carriers has been attained by either covalent or non-covalent interaction methods. Various anticancer drugs, like 10-hydroxycamptothecin, cisplatin, and methotrexate, can be conjugated covalently to the sidewalls of CNTs. However, these chemical conjugations might alter the chemical entities of the loaded drugs, especially if they are covalently associated by nonbiodegradable linkages, resulting in lower drug efficacy and relevant side effects. Figure 12.9 depicts some of the pathways used to conjugate a drug to the surface of CNTs.

CNTs have emerged as a promising drug carrier for many anticancer agents as well. For instance, Chetana et al. studied the latencies of MWCNTs as a carrier for efficient targeting of methotrexate to cancerous tissues. They functionalized MWCNTs using DSPE-mPEG 2000 and then reacted with methotrexate (MTX) to produce MWCNT-mPEG-MTX conjugate, and these were assessed for drug loading and *in vitro* diffusion in acidic media [63]. Similarly, for loading of the anticancer drug, epirubicin hydrochloride (EPI), Chen et al. synthesized carboxylated multiwalled carbon nanotubes (c-MWNTs) with nitric acid treatment, further evaluating the adsorption ability of c-MWNTs as adsorbents. The results implied that CNTs were able to form supramolecular complexes with EPI via π–π-stacking and possessed favorable loading properties as drug carriers that increased on increasing the pH. Lowering the temperature facilitated the adsorption [63]. Large surface area exists for supramolecular chemistry on SWNTs functionalized non-covalently or covalently via a regular surfactant or acid oxidation route. Water-solvable SWNTs functionalized with PEG via these routes allow for higher degrees of π stacking of aromatic molecules, including a cancer drug (DOX) with ultrahigh loading capacity, a widely used fluorescence molecule (fluorescein), and combinations of molecules. DOX is a popularly used chemotherapeutic drug for treating various cancers [29]. Similarly, Wu et al. used a solvothermal method to prepare MWCNT/cobalt ferrite ($CoFe_2O_4$) magnetic hybrids and DOX loaded onto the hybrids that subsequently released in a sustained and pH-responsive manner [64].

A unique plasma polymerization technique was employed to prepare CNTs containing an ultrathin poly(lactic-co-glycolic acid) (PLGA) coating. Paclitaxel was loaded on PLGA-coated CNTs, and its efficiency was measured by high-performance liquid chromatography (HPLC) [65]. In further studies, drug paclitaxel (PTX) was linked with a nontoxic lipid molecule docosanol for further functionalization with CNT surfaces and loading via hydrophobic interactions. High level of drug loading onto SWNT was achieved by lipid

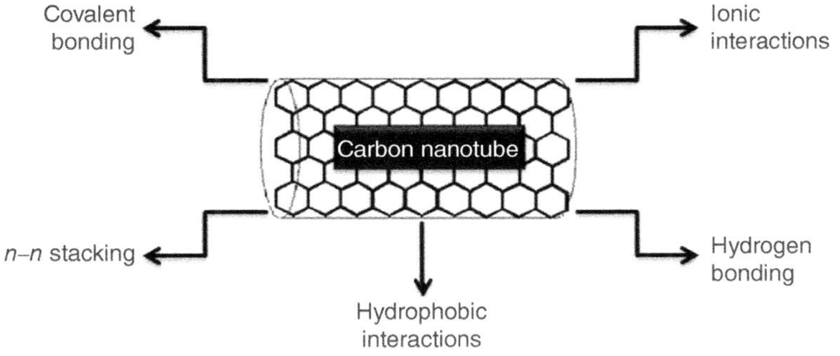

Figure 12.9 Methods to conjugate a drug to the surface of CNTs.

drug approach. Chemotherapeutic drugs are loaded into the interior of SWNT by a general capillarity-induced filling [66]. Li et al. studied carboxyl group-functionalized multiwalled carbon nanotubes (MWNTs) as set for loading water-insoluble drug, carvedilol (CAR), to increase the drug loading capacity using three different techniques, namely, the fusion method, the incipient wetness impregnation method, and the solvent method. The methods employed to load the drug had a considerable effect on the drug loading dissolution, and physical state of the drug along with its distribution. In addition, the solvability of the drug was increased when carried by both MWNTs and functional MWNTs that might help improve the bioavailability [67]. Moreover, a dual-targeting drug delivery system based on PEGylated oxidized multiwalled carbon nanotubes (O-MWNTs) modified with angiopep-2 (O-MWNTs-PEG-ANG) was developed by Ren et al. for the treatment of brain glioma. Studies showed that O-MWNTs can not only spread in brains but also accumulate in tumors and have ultrahigh surface area with exceptionally high loading anticancer drug of DOX, selected as drug carrier [68]. To culminate, the alternative of drug loading mechanism and location reveals a fragile balance between drug loading efficiency and therapeutic efficacy of pharmaceutical drugs (Table 12.1).

Table 12.1 Various drug-loaded functionalized carbon nanotubes (FCNTs).

S. No	Type of carbon nanotubes	Drug loaded	Functionalized molecule	References
1	Multiwalled carbon nanotube (MWCNT)	Methotrexate (MXT)	MWCNT-mPEG-MTX	[63]
2	PEGylated single-walled carbon nanotubes (SWNTs)	Doxorubicin (DOX)	PL-SWNT-DOX	[29]
3	Multiwalled carbon nanotube (MWCNT)	Doxorubicin (DOX)	DOX-loaded MWCNT/ $CoFe_2O_4$-180	[64]
4	Carbon nanotubes (CNT)	Paclitaxel (PTX)	PTX-loaded PLGA-coated CNTs	[65]
5	Carboxyl multiwalled carbon nanotubes (MWNTs)	Carvedilol (CAR)	CAR-loaded MWNTs	[67]
6	Polypyrrole-sealed carbon nanotubes (PPy/ CNTs)	Dexamethasone (DEX)	DEX-loaded PPy/CNTs	[69]
7	Oxidized multiwalled carbon nanotubes (O-MWNTs)	Angiopep-2 (ANG)	DOX-O-MWNTs-PEG-ANG	[68]
8	Single-walled carbon nanotubes (SWNTs)	Paclitaxel (PTX)	SWNT-lipid-PTX	[66]
9	Multiwalled carbon nanotubes (MWCNTs)	Doxorubicin (DOX)	DOX-FA-MN-MWCNT	[70]
10	Carboxylated multiwalled carbon nanotubes (c-MWNTs)	Epirubicin hydrochloride (EPI)	c-MWNTs-EPI	[63]

12.10.3 FCNTs in Drug Targeting

The very focus of scientific studies and research has been realigned to create mechanisms that could achieve higher levels of efficient drug delivery. In modern era, this is being worked on due to the potential of targeted drug delivery, in which drug is delivered to the target site by the help of a carrier. This method enhances the therapeutic effect of the administered drug as it accumulates only at the target site in maximum concentrations. Drug targeting can be efficiently executed by employing functionalized CNTs, which can even lead to controlled release of drugs. Drug targeting is known to take place when a specified part is targeted with a live therapeutic drug for extended period of times. CNTs are often used in drug targeting because they have the ability to stay in circulatory systems for long durations, owing to their hydrophobic nature. Generally they are used to target various malignant disorders including several types of cancers such as breast, prostate, brain, and testicular cancers [71]. There are various stimuli whose application can result in accomplishment of controlled drug delivery using CNTs such as temperature, magnetic, electric, etc. Several biomolecules such as lectins, RNA, DNA, and proteins can also be transported via CNTs [72, 73]. Application of MWCNTs and pH-responsive gel has been demonstrated to achieve effective delivery of DOX. MWCNTs and pH-responsive gel were together employed to form nano-hybrid system. Nanocomposites were coated with chitosan. U-87 glioblastoma cell was targeted, and an efficient reduction in cancer proliferation was observed when drug was delivered via this nanosystem [74]. To access electrical properties, a transdermal patch has been created using polymer/MWCNTs. It has been demonstrated that the release of drugs can be made rapid through the effect of heating [75]. These are some of the examples focused on studies that explore targeted drug delivery through MWCNTs. Their role toward specific disorders is discussed as follows.

12.10.3.1 Cancer Targeting

Cancer has always been one of the most challenging diseases to treat across the world. The greatest need for an effective treatment is to eliminate malignant tumors without affecting healthy cells in the vicinity. Conventional methodology of treatments such as chemotherapy is not yet efficient for using this approach, as in most of the cases nonspecific accumulation of drugs takes place within the body, leading to harmful consequences. CNTs have been studied and researched upon for achieving *in vivo* biodistribution and improved tumor targeting in mice. Also, phospholipid-functionalized SWCNTs containing PEG are very stable *in vivo*. CNTs loaded with chemotherapeutic agents have demonstrated higher cellular uptake in tumor cells in comparison with healthy cells. It has also been found that the tumor-eliminating activity of MWCNTs is enhanced when they are functionalized with siRNA and DNA. Moreover efficient and sustained suppression in tumor growth has been observed by employing SWCNT–siRNA complexes. You et al. designed recently a cell-penetrating cancer-targeted molecule-functionalized MWCNTs for orthotopicglioma therapy. This system exhibited decreased cytotoxicity along with efficient cancer recognition and attack through reactive oxygen species (ROS) generation. Trans-activating transcriptional (TAT) activator–polyethylenimine (PEI)–biotin copolymer was used to functionalize MWCNTs. This system was able to enhance cytotoxicity of oxaliplatin toward glioma cells through the excessive production of ROS, penetrate blood–brain barrier, and show significant anticancer activity against orthotopicglioma [76]. An anticancer drug DOX hydrochloride has been employed to achieve its intracellular delivery through the use of PEGylated CNTs [77]. The surface of CNTs was effectively coated with polydopamine. These PEGylated CNTs demonstrated enhanced biocompatibility toward cancerous cells along with efficient delivery of the drug into the cells. A supramolecular MWCNT–DOX–copolymer was investigated by Boucetta and coworkers to evaluate anticancer activities of the complex. Fluorescent spectrophotometry was used to analyze the interaction of DOX with nanotubes and its cellular uptake. The complex reported elevated levels of cytotoxicity and thereby higher cell eradication capability [78].

12.10.3.2 Brain Targeting

The impermeable blood–brain barrier disrupts the targeting of brain by drug molecules to preserve the milieu of the brain. Due to this, entry of various substances is not allowed. Even if a few neuropharmaceutical drugs or agents manage to pass through the barrier, they are degraded by the enzymes that are present in the brain. Taking these concepts into account, it can be understood why conventional drug delivery systems are unable to transport drug into the brain and produce the required therapeutic effect. CNTs have been studied and researched on, regarding their drug delivery capabilities as they possess the ability to cross the blood–brain barrier [79]. Therefore, agents can be efficiently transported to the microglial cells of the brain by employing MWCNTs. In a recent study, surface modification of CNTs and unique electronic properties along with biocompatibility of gold nanoparticles (AuNPs) have been productively employed to develop Au/CNT hybrid nanomaterials that could detect ascorbic acid and dopamine [80]. Moreover, efforts have been made to treat neurodegenerative disorders through the exploitation of magnetic properties of the nanotubes. CNTs also cause cellular differentiation into neurons when they are combined with nerve growth factors and, thus, result in an efficient framework for treating neurodegenerative disorders.

12.10.3.3 Lymphatic Targeting

Several malignancies can metastasize through the lymphatic system. Metastasis of cancer can be blocked effectively using drug delivery carriers like CNTs by targeting to the lymphatic canal. Radical polymerization can be used to conjugate polyacrylic acid (PAA) onto CNTs. Fe_3O_4-based magnetic nanoparticles can be further adsorbed on the PAA-CNT surface and additionally be loaded with gemcitabine. Post three hours of subcutaneous injection, it was studied that CNTs were found only in the local lymphatic nodes and were absent in major organs such as the spleen, liver, heart, kidney, and lungs. Gemcitabine alone cannot distribute in the lymphatic system without the help of such nanoformulations [79]. Feng Yang and coworkers further investigated this novel lymphatic-targeting drug delivery system using functionalized mMWNTs *in vivo*. mACs (micro activated carbons) and mMWNTs have been studied to inhibit cancer lymph node metastasis as magnetic lymphatic-targeting drug delivery systems. Using implanted *in vivo* magnets and high-dose agents in addition with mACs and mMWNTs inhibited lymph node metastasis and enhanced gemcitabine (GEM) cytotoxicity [81].

Jian Ji et al. investigated magnetic multiwalled carbon nanotube–doxorubicin (DOX-O-mMWNT) as magnetic drug delivery systems targeting lymphatic system. The studies showed that DOX-O-mMWNT inhibited lymph node metastasis and breast cancer cells. It was observed that DOX and magnetic particle conjugate showed stronger inhibitory effects on tumors of mice and lymph node metastasis in comparison with simply using DOX. Moreover, the influence of magnetic field had enhanced the antitumor effects of DOX-O-mMWNT [82].

12.10.3.4 Tuberculosis Targeting

Tuberculosis is a chronic illness that is caused by *Mycobacterium tuberculosis*. It is a rod-shaped bacterium that seizes the walls of pulmonary concavity and divides slowly. This slow growth is the cause of long-term drug therapies that target larger bacterial populations. The TB therapy pertains to a combination of four first-line drug regime of isoniazid (INH), pyrazinamide (PZA), rifampicin, and ethambutol. A major blockade for antibiotics in targeting mycobacterium is because of low permeability of the mycobacteria's cell wall. Surface FCNTs have emerged as effective nanocarriers of drug. Recently, in their experiment, Bianco et al. demonstrated that SWCNTs functionalized by the 1,3-dipolar cycloaddition of azomethineylides are apt for negotiating mammalian and prokaryotic cells [83] as well as handling traffic via different cellular barriers. For instance, we know that PZA is metabolized into its active state, i.e. pyrazoic acid by the amidase activity of *M. tuberculosis* nicotinamidase/pyrazinamidase (MtPncA) that is encoded by the pncA gene [84]. FCNTs can be seen as facilitators in the loading and delivery of a PZA onto the active undertaking pathway of pncA without the nanotube influencing the structural [85]. Theoretical investigations on the interaction of non-steroid anti-inflammatory drug nimesulide along with pristine and silicon (Si)-doped capped SWCNT were performed by Zanella and coworkers. The robust interaction of nimesulide with Si-doped SWCNT provided as better drug delivery carriers in comparison with pristine capped SWCNT [86] (Table 12.2).

12.11 Biocompatibility and Toxicity Considerations of FCNTs

The biocompatibility and toxicological issues of FCNTs should be thoroughly analyzed before they are employed as drug delivery systems. Metallic impurities and hydrophobic aggregation of CNTs are mainly responsible for causing unusual toxicity issues. Fortunately, the toxicity caused by CNTs can be reduced by modifying degree of functionalization associated with their bio-distribution profile. Several studies are available concerning the safety of CNTs. However, there is contradiction in the available reports on CNTs, and they are found to be disrupted too. Functionalized CNTs are considered to be a promising nanovector for biomedical implementation [79, 92]. Yang et al. demonstrated that PEGylated SWCNTs that were covalently functionalized were easily defunctionalized by radical attack in the liver or by enzymatic hydrolysis and were transformed into oxidized CNTs. So the possibility of long-term *in vivo* toxicity of covalently PEGylated CNTs was directed to the oxidized CNTs [93]. There might be different toxicological issues caused by different classes of CNTs (MWCNTs and SWCNTs), owing to the variability in their surface areas. SWCNTs are more prone to aggregation/bundling as they possess greater surface area, due to the van der Waals forces, and the bundling is known to be

Table 12.2 Drug targeting via functionalized carbon nanotubes (FCNTs).

S. No	Type of carbon nanotube	Functionalizing agent	Characteristics	References
1	MWCNTs	TAT-PEI biotin copolymer	Improvement in cytotoxicity of oxaliplatin along with anticancer activity	[76]
2.	MWCNTs	PEG and polydopamine	Successful delivery of doxorubicin into the cells and improved biocompatibility	[77]
3.	SWCNTs	Hyaluronic acid	Significant anticancer activity and induction of apoptosis in tumor cells	[87]
4	SWCNTs	Palmitoyl chitosan and carboxymethyl chitosan	Efficient drug loading along with controlled release at reduced pH	[88]
5	SWCNTs	Polyethylenimine	Enhanced tumoricidal activity	[89]
6	MWCNTs	Pluronic F127, carboxyl group, and amino group	Efficient cellular uptake in microglial cells	[90]
7	MWCNTs	Amino group	System was able to cross blood–brain barrier *in vitro* and showed significant brain uptake *in vivo*	[91]
8	MWCNTs	DOX/PEG-FA-MWCNTs and DOX/MWCNTs	Drug delivery systems by targeting MCF-7 breast cancer cell line. Higher uptake and prolonged circulation time for DOX/PEG-FA-MWCNT nanoformulations	[59]
9	MWCNTs	PAA-CNT	Lymph node metastasis was inhibited using lymphatic targeting	[79]
10	mACs and mMWCNTs	PAA-CNT	Lymph node metastasis was inhibited, and GEM cytotoxicity was enhanced using magnetic lymphatic targeting	[81]
11	MWCNTs	O-mMWNT–DOX	Inhibited lymph node metastasis and breast cancer cells	[82]
12	SWCNTs	Polyethylene glycol (PEG) oligomer	Cellular uptake of drug and degrade the therapeutic activity of drug molecules	[29]

Table 12.2 (Continued)

S. No	Type of carbon nanotube	Functionalizing agent	Characteristics	References
13	SWCNTs	Boron nitride nanotubes (BNNTs)	Enhances nanotube chirality and electronic properties	[85]
14	SWCNTs	Silicon (Si)-doped capping	Higher binding energy, better drug delivery carrier	[86]

harmful for the living cells, tissues, and organs. Aggregation in MWCNTs might be avoided, owing to the fact that the surface area of MWCNTs is slightly lower due to many active defect sites. Thus, in our opinion SWCNTs are more toxic as compared with MWCNTs. Both types of CNTs should be hydrophilic to be able to serve as nanocarriers. Thus, modification in surface chemistry plays a vital role for enhancing the biocompatibility of the nanotubes [94]. Till date available, studies have demonstrated that the decrease in the toxicity of nanotubes is achievable through high degree of functionalization and suggest that the f-CNTs are less toxic [75, 94, 95].

12.12 Conclusion and Future Perspective

CNTs have demonstrated to be one of the novel functional materials expanding the experimental and theoretical horizons to explore its unique properties. The electronic properties of CNTs can be modified to fit the desired functionality on the basis of its length, size, and indices (n, m). CNTs are highly reactive due to the curvature and conjugated π-electron backbone; they have properties analogous to graphene and fullerene. There are some major disadvantages that restricts the applicability of CNTs: first, the preparation of CNTs yields a combination of both semiconducting and metallic tubes, varying in chirality and diameter, and purification and separation of these nanotubes is challenging. Also, CNTs are hydrophobic and tend to agglomerate in organic and aqueous solvents. Hence, a sought-after approach like covalent and non-covalent surface functionalizations can improve dispersibility and reduce hydrophobicity and bundling effect. Even though CNTs proved their practical applicability in all aspects of science, its long-term applications need to be evaluated from the prospects of environmental risks to human health [96–100]. Appropriate evaluation and in-depth research are necessary to render nanotubes useful for diverse and environmentally benign applications (Hussain [101, 102]; Mallakpour, Handbook of Polymer Nanocomposites for Industrial Applications). *In vivo* and *in vitro* cytotoxicity can be reduced via surface functionalization of CNTs. Clearly, continual attention of toxicologists for CNT technology is much needed before it can be used for treatment of diseases. CNT is certainly worth further research and is an extremely promising application of nanotechnology, as the existing approaches are extensively detrimental and only partially effective.

Important Websites

https://www.nanowerk.com/nanotechnology/introduction/introduction_to_nanotechnology_22.php.
https://www.nanotech-now.com/nanotube-buckyball-sites.htm.
https://technology.nasa.gov/public_domain/GSC-14435-1.
https://nanografi.com/carbon-nanotubes.
https://www.britannica.com/science/carbon-nanotube.
https://www.sciencedaily.com/terms/carbon_nanotube.htm.
https://greenchemistry.yale.edu/environmentally-and-economically-sustainable-carbon-nanotube-manufacturing.

References

1 Abdolmaleki, A., Mallakpour, S., and Karshenas, A. (2017). Facile synthesis of glucose-functionalized reduced graphene oxide (GFRGO)/poly(vinyl alcohol) nanocomposites for improving thermal and mechanical properties. *Mater. Sci. Eng. B* 217: 26–35.

2 Mallakpour, S. and khodadadzadeh, L. (2018). Chapter 7 biocompatible and biodegradable chitosan nanocomposites loaded with carbon nanotubes. In: *Biodegradable and Biocompatible Polymer Composites Processing, Properties and Applications* (ed. N.G. Shimpi), 187–221. Sawston, Cambridge, UK: Elsevier, Woodhead Publishing https://doi.org/10.1016/B978-0-08-100970-3.00007-9.

3 Mallakpour, S. and Khadem, E. (2016). Carbon nanotube–metal oxide nanocomposites: fabrication, properties and applications. *Chem. Eng. J.* 302: 344–367.

4 Abdolmaleki, A., Mallakpour, S., and Karshenas, A. (2017). Synthesis and characterization of new nanocomposites films using alanine-Cu-functionalized graphene oxide as nanofiller and PVA as polymeric matrix for improving of their properties. *J. Solid State Chem.* 253: 398–405.

5 Abdolmaleki, A., Mallakpour, S., and Mahmoudian, M. (2017). Preparation and evaluation of edge selective sulfonated graphene by chlorosulfuric acid as an active metal-free electrocatalyst for oxygen reduction reaction in alkaline media. *ChemistrySelect* (34): 11211–11217.

6 Abdolmaleki, A., Mallakpour, S., and Borandeh, S. (2016). Improving interfacial interaction of l-phenylalanine-functionalized graphene nanofiller and poly(vinyl alcohol) nanocomposites for obtaining significant membrane properties: morphology, thermal, and mechanical studies. *Polym. Compos.* 37: 1924–1935.

7 Mallakpour, S., Abdolmaleki, A., and Karshenas, A. (2017). Graphene oxide supported copper coordinated amino acids as novel heterogeneous catalysts for epoxidation of norbornene. *Catal. Commun.* 92: 109–113.

8 Mallakpour, S., Abdolmaleki, A., Mahmoudian, M. et al. (2017). Synergetic effect of synthesized sulfonated polyaniline/quaternized graphene and its application as a high-performance supercapacitor electrode. *J. Mater. Sci.* 52: 9683–9695.

9 Mallakpour, S., Abdolmaleki, A., and Borandeh, S. (2017). Fabrication of amino acid-based graphene-zinc oxide (ZnO) hybrid and its application for poly(ester–amide)/graphene-ZnO nanocomposite synthesis. *J. Thermoplast. Compos. Mater.* 30: 358–380.

10 Mallakpour, S. and Khadem, E. (2018). Construction of crosslinked chitosan/nitrogen-doped graphene quantum dot nanocomposite for hydroxyapatite biomimetic mineralization. *Int. J. Biol. Macromol.* 120: 1451–1460.

11 Mallakpour, S. and Khadem, E. (2019). Linear and nonlinear behavior of crosslinked chitosan/N-doped graphene quantum dot nanocomposite films in cadmium cation uptake. *Sci. Total Environ.* 690: 1245–1253.

12 Mallakpour, S., Behranvand, V., and Mallakpour, F. (2019). Synthesis of alginate/carbon nanotube/carbon dot/fluoroapatite/TiO_2 beads for dye photocatalytic degradation under ultraviolet light. *Carbohydr. Polym.* 224: 115138.

13 Mallakpour, S. and Behranvand, V. (2018). Synthesis of mesoporous recycled poly(ethylene terephthalate)/MWNT/carbon quantum dot nanocomposite from sustainable materials using ultrasonic waves: application for methylene blue removal. *J. Clean. Prod.* 190: 525–537.

14 Bharadwaj, L. and Bhalla, V. (2007). Carbon nanotubes – role in health care. In: *Nanoparticles for Pharmaceutical Applications* (ed. A.J. Domb), 177–190. American Scientific Publishers.

15 De Rosa, A.M., Greco, K., Rajamani, S., and Sitharaman, B. (2010). Recent patents on single-walled carbon nanotubes for biomedical imaging, drug delivery and tissue regeneration. *Recent Pat. Biomed. Eng.* 3: 86–94.

16 Beg, S., Rizwan, M., Sheikh, A.M. et al. (2011). Advancement in carbon nanotubes: basics, biomedical applications and toxicity. *J. Pharm. Pharmacol.* 63: 141–163.

17 Ren, Z.F., Huang, Z.P., Wang, D.Z. et al. (1999). Growth of a single free standing multiwall carbon nanotube on each nanonickel dot. *Appl. Phys. Lett.* 75: 1086–1089.

18 Ren, Z.F., Huang, Z.P., Xu, J.W. et al. (1998). Synthesis of large arrays of well-aligned carbon nanotubes on glass. *Science* 282: 1105–1107.

19 Grobert, N. (2007). Carbon nanotubes-becoming clean. *Mater. Today* 10 (1–2): 28–35.

20 Mallakpour, S. and Khadem, E. (2019). Chapter 8 carbon nanotubes for heavy metals removal. In: *Composite Nanoadsorbents* (eds. G. Kyzas and A.C. Mitrpoulos), 181–210. Amsterdam, Netherlands: Elsevier https://doi.org/10.1016/B978-0-12-814132-8.00009-5, eBook ISBN: 9780128141335.

21 Mallakpour, S. and Rashidimoghadam, S. (2019). Chapter 9 carbon nanotubes for dyes removal. In: *Composite Nanoadsorbents* (eds. G. Kyzas and A.C. Mitrpoulos), 211–244. Amsterdam, Netherlands: Elsevier https://doi.org/10.1016/B978-0-12-814132-8.00010-1. eBook ISBN: 9780128141335.

22 Mehra, N.K., Jain, A.K., Lodhi, N. et al. (2008). Challenges in the use of carbon nanotubes for biomedical application. *Crit. Rev. Ther. Drug Carrier. Syst.* 25 (2): 169–206.

23 Mallakpour, S. and Soltanian, S. (2016). Surface functionalization of carbon nanotubes: fabrication and applications. *RSC Adv.* 6: 109916–109935.

24 Kesharwani, P., Ghanghoria, R., and Jain, N.K. (2012). Carbon nanotubes exploration in cancer cell lines. *Drug Discov. Today* 17 (17–18): 1023–1030.

25 Mehra, N.K. and Jain, N.K. (2013). Development, characterization of surface engineered carbon nanotubes for anticancer targeting potential. *J. Drug Target.* 21 (8): 745–758.

26 Karousis, N. and Tagmatarchis, N. (2010). Current progress on the chemical modification of carbon nanotubes. *Chem. Rev.* 110: 5366–5397.

27 Jain, A.K., Mehra, N.K., Lodhi, N. et al. (2007). Carbon nanotubes and their toxicity. *Nanotoxicology* 1 (3): 167–197.

28 Tasis, D., Tagmatarchis, N., Bianco, A., and Prato, M. (2006). Chemistry of carbon nanotubes. *Chem. Rev.* 106 (3): 1105–1136.
29 Liu, Z., Winters, M., Holodniy, M., and Dai, H. (2007). siRNA delivery into human T cells and primary cells with carbon-nanotube transporters. *Angew. Chem. Int. Ed.* 46: 2023–2027.
30 Huang, H., Yuan, Q., Shah, J.S., and Misra, R.D.K. (2011). A new family of folate-decorated and carbon nanotubes-mediated drug delivery system: synthesis and drug delivery response. *Adv. Drug Deliv. Rev.* 63: 1332–1339.
31 Zhang, X., Meng, L., Lu, Q. et al. (2009). Targeted delivery and controlled release of doxorubicin to cancer cells using modified single wall carbon nanotubes. *Biomaterials* 30 (30): 6041–6047.
32 Kam, N.W.S., Liu, Z., and Dai, H. (2005). Functionalization of carbon nanotubes via cleavable disulfide bonds for efficient intracellular delivery of siRNA and potent gene silencing. *J. Am. Chem. Soc.* 127: 12492–12493.
33 Liu, Z., Sun, X., Nakayama-Ratchford, N., and Dai, H. (2007). Supramolecular chemistry on water-soluble carbon nanotubes for drug loading and delivery. *ACS Nano* 1 (1): 50–56.
34 Geim, A.K. and Novoselov, K.S. (2007). The rise of graphene. *Nat. Mater.* 6: 183–191.
35 Prasad, P.V., Kumar, A., Pal, P.C. et al. (2011). Bird's eye view on the recent advances in drug delivery systems. *J. Biomater. Nanobiotechnol.* 2: 544–556.
36 Liu, L., Guo, G.Y., Jayanthi, C.S., and Wudate, S.Y. (2002). Colossal paramagnetic moments in metallic carbon nanotori. *Phys. Rev. Lett.* 88: 217206.
37 Novoselov, K.S., Geim, A.K., Morozov, S.V. et al. (2004). Electric field effect in atomically thin carbon films. *Science* 306: 666–669.
38 Dresselhaus, M.S., Dresselhaus, G., and Jorio, A. (2004). Unusual properties and structures of carbon nanotubes. *Annu. Rev. Mater. Res.* 34: 247.
39 Pantarotto, D., Briand, J.-P., Prato, M., and Bianco, A. (2004). Translocation of bioactive peptides across cell membranes by carbon nanotubes. *Chem. Commun.* 1: 16–17.
40 Kam, N.W.S., Jessop, T.C., Wender, P.A., and Dai, H. (2004). Nanotube molecular transporters: internalization of carbon nanotube-protein conjugates into mammalian cells. *J. Am. Chem. Soc.* 126: 6850–6851.
41 Pantarotto, D., Singh, R., McCarthy, D. et al. (2004). Functionalized carbon nanotubes for plasmid DNA gene delivery. *Angew. Chem. Int. Ed. Engl.* 43: 5242.
42 Liu, Z., Chen, K., Davis, C. et al. (2008). Drug delivery with carbon nanotubes for in vivo cancer treatment. *Cancer Res.* 68 (16): 6652–6660.
43 Cai, D., Doughty, C.A., Potocky, T.B. et al. (2007). Carbon nanotube-mediated delivery of nucleic acids does not result in non-specific activation of B lymphocytes. *Nanotechnology* 18 (365): 101–110.
44 Chin, S.F., Baughman, R.H., Dalton, A.B. et al. (2007). Amphiphilic helical peptide enhances the uptake of single-walled carbon nanotubes by living cells. *Exp. Biol. Med.* 232 (9): 1236–1244.
45 Kostarelos, K., Lacerda, L., Pastorin, G. et al. (2007). Cellular uptake of functionalized carbon nanotubes is independent of functional group and cell type. *Nat. Nanotechnol.* 2 (2): 108–113.
46 Kam, N.W.S. and Dai, H.J. (2005). Carbon nanotubes as intracellular protein transporters: generality and biological functionality. *J. Am. Chem. Soc.* 127 (16): 6021–6026.

47 Chen, J.Y., Chen, S.Y., Zhao, X.R. et al. (2008). Functionalized single-walled carbon nanotubes as rationally designed vehicles for tumor targeted drug delivery. *J. Am. Chem. Soc.* 130 (49): 16778–16785.

48 Prato, M., Kostarelos, K., and Bianco, A. (2008). Functionalized carbon nanotubes in drug design and discovery. *Acc. Chem. Res.* 41: 60–68.

49 Zeineldin, R., Al-halik, M., and Hudson, L.G. (2009). Role of polyethyleneglycol integrity in specific receptor targeting of carbon nanotubes to cancer cells. *Nano Lett.* 9 (2): 751–757.

50 Heister, E., Neves, V., Tilmaciu, C. et al. (2009). Triple functionalisation of single-walled carbon nanotubes with doxorubicin, a monoclonal antibody, and a fluorescent marker for targeted cancer therapy. *Carbon* 47 (9): 2152–2160.

51 Mu, Q.X., Broughton, D.L., and Yan, B. (2009). Endosomal leakage and nuclear translocation of multiwalled carbon nanotubes: developing a model for cell uptake. *Nano Lett.* 9 (12): 4370–4375.

52 Modi, C.D., Patel, S.J., Desai, A.B., and Murthy, R.S.R. (2011). Functionalization and evaluation of PEGylated carbon nanotubes as novel drug delivery for methotrexate. *J. Appl. Pharm. Sci.* 1 (5): 103–108.

53 Dolatabadi, J.E.N., Jamali, A.A., and Hasanzadeh, M. (2011). Quercetin delivery into cancer cells with single walled carbon nanotubes. *Int. J. Biosci. Biochem. Bioinforma.* 1 (1): 21–25.

54 Benincasa, M., Pacor, S., Wu, W. et al. (2011). Antifungal activity of amphotericin B conjugated to carbon nanotubes. *ACS Nano* 5: 199–208.

55 Brady-Estévez, A.S., Kang, S., and Elimelech, M. (2008). A single-walled-carbon-nanotube filter for removal of viral and bacterial pathogens. *Small* 4: 481–484.

56 Abarrategi, A., Gutierrez, M.C., Moreno-Vicente, C. et al. (2008). Multiwall carbon nanotube scaffolds for tissue engineering purposes. *Biomaterials* 29 (1): 94–102.

57 Long, D.W., Wu, G.Z., and Zhu, G.L. (2008). Noncovalently modified carbon nanotubes with carboxymethylated chitosan: a controllable donor-acceptor nanohybrid. *Int. J. Mol. Sci.* 9 (2): 120–130.

58 Zhang, X.K., Wang, X.F., Lu, Q.H., and Fu, C.L. (2008). Influence of carbon nanotube scaffolds on human cervical carcinoma HeLa cell viability and focal adhesion kinase expression. *Carbon* 46 (3): 453–460.

59 Sharma, S., Mehra, N.K., Jain, K., and Jain, N.K. (2016). Effect of functionalization on drug delivery potential of carbon nanotubes. *Artif. Cell. Nanomed. Biotechnol.* 44 (8): 1851–1860.

60 Mehra, N.K., Verma, A.K., Mishra, P.R., and Jain, N.K. (2014). The cancer targeting potential of D-a-tocopheryl polyethylene glycol 1000 succinate tethered multi walled carbon nanotubes. *Biomaterials* 35: 4573–4588.

61 Heister, E., Neves, V., Lamprecht, C. et al. (2012). Drug loading, dispersion stability, and therapeutic efficacy in targeted drug delivery with carbon nanotubes. *Carbon* 50: 622–632.

62 Tsai, H.-C., Lin, J.-Y., Maryani, F. et al. (2013). Drug-loading capacity and nuclear targeting of multiwalled carbon nanotubes grafted with anionic amphiphilic copolymers. *Int. J. Nanomed.* 8: 4427–4440.

63 Chen, Z., Pierre, D., He, H. et al. (2011). Adsorption behavior of epirubicin hydrochloride on carboxylated carbon nanotubes. *Int. J. Pharm.* 405: 153–161.

64 Wu, H., Liu, G., Wang, X. et al. (2011). Solvothermal synthesis of cobalt ferrite nanoparticles loaded on multiwalled carbon nanotubes for magnetic resonance imaging and drug delivery. *Acta Biomater.* 7: 3496–3504.

65 Guo, Y., Shi, D., Cho, H. et al. (2008). In vivo imaging and drug storage by quantum-dot-conjugated carbon nanotubes. *Adv. Funct. Mater.* 18: 2489–2497.

66 Shao, W., Paul, A., Zhao, B. et al. (2013). Carbon nanotube lipid drug approach for targeted delivery of a chemotherapy drug in a human breast cancer xenograft animal model. *Biomaterials* 34 (38): 1–11.

67 Li, Y., Wang, T., Wang, J. et al. (2011). Functional and unmodified MWNTs for delivery of the water-insoluble drug Carvedilol – a drug-loading mechanism. *Appl. Surf. Sci.* 257: 5663–5670.

68 Ren, J., Shen, S., Wang, D. et al. (2012). The targeted delivery of anticancer drugs to brain glioma by PEGylated oxidized multi-walled carbon nanotubes modified with angiopep-2. *Biomaterials* 33: 3324–3333.

69 Luo, X., Matranga, C., Tan, S. et al. (2011). Carbon nanotube nanoreservior for controlled release of anti-inflammatory dexamethasone. *Biomaterials* 32: 6316–6323.

70 Lu, Y.-J., Wei, K.-C., Ma, C.-C.M. et al. (2012). Dual targeted delivery of doxorubicin to cancer cells using folate-conjugated magnetic multi-walled carbon nanotubes. *Colloids Surf. B. Biointerfaces* 89: 1–9.

71 Thakare, V.S., Das, M., Jain, A.K. et al. (2010). Carbon nanotubes in cancer theragnosis. *Nanomedicine (Lond.)* 5: 12771301.

72 Che, J., Cagin, T., and Goddard, W.A. III (2000). Thermal conductivity of carbon nanotubes. *Nanotechnology.*

73 Khan, M.U., Reddy, K.R., Snguanwongchai, T. et al. (2016). Polymer brush synthesis on surface modified carbon nanotubes via in situ emulsion polymerization. *Colloid Polym. Sci.* 294(10): 1599–1610.

74 Seyfoori, A., Sarfarazijami, S., and Seyyed Ebrahimi, S.A. (2019). pH-responsive carbon nanotube-based hybrid nanogels as the smart anticancer drug carrier. *Artif. cell. Nanomed. Biotechnol.* 47(1): 1437–1443.

75 Im, J.S., Bai, B.C., and Lee, Y.S. (2010). The effect of carbon nanotubes on drug delivery in an electro-sensitive transdermal drug delivery system. *Biomaterials* 31(6): 1414–1419.

76 You, Y., Wang, N., He, L. et al. (2019). Designing dual-functionalized carbon nanotubes with high blood–brain-barrier permeability for precise orthotopicglioma therapy. *Dalton Trans.* 48: 1569–1573.

77 Xu, H., Liu, M., Lan, M. et al. (2016). Mussel inspired PEGylated carbon nanotubes: biocompatibility evaluation and their drug delivery applications. *Toxicol. Res.* 5: 1371–1379.

78 Ali-Boucetta, H., Al-Jamal, K.T., McCarthy, D. et al. (2008). Multiwalled carbon nanotube-doxorubicin supramolecular complexes for cancer therapeutics. *Chem. Commun.* (4): 459–461.

79 Yang, Z., Zhang, Y., Yang, Y. et al. (2010). Pharmacological and toxicological target organelles and safe use of single-walled carbon nanotubes as drug carriers in treating Alzheimer disease. *Nanomed. Nanotechnol. Biol. Med.* 6 (3): 427–441.

80 Vinoth, V., Wu, J.J., Asiri, A.M., and Anandan, S. (2015). Simultaneous detection of dopamine and ascorbic acid using silicate network interlinked gold nanoparticles and multiwalled carbon nanotubes. *Sensors Actuators B Chem.* 210: 731–741.

81 Yang, F., Jin, C., Dong, Y. et al. (2011). Magnetic functionalised carbon nanotubes as drug vehicles for cancer lymph node metastasis treatment. *Eur. J. Cancer* 4 (7): 1873–1882.

82 Ji, J., Liu, M., Meng, Y. et al. (2016). Experimental study of magnetic multi-walled carbon nanotube-doxorubicin conjugate in a lymph node metastatic model of breast Cancer. *Med. Sci. Monit.* 22: 2363–2373.

83 Bianco, K., Kostarelos, C.D., and Partidos, M.P. (2005). Biomedical applications of functionalised carbon nanotubes. *Chem. Commun.* 5: 571.

84 Petrella, S., Gelus-Ziental, N., Maudry, A. et al. (2011). Crystal structure of the pyrazinamidase of *Mycobacterium tuberculosis*: insights into natural and acquired resistance to pyrazinamide. *PLoS One* 6: e15785.

85 Saikia, N., Jha, A.N., and Deka, R.C. (2013). Interaction of pyrazinamide drug functionalized carbon and boron nitride nanotubes with pncA protein: a molecular dynamics and density functional approach. *RSC Adv.* 3: 15102–15107.

86 Zanella, I., Fagan, S.B., Mota, R., and Fazzio, A. (2007). Ab initio study of pristine and Si-doped capped carbon nanotubes interacting with nimesulide molecules. *Chem. Phys. Lett.* 439: 348–353.

87 Yao, H.-j., Zhang, Y.-g., Sun, L., and Liu, Y. (2014). The effect of hyaluronic acid functionalized carbon nanotubes loaded with salinomycin on gastric cancer stem cells. *Biomaterials* 35: 9208–9223.

88 Mohammadi, Z.A., Aghamiria, S.F., Zarrabi, A., and Talaiea, M.R. (2015). A comparative study on non-covalent functionalization of carbonnanotubes by chitosan and its derivatives for delivery of doxorubicin. *Chem. Phys. Lett.* 642: 22–28.

89 Taghavi, S., Hashem, A., Abnous, N.K., and Ramezani, M. (2017). Polyethylenimine-functionalized carbon nanotubes tagged with AS1411 aptamer for combination gene and drug delivery into human gastric cancer cells. *Int. J. Pharm.* 516 (1–2): 301–312.

90 Bussy, C., Al-Jamal, K.T., Boczkowski, J. et al. (2015). Microglia determine brain region-specific neurotoxic responses to chemically functionalized carbon nanotubes. *ACS Nano* 9 (8): 7815–7830.

91 Kafa, H., Wang, J.T.-W., Rubio, N. et al. (2015). The interaction of carbon nanotubes with an in vitro blood–brain barrier model and mouse brain in vivo. *Biomaterials* 53: 437–452.

92 Bottini, M., Rosato, N., and Bottini, N. (2011). PEG-modified carbon nanotubes in biomedicine current status and challenges ahead. *Biomacromolecules* 12 (10): 3381–3393.

93 Yang, S.T., Fernando, K.A., Liu, J.H. et al. (2008). Covalently PEGylated carbon nanotubes with stealth character in vivo. *Small* 4 (7): 940–944.

94 Vashist, S.K., Zheng, D., Pastorin, G. et al. (2011). Delivery of drugs and biomolecules using carbon nanotubes. *Carbon* 49 (13): 4077–4097.

95 Mehra, N.K., Mishra, V., and Jain, N.K. (2013). Receptor based therapeutic targeting. *Ther. Deliv.* 4 (3): 369–394.

96 Hussain, C.M. (2019). *Handbook of Environmental Materials Management*, 1e. Springer International Publishing.

97 Hussain, C.M. and Kharisov, B. (2017). *Advanced Environmental Analysis-Application of Nanomaterials*. The Royal Society of Chemistry.

98 Hussain, C.M. (2018). *Handbook of Nanomaterials for Industrial Applications*. Elsevier.

99 Hussain, C.M. and Keçili, R. (2019). *Modern Environmental Analysis Techniques for Pollutants*, 1e. Elsevier.

100 Hussain, C.M. and Mishra, A.K. (2019). *Nanotechnology in Environmental Science*, vol. 2. Wiley.

101 Hussain, C.M. (2020). *Handbook of Functionalized Nanomaterials for Industrial Applications*. Elsevier.

102 Hussain, C.M. (2020). *Handbook of Manufacturing Applications of Nanomaterials*. Elsevier.

13

Adsorptive Removal of Fluoride by Carbon Nanomaterials

Tanvir Arfin

Environmental Materials Division, CSIR-National Environmental Engineering Research Institute, Nagpur 440020, India

13.1 Introduction

The material science is considered as the primary sector in life of human being generating growth elongation of new innovative material by union processing of varied sources [1–5]. At present, the lifestyle of the people has given way for the development of novel material [6–11]. Such material is a primary source for the existence of an individual and survival of life; without it could be a difficult task [12–16]. Today, life depends on different materials [17–22]. It has enabled for comfortable and better sustaining life and even imparted the comforts and a satisfactory level of the single person [23–27]. It is also aiding for the progress and development of the country [28–33]. Fluoride is nonbiodegradable material present in Earth's crust. Natural source of fluoride in the human body is calcium fluoride. The bones and teeth are a storehouse of fluoride in the body. It is required in a minimal amount to reduce tooth decay and avoid dental problems [34, 35]. According to the studies, the fluoride present in drinking water helps to reduce 50–60% dental problems in children and maintain bone strength. It is required in a particular amount: if it is below the prescribed limit, then a dental crisis occurs, and if it exceeds the threshold, then it leads to three medical conditions, i.e. non-skeletal, skeletal, and dental fluoroses [36, 37]. Other sources of fluoride include soft drinks, various foods, and other products, and their intake increases fluoride level in the body. While consuming food sources, which are essential in nutrients like vitamin D, E, and C, calcium and various anti-oxidants will assist in reducing the toxic effects of fluoride intake.

The level of fluoride present in our food depends on various factors such as the geographical condition of the area, water utilized in agricultural fields and amount of precipitate in the particular area, and water used in cooking and washing fruits and vegetables and for drinking purposes. Malnutrition adds fuel to fire by increasing fluoride level; it is due to insufficient knowledge regarding nutritive content of foods that plays an essential role in reducing fluoride content in the body. Poverty and ignorance are also the reasons of malnutrition [38].

Environmental Applications of Carbon Nanomaterials-Based Devices, First Edition.
Edited by Shadpour Mallakpour and Chaudhery Mustansar Hussain.

13.2 Geochemistry of Fluoride

Fluoride due to its electronegativity is very reactive, so it reacts with all organic and inorganic substances. It is included in the halogen family with valence 1 and atomic number and atomic weight of 9 and 18.998, respectively. It occurs in combined form in nature. Average fluorine amount is 0.05–0.1% or 500–1000 mg/kg in Earth's crust. It occurs in association with chemicals like phosphate minerals, clay material, and tremolite. Important fluorosis materials that are essential mineral in various rock type are apatite $Ca(PO_4)_3(F/OH/Cl)$, fluorapatite $Ca_5(PO_4)_3F$, and fluorite (CaF_2). By weight, half the amount of fluorine is present inside fluorite. Micas, pyroxenes, and clay minerals have fluorine in their composition. Sorption capacity of soil increases due to a large amount of clay mineral in it, and it also acts as a natural barrier by preventing pollutants from being mixed with groundwater. Fluoride gets adsorbed on clay particles in the soil where ionic exchange occurs and replacement of OH^- the group with fluoride ions also takes place. Due to the fast weathering process in the case of mica and amphiboles, mobility of fluoride in soil increases [39]. Various types of rock and its fluoride amount are listed in Table 13.1 [40].

13.3 Fluoride in Water

Water is an essential source for the environment and sustainable development. Among the total available water on Earth, only 1% is available for our requirement. Water distribution is categorized into surface water and groundwater in which 0.6% part is occupied by groundwater. India depends on groundwater for its entire water requirement like irrigation in agricultural fields, animal husbandry, washing, cooking, etc. From contamination point of view, groundwater is considered as the safest source of water. Still, its purity is disturbed

Table 13.1 Nature of rock and its fluoride amount.

S. No.	Type of rock	F^- (in mg/kg)
1.	Conglomerate	963
2.	Igneous	1043
3.	Killas	1873
4.	Gneiss	1563
5.	Metamorphic	1503
6.	Silexite	982
7.	Sandstone	903
8.	Sedimentary	1175
9.	Granite	1043
10.	Schist	1703

Source: Based on He et al. [40].

due to leaching of various pollutants such as sulfuric acid, fluoride, mercury, etc., which are released from multiple industries and other chemical sources.

13.3.1 Dynamics of Fluoride in Groundwater

Emancipation of fluoride depends on the following factors:

- Geochemistry of water
- Mineralogical composition
- Extend of alteration of sediments

Fluoride-contributing minerals in groundwater are aegirine, hornblende, biotite, etc., and other minerals include epidote, apatite, sphene, etc. Geogenic minerals release fluoride in groundwater inconsiderable amount. CaF_2 solubility decides fluoride amount in groundwater while the dissolution of calcite and fluorite governs Ca^{2+} solubility in groundwater. Therefore chemical composition with regard to Na-HCO_3 variation occurs due to the interaction between permeable rock and groundwater [41].

The reduced form of fluorine is fluoride, and it is highly reactive. Based on weight, its abundance is 0.3 g/kg in Earth's crust, and it thus occupies 13th position. In India, Topaz is the main source of fluoride. It is found in −1 oxidation state in combination with various chemicals. Common associates of fluoride ion are KF, NaF, HF, etc. Fluorspar (CaF_2), fluorapatite [$3Ca_3(PO_4)_2 \cdot Ca(F, Cl_2)$], and cryolite ($Na_3AlH_6$) are sources of fluoride.

The impact of fluoride on environment and living being depends on its concentration. Beneficial fluoride range for the bones and teeth is 0.4–1.0 mg/l. Toxicity and harmful nature of fluoride depends on its oxidation state. Fluoride shows its effects in elemental form. In India, the beneficial limit for fluoride is 1 mg/l in the drinking water. According to the different organization like Bureau of Indian Standards (BIS), Indian Council of Medical Research (ICMR), World Health Organization (WHO), Indian Standard Institution (ISI), and others, range of fluoride amount varies. The concentration range of fluoride is listed in Table 13.2.

Table 13.2 Various organization with limit.

S. No.	Organization	F^- (desirable limit) (ppm)
1.	Bureau of Indian Standards (BIS)	0.6–1.2
2.	The United States Public Health (USPH)	0.8
3.	Indian Standard Institution (ISI)	1.5
4.	Indian Council of Medical Research (ICMR)	1
5.	World Health Organization (WHO)	1.5
6.	The Committee on Public Health Engineering Manual and Code of Practice, Government of India	1
7.	The International Standard for Drinking Water	0.5

13.4 Fluoride Solubility and Temperature

Decreased solubility and dissolution rate in the case of CaF_2 was seen in granitic terrains. In the case of fluoride, equilibrium constant depends on temperature, and about 30% fluoride solubility is seen when stability constant increment occurs from $10^{-10.8}$ to $10^{-10.57}$. Many-fold increment in the case of fluoride solubility is observed when the temperature increases from 273.15 to 373.15 K [42].

13.5 Sources of Fluoride in the Environment

Human-made activities and natural calamities contribute toward fluoride addition in the environment. Natural contributor to fluoride source involves soil and rock. Natural leaching contributes toward contamination of groundwater occurred during percolation of water through rocks and soil. The indirect cause of groundwater contamination includes irrigation system for long duration, evaporation water, and long-term storage of water between rocks and soil. Table 13.3 shows the percentage content of fluoride in mineral rocks [43].

13.6 Health Effects of Fluoride

If fluoride is present in the diet regularly, then such a long exposure to fluoride leads to diseases like skeletal fluorosis and dental fluorosis. Ion attraction between the fluoride ion and the teeth is the reason for fluorosis where the negatively charged fluoride ions attracted toward the positively charged ions present in the teeth. Fluoride will not kill immediately, but it will accumulate in the brain and degenerates the body slowly. Fluoride affects in the following ways as shown in Table 13.4.

13.7 Removal Technologies

For the removal of fluoride, different methods are available such as precipitation/coagulation, adsorption, ion exchange, reverse osmosis, and electrodialysis. Among this method,

Table 13.3 Mineral rock showing percentage content of fluoride.

S. No.	Minerals	Chemical formula	F (%)
1.	Sellaite	MgF_2	61
2.	Villianmite	NaF	55
3.	Fluorspar	CaF_2	49
4.	Cryolite	Na_3AlF_6	45
5.	Bastnaesite	(Ce, La) $(CO_3)F$	9
6.	Fluorapatite	$Ca_3(PO_4)_3F$	3–4

Source: Based on Vithanage and Bhattacharya [43].

Table 13.4 Diseases occurred due to fluoride.

Cancer	Lowers IQ	Memory loss	Kills brain cells
Bone damage in the unborn fetus	Inhibits melatonin production	Kidney stones and kidney failure	Interferes with white blood cell (WBC) production
Bone weakness	Impairs immune system	Affects heart circulation	Promotes mental disturbance
Dental fluorosis	Apathy and passivity	Accelerates the ageing process	Attacks hypothalamus
Joint problems	Autism	Attacks thyroid	Alzheimer's disease
Damage stomach	DNA damage	Shortens life span	Skeletal fluorosis

precipitation/coagulation is the most recognized one as it is cost-effective, showing less efficiency for the removal of fluoride below the concentration 5 mg/l. Such a process needs a certain quantity of chemicals for enhancing the costing and sludge generation, as well as problem based on the disposal. This process also requires acid neutralization since few substances such as SO_4^{2-} and aluminium fluoride (AIF) complex are still present in treated water as well. Ion exchange can be an efficient method, but the methods possess quite less selectivity, and it is even a costly method. The processes such as reverses osmosis and electrodialysis reduce the problems faced by precipitation/coagulation and ion exchange because they involve no chemicals and even show extraordinary response toward fluoride removal. However, they require substantial capital uses, and the cost of the operating system is also high.

In comparison with other mentioned techniques, the adsorption process is cost-effective and viable. The adsorption process is employed to large extend due to various factors such as effectiveness, convenience, easy operational technique, simple designing, and economic and environmental benefits. The hazardous effect of fluoride on the human being is reduced because of the defluoridation method [44].

13.8 Classification of Adsorbents

The attempt has been taken to show a different range of absorbent employed for fluoride removal from the drinking water. The adsorbents, namely, clays; geomaterials; minerals; low-cost materials and zeolite [45]; chitin–chitosan–alginate–polymers–cellulose–resins–algae and fungi [46]; industrial, agricultural, and biomass wastes [47]; carbonaceous materials [48]; aluminum [49]; calcium [50]; mixed metal oxide [51]; nanomaterials; building materials; and other miscellaneous [52], have been developed earlier and are used on a large scale for fast defluoridation in an aqueous solution. The types of adsorbents used are given in the Figure 13.1.

13.9 Carbon-Based Adsorbents

At present, carbon from various species are made and employed for the removal of fluoride and different application as well. The raw material for manufacturing carbon is

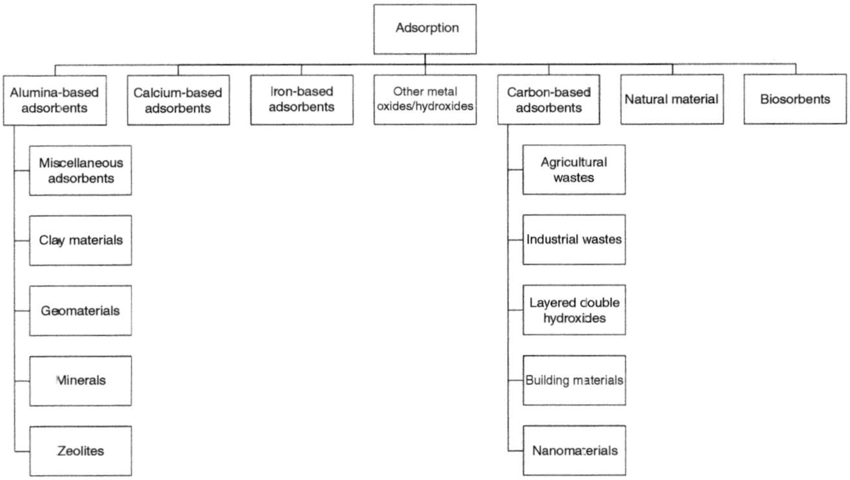

Figure 13.1 Classification of adsorbents.

available easily. The carbonaceous substances such as plants, animal, mineral posing large concentration of carbon are changed easily in an activated form by employing gas-activated method. Activated carbon is prepared from raw material, namely, nut shells, wood, and charcoal. The activated carbon prepared by the hardwood is given preference for the process of adsorption as the charcoal attained by softwood is unstable and it gets crumbles quickly. Researchers have even reported about the use of some carbon-based sorbet for the removal of fluoride. Bhargava and Killedar [53] have conducted the process of fluoride adsorption on fishbone charcoal in the system as the mobile media. The ratio of equilibrium or sorbate concentration with that of initial concentration and the defluoridation capacity were observed to show varied response. It was inverse to the mass input rate of sorbent and also exhibited various direct proportion for a flow rate of the sorbent. The initial concentration of sorbate was observed to work as sorbent–sorbate mass input ratio.

13.9.1 Carbon Nanomaterials (CNM)

Nanotechnology plays a significant role in the technological field [54–58], where nanomaterial is the primary substance in the advancement of such technology [59–64]. According to the name itself, it is clear that nanomaterial has a dimension of 1–100 nm in the structural unit [65, 66]. The nanoscale working licenses for cautiously controlling the significant material accomplishing the interesting features that generally change from that are available in subatomic level or mass structure [67–71]. Even at present, some of the material is being synthesized, namely, carbon nanotube, graphene, fullerenes, etc. With the development in the nanotechnological sector, the interest of the researcher has also enhanced to use the novel properties and features in different application like treatment of water, production of energy, contamination sensing, and so on [72–76].

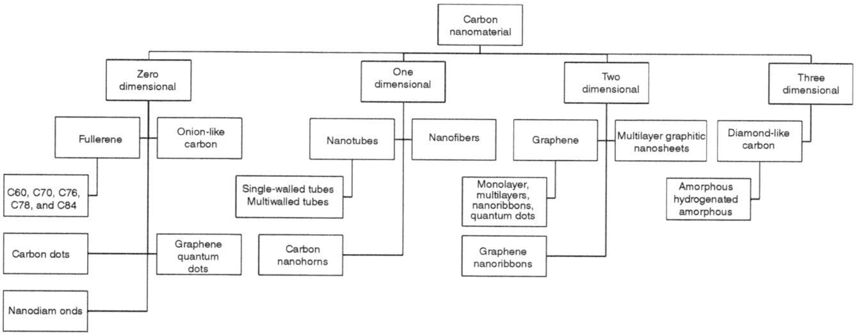

Figure 13.2 Different types of carbon nanomaterials.

In contrast to the other nanomaterial, the carbon nanomaterial (CNM) and its derivatives have gathered lots of interest as they possess excellent character and it has the potential to be applied in various application. Carbon is capable of bonding itself in an extraordinary architectural way and helps in forming exclusively low-dimension structure. Different types of CNM are shown in Figure 13.2.

13.9.1.1 Carbon Nanotube (CNT)

Carbon nanotube (CNT) has gained a lot of attention in the last 20 years and is widely used in the treatment of wastewater as it has potential in pollution removal ability [77–82]. It has been already reported before about the removal of fluoride by CNT-based composite in literature and is exhibited in Table 13.5.

As per the structure, CNTs are categorized into two classes, such as single-walled nanotubes (SWCNTs) and multiwalled nanotubes (MWCNTs). Single-walled nanotubes (SWNTs) are rolled graphene sheet forming a tube of diameter ranging between 0.7 and 1.4 nm. Multiwalled nanotubes (MWNT) has numerous layers rolled up into a diameter of 100 nm [88–91]. SWCNT- and MWCNT-based materials for fluoride removal are listed in Table 13.6.

13.9.1.2 Graphene

When we go through the new-generation carbon-based nanomaterial, graphene is the most favorable developed and engineered material in various applications [96–107]. It is a 2D

Table 13.5 Study of fluoride adsorption onto various CNT-based materials.

S.No.	Adsorbent	Q_{max} (mg/g)	References
1.	Aligned CNTs	3.0	[83]
2.	Al_2O_3-supported CNTs	28.7	[84]
3.	Coconut-shell-derived carbon/CNT composite	0.36	[85]
4.	CNT-HAP	11.05	[86]
5.	Al_2O_3/CNTs	45.32	[87]

Table 13.6 Study of fluoride adsorption onto various CNT-based materials.

S.No.	Adsorbent	Q_{max} (mg/g)	References
1.	SWCNTs	16.63	[92]
2.	HA-MWCNTs	39.22	[93]
3.	MWCNTs	2.83	[94]
4.	SWCNTs	2.40	[94]
5.	MWCNTs	3.5	[95]

Table 13.7 Study of fluoride adsorption onto various graphene-based materials.

S.No.	Adsorbent	Q_{max} (mg/g)	References
1.	Graphene oxide/alumina	4.68	[112]
2.	Zr-chitosan (CTS) / (slash) graphene oxide (GO)	29.0588	[113]
3.	FeOOH/GO	17.65	[114]
4.	$FeOOH^{+}Ac/GO$	19.82	[115]
5.	AlOOH@reduced graphene oxide	118.7	[116]

carbon in allotrope form and contains carbon allotropes, namely, graphite, fullerenes, etc. [108–111]. Graphene-based material for fluoride removal is listed in Table 13.7.

13.10 Conclusion

Nanotechnology had gathered attention in the field of research and brought about novel modernization in science and technology [117–121]. The nanomaterial, namely, CNM, can be now manufactured in large industrial scale and put in use in wide extend. In a short duration, the extraordinary characteristic of CNM has gathered interest by the invention of fullerenes (1985), CNTs (1991), and graphene (2004). This has led to the influence and development of industrial production on a large scale. The contamination of drinking water in the form of fluoride is caused by natural and anthropogenic activities. It is considered as severe problems posing a threat to the health issue of human being global. Adsorption is the highly recognized process for treatment and removal of fluoride, and it shows favorable results in the form of CNM adsorbent. In this chapter, the lists of different CNM adsorbent are formulated, and the adsorption capacities of the removal of fluoride are highlighted and briefly discussed. It is clear from the literature survey that adsorbent is responsible for exhibiting excellent potential for defluoridation. However, it requires a particular measure to find the practical application of the adsorbent in the commercial field, which can control the pollution aspect.

Acknowledgment

Authors acknowledge the Knowledge Resource Centre, CSIR-NEERI (CSIR-NEERI/KRC/2020/JUNE/EMD/3), for their support.

Important Websites

https://www.google.com/search?q=Adsorptive+removal+of+fluoride+by+Carbon+Nanomaterials&rlz=1C1CHBF_enUS858US858&oq=Adsorptive+removal+of+fluoride+by+Carbon+Nanomaterials&aqs=chrome..69i57.823j0j7&sourceid=chrome&ie=UTF-8.

https://www.google.com/search?q=Adsorptive+removal+of+fluoride+by+Carbon+Nanomaterials&newwindow=1&rlz=1C1CHBF_enUS858US858&sxsrf=ALeKk03d8nRKXdgI5UOyqn6TlsqcOD-9rw:1603717995525&source=lnms&tbm=bks&sa=X&ved=2ahUKEwiDrsWLq9LsAhWSr3EKHQLNDDQQ_AUoAXoECAQQAw&biw=1368&bih=802.

https://www.google.com/search?q=Adsorptive+removal+of+fluoride+by+Carbon+Nanomaterials&newwindow=1&rlz=1C1CHBF_enUS858US858&tbm=isch&sxsrf=ALeKk0309N9TVuX8PhpxhX3ZODHijYhqPw:1603718177201&source=lnms&sa=X&ved=0ahUKEwibhJbiq9LsAhWzo3EKHcJgAPUQ_AUICygB&biw=1368&bih=802&dpr=2.

References

1 Arfin, T. (2015). Applications of bentonite and its role as a base of engineering chemistry. In: *Bentonite: Characteristics, Uses and Implications for the Environment*, Ch. 5 (ed. A.K. Mishra), 127–135. New York: Nova Science Publishers.

2 Mohammad, F., Arfin, T., and Al-Lohedan, H.A. (2017). Enhanced biological activity and biosorption performance of trimethyl chitosan-loaded cerium oxide particles. *J. Ind. Eng. Chem.* 45: 33–43.

3 Arfin, T., Tarannum, A., and Sonawane, K. (2018). Green and sustainable advanced materials: an overview. In: *Green and Sustainable Advanced Materials: Processing and Characterization*, vol. I, Ch. 1 (eds. S. Ahmed and C.M. Hussain), 1–34. Beverly, MA: Scrivener Publishing LLC.

4 Arfin, T. (2019). Bionanoceramic and bionanocomposite: concepts, processing, and applications. In: *Handbook of Polymer and Ceramic Nanotechnology*, Ch. 7 (eds. C.M. Hussain and S. Thomas), 1–19. Switzerland: Springer International Publishing.

5 Arfin, T. (2019). Reactive and functional polymers. In: *Advanced Functional Textiles and Polymers: Fabrication, Processing and Applications*, Ch. 10 (eds. S. Ul-Islam and B.S. Butola), 1–19. Beverly, MA: Scrivener Publishing LLC.

6 Arfin, T. and Sonawane, K. (2018). An excellence method on starch-based materials: a promising stage for environmental application. In: *Green and Sustainable Advanced Materials: Applications*, Vol. II, Ch. 8 (eds. S. Ahmed and C.M. Hussain), 177–208. Beverly, MA: Scrivener Publishing LLC.

7 Arfin, T. and Mogarkar, P.R. (2018). Bio-based material protein and its novel applications. In: *Biocomposites: Biomedical and Environmental Applications*, Ch.15 (eds. S. Ahmed, S. Ikram, S. Kanchi and K. Bisetty), 405–432. Singapore: Pan Stanford Publishing.

8 Athar, S. and Arfin, T. (2017). Commercial and prospective applications of gelatin. In: *Natural Polymers: Derivatives, Blends and Composites*, vol. II, Ch. 11 (eds. S. Ikram and S. Ahmed), 199–216. New York: Nova Science Publishers.

9 Mogarkar, P.R. and Arfin, T. (2017). Chemical and structural importance of starch-based derivative and its applications. In: *Natural Polymers: Derivatives, Blends and Composites*, vol. II, Ch. 5 (eds. S. Ikram and S. Ahmed), 73–87. New York: Nova Science Publishers.

10 Arfin, T. (2019). Marine polysaccharides: an overview. In: *Marine Polysaccharides: Advances and Multifaceted Applications*, Ch. 1 (eds. S. Ahmed and A. Soundararajan), 1–11. Singapore: Pan Stanford Publishing.

11 Arfin, T., Bushra, R., and Kriek, R.J. (2013). Ionic conductivity of alkali halides across a polyaniline-zirconium (IV)-arsenate membrane. *Anal. Bioanal. Electrochem.* 5: 206–221.

12 Arfin, T. and Yadav, N. (2012). Impedance characteristics and electrical double-layer capacitance of composite polystyrene-cobalt-arsenate membrane. *J. Ind. Eng. Chem.* 19: 256–262.

13 Arfin, T., Falch, A., and Kriek, R.J. (2012). Evaluation of charge density and the theory for calculating membrane potential for a nano-composite nylon-6,6 nickel phosphate membrane. *Phys. Chem. Chem. Phys.* 14: 16760–16769.

14 Arfin, T. and Rafiuddin (2011). An electrochemical and theoretical comparison of ionic transport through a polystyrene-based cobalt arsenate membrane. *Electrochim. Acta* 56: 7476–7483.

15 Arfin, T. and Fatima, S. (2014). Anticipating behaviour of advanced materials in healthcare. In: *Advanced biomaterials and biodevices*, Ch. 7 (eds. A. Tiwari and A.N. Nordin), 243–287. Beverly, MA: Scrivener Publishing LLC.

16 Arfin, T. and Fatima, S. (2014). Conductometric studies with polystyrene calcium phosphate membrane. *Asian J. Adv. Basic Sci.* 2: 1–14.

17 Onwudiwe, D.C., Arfin, T., Strydom, C.A., and Kriek, R.J. (2013). Synthesis, spectroscopic characterization and behaviour of AC impedance spectroscopy of Cd(II) bis (*N*-para-methylphenyl dithiocarbamate). *Electrochim. Acta* 104: 19–25.

18 Arfin, T. and Rafiuddin (2009). Transport studies of nickel arsenate membrane. *J. Electroanal. Chem.* 636: 113–122.

19 Arfin, T., Jabeen, F., and Kriek, R.J. (2011). An electrochemical and theoretical comparison of ionic transport through a polystyrene based titanium-vanadium (1:2) phosphate membrane. *Desalination* 274: 206–211.

20 Arfin, T., Athar, S., and Rangari, S. (2019). Proteins and their novel applications. In: *Handbook of Biopolymers: Advances and Multifaceted Applications*, Ch. 4 (eds. S. Ahmed, S. Kanchi and G. Kumar), 75–93. Singapore: Pan Stanford Publishing.

21 Arfin, T. and Fatma, S. (2014). Synthesis, influence of electrolyte solutions on impedance properties and in-vitro antibacterial studies of organic-inorganic composite membrane. *Adv. Ind. Eng. Manag.* 3: 19–30.

22 Arfin, T. and Kumar, C. (2014). Synthesis, characterization, conductivity and antibacterial activity of ethyl cellulose manganese (II) hydrogen phosphate. *Anal. Bioanal.Chem.* 6: 403–421.

23 Arfin, T. and Sonawane, K. (2018). Bio-based materials: past to future. In: *Bio-based Materials for Food Packaging*, Ch. 1 (ed. S. Ahmed), 1–32. Singapore: Springer International Publishing.

24 Arfin, T. and Tarannum, A. (2017). Polymer materials: from the past to the future. In: *Green Polymeric Materials*, Ch. 2 (eds. S. Ahmed and S.I. Annu), 35–52. New York: Nova Science Publishers.

25 Arfin, T. and Mohammad, F. (2016). Chemistry and structural aspects of chitosan towards biomedical. In: *Natural Polymers: Derivatives, Blends and Composite*, vol. 1, Ch. 13 (eds. S. Ikram and S. Ahmed), 265–280. New York: Nova Science Publishers.

26 Arfin, T. and Mohammad, F. (2015). Electrical conductivity, mechanical stability, antibacterial and anticancer activities of ethyl cellulose-tin (II) hydrogen phosphate. *Adv. Mater. Lett.* 6: 1058–1065.

27 Arfin, T. and Mohammad, F. (2014). Electrochemical, dielectric behaviour and in vitro antimicrobial activity of polystyrene-calcium phosphate. *Adv. Ind. Eng. Manag.* 3: 25–38.

28 Arfin, T. and Mohammad, F. (2015). Dendrimer and its role for the advancement of nanotechnology and bioengineering. In: *Advances in Materials Science Research*, vol. 21, Ch. 7 (ed. M.C. Wythers), 157–174. New York: Nova Science Publishers.

29 Arfin, T., Mohammad, F., and Yusof, N.A. (2015). Applications of polystyrene and its role as a base in industrial chemistry. In: *Polystyrene: Synthesis, Characteristics and Applications*, Ch. 10 (ed. C. Lynwood), 269–280. New York: Nova Science Publishers.

30 Mohammad, F. and Arfin, T. (2013). Cytotoxic effects of polystyrene-titanium-arsenate composite in cultured H9c2 cardiomyoblasts. *Bull. Environ. Contam. Toxicol.* 91: 689–696.

31 Arfin, T. and Mohammad, F. (2013). DC electrical conductivity of nano-composite polystyrene-titanium-arsenate membrane. *J. Ind. Eng. Chem.* 19: 2046–2051.

32 Arfin, T. and Rafiuddin (2009). Electrochemical properties of titanium arsenate membrane. *Electrochim. Acta* 54: 6928–6934.

33 Arfin, T., Singh, B., and Varshney, N. Biological adhesion behavior of superhdrophobic polymer coating. In: *Superhydrophobic Polymer Coatings: Fundamentals, Design, Fabrication, and Applications*, Ch.7 (eds. S.K. Samal, S. Mohanty and S.K. Nayak), 161–177. Amsterdam: Elsevier.

34 Waghmare, S.S., Arfin, T., Lataye, D. et al. (2015). Adsorption behaviour of eggshell modified polyalthia longifolia leaf based alumina as a novel adsorbents for fluoride removal from drinking water. *Int. J. Adv. Res. Innov. Ideas Educ.* 1: 904–925.

35 Waghmare, S.S., Arfin, T., Rayalu, S. et al. (2015). Adsorption behaviour of modified zeolite as novel adsorbents for fluoride removal from drinking water: surface phenomena, kinetics and thermodynamics studies. *Int. J. Sci. Eng. Technol. Res.* 4: 4114–4124.

36 Waghmare, S., Arfin, T., Manwar, N. et al. (2015). Preparation and characterization of polyalthia longifolia based adsorbent for removing fluoride from drinking water. *Asian J. Adv. Basic Sci.* 4: 12–24.

37 Waghmare, S.S. and Arfin, T. (2015). Fluoride induced water pollution issue and its health efficacy in India- a review. *Int. J. Eng. Res. General Sci.* 3: 345–358.

38 Susheela, A.K. and Toteja, G.S. (2018). Prevention & control of fluorosis & linked disorders: developments in the 21st century-reaching out to patients in the community & hospital settings for recovery. *Indian J. Med. Res.* 148: 539–547.

39 Ayoob, S. and Gupta, A.K. (2006). Fluoride in drinking water: a review on the status and stress effects. *Crit. Rev. Env. Sci. Technol.* 36: 433–487.

40 He, J., An, Y., and Zhang, F. (2013). Geochemical characteristics and fluoride distribution in the ggroundwater of the Zhangye basin in the North Western China. *J. Geochem. Explor.* 135: 22–30.

41 Chenini, I., Farhat, B., and Mammou, A.B. (2010). Identification of major sources controlling groundwater chemistry from a multi-layered aquifer system. *Chem. Speciation Bioavailability* 22: 183–189.

42 Reynolds, J.G. and Belsher, J.D. (2017). A review of sodium fluoride solubility in water. *J. Chem. Eng. Data* 62: 1743–1748.

43 Vithanage, M. and Bhattacharya, P. (2015). Fluoride in the environment: sources, distribution and defloridation. *Environ. Chem. Lett.* 13: 131–147.

44 Waghmare, S.S. and Arfin, T. (2015). Fluoride removal from water by various techniques: review. *Int. J. Innov. Sci. Eng. Technol.* 2: 560–571.

45 Waghmare, S.S. and Arfin, T. (2015). Fluoride removal by clays, geomaterials, minerals, low cost materials and zeolites by adsorption: a review. *Int. J. Sci. Eng. Technol. Res.* 4: 3663–3676.

46 Waghmare, S.S. and Arfin, T. (2015). Defluoridation by adsorption with chitin-chitosan-alginate-polymers-cellulose-resins-algae and fungi: a review. *Int. Res. J. Eng. Technol.* 2: 1179–1197.

47 Waghmare, S.S. and Arfin, T. (2015). Fluoride removal by industrial, agricultural and biomass wastes as adsorbents: review. *Int. J. Adv. Res. Innov. Ideas Educ.* 1: 628–653.

48 Waghmare, S.S. and Arfin, T. (2015). Fluoride removal from water by carbonaceous materials: review. *Int. J. Modern Trends Eng. Res.* 2: 355–361.

49 Waghmare, S.S. and Arfin, T. (2015). Fluoride removal from water by aluminium based adsorption: a review. *J. Biol. Chem. Chron.* 1: 1–11.

50 Waghmare, S.S. and Arfin, T. (2015). Fluoride removal from water by calcium materials: a state-of-the-art review. *Int. J. Innov. Res. Sci. Eng. Technol.* 4: 8090–8102.

51 Waghmare, S.S. and Arfin, T. (2015). Fluoride removal from water by mixed metal oxide adsorbent materials: a state-of-the-art review. *Int. J. Eng. Sci. Res. Technol.* 4: 519–536.

52 Waghmare, S.S., Arfin, T., Lataye, D., and Rayalu, S. (2015). Defluoridation by nanomaterials, building materials and other miscellaneous materials: a systematic review. *Int. J. Innov. Res. Sci. Eng. Technol.* 4: 11998–12010.

53 Bhargava, D.S. and Killedar, D.J. (1992). Fluoride adsorption on fishbone charcoal through a moving media adsorber. *Water Res.* 26: 781–788.

54 Khan, U., Malik, N., and Arfin, T. (2017). Nanofibrillated cellulose and copoly(amino acid) hydrogel matrics in biotechnology and biomedicine. In: *Nanocellulose and Nanohydrogel Matrices: Biotechnological and Biomedical Applications*, Ch. 12 (eds. M. Jawaid and F. Mohammad), 331–352. Weinheim: Wiley-VCH.

55 Athar, S., Bushra, R., and Arfin, T. (2017). Cellulose nanocrystals and PEO/PET hydrogel material in biotechnology and biomedicine: current status and future prospects. In: *Nanocellulose and Nanohydrogel Matrices: Biotechnological and Biomedical Applications*, Ch. 7 (eds. M. Jawaid and F. Mohammad), 139–173. Weinheim: Wiley-VCH.

56 Borkar, R., Waghmare, S.S., and Arfin, T. (2017). Bacterial cellulose and polyester hydrogel matrics in biotechnology and biomedicine: current status and future prospects. In:

Nanocellulose and Nanohydrogel Matrices: Biotechnological and Biomedical Applications, Ch.2 (eds. M. Jawaid and F. Mohammad), 21–46. Weinheim: Wiley-VCH.

57 Arfin, T. and Tarannum, A. (2018). Engineered nanomaterials for industrial application: an overview. In: *Handbook of Nanomaterials for Industrial Applications*, Ch. 6 (ed. C.M. Hussain), 127–134. Amsterdam: Elsevier.

58 Arfin, T. (2017). Chitosan and its derivatives: overlook of commercial application in diverse field. In: *Chitosan: derivatives, composites and applications, Ch. 5* (eds. S. Ahmed and S. Ikram), 115–150. Beverly, MA: Scrivener Publishing LLC.

59 Hussain, C.M. and Kharisov, B. (2017). *Advanced Environmental Analysis-Application of Nanomaterials*, vol. 1. Cambridge: The Royal Society of Chemistry.

60 Hussain, C.M. (2018). *Handbook of Nanomaterials for Industrial Applications*. Amsterdam: Elsevier.

61 Hussain, C.M. and Mishra, A.K. (2019). *Nanotechnology in Environmental Science*, vol. 2. Hoboken, NJ: Wiley.

62 Hussain, C.M. (2019). *Handbook of Environmental Materials Management*, 1e. Switzerland: Springer International Publishing.

63 Hussain, C.M. and Keçili, R. (2019). *Modern Environmental Analysis Techniques for Pollutants*, 1e. Amsterdam: Elsevier.

64 Mallakpour, S. and Rashidimoghadam, S. (2020). Utilization of starch and starch/carbonaceous nanocomposites for removal of pollutants from wastewater. In: *Handbook of Polymer Nanocomposites for Industrial Applications*, 1e, Ch. 29 (ed. C.M. Hussain). Amsterdam: Elsevier, 477–502.

65 Hussain, C.M. (2020). *Handbook of Functionalized Nanomaterials for Industrial Applications*. Amsterdam: Elsevier.

66 Hussain, C.M. (2020). *Handbook of Manufacturing Applications of Nanomaterials*. Amsterdam: Elsevier.

67 Arfin, T., Varshney, N., and Singh, B. (2020). Ionic liquid modified activated carbon for the treatment of textile wastewater. In: *Green Materials For Wastewater Treatment*, Ch.11 (eds. Mu.Naushad and E. Lichtfouse), 257–275. Switzerland: Springer International Publishing.

68 Arfin, T., Sonawane, K., and Tarannum, A. (2019). Review on detection of phenol in water. *Adv. Mater. Lett.* 10: 753–785.

69 Onwudiwe, D.C., Arfin, T., and Strydom, C.A. (2015). Surfactant mediated synthesis of ZnO nanospheres at elevated temperature, and their dielectric properties. *Superlattices Microstruct.* 81: 215–225.

70 Mohammad, F., Arfin, T., and Al-Lohedan, H.A. (2019). Biocompatible polylactic acid-reinforced nickel-arsenate composite: studies of electrochemical conductivity, mechanical stability, and cell viability. *Mater. Sci. Eng., C* 102: 142–149.

71 Mohammad, F., Arfin, T., Bwatanglang, I.B., and Al-Lohedan, H.A. (2019). Starch-based nanocomposites: types and industrial applications. In: *Bio-based Polymers and Nanocomposites: Preparation, Processing, Properties & Performance*, Ch.8 (eds. M.L. Sanyang and M. Jawaid), 157–181. Switzerland: Springer International Publishing.

72 Mohammad, F., Arfin, T., Saba, N. et al. (2018). Electrical conductivity and biological efficacy of ethyl cellulose and polyaniline-based composites. In: *Electrically Conductive Polymers and Polymer Composites: From Synthesis to Biomedical Applications*, Ch. 9 (eds. A. Khan, M. Jawaid, A.A.P. Khan and A.M. Asiri), 181–197. Weinheim: Wiley-VCH.

73 Sophia, A.C., Arfin, T., and Lima, E.C. (2019). Recent developments in adsorption of dyes using graphene-based nanomaterials. In: *A New Generation Material Graphene: Applications in Water Technology*, Ch. 18 (ed. Mu.Naushad), 439–471. Switzerland: Springer International Publishing.

74 Arfin, T. (2018). Current innovative chitosan-based water treatment of heavy metals: a sustainable approach. In: *Handbook of Biopolymers: Advances and Multifaceted Applications*, Ch. 7 (eds. S. Ahmed, S. Kanchi and G. Kumar), 167–182. Singapore: Pan Stanford Publishing.

75 Arfin, T. and Tarannum, A. (2019). Rapid determination of lead ions using polyaniline-zirconium (IV) iodate-based ion selective electrode. *J. Environ. Chem. Eng.* 7: 102811.

76 Arfin, T. and Rafiuddin (2012). Metal ion transport through a polystyrene-based cobalt arsenate membrane: application of irreversible thermodynamics and theory of absolute reaction rates. *Desalination* 284: 100–105.

77 Zhang, Q., Huang, J.Q., Qian, W.Z. et al. (2013). The road for nanomaterials industry: a review of carbon nanotube production, post-treatment, and bulk applications for composites and energy storage. *Small* 9: 1237–1265.

78 Mallakpour, S. and Soltanian, S. (2016). Surface functionalization of carbon nanotubes: fabrication and applications. *RSC Adv.* 6: 109916–109935.

79 Mallakpour, S., Behranvand, V., and Mallakpour, F. (2019). Synthesis of alginate/carbon nanotube/carbon dot/fluoroapatite/TiO_2 beads for dye photocatalytic degradation under ultraviolet light. *Carbohydr. Polym.* 224: 115138.

80 Mallakpour, S. and Khodadadzadeh, L. (2018). Biocompatible and biodegradable Chitosan nanocomposites loaded with carbon nanotubes. In: *Biodegradable and Biocompatible Polymer Composites Processing, Properties and Applications*, Ch. 7 (ed. N.G. Shimpi), 187–221. Sawston, Cambridge, United Kingdom: Elsevier, Woodhead Publishing.

81 Mallakpour, S. and Khadem, E. (2019). Carbon nanotubes for heavy metals removal. In: *Composite Nanoadsorbents*, Ch.8 (eds. G. Kyzas and A.C. Mitrpoulos), 181–210. Amsterdam: Elsevier.

82 Mallakpour, S. and Rashidimoghadam, S. (2019). Carbon nanotubes for dyes removal. In: *Composite Nanoadsorbents*, Ch.9 (eds. G. Kyzas and A.C. Mitrpoulos), 211–244. Amsterdam: Elsevier.

83 Li, Y.-H., Wang, S., Zhang, X. et al. (2003). Adsorption of fluoride from water by aligned carbon nanotubes. *Mater. Res. Bull.* 38: 469–476.

84 Li, Y.-H., Wang, S., Cao, A. et al. (2001). Adsorption of fluoride from water by amorphous alumina supported on carbon nanotubes. *Chem. Phys. Lett.* 350: 412–416.

85 Araga, R., Kali, S., and Sharma, C.S. (2019). Coconut-shell-derived carbon/carbon nanotube composite for fluoride adsorption from aqueous solution. *Clean Soil Air Water* 47: 1800286.

86 Tang, Q., Duan, T., Li, P. et al. (2018). Enhanced defluoridation capacity from aqueous media via hydroxyapatite decorated with carbon nanotube. *Front Chem.* 6: 104.

87 Li, Y.H., wang, S., Zhang, X. et al. (2003). Removal of fluoride from water by carbon nanotube supported alumina. *Environ. Technol.* 24: 391–398.

88 Bushra, R., Arfin, T., Oves, M. et al. (2016). Development of PANI/MWCNTs decorated with cobalt oxide nanoparticles towards multiple electrochemical, photocatalytic and biomedical application sites. *New J. Chem.* 40: 9448–9459.

89 Arfin, T. (2018). MWCNT polymer composites: environmental applications. In: *Handbook of Bionanocomposites: Green and Sustainable Materials*, Ch.9 (eds. S. Ahmed and S. Kanchi), 235–245. Singapore: Pan Stanford Publishing.

90 Mallakpour, S. and Behranvand, V. (2018). Synthesis of mesoporous recycled poly(ethylene terephthalate)/MWNT/carbon quantum dot nanocomposite from sustainable materials using ultrasonic waves: application for methylene blue removal. *J. Cleaner Prod.* 190: 525–537.

91 Mallakpour, S. and Khadem, E. (2016). Carbon nanotube–metal oxide nanocomposites: fabrication, properties and applications. *Chem. Eng. J.* 302: 344–367.

92 Balarak, D., Esfandyari, Y., Bazrafshan, E. et al. (2016). Adsorption of fluoride from aqueous solutions by carbon nanotubes: determination of equilibrium, kinetic and thermodynamic parameters. *Fluoride* 49: 71–83.

93 Ruan, Z., Tian, Y., Ruan, J. et al. (2017). Synthesis of hydroxyapatite/multi-walled carbon nanotubes for the removal of fluoride ions from solution. *Appl. Surf. Sci.* 412: 578–590.

94 Deghani, M.H., Haghighat, G.A., Yetilmezsoy, K. et al. (2016). Adsorptive removal of fluoride from aqueous solution using single-and multi-walled carbon nanotubes. *J. Mol. Liq.* 216: 401–410.

95 Ansari, M., Kazemipour, M., Dehghani, M., and Kazemipour, M. (2011). The defluoridation of drinking water using multi-walled carbon nanotubes. *J. Fluorine Chem.* 132: 516–520.

96 Mohammad, F., Arfin, T., and Al-lohedan, H.A. (2019). Enhanced biosorption and electrochemical performance of sugarcane bagasse derived a polylactic acid-graphene oxide-CeO_2 composite. *Mater. Chem. Phys.* 229: 117–123.

97 Mohammad, F., Arfin, T., and Al-lohedan, H.A. (2019). Development of graphene-based nanocomposites as potential materials for supercapacitors and electrochemical cells. In: *Graphene-based Nanotechnologies for Energy and Environmental Applications*, Ch.8 (eds. M. Jawaid, A. Ahmad and D. Lokhat), 145–154. Amsterdam: Elsevier.

98 Malik, N., Arfin, T., and Khan, A.U. (2019). Graphene nanomaterials: chemistry and pharmaceutical perspectives. In: *Nanomaterials for Drug Delivery and Therapy*, Ch.13 (ed. A.M. Grumezescu), 373–402. Amsterdam: Elsevier.

99 Mallakpour, S. and Khadem, E. (2019). Linear and nonlinear behavior of crosslinked chitosan/N-doped graphene quantum dot nanocomposite films in cadmium cation uptake. *Sci. Total Environ.* 690: 1245–1253.

100 Mallakpour, S., Abdolmaleki, A., and Borandeh, S. (2017). Fabrication of amino acid-based graphene-zinc oxide (ZnO) hybrid and its application for poly(ester–amide)/graphene-ZnO nanocomposite synthesis. *J. Thermoplast. Compos. Mater.* 30: 358–380.

101 Mallakpour, S., Abdolmaleki, A., and Karshenas, A. (2017). Graphene oxide supported copper coordinated amino acids as novel heterogeneous catalysts for epoxidation of norbornene. *Catal. Commun.* 92: 109–113.

102 Mallakpour, S., Abdolmaleki, A., Mahmoudian, M. et al. (2017). Synergetic effect of synthesized sulfonated polyaniline/quaternized graphene and its application as a high-performance supercapacitor electrode. *J. Mater. Sci.* 52: 9683–9695.

103 Mallakpour, S. and Khadem, E. (2018). Construction of crosslinked chitosan/nitrogen-doped graphene quantum dot nanocomposite for hydroxyapatite biomimetic mineralization. *Int. J. Biol. Macromol.* 120: 1451–1460.

104 Abdolmaleki, A., Mallakpour, S., and Karshenas, A. (2017). Synthesis and characterization of new nanocomposites films using alanine-Cu-functionalized graphene oxide as nanofiller and PVA as polymeric matrix for improving of their properties. *J. Solid State Chem.* 253: 398–405.

105 Abdolmaleki, A., Mallakpour, S., and Borandeh, S. (2016). Improving interfacial interaction of l-phenylalanine-functionalized graphene nanofiller and poly(vinyl alcohol) nanocomposites for obtaining significant membrane properties: Morphology, thermal, and mechanical studies. *Polym. Compos.* 37: 1924–1935.

106 Abdolmaleki, A., Mallakpour, S., and Mahmoudian, M. (2017). Preparation and evaluation of edge selective sulfonated graphene by chlorosulfuric acid as an active metal-free electrocatalyst for oxygen reduction reaction in alkaline media. *ChemistrySelect* (34): 11211–11217.

107 Abdolmaleki, A., Mallakpour, S., and Karshenas, A. (2017). Facile synthesis of glucose-functionalized reduced graphene oxide (GFRGO)/poly(vinyl alcohol) nanocomposites for improving thermal and mechanical properties. *Mater. Sci. Eng., B* 217: 26–35.

108 Arfin, T. and Athar, S. (2018). Graphene of advanced organic photovoltaics. In: *Nanomaterials: Biomedical, Environmental and Engineering Applications* (eds. S. Kanchi, S. Ahmed, M.I. Sabela and C.M. Hussain), Ch.3, 93–104. Beverly, MA: Scrivener Publishing LLC.

109 Arfin, T. and Rangari, S.N. (2018). Graphene oxide-ZnO nanocomposite modified electrode for the detection of phenol. *Anal. Methods* 10: 347–358.

110 Arfin, T., Bushra, R., and Mohammad, F. (2016). Electrochemical sensor for the sensitive detection of o-nitrophenol using graphene oxide-poly(ethyleneimine) dendrimer-modified glassy carbon electrode. *Graphene Technol.* 1: 1–15.

111 Arfin, T. (2020). Functional graphene-based nanodevices: emerging diagnostic tool. In: *Nanomaterials in Diagnostic Tools and Devices* (eds. S. Kanchi and D. Sharma), Ch.3, 85–112. Amsterdam: Elsevier.

112 Xu, N., Li, S., Li, W., and Liu, Z. (2020). Removal of fluoride by graphene oxide/alumina nanocomposite: adsorbent preparation, characterization, adsorption performance and mechanisms. *ChemistrySelect* 5: 1818–1828.

113 Zhang, J., Chen, N., Su, P. et al. (2017). Fluoride removal from aqueous solution by zirconium-chitosan/graphene oxide membrane. *React. Funct. Polym.* 114: 127–135.

114 Kuang, L., Liu, Y., Fu, D., and Zhao, Y. (2017). FeOOH-graphene oxide nanocomposites for fluoride removal from water: acetate mediated nano FeOOH growth and adsorption mechanism. *J. Colloid Interface Sci.* 490: 259–269.

115 Sun, R., Zhang, H.-B., Qu, J. et al. (2016). Supercritical carbon dioxide fluid assisted synthesis of hierearchical AlOOH@reduced graphene oxide hybrids for efficient removal of fluoride ions. *Chem. Eng. J.* 292: 174–182.

116 Mohammad, F. and Arfin, T. (2014). Thermodynamics and electrochemical characterization of core-shell type gold-coated superparamagnetic iron oxide nanoparticles. *Adv. Mater. Lett.* 5: 315–324.

117 Mohammad, F., Arfin, T., and Al-Lohedan, H.A. (2017). Sustained drug release and electrochemical performance of ethyl cellulose-magnesium hydrogen phosphate composite. *Mater. Sci. Eng., C* 71: 735–743.

118 Arfin, T. and Mohammad, F. (2016). Electrochemical, antimicrobial and anticancer effects of ethyl cellulose-nickel (ii) hydrogen phosphate. *Innov. Corros. Mater. Sci.* 6: 10–18.

119 Arfin, T. and Rafiuddin (2010). Thermodynamics of ion conductivity of alkali halides across a polystyrene-based titanium arsenate membrane. *Electrochim. Acta* 55: 8628–8631.

120 Arfin, T. and Yadav, N. (2012). Impedance characteristics and electrical double layer capacitance of polystyrene based nickel arsenate membrane. *Anal. Bioanal. Electrochem.* 4: 135–152.

121 Arfin, T. (2020). Cellulose and hydrogel matrices for environmental applications. In: *Sustainable Nanocellulose and Nanohydrogels from Natural Sources*, Ch. 12 (eds. F. Mohammad, H.A. Al-Lohedan and M. Jawaid), 255–274. Amsterdam: Elsevier.

14

Functionalized Carbon Nano-Membranes Based Devices for Water Purification Technology

Lindomar Cordeiro A. de Araújo[1] *and Luiz Pereira da Costa*[1,2]

[1] *Federal University of Amazonas – UFAM, Graduate Program in Science and Technology for Amazon Resources (PPGCTRA), Institute of Exact Science and Technology (ICET), Rua: Nossa Senhora do Rosário, 3863, Tiradentes. CEP: 69103-128, Itacoatiara, AM, Brazil*
[2] *Federal University of Amazonas – UFAM, Graduate Program in Chemistry (PPGQ), Avenida Gal. Rodrigo Otávio J. Ramos, no. 6.200, Coroado, CEP: 69080-900, University Campus "Arthur Virgílio Filho", Manaus, AM, Brazil*

14.1 Introduction

With industrial and technological growth, the generation of waste has increased significantly. Conventional techniques previously effective in the treatment or recovery of areas impacted by the waste generated (in case of accidents) are not always effective. Also, the need to use water, considered a "universal solvent" and "vital liquid" for maintaining the life of living beings, becomes increasingly essential for its use in various activities. In some regions, even those considered developed, water is rationed due to its misuse or little rainfall. This motivated the search for technologies that solve the lack of water [1–3].

Research in nanotechnology has been developed and adapted according to the needs of the modern world to minimize the lack of water that traditional methods have no good results. As a science that proposes to study materials with microscopic dimensions on a nanometric scale, it has been gaining space in the scientific and technological world to solve simple or complex problems and provide new products essential for development. Compounds or devices manufactured by this field of knowledge are called nanomaterials [4].

Some authors describe nanomaterials as "nano-tools" with characteristics that differ from bulk materials [4–6]. These nanomaterials have been developed when the need for pure water in some regions of the world has increased and contributed to improved methods of solving this problem. Due to the increase in waste generation, it also aims to contribute to processes that reduce the generation of pollutants or that are capable of removing them [7, 8]. As a result, nanomaterials are seen as an alternative to assist in the clean-up process or, as a last resort, ease the "problem situation" of pollution.

In this sense, nanomaterials, including those based on carbon, have been used. Its properties are so exceptional that other "nano-objects" cannot reproduce their characteristics or perform their duties [9, 10]. Due to the discovery of the characteristics of these materials,

Environmental Applications of Carbon Nanomaterials-Based Devices, First Edition.
Edited by Shadpour Mallakpour and Chaudhery Mustansar Hussain.

research has been developing new methods or existing techniques perfected to obtain water with a lower concentration of salts (desalination) and remove pollutants that harm health and the environment (heavy metals and microorganisms).

Among the ways to decrease the removal of salts or pollutants, nanomembranes have often been manufactured. Nanomembrane is a synthetic nanotechnological material commonly made with nanocomposites, with nanometric thickness. It can reach macroscopic surface areas and contain good physical, chemical, and antimicrobial characteristics [11, 12]. Particularly those based on carbon nanotubes (CNTs) have been studied and improved in recent years. Since its discovery in 1991 [13], CNTs have aroused the interest of researchers to develop new techniques or enhance existing materials. This occurs due to these materials' characteristics, as good mechanical and physical–chemical properties [14, 15]. CNTs are nanometric structures with measurements in nanometers and formed by graphene sheets "rolled" in a cylinder shape with a hollow internal cavity [16]. According to some authors, CNTs can be divided into two groups: (i) single-walled carbon nanotube (SWCNT) that are formed by only one layer of rolled graphene, and (ii) multi-walled or multilayer carbon nanotubes (MWCNT) that have two or more (several) rolled graphene layers [17–21].

CNTs are extensively used for their excellent characteristics, which have been reported in several studies. They are considered hard materials because of their tubular network and sp^2 hybridization of carbon [14, 22, 23]. They are regularly applied in filtration devices because they have good water permeability and are selective for different substances [14, 24–26]. Other characteristics include high elasticity/flexibility [27], good characteristics of stiffness and resistance from to carbon–carbon bonds [28], excellent permeability, selectivity, electrical and thermal properties, and mechanical resistance [28–33]. Also, these nanomaterials can modify the characteristics of other materials [34, 35].

Although having good characteristics, CNTs and other carbon-based materials such as graphene are rarely applied in their pure forms. They are commonly used, dispersed, or modified with other elements in polymeric matrices [36]. Thus, according to Ahn et al., membranes formed with CNTs can be classified in two ways: (i) vertical CNT membranes and (ii) mixed CNT membrane with other nanocomposites [37].

Recent studies have demonstrated an improvement in the performance of these membranes, such as an increase in the flow of water vapor, in the desalination process [38], improvement in the performance of reverse osmosis [39], efficiency in the removal of heavy metals [40] and others.

The ability to remove salts and other membrane properties based on CNTs can be further enhanced by modifying their surface by functionalization processes. These processes aim to assign new characteristics or enhance existing ones in a nanomaterial, such as CNTs, via alteration of the surface with adsorption of chemical groups or elements ions, atoms, or molecules [17, 20, 41].

Carbon-based membranes such as CNTs are quite effective in the purification process. In this way, studies must be developed to improve purifying water mechanisms, such as desalination processes, removing heavy metals and microorganisms, and using these materials [42].

14.2 Desalination

Desalination is a process of removing salts from water, making it suitable for human consumption or other activities [43, 44]. This process of reducing water salts has gained importance and has been improved in recent decades due to the lack of water in several regions of the planet [2]. Some countries such as Australia, Israel, Spain, and South Africa obtain drinking water from the sea or another source through desalination methods [45]. Through the problem of obtaining drinking water, several studies were directed to the development of mechanisms involving membranes to turn saline water (e.g. seawater) into drinking water [45]. These mechanisms involving desalination methods can be divided into two groups: (i) thermal processes, indicating phase change in the liquid (distillation, freezing), and (ii) salt removal processes, with membranes and filtration methods, such as reverse osmosis and nanofiltration [46–49].

A graphene membrane with a thickness of about 0.3 nm has good liquid flow results compared to other traditional membranes [50–56]. Compared to other materials, graphene is considered an excellent material for desalination processes. It can promote a flow rate inversely proportional to its thickness, which is advantageous since it has nanometric measurements [54, 57].

According to Ma et al., the composite membranes based on CNTs have been considered suitable for water purification because they combine the characteristics of this nanomaterial with existing technologies. The seawater desalination process has been enhanced using reverse osmosis purification processes concomitantly with functionalized CNT-based membranes [58]. The functionalized CNTs improved the characteristics of the membranes, such as hydrophilicity, porosity, and water flow barrier control. Son et al. were able to confirm the effectiveness of functionalized CNTs in a thin film on the membrane of a reverse osmosis device used to decrease the salt content of the water [59].

Wang et al. evaluated the ability of CNTs to interfere in the properties of a mixed matrix membrane (CNT/PES (polyethersulfone). It was observed that the greater the water flow, the greater the pressure in the presence of CNTs. The study data revealed that the removal of salts was also greater with CNTs in the membrane. This study demonstrated that with the presence of CNTs, the synthesized membranes showed greater water flow and salt removal when compared to PES membranes [60]. Bhadra et al. evaluated the performance of a membrane with carboxylated (functionalized) CNTs for the desalination process. The authors found that CNTs interfered with the interaction of CNTs with water vapor and increased desalination efficiency. Figure 14.1 shows the proposed mechanism of action of the synthesized membrane. The interaction of water vapors with functionalized CNTs optimizes the flow of water [61].

The functionalization of CNTs demonstrates the effectiveness of this nanomaterial in reverse osmosis devices. In a study using CNTs functionalized with zwitterion (Z-CNTs) in a membrane, good results were observed in the process of removing salts. The authors observed a reduction in the reverse flow of extraction solutes, including NaCl (55.5%), $NH_4H_2PO_4$ (83.8%), $(NH_4)_2HPO_4$ (74.5%), NH_4Cl (70.8%), NH_4HCO_3 (61.9%). A significant reduction in the direct flow of the solute was also detected using a membrane coated with fewer pollutants leaked into the extraction solution [62]. Sabir et al. report that the

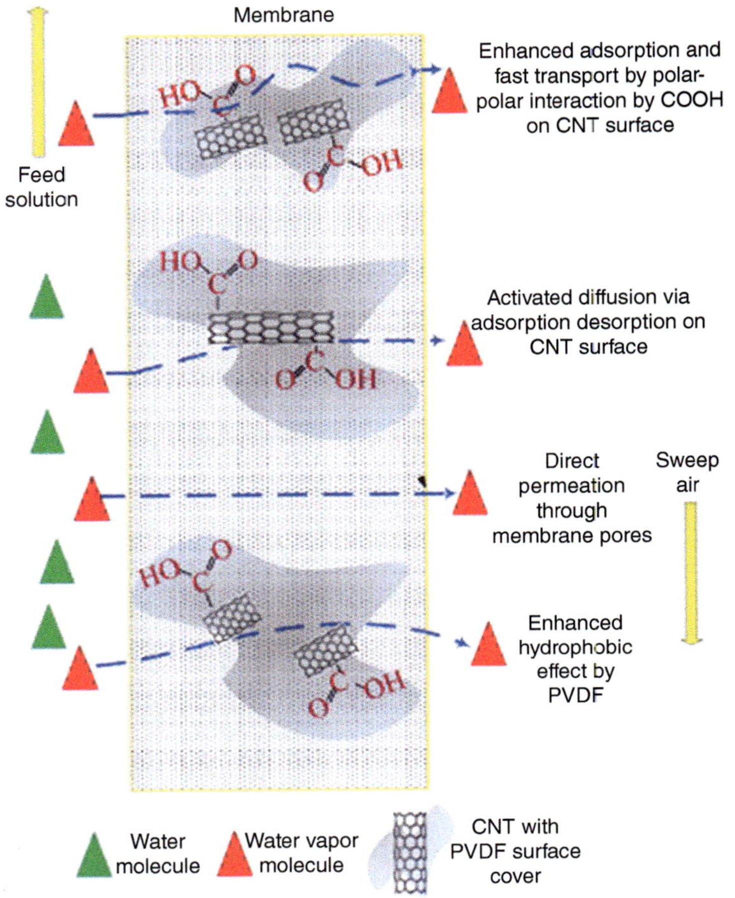

Figure 14.1 Mechanism of action of the membrane containing carboxylated CNTs. Source: Bhadra et al. [61]. Elsevier.

permeability of a membrane composed of cellulose acetate/polyethylene glycol-400 for a reverse osmosis device showed a better performance when CNTs (functionalized) were incorporated into the membrane, showing better permeability and salt rejection (99.8%) as greater was the number of incorporated CNTs [63]. Zhao et al. reported an improvement in the permeability and selectivity of CO_2/H_2 of a mixed matrix membrane of functionalized CNTs [64].

Several studies on using CNTs applied to water desalination based on using a membrane containing functionalized CNTs are presented in Table 14.1.

In all the works shown, their properties were improved with the presence of functionalized CNTs. This denotes the importance that this nanomaterial presents in water desalination processes. It also ratifies its contribution to humanity that needs water that is suitable for consumption.

Table 14.1 Membranes based on functionalized CNTs applied in water desalination processes.

Study	Results	References
Fabrication of tethered carbon nanotubes in cellulose acetate/ polyethylene glycol-400 composite membranes for reverse osmosis	• The number of MWCNTs (functionalized) showed a direct relationship in permeability and salt rejection (99.8%)	[63]
Integrating seawater desalination and wastewater reclamation forward osmosis process using thin-film composite mixed matrix membrane with functionalized carbon nanotube blended polyethersulfone support layer.	• MMPF showed improvement in water flow • Optimized hydrophilicity and selectivity • Decrease in alginate encrustation	[65]
Enhanced desalination via functionalized carbon nanotube immobilized membrane in direct contact membrane distillation	• The carboxylic acid (–COOH) functional group causes interference in the membrane surface properties • Salt rejection (99.9) • Better thermal stability of the membrane	[66]
Improvement in desalination performance of thin-film nanocomposite nanofiltration membrane using amine-functionalized multi-walled carbon nanotube	• Improvement in hydrophilicity and smooth surfaces for modified membranes • The presence of NH_2-MWCNT (0.005%) caused good rejection of Na_2SO_4 (36.71%) and NaCl (95.72%) • Membranes containing functionalized MWCNT showed better resistance to incrustation	[67]

MMPF: Thin-film mixed matrix membrane; CNTs: Carbon Nanotubes; MWCNT: multi-walled carbon nanotube; NF: Nanofiltration.

14.3 Removal of Particles (Ions, Heavy Metals)

The industrial and technological advances of the last four decades have generated increasing amounts of waste, which can impact the environment when they are not correctly destined and treated. The consequences of incorrect waste disposal can cause pollution of rivers, seas, forests, and groundwater that will affect living beings in general, directly or indirectly. In this sense, the use of nanotechnology for wastewater treatment purposes is becoming frequent and improving. Research on the development of effective nanomaterials for removing heavy metal ions from wastewater has made great strides.

CNTs have been considered effective and promising for wastewater treatment and the removal of pollutants such as Ni^{2+}, Cu^{2+}, Cd^{2+}, Cr^{3+}, Pb^{2+}, Cd^{2+}, among others. They have good characteristics, such as their large surface area, and the presence of mesopores, particle/ion retention capacity [25, 68]. Removing this type of pollutant is of paramount importance since they are extremely harmful to health [68].

Chen et al. evaluated the efficiency of a poly(vinylidene fluoride) membrane (PVDF) with multifunctional active carbon and CNT surface for removing oil and heavy metals. It was noted that the membrane showed good antifouling results, effective in removing toxic pollutants and heavy metals like Ni^{+}, Cu^{2+}, and Cr^{3+} [69]. It should be noted that the removal of pollutants. It is related to the surface conditions of the used nanomaterial.

Functionalization is a way to improve the characteristics of CNTs, enhancing their effectiveness in purification processes, increasing the absorption capacity via electrostatic attractions [68]. CNTs used to remove heavy metals from wastewater are sometimes oxidized (functionalized) to increase their effectiveness or used concurrently with other metal ions linked to organic substances [68, 70, 71]. The presence of functional groups containing oxygen in CNTs makes them good metal removers, such as Zn^{2+} and Cd^{2+}. CNTs functionalized with carboxyl, hydroxyl, and phenol groups are ion capture sites that suggest the efficacy of water purification and improvement in removing ions from heavy metals [68, 72, 73]. Studies report that functionalization with hydroxyquinolines optimizes the removal processes, making CNTs excellent removers of other metal ions such as Pb^{2+}, Cd^{2+}, Cu^{2+}, and Zn^{2+} [74–78].

Deng et al. used MWCNTs and electrospinning nanofibers to produce a nanocomposite membrane for removing heavy metal ions. The MWCNTs were modified via the interaction of MWCNTs-COOH with polyethyleneimine for incorporation into the nanofiber. The results showed that the membrane produced presented greater mechanical resistance and hydrophilicity, high permeation, and filtration. The adsorption results for Pb^{2+} and Cu^{2+} were higher than other membranes, showing excellent membrane capacity for removing heavy metals [79].

Ali et al. studied the removal of Zn^{2+} and the regeneration capacity of a functionalized MWCNT membrane. The data revealed good regenerative capacity of the membrane. The authors observed that the functionalized MWCNT membrane effectively removed Zn^{2+} regardless of the liquid they used in the experiments (synthetic water or wastewater). The removal efficiency was attributed to MWCNT with functional groups containing oxygen (–OH, –COOH) in addition to high potential and hydrophilicity [80].

In another study, a filtration membrane for removing zinc ions was improved by the functionalization of MWCNTs by an Ar/O_2 plasma. After treatment with plasma, the number of functional groups containing oxygen on the surface increased. As a result, total removal of zinc ions in the pre-prepared solution in the laboratory and 80% removal of Zn^{2+} ions in wastewater were observed, attributed to complexation reactions on the membrane surface [81].

Usman Farid et al. had also successfully removed zinc ions using a membrane with CNTs functionalized with Ar/O_2 plasma. The membranes synthesized with functionalized MWCNTs had good values in removing Zn^{2+} (90% removal). The authors also observed that other ions in the experiments (Na^{+}, Ca^{2+}, Mg^{2+}, Mn^{2+}, and Fe^{2+}) did not influence zinc removal [82]. These studies reveal that, as reported, CNTs functionalized with groups containing oxygen on the surface corroborate to remove heavy metals (such as Zn^{2+}) from wastewater.

14.4 Removal of Microorganisms

With disordered and inadequate use of natural resources, pollution has increased considerably in the last decades, resulting in diseases in various parts of the world.

The search for alternatives for reducing these harmful effects has been constant, mainly to have clean water for consumption. Due to the ubiquity of the microorganisms, various water source devices and systems are contaminated, either by planktonic cells or the formation of biofilms via sessile cells [83]. Nanomaterials are used to remedy problems caused by microorganisms, including water pollution. Several nanomaterials are being studied in search of antimicrobial properties. They are used to remove and inactivate bacteria and viruses, in addition to the generation of other biomaterials [36]. As a result, research with CNTs is being conducted, and other important aspects of membranes containing CNTs have been discovered, including the ability to act against microorganisms [84–86].

Understanding the antimicrobial effect of nanomaterials is an important step for its application, for the production of another material, or for determining its efficiency using functionalization processes. CNTs are as effective against microorganisms and viruses as other types of nanomaterials [86–88]. The inactivation of microorganisms is due to the ability to damage the microbial cell wall by promoting internal cell material release. The capacity of inactivation of CNTs is due to some characteristics of nanomaterials, such as nanometric dimension, morphology, and the largest surface area [89, 90]. Another important fact is the ability to interact with the bacterial cell at specific sites, so the smaller the size of the CNTs, the greater the likelihood of interaction with the microorganisms [91]. Chen et al. reported MWCNTs 50 μm in length with antimicrobial properties [92].

The inactivation efficiency of microorganisms by nanomaterials may contain other variables that must be taken into account. Therefore, characteristics such as the dimensions of the CNTs (SWCNT or MWCNT), their organization in a matrix (dispersed or superficial), presence of other chemical elements, concentrations of CNTs, contact time, among others, are factors that should be evaluated when applying these types of nanomaterial [93–95].

Several studies report the efficiency of CNTs against microorganisms. Brady-Estévez observed that the SWCNT filter was able to retain and inactivate bacteria and viruses [86]. In another study by the same research group, Brady-Estévez et al. observed an increase in viral inactivation when using salts (NaCl, $CaCl_2$) with SWCNT. The presence of NaCl promoted the formation of a cationic interaction, which resulted in virus adsorption in the SWCNT. The results showed that the thin layer of CNTs efficiently removed microorganisms present in the system [96].

Rahaman et al. also found a way to inactivate and remove viruses by evaluating the efficiency of an MWCNT filter in an electrochemical device in the presence of organic substances. When a potential of 2–3 V was applied, the complete inactivation results of the virus in alginate were obtained. Virus removal was related to the formation of an external electric field, while inactivation was explained by direct oxidation on the surface of the MWCNT [97]. In the study by Wang et al., CNTs inserted in a membrane were functionalized with sodium lignosulfonate before dispersion in the membrane. It improved its properties, such as water flow, hydrophilicity, and antifouling performance. The analysis revealed the efficiency of the membrane against microorganisms through the application of an electric field, showing antimicrobial activity against the bacterium *Escherichia coli* [98].

Some studies suggest the efficiency of doping CNTs with some metal in enhancing their properties and, consequently, of the membranes in which they are incorporated, such as silver nanoparticles (AgNPs). This can result in a nanomaterial with interesting characteristics when inserted in certain devices. The literature data confirm that the antimicrobial activity of CNTs is very consistent for application in the inactivation of microorganisms [84, 86, 93, 96, 99, 100]. Silver nanoparticles are already well-known for having good antimicrobial activity for binding to proteins causing enzymatic damage (causes changes in the cell membrane, interference with DNA replication and electron transport, among others), and low toxic effect against humans. It makes this material promising in the area of nanotechnology [101–104].

Combining these two nanomaterials (AgNPs and CNTs) and their characteristics form an interesting and effective nanocomposite material in eliminating microorganisms. It represents a good alternative for treating and purifying water for human consumption, free of microorganisms.

Ihsanullah et al. investigated a new membrane with antimicrobial characteristics for the pretreatment of desalination with AgNPs doped CNTs. In this study, the presence of AgNPs caused changes in the membrane properties based on CNTs. In addition to the hydrophilic properties observed, the membrane could inactivate all microorganisms (*E. coli*) used in the experiments, corroborating that the junction of AgNPs and CNTs can result in a promising material against microorganisms [105].

In general, it is observed that the processes of removing microorganisms using membranes can be significantly improved by adopting different strategies. Table 14.2 presents some references with succinct information from these studies using functionalized or doped CNTs against microorganisms.

Table 14.2 Membranes containing CNTs against microorganisms.

Microorganisms	Observation	References
Escherichia coli	• The CNTs functionalized membrane showed antibacterial activity.	[106]
Escherichia coli	• Biocidal properties obtained via covalent bonding of SWCNT to the membrane surface • SWCNT modified by purification and ozonolysis increasing cytotoxic characteristics and dispersion • After 1 h of contact, SWCNT membranes showed inactivation of 60% of bacterial cells	[107]
Pseudomonas aeruginosa	• Membrane/functionalized MWCNTs and Ag nanoparticles showed anti-adhesive properties and antibacterial activity • Improved permeability and anti-biofouling property	[108]
Escherichia coli	• Study conducted on a hollow fiber membrane with a deposit on the surface of a porous network of Ag/MWCNTs • MWCNT carboxylated and Ag nanoparticles were loaded onto the nanotube surface via reduction of Ag ion by –COOH • Possibility of adaptation to other devices/membrane for better antimicrobial and antifouling performance	[109]

14.5 Final Considerations

Nanotechnology is a field of knowledge that must be explored and has a lot to offer humanity. It should be noted that although clean-up processes have been developed over the past few decades, there is still a need for the development and application of new technologies for water purification in various regions of the world that require water for consumption. Essentially in the case of manufacturing carbon-based nanomembranes, such as the use of nanotubes (CNTs), several mechanisms can still be improved and extended to a wider application spectrum. The incorporation of CNTs into polymeric matrices in obtaining membranes is constantly being improved to purify water for human consumption. This nanomaterial contributes significantly to improving the properties of the membranes produced, such as hydrophilicity and selectivity. Membrane properties are further improved when CNTs undergo functionalization methods [110–115].

Desalination methods, removal of heavy metals, removal/inactivation of microorganisms are optimized with functionalized CNTs embedded in membranes in these processes. However, despite the good results that research has shown, several barriers must be overcome, ranging from more efficient and low-cost ways of producing nanomembranes to improving existing technologies to develop devices with low impact on the environment.

Important Websites

http://www2.mst.dk/Udgiv.
http://ec.europa.eu/environment.
http://www.carbon-nano.com.
www.timesnano.com.
www.discovermagazine.com.
www.understandingnano.com.
www.sciencedaily.com.
www.ohsrep.org.au.
www.nano.gov.
www.bccresearch.com.

References

1 Elimelech, M. and Phillip, W.A. (2011). The future of seawater desalination: energy, technology, and the environment. *Science* 333: 712–717.
2 Shannon, M.A. et al. (2008). Science and technology for water purification in the coming decades. *Nature* 452: 301–310.
3 Tundisi, J. and Water, G. (2008). Resources in the future: problems and solutions. *Estud. Avançados* 22: 7–16.
4 Abdo, T.S. and Kumar, V.G. (2016). Synthesis, classification, and properties of nanomaterials. *Nanomater. Polym. Membr.* 101: 83–133.

5 Santhosh, C. et al. (2016). Role of nanomaterials in water treatment applications: a review. *Chem. Eng. J.* 306: 1116–1137.
6 Cesaria, M. and Di Bartolo, B. (2017). Nanomaterials: basic concepts and quantum models. In: *Nano-Optics: Principles Enabling Basic Research and Applications*, vol. 104302 (eds. B. Di Bartolo, J. Collins and L. Silvestri), 43–105. Netherlands: Springer.
7 Bottero, J.Y., Rose, J., and Wiesner, M.R. (2006). Nanotechnologies: tools for sustainability in a new wave of water treatment processes. *Integr. Environ. Assess. Manage.* 2: 391–395.
8 Amin, M.T., Alazba, A.A., and Manzoor, U. (2014). A review of removal of pollutants from water/wastewater using different types of nanomaterials. *Adv. Mater. Sci. Eng.* 2014: 1–24.
9 Rogers, J.A., Lagally, M.G., and Nuzzo, R.G. (2011). Synthesis, assembly and applications of semiconductor nanomembranes. *Nature* 477: 45–53.
10 Turchanin, A. and Gölzhäuser, A. (2016). Carbon nanomembranes. *Adv. Mater.* 28: 6075–6103.
11 Agboola, O., Sadiku, E.R., and Mokrani, T. (2016). Nanomembrane materials based on polymer blends. In: *Design and Applications of Nanostructured Polymer Blends and Nanocomposite Systems*. 1–442. New York: Elsevier. https://doi.org/10.1016/B978-0-323-39408-6.00006-6.
12 Angelova, P. and Gölzhäuser, A. (2017). Carbon nanomembranes. *Phys. Sci. Rev.* 2: 1–19.
13 Lijima, S. (1991). Helical microtubules of graphitic carbon. *Nature* 354: 56–58.
14 Yu, M.F. et al. (2000). Strength and breaking mechanism of multiwalled carbon nanotubes under tensile load. *Science* 287: 637–640.
15 Moniruzzaman, M. and Winey, K.I. (2006). Polymer nanocomposites containing carbon nanotubes. *Macromolecules* 39: 5194–5205.
16 Usui, Y., Haniu, H., Tsuruoka, S., and Saito, N. (2012). Carbon nanotubes innovate on medical technology. *Med. Chem.* 02: 1–6.
17 Fonseca e Trindade, J. (2011). *Nanotubos De Carbono: Propriedades E Funcionalização*, 14. Instituto Politecnico da Guarda.
18 Ojha, A. (2020). Nanomaterials for removal of waterborne pathogens. In: *Waterborne Pathogens*. 385–432. New York: Elsevier. https://doi.org/10.1016/b978-0-12-818783-8.00019-0.
19 Wadhawan, S., Jain, A., Nayyar, J., and Mehta, S.K. (2020). Role of nanomaterials as adsorbents in heavy metal ion removal from waste water: a review. *J. Water Process Eng.* 33: 101038.
20 De Souza Filho, A.G. and Fagan, S.B. (2007). Funcionalização de nanotubos de carbono. *Quim. Nova* 30: 1695–1703.
21 Zarbin, A.J.G. and Oliveira, M.M. (2013). Nanoestruturas de carbono (nanotubos, grafeno): Quo Vadis? *Quim. Nova* 36: 1533–1539.
22 Lee, C.H., Johnson, N., Drelich, J., and Yap, Y.K. (2011). The performance of superhydrophobic and superoleophilic carbon nanotube meshes in water-oil filtration. *Carbon* 49: 669–676.
23 Demczyk, B.G., Wang, Y.M., Cumings, J. et al. (2002). Direct mechanical measurement of the tensile strength and elastic modulus of multiwalled carbon nanotubes. *Mater. Sci. Eng., A* 334: 173–178.
24 Herbst, M.H., Fernandes Macêdo, M.I., and Rocco, A.M. (2004). Tecnologia dos nanotubos de carbono: Tendências e perspectivas de uma área multidisciplinar. *Quim. Nova* 27: 986–992.

25 Mubarak, N.M., Sahu, J.N., Abdullah, E.C., and Jayakumar, N.S. (2014). Removal of heavy metals from wastewater using carbon nanotubes. *Sep. Purif. Rev.* 43: 311–338.

26 Majumder, M., Chopra, N., Andrews, R., and Hinds, B.J. (2005). Nanoscale hydrodynamics: enhanced flow in carbon nanotubes. *Nature* 438: 44–44.

27 Yakobson, B.I., Brabec, C.J., and Bernholc, J. (1996). Nanomechanics of carbon tubes: instabilities beyond linear response. *Phys. Rev. Lett.* 76: 2511–2514.

28 Shi, D., Guo, Z., and Bedford, N. (2015). Carbon nanotubes. *Nanomaterials and Devices*. In: New York: Elsevier, 49–82. https://doi.org/10.1016/B978-1-4557-7754-9.00003-2.

29 Ma, P.C., Siddiqui, N.A., Marom, G., and Kim, J.K. (2010). Dispersion and functionalization of carbon nanotubes for polymer-based nanocomposites: a review. *Compos. Part A: Appl. Sci. Manuf.* 41: 1345–1367.

30 Aroon, M.A., Ismail, A.F., Montazer-Rahmati, M.M., and Matsuura, T. (2010). Morphology and permeation properties of polysulfone membranes for gas separation: effects of non-solvent additives and co-solvent. *Sep. Purif. Technol.* 72: 194–202.

31 Ismail, A.F., Goh, P.S., Sanip, S.M., and Aziz, M. (2009). Transport and separation properties of carbon nanotube-mixed matrix membrane. *Sep. Purif. Technol.* 70: 12–26.

32 Sanip, S.M. et al. (2011). Gas separation properties of functionalized carbon nanotubes mixed matrix membranes. *Sep. Purif. Technol.* 78: 208–213.

33 Skoulidas, A.I., Ackerman, D.M., Johnson, J.K., and Sholl, D.S. (2002). Rapid transport of gases in carbon nanotubes. *Phys. Rev. Lett.* 89: 13–16.

34 Allaoui, A., Bai, S., Cheng, H.M., and Bai, J.B. (2002). Mechanical and electrical properties of a MWNT/epoxy composite. *Compos. Sci. Technol.* 62: 1993–1998.

35 Loos, M.R., Coelho, L.A.F., Pezzin, S.H., and Amico, S.C. (2008). Effect of carbon nanotubes addition on the mechanical and thermal properties of epoxy matrices. *Mater. Res.* 11: 347–352.

36 Smith, S.C. and Rodrigues, D.F. (2015). Carbon-based nanomaterials for removal of chemical and biological contaminants from water: a review of mechanisms and applications. *Carbon* 91: 122–143.

37 Ahn, C.H. et al. (2012). Carbon nanotube-based membranes: fabrication and application to desalination. *J. Ind. Eng. Chem.* 18: 1551–1559.

38 Ragunath, S., Roy, S., and Mitra, S. (2018). Carbon nanotube immobilized membrane with controlled nanotube incorporation via phase inversion polymerization for membrane distillation based desalination. *Sep. Purif. Technol.* 194: 249–255.

39 Farahbakhsh, J., Delnavaz, M., and Vatanpour, V. (2019). Simulation and characterization of novel reverse osmosis membrane prepared by blending polypyrrole coated multiwalled carbon nanotubes for brackish water desalination and antifouling properties using artificial neural networks. *J. Membr. Sci.* 581: 123–138.

40 Tofighy, M.A. and Mohammadi, T. (2011). Adsorption of divalent heavy metal ions from water using carbon nanotube sheets. *J. Hazard. Mater.* 185: 140–147.

41 Gupta, V.K. et al. (2015). Nanoparticles as adsorbent; a positive approach for removal of noxious metal ions: a review. *Sci. Technol. Dev.* 34: 195–214.

42 Ashbolt, N.J. (2004). Microbial contamination of drinking water and disease outcomes in developing regions. *Toxicology* 198: 229–238.

43 Shaffer, D.L., Yip, N.Y., Gilron, J., and Elimelech, M. (2012). Seawater desalination for agriculture by integrated forward and reverse osmosis: improved product water quality for potentially less energy. *J. Membr. Sci.* 415–416: 1–8.

44 Kadhom, M. and Deng, B. (2019). Thin film nanocomposite membranes filled with bentonite nanoparticles for brackish water desalination: a novel water uptake concept. *Microporous Mesoporous Mater.* 279: 82–91.

45 Blersch, C.L. and Du Plessis, J.A. (2017). Planning for desalination in the context of the Western Cape water supply system. *J. South Afr. Inst. Civil Eng.* 59: 11–21.

46 Kucera, J. (2014). Introduction to desalination. In: *Desalination*, 1–37. Wiley. https://doi.org/10.1002/9781118904855.ch1.

47 Zhao, S., Zou, L., Tang, C.Y., and Mulcahy, D. (2012). Recent developments in forward osmosis: opportunities and challenges. *J. Membr. Sci.* 396: 1–21.

48 Fritzmann, C., Löwenberg, J., Wintgens, T., and Melin, T. (2007). State-of-the-art of reverse osmosis desalination. *Desalination* 216: 1–76.

49 Khawaji, A.D., Kutubkhanah, I.K., and Wie, J.M. (2008). Advances in seawater desalination technologies. *Desalination* 221: 47–69.

50 Humplik, T. et al. (2011). Nanostructured materials for water desalination. *Nanotechnology* 22: 1–19.

51 Geise, G.M. et al. (2010). Water purification by membranes: the role of polymer science. *J. Polym. Sci., Part B: Polym. Phys.* 48: 1685–1718.

52 Suk, M.E. and Aluru, N.R. (2010). Water transport through ultrathin graphene. *J. Phys. Chem. Lett.* 1: 1590–1594.

53 Mishra, A.K. and Ramaprabhu, S. (2011). Functionalized graphene sheets for arsenic removal and desalination of sea water. *Desalination* 282: 39–45.

54 Celebi, K. et al. (2014). Ultimate permeation across atomically thin porous graphene. *Science* 344: 289–292.

55 Cohen-Tanugi, D. and Grossman, J.C. (2012). Water desalination across nanoporous graphene. *Nano Lett.* 12: 3602–3608.

56 Sint, K., Wang, B., and Král, P. (2008). Selective ion passage through functionalized graphene nanopores. *J. Am. Chem. Soc.* 130: 16448–16449.

57 O'Hern, S.C. et al. (2012). Selective molecular transport through intrinsic defects in a single layer of CVD graphene. *ACS Nano* 6: 10130–10138.

58 Ma, L., Dong, X., Chen, M. et al. (2017). Fabrication and water treatment application of carbon nanotubes (CNTs)-based composite membranes: a review. *Membranes* 7(1): 1–21.

59 Son, M. et al. (2015). Efficacy of carbon nanotube positioning in the polyethersulfone support layer on the performance of thin-film composite membrane for desalination. *Chem. Eng. J.* 266: 376–384.

60 Wang, L. et al. (2015). Fabrication and characterization of polyethersulfone/carbon nanotubes (PES/CNTs) based mixed matrix membranes (MMMs) for nanofiltration application. *Appl. Surf. Sci.* 330: 118–125.

61 Bhadra, M., Roy, S., and Mitra, S. (2013). Enhanced desalination using carboxylated carbon nanotube immobilized membranes. *Sep. Purif. Technol.* 120: 373–377.

62 Zou, S., Smith, E.D., Lin, S. et al. (2019). Mitigation of bidirectional solute flux in forward osmosis via membrane surface coating of zwitterion functionalized carbon nanotubes. *Environ. Int.* 131: 104970.

63 Sabir, A. et al. (2015). Fabrication of tethered carbon nanotubes in cellulose acetate/polyethylene glycol-400 composite membranes for reverse osmosis. *Carbohydr. Polym.* 132: 589–597.

64 Zhao, Y., Jung, B.T., Ansaloni, L., and Ho, W.S.W. (2014). Multiwalled carbon nanotube mixed matrix membranes containing amines for high pressure CO_2/H_2 separation. *J. Membr. Sci.* 459: 233–243.

65 Choi, H.-G., Son, M., and Choi, H. (2017). Integrating seawater desalination and wastewater reclamation forward osmosis process using thin-film composite mixed matrix membrane with functionalized carbon nanotube blended polyethersulfone support layer. *Chemosphere* 185: 1181–1188.

66 Roy, S., Bhadra, M., and Mitra, S. (2014). Enhanced desalination via functionalized carbon nanotube immobilized membrane in direct contact membrane distillation. *Sep. Purif. Technol.* 136: 58–65.

67 Zarrabi, H., Yekavalangi, M.E., Vatanpour, V. et al. (2016). Improvement in desalination performance of thin film nanocomposite nanofiltration membrane using amine-functionalized multiwalled carbon nanotube. *Desalination* 394: 83–90.

68 Rao, G.P., Lu, C., and Su, F. (2007). Sorption of divalent metal ions from aqueous solution by carbon nanotubes: a review. *Sep. Purif. Technol.* 58: 224–231.

69 Chen, Z. et al. (2020). Hierarchical poly(vinylidene fluoride)/active carbon composite membrane with self-confining functional carbon nanotube layer for intractable wastewater remediation. *J. Membr. Sci.* 603: 118041.

70 Afzali, D., Jamshidi, R., Ghaseminezhad, S., and Afzali, Z. (2012). Preconcentration procedure trace amounts of palladium using modified multiwalled carbon nanotubes sorbent prior to flame atomic absorption spectrometry. 1st Nano Update. *Arabian J. Chem.* 5: 461–466.

71 Gupta, V.K., Agarwal, S., and Saleh, T.A. (2011). Synthesis and characterization of alumina-coated carbon nanotubes and their application for lead removal. *J. Hazard. Mater.* 185: 17–23.

72 Bhanjana, G., Dilbaghi, N., Kim, K.H., and Kumar, S. (2017). Carbon nanotubes as sorbent material for removal of cadmium. *J. Mol. Liq.* 242: 966–970.

73 Bystrzejewski, M., Pyrzyńska, K., Huczko, A., and Lange, H. (2009). Carbon-encapsulated magnetic nanoparticles as separable and mobile sorbents of heavy metal ions from aqueous solutions. *Carbon* 47: 1201–1204.

74 Kosa, S.A., Al-Zhrani, G., and Abdel Salam, M. (2012). Removal of heavy metals from aqueous solutions by multi-walled carbon nanotubes modified with 8-hydroxyquinoline. *Chem. Eng. J.* 181–182: 159–168.

75 Cho, H.H. et al. (2010). Sorption of aqueous Zn[II] and Cd[II] by multiwall carbon nanotubes: the relative roles of oxygen-containing functional groups and graphenic carbon. *Langmuir* 26: 967–981.

76 Yu, F., Wu, Y., Ma, J., and Zhang, C. (2013). Adsorption of lead on multi-walled carbon nanotubes with different outer diameters and oxygen contents: kinetics, isotherms and thermodynamics. *J. Environ. Sci.* 25: 195–203.

77 Li, Y.H. et al. (2003). Competitive adsorption of Pb^{2+}, Cu^{2+} and Cd^{2+} ions from aqueous solutions by multiwalled carbon nanotubes. *Carbon* 41: 2787–2792.

78 Chandrashekhar Nayak, M., Isloor, A.M., Inamuddin, L.B. et al. (2019). Polyphenylsulfone/multiwalled carbon nanotubes mixed ultrafiltration membranes: fabrication, characterization and removal of heavy metals Pb^{2+}, Hg^{2+}, and Cd^{2+} from aqueous solutions. *Arabian J. Chem.* 13(3): 4661–4672. https://doi.org/10.1016/j.arabjc.2019.10.007.

79 Deng, S., Liu, X., Liao, J. et al. (2019). PEI modified multiwalled carbon nanotube as a novel additive in PAN nanofiber membrane for enhanced removal of heavy metal ions. *Chem. Eng. J.* 375: 122086.

80 Ali, S. et al. (2019). Efficient removal of zinc from water and wastewater effluents by hydroxylated and carboxylated carbon nanotube membranes: behaviors and mechanisms of dynamic filtration. *J. Hazard. Mater.* 365: 64–73.

81 Ali, S., Shah, I.A., Ahmad, A. et al. (2019). Ar/O_2 plasma treatment of carbon nanotube membranes for enhanced removal of zinc from water and wastewater: a dynamic sorption-filtration process. *Sci. Total Environ.* 655: 1270–1278.

82 Usman Farid, M. et al. (2017). Increased adsorption of aqueous zinc species by Ar/O_2 plasma-treated carbon nanotubes immobilized in hollow-fiber ultrafiltration membrane. *Chem. Eng. J.* 325: 239–248.

83 Pandey, P.K., Kass, P.H., Soupir, M.L. et al. (2014). Contamination of water resources by pathogenic bacteria. *AMB Express* 4: 1–16.

84 Hussein, M.A., El-Shishtawy, R.M., Alamry, K.A. et al. (2019). Efficient water disinfection using hybrid polyaniline/graphene/carbon nanotube nanocomposites. *Environ. Technol.* 40: 2813–2824.

85 Baoukina, S., Monticelli, L., and Tieleman, D.P. (2013). Interaction of pristine and functionalized carbon nanotubes with lipid membranes. *J. Phys. Chem. B* 117: 12113–12123.

86 Brady-Estévez, A.S., Kang, S., and Elimelech, M. (2008). A single-walled-carbon-nanotube filter for removal of viral and bacterial pathogens. *Small* 4: 481–484.

87 Akasaka, T. and Watari, F. (2009). Capture of bacteria by flexible carbon nanotubes. *Acta Biomater.* 5: 607–612.

88 Upadhyayula, V.K.K., Deng, S., Smith, G.B., and Mitchell, M.C. (2009). Adsorption of *Bacillus subtilis* on single-walled carbon nanotube aggregates, activated carbon and NanoCeram™. *Water Res.* 43: 148–156.

89 Upadhyayula, V.K.K., Deng, S., Mitchell, M.C., and Smith, G.B. (2009). Application of carbon nanotube technology for removal of contaminants in drinking water: a review. *Sci. Total Environ.* 408: 1–13.

90 Narayan, R.J., Berry, C.J., and Brigmon, R.L. (2005). Structural and biological properties of carbon nanotube composite films. *Mater. Sci. Eng., B* 123: 123–129.

91 Aslan, S. et al. (2010). Antimicrobial biomaterials based on carbon nanotubes dispersed in poly(lactic-co-glycolic acid). *Nanoscale* 2: 1789–1794.

92 Chen, H. et al. (2013). Broad-spectrum antibacterial activity of carbon nanotubes to human gut bacteria. *Small* 9: 2735–2746.

93 Kang, S., Herzberg, M., Rodrigues, D.F., and Elimelech, M. (2008). Antibacterial effects of carbon nanotubes: size does matter! *Langmuir* 24: 6409–6413.

94 Li, Q. et al. (2008). Antimicrobial nanomaterials for water disinfection and microbial control: potential applications and implications. *Water Res.* 42: 4591–4602.

95 Wick, P. et al. (2007). The degree and kind of agglomeration affect carbon nanotube cytotoxicity. *Toxicol. Lett.* 168: 121–131.

96 Brady-Estévez, A.S., Nguyen, T.H., Gutierrez, L., and Elimelech, M. (2010). Impact of solution chemistry on viral removal by a single-walled carbon nanotube filter. *Water Res.* 44: 3773–3780.

97 Rahaman, M.S., Vecitis, C.D., and Elimelech, M. (2012). Electrochemical carbon-nanotube filter performance toward virus removal and inactivation in the presence of natural organic matter. *Environ. Sci. Technol.* 46: 1556–1564.

98 Wang, W. et al. (2018). Preparation and characterization of SLS-CNT/PES ultrafiltration membrane with antifouling and antibacterial properties. *J. Membr. Sci.* 548: 459–469.

99 Koo, J. (2016). An overview of nanomaterials. In: *Fundamentals, Properties, and Applications of Polymer Nanocomposites*, 22–108. Cambridge University Press. https://doi.org/10.1017/CBO9781139342766.003.

100 Kang, S., Pinault, M., Pfefferle, L.D., and Elimelech, M. (2007). Single-walled carbon nanotubes exhibit strong antimicrobial activity. *Langmuir* 23: 8670–8673.

101 Liau, S.Y., Read, D.C., Pugh, W.J. et al. (1997). Interaction of silver nitrate with readily identifiable groups: relationship to the antibacterial action of silver ions. *Lett. Appl. Microbiol.* 25: 279–283.

102 Sondi, I. and Salopek-Sondi, B. (2004). Silver nanoparticles as antimicrobial agent: a case study on *E. coli* as a model for gram-negative bacteria. *J. Colloid Interface Sci.* 275: 177–182.

103 Feng, Q.L., Wu, J., Chen, G.Q. et al. (2000). A mechanistic study of the antibacterial effect of silver ions on *Escherichia coli* and *Staphylococcus aureus*. *J. Biomed. Mater. Res.* 52: 662–668.

104 Zodrow, K. et al. (2009). Polysulfone ultrafiltration membranes impregnated with silver nanoparticles show improved biofouling resistance and virus removal. *Water Res.* 43: 715–723.

105 Ihsanullah, L.T., Al-Amer, A.M., Khalil, A.B. et al. (2015). Novel anti-microbial membrane for desalination pretreatment: a silver nanoparticle-doped carbon nanotube membrane. *Desalination* 376: 82–93.

106 Farid, M.U., Khanzada, N.K., and An, A.K. (2019). Understanding fouling dynamics on functionalized CNT-based membranes: mechanisms and reversibility. *Desalination* 456: 74–84.

107 Tiraferri, A., Vecitis, C.D., and Elimelech, M. (2011). Covalent binding of single-walled carbon nanotubes to polyamide membranes for antimicrobial surface properties. *ACS Appl. Mater. Interfaces* 3: 2869–2877.

108 Kim, E., Hwang, G., El-din, M.G., and Liu, Y. (2012). Development of nanosilver and multi-walled carbon nanotubes thin-film nanocomposite membrane for enhanced water treatment. *J. Membr. Sci.* 394–395: 37–48.

109 Yoosefi, A., Wang, R., and Xu, R. (2013). The effect of re-generable silver nanoparticles / multi-walled carbon nanotubes coating on the antibacterial performance of hollow fiber membrane. *Chem. Eng. J.* 230: 251–259.

110 Mallakpour, S. and Khadem, E. (2019). Carbon nanotubes for heavy metals removal. In: *Composite Nanoadsorbents*, Chapter 8 (eds. G. Kyzas and A.C. Mitrpoulos), 181–210. Amsterdam, Netherlands: Elsevier.

111 Mallakpour, S. and Rashidimoghadam, S. (2019). Carbon nanotubes for dyes removal. In: *Composite Nanoadsorbents*, Chapter 9 (eds. G. Kyzas and A.C. Mitrpoulos), 211–244. Amsterdam, Netherlands: Elsevier.

112 Abdolmaleki, A., Mallakpour, S., and Karshenas, A. (2017). Facile synthesis of glucose-functionalized reduced graphene oxide (GFRGO)/poly(vinyl alcohol) nanocomposites for improving thermal and mechanical properties. *Mater. Sci. Eng., B* 217: 26–35.

113 Abdolmaleki, A., Mallakpour, S., and Karshenas, A. (2017). Synthesis and characterization of new nanocomposites films using alanine-Cu-functionalized graphene oxide as nanofiller and PVA as polymeric matrix for improving of their properties. *J. Solid State Chem.* 253: 398–405.

114 Mallakpour, S., Behranvand, V., and Mallakpour, F. (2019). Synthesis of alginate/carbon nanotube/carbon dot/fluoroapatite/TiO_2 beads for dye photocatalytic degradation under ultraviolet light. *Carbohydr. Polym.* 224: 115–138.

115 Mallakpour, S. and Behranvand, V. (2018). Synthesis of mesoporous recycled poly(ethylene terephthalate)/MWNT/carbon quantum dot nanocomposite from sustainable materials using ultrasonic waves: application for methylene blue removal. *J. Cleaner Prod.* 190: 525–537.

15

Functionalized Bio-carbon Nanomaterials for Environmental Utilizations

Mahtabin R. Rozbu, Ahmedul Kabir, and Paulraj M. Selvakumar

Asian University for Women, Science and Math Program, M.M. Ali Road, Chittagong, 4000, Bangladesh

15.1 Introduction

Humanity's ability to innovate, pushing, and exploring science and technology has always brought revolutionary changes throughout history. Although yet rising, Nanomaterial and Nanotechnology, coupled with Biotechnology, demonstrate such a revolution rooted in almost all scientific arenas by connecting physicists, biologists, engineers, environmentalists, chemists, material scientists, and many more. Expanding as an interdisciplinary zone and serving applications in diverse fields, nanoparticles (NPs) and nanomaterials form the basis of future technologies. However, various scientific and technological advancements have had a continual detrimental impact on the environment, making it utterly important to bring about innovations that support green technology for a sustainable environment and thriving biodiversity. Therefore, nanomaterials, their interaction with the environment, and their synthesis must be non-hazardous, biocompatible, and eco-friendly.

Starting with the fundamental question, what is a nanoparticle? Nanoparticles, organic or inorganic, are constituents of three dimensions being less than 100 nm or (1 billionth of a meter). The reason why nanoparticles gained vast attention over the past decade was their weird or incredible ability to behave differently from their precursor parent body, solely based on their size. A particle demonstrating different chemical and physical characteristics, at different sizes, despite being made from the same atoms and electrons, invites a new realm of discovery and research. For example, a gold block has different physical and chemical traits than gold nanoparticles of different sizes. Gold nanoparticles display a range of colors from blue to red when immersed in a gold colloidal solution. For instance, the blue solution consists of the smallest gold particles. Such discovery of significant change of properties at the nanoscale leads toward developing various devices, with recent stretches being on environmentally friendly applications.

Environmental Applications of Carbon Nanomaterials-Based Devices, First Edition.
Edited by Shadpour Mallakpour and Chaudhery Mustansar Hussain.

Prominently, there are six types of nanoparticles categorized based on distinct morphology, size, physical and chemical properties [1–3]:

(1) Metal nanoparticles
(2) Ceramic nanoparticles
(3) Semiconductor nanoparticles
(4) Polymeric nanoparticles
(5) Lipid-based nanoparticles
(6) Carbon-based nanoparticles

Metal nanoparticles are usually synthesized by chemical, electrochemical, or photochemical methods from metal forerunners [1]. A prime example of metal nanoparticles is Gold nanoparticles. They have various applications such as adsorbents, bioimaging materials, and functional moieties to synthesize other nanomaterials. Other considerable metal nanoparticles include the silver nanoparticle, lead nanoparticles, and many more. Metal nanoparticles tend to have higher surface energy and can be toxic to the environment.

Inorganic solids of oxides, carbides, carbonates, and phosphates are the precursors of *ceramic nanomaterials* with high thermal resistance. Despite disposal demerits though completely non-toxic to the environment, ceramic nanoparticles have primary applications in photocatalysis, photodegradation of dyes, and imaging. However, its nature of being chemically inert and scope of manipulation in terms of size and porosity makes ceramic nanoparticles suitable to form effective drug delivery systems [4].

Semiconductor nanoparticles can display metallic and non-metallic properties that vary depending on size, especially at the Bohr diameter for exciton [5]. Semiconductor nanoparticles having wider band gaps are employed to make fluorescent materials and are known for their diminishing photobleaching effect [6]. Utilized in the process of water splitting, photocatalysis, and in the development of electronic devices, as surface coatings, semiconductor nanoparticles have a diverse range of applications [1].

Polymeric nanoparticles are obtained from organic materials; they tend to form nano-capsular or nano-spherical shapes depending on the synthesis method [1]. Nano-capsule-shaped polymeric nanoparticles consist of a core shape, whereas the nano-spherical nanoparticles consist of a matrix [1]. The dispersion properties of polymeric nanoparticles significantly depend on their morphology. However, both the polymeric nanoparticles are comparatively quite biocompatible and biodegradable. They have significant applications in the pharmaceutical industries and biomedical fields for drug delivery, diagnostics, imaging, specific cell targeting, and many more [1, 7, 8].

Lipid nanoparticles can be considered bio-nanomaterial made of a lipid core and a matrix of soluble lipophilic molecules. The external layer is held together by emulsifiers and surfactants. The diameter is about 10–100 nm [1]. Lipid-based nanoparticles, as bioparticles, lay an important role in drug delivery applications, particularly in the field of biomedicine.

Carbon nanoparticles are among the most popular nanoparticles with many applications in environmental science, engineering, biomedicine, and electronics [9–13]. One of the most significant nanomaterials precursing carbon nanoparticles is carbon nanotubes (CNTs) and graphenes, structurally more tensile and stronger than steel [14, 15]. The latest research in material science looks into carbon nanoparticles of organic origin to significantly reduce any possible toxicological environmental, or health effects.

In contrast to nanoparticles, nanomaterials are structural components consisting of only one dimension is less than 100 nm. Nano-objects have two dimensions, less than 100 nm [16]. Like nanoparticles, nanomaterials have the intrinsic property of having different chemical and physical characteristics depending on their size [17]. Different nanomaterials have been synthesized, with popular nanomaterials being fullerenes, graphene nanostructures, nanozymes, and nanotubes [18–20]. CNTs had established prominent popularity. However, with the growing concern of environmental effects, nanomaterials are constructed and engineered to be significantly biodegradable and least toxic. Such a notion of creating a sustainable "green" environment and eco-friendly nanoparticles/nanomaterials developed the concept of bio-nanomaterials. Bio-nanomaterials, still under research and study, tend to be polymers of biomolecules, such as cellulose, lipids, DNA, RNA, and amino acids, mainly of organic origin.

In this chapter, a wide range of materials, from carbon nanomaterials to particular bio-carbon nanomaterials, has been observed based on their scope of functionalization and advancements to develop "green" technologies. Economic viability was also considered with high importance.

15.2 Carbon Nanomaterial

Carbon's four-electron valency makes it a special element, enabling it to form polymers at the nanoscale undergoing single, double, or even triple covalent bonding [21]. Such intrinsic properties of carbon allowed the possibility of synthesizing various carbon-based nanomaterials with remarkable unique properties such as increased tensile strength, durability, and immense thermal resistivity. Three unique carbon nanostructures can be classified based on their shapes.

(1) Fullerenes – spherical
(2) Carbon nanotubes (CNTs) – cylindrical
(3) Graphenes – unfolded sheets

15.3 Synthesis of Fullerenes

Fullerenes were first synthesized from carbon vapors, in 1990, by W. Krätschmer and D.R. Huffman, using evaporating graphite electrodes in the presence of helium gas [21–23]. Today, there are many techniques for synthesizing different fullerenes (C_{60}, C_{70}) by adjusting the synthesis parameters and purification methodology in organic solvents such as benzene and toluene [21, 24]. However, the production of fullerenes is not quite economically viable due to the absence of a better and cheaper alternative synthetic technique.

15.4 Synthesis of CNTs

CNTs are one of the most commonly synthesized carbon nanomaterials. It is an exceptional allotrope with a wide range of applications and "wonder" properties [25, 26]. Chemical Vapor Deposition (CVD) technique that enhances catalytic arc discharge under milder conditions (temperature and pressure) is used for the synthesis of CNTs [21, 27, 28]. Hydrocarbons decompose in a reactant chamber consisting of inert gas to form carbons that develop into non-tubular structures upon the dissociation of organic precursors consisting of nickel, iron, or cobalt nanoparticle catalysts [16, 21]. The non-tubular carbons, although structurally porous like CNTs, are not as adsorbent. Hence, ensuring the purity of CNT is vital to have efficient adsorption properties within the synthesized CNTs. The synthesis mechanism encourages non-tubular carbons to diffuse into CNTs that reduce active adsorbent sites to the sorbate ([29], section 5.2; [30]). Eventually, the CNTs are purified via acid and gas-phase oxidation treatment [16, 31], which results in oxygen atom groups that further increases hydrophilicity, polarity, and ion exchange characteristics ([29], section 5.2). Due to higher yield and purity, CVD is an economically better option than techniques such as CNTs arc discharge or laser ablation [21, 32, 33].

15.5 Synthesis of Graphenes

Similar to CNTs, CVD is also considered the most efficient method for large scale graphene production. The basic synthesis operation ensures graphene or graphene oxides as sheets, usually of varying qualities depending on specific applications [21]. Certain physical and chemical methods or the laser ablation technique can also split/cut graphite and nanotubes to obtain graphene sheets [34, 35].

Industry and laboratory synthesis of carbon nanomaterials like CNTs, graphenes, and fullerenes always require intense chemical and physical mechanisms that do not completely support environmental sustainability. Although carbon nanomaterials can have vast applications in cleaning environmental pollutants, their synthesis is heavy on the energy demand [25, 26, 36]. Hence, researchers are looking for a further "green" and cost-effective way of synthesizing these nanomaterials. They are also interested in alternative nanomaterials in nature. Interestingly, carbon-based nanoparticles occur in nature [21, 37, 38]. For example, CNTs in a coal-petroleum mix were reported, and the synthesis of single-walled CNT by CVD method, using volcanic lava (substrate) and a catalyst (consisting of metal oxides) was possible [39, 40]. It demonstrated that CNTs could form in nature at extremely high temperatures [39]. Another speculation of carbon-based nanomaterials was the presence of fullerenes in particular geological materials and meteorites [21, 41, 42].

Furthermore, particular fullerenes were found in Chinese hibiscus pollens, which played a significant biological and evolutionary role in pollination [21, 43]. The incorporation of carbon nanomaterials found in nature still requires further research and mining/extraction technologies to be used as substances for greater applications. However, whether from nature or manufactured artificially, carbon-based nanoparticles raise questions and concerns on their possible effect on the environment. The consequences of nanomaterials entering the food chain and interacting with biological systems are yet unknown. The need for figuring methods and discovering techniques of synthesizing and generating bio-based carbon nanomaterials (preferably of organic origin) is the new era in nanotechnology [21].

15.6 Bio-carbon Nanomaterials

Carbon nanoparticles synthesized from bio-based sources are better substitutes for metal nanoparticles because of their inherent quality of biodegradability, biocompatibility, and eco-friendly nature [44–46]. The carbon of bio-carbon nanomaterials is usually derived from biomass and hence is of organic origin. Before acknowledging some prominent bio-carbon nanomaterials and their subsequent applications, it is important to understand bio-carbon nanoparticles/nanomaterials and their corresponding necessity for commercialization.

15.7 Functionalization of Nanomaterials

Tailoring a nanomaterial by adding ligands, amino acids, polymers, organic or inorganic moieties to have certain physicochemical properties for particular applications is essentially functional. A nanomaterial can be functionalized directly during or after synthesis, depending on the application. Targeting chemical groups on nanomaterials' surface creates specific surface-active sites identified as surface functionalization. Surface functionalization allows for the control of nanoparticles and associated photonic, catalytic, and hybrid reaction systems giving rise to enhanced unique functionality. The addition of a functional group(s) enables producing a customized range of nanomaterials with different functional properties [8, 47–49]. Functionalization effectively provides nanomaterials with active sites, where the functional entity binds with the nanomaterial using covalent or noncovalent bonding through simple coupling reactions ([29], section 10.2; [50–52]). For example, upon functionalization, certain biocarbon nanomaterials can be modified to attain better adsorbent capacity, which can be used to remove pollutants and wastewater treatment ([29, 53], section 10.3). Another example in bioengineering is the biofunctionalization of carbon nano-networks, which stimulates efficient cell adhesion and proliferation [54].

Functionalization can be achieved through several grafting techniques. Ligands and moieties are grafted in the nanomaterial, each requiring unique methodologies with their own merits and demerits ([29, 55, 56], section 10.4). Another functionalization method involves the direct addition of a functional group during nanomaterial synthesis, usually *in situ* and co-condensation. Various functionalization techniques bind functional groups to a specific nanomaterial via strong covalent or noncovalent bonds (i.e. van der Waals' forces). Functionalization of CNTs and graphenes tend to utilize noncovalent bonds. The interaction between π bonds and the functional moieties (usually aromatic organic groups) gives new functional traits without altering the electronic properties ([29, 57, 58], section 8.1).

15.7.1 Importance of Functionalization

The addition of extra organic or inorganic moieties is an effective approach to control the physicochemical and toxicological properties of nanomaterials. This enables a tailored functionality toward the desired application. Moreover, functionalization has been observed to provide certain benefits, especially in diminishing environmental hazards as well as to enhance nanoparticle stability [59]. Higher degrees of functionalization in nanomaterials leads to the decrease of toxicity of the nanoparticles. Furthermore, functionalization opened the realm of innovative treatment methodologies such as improving the adsorption of pollutants from water, removing air pollution, and various other environmental clean-up techniques [60]. Functionalization also allowed tuning the sensitivity of nanosensors, using single-stranded DNA, polymers, noble metals, and biomolecules. They act as sensitizing agents, giving rise to various bio-fabricated nanomaterials [61–63].

The recent advancements in nanoscience, nanoengineering, and nanotechnology, originating from novel innovations, indicate that functionalization/surface functionalization of nanomaterials provides a ladder to surpass challenges and find solutions to current limitations. Through property changes using functionalization, bio-carbon nanomaterials have found several applications in electronics, waste management (particular for water treatment plants and applications), biomedicine [8], and many more. With the increasing popularity of organic materials in the market, the future of bio-carbons looks promising. Some of the remarkable bio-carbon nanomaterials and their applications are discussed below.

15.8 Nanocellulose

A promising bio-carbon nanomaterial on the rise, with several applications in technical fields and "environment clean up," is the nanocellulose (NC). A polysaccharide with individualized nanofiber, the polymer of β-D-glucopyranose is a bio-carbon nanomaterial that gained popularity for its abundance, biocompatibility, and sustainability (Figure 15.1) ([29, 65, 66], section 18.4). Cellulose derivatives with properties like hydrophilicity, biodegradability, chirality, easy surface modification enhance functionalization for different applications in various fields ([29, 67, 68], section 18.4). For centuries, cellulose derivatives were signature materials in the paper, food, and pharmaceutical industries, making a recent emergence to be particularly named a "bio" nanomaterial and be considered for especially "green" engineering advancements, alongside CNTs.

15.8.1 Synthesis of Nanocellulose (NC)

The synthesis of NC follows different methodology, chemical reagent, and procedures depending on the type of NC synthesized. Microcrystalline cellulose (MCC), pure cellulose material, usually acts as a precursor for most NC synthesis processes ([29], section 18.4). Generally, the synthesis method comprises an extraction procedure using an alkali or bleaching agent, involving the oxidation of primary alcohols to carboxylic groups using 2,2,6,6-tetramethylpiperidinyloxy (TEMPO) radical procedures or strong acid hydrolysis (usually 65% of sulfuric acid at 45 °C). The property of NC to swell up in aqueous solutions

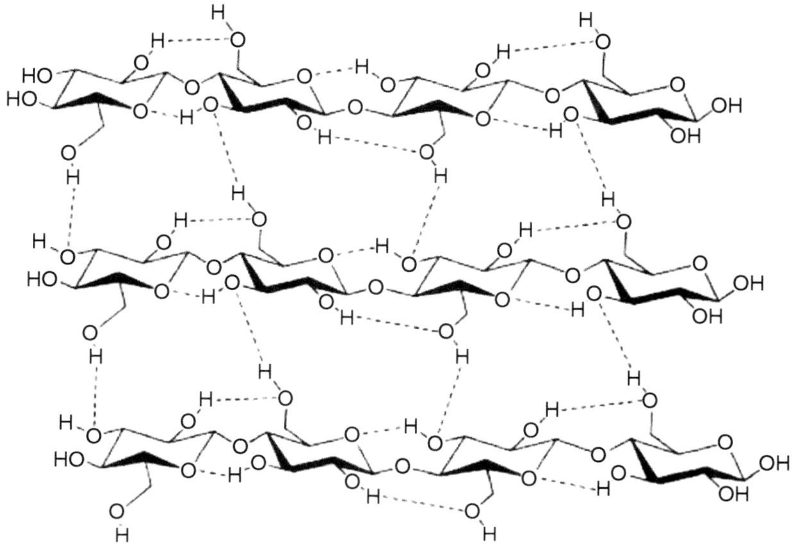

Figure 15.1 Intra- and intermolecular hydrogen bonds in a nanocellulose polymer [64]. Source: Jin et al. [65], Soni et al. [66], Mishra et al. [29].

enables the formation of nanocrystals via the acid hydrolysis system. It is followed by an ultrasonication for 24 hours, separating the nanofibrils, which form the crystalline, nanowhisker, or a different tailored structure. Particular procedures and reagents can result in different NC types, which can be further functionalized to meet specific structural requirements, enabling the synthesis of a range of NC at a relatively low cost.

Primarily, there are four types of cellulosic nanomaterials, each requiring unique specific synthesis procedures and methodology ([29], section 18.4):

(1) Cellulose nanocrystals (CNCs)
(2) Cellulose nanofibrils (CNFs)
(3) Hairy NC (2,3-dicarboxylic cellulose nanofibrils, DCCs)
(4) Bacterial nanocellulose (BNC)

15.8.2 Synthesis of CNCs

CNC is called nanocrystalline cellulose and cellulose nanowhiskers, which are prepared using enzymatic reactions ([29, 69, 70], section 18.4) or strong acid hydrolysis [71, 72], resulting in nanostructures. The highly crystalline structures are targeted diameters 5–30 nm, having lengths of 100–500 nm range ([29], section 18.4).

15.8.3 Synthesis of CNFs

Cellulose nanofiber or nanofibrillated cellulose and CNF are synthesized by the disintegration of cellulose micro/macro fibers. Both physical and chemical methods (consisting of enzymatic reactions) produce nanofibrils that are usually a length of more than 1 μm in diameter and 10–100 nm in range ([29], section 18.4). Particular synthetic processes and the source of cellulose used can bring variations in the diameter and dimensions of the CNFs synthesized [73]. Unlike CNCs, CNFs consist of crystallized and unshaped regions [74].

15.8.4 Synthesis of DCCs

Commonly known as Hairy Nanocellulose, De-carboxylated cellulose (DCC) are electrostatically stabilized nanocrystalline cellulose. Being structurally similar to CNCs, DCC consists of crystalline and amorphous regions [75] having polymer chains at both ends [75, 76].

15.8.5 Synthesis of BNC

Synthesized from the extracellular constituents of microbial or bacterial cellulose (usually of different strains), growing glucose chains inside the microorganism bacteria utilize a bottom-up methodology ([29, 77], section 18.4). *Acetobacter xylinum* tends to be the most common bacteria used in the synthesis. Once the optimum quantity of BNC is generated, further enzymatic or acid hydrolysis enhances the crystallinity of the nanoparticles of size between 20 and 100 nm ([29], section 18.4).

15.8.6 Applications

15.8.6.1 NC in Purification Technology as Films and Foams

One significant application of NCs is as sorbents, acting as bioremediators by absorbing pollutants from biowaste, sludge, and bacteria. Having low density but great surface area, NCs make excellent membranes, solid fibers, and films and improvise absorbent materials for separation techniques. Upon functionalization, lanthanum-doped NC-based hybrid films act as efficient phosphate absorbents compared to graphene nanosheets. It is also biocompatible ([29, 78], section 18.4). Therefore, such use of NC crystals for hybrid films stands to be cost-effective and efficient, maintaining reduced pollution throughout the process. Secondly, numerous toxic inorganic species and entities, radioactive heavy metals, silver nanoparticles from dyes, chemical industries effluents, and various other sources, runoff into water, air, and the environment are hard to retrieve. Functionalizing NC materials resolves issues as it improves efficacy rates of absorption binding with pollutant substrates ([29, 79–81], section 18.5; [82–84]):

(1) *Radioactive and heavy metals*: Proper surface modification of NC nanocrystals enables the design of selective adsorbents. For example, negatively charged phosphate has been observed to prevent cellulose nanofibers aggregation and increase hazardous metals adsorption by up to 100%, representing the efficacy of CNCs upon specific functionalization [85]. Secondly, sulfonation and peroxidate oxidation have been observed to make NCs effective Lead adsorbents [86]. Since sulfur atoms have a high affinity toward Pb(II), together with an increased surface area at the nanoscale, the cellulose nanofibers hold better adsorbent capacities when

compared with unfunctionalized or general NCs ([29, 83], section 18.5). Cellulose nanomaterials can be modified with various functionalization and modification to remove specific metals from synthetic water or other media, disconcerting the pH, being better filters [87–90].

(2) *Viruses, bacterias, mycotoxins*: Functionalization by carboxyl and positive amine ending groups enabled NCs to be good adsorbents for even viruses and bacterias via electrostatic interactions. Moreover, modification using grafting fatty acids allows the removal of mycotoxins, once again demonstrating NCs' efficiency as membrane filters and reflecting diverse adsorbent capability [90–92]
(3) *Organic substances, oil spillage, herbicides*: Hydrophobic aerogels, organogels, hydrogel-based NCs acts as sponges for organic solvents and oils. This property resolves dilemmas such as oil spillage via bioremediation at a cheaper rate.
(4) *Air pollution*: Various porous morphology of NCs make them excellent applicants for air pollutant removal. Moreover, depending on appropriate functionalization, NCs can be used to absorb CO_2 in the atmosphere. For instance, using aerogels and amine groups, preferably polyethylenimine, with nanofibrils of cellulose captures carbon dioxide, avoiding urea to be produced as a byproduct of the reaction. Additionally, exposure of NC foams to 0.1 M of HCl increases permeability to hydrogen in air.

15.8.7 NC as Solar Cells

Cellulose nanofibers with particular polymers can serve as solar cells and mechanical support for solar cells. The application of NC in solar cells had promising solar conversion efficiency, further developed upon research [93–95]. With global technology seeking efficient renewable sources of energy, the application of NC as solar cells is of relevance, having a promising future, solving both the economy of production and environmental sustainability.

15.8.8 NC as Stabilizing Agent

Paper production requires strong adhesives and several chemicals to attain a smooth surface and a dense, strong fine paper that can enhance printing quality. BNC, due to its fine porous structure, having optical, lightweight, biodegradable, and recyclable qualities, acts as a good stabilizing agent for paper production, without the necessity of extra hazardous chemical adhesives [96]. This makes the paper smoother and transparent, yet strong and durable, making it useful for further applying solar cells, touch sensors, and mobile devices, where transparent NC papers are growing in demand [97].

15.8.9 NC in Biomedicine

Acting as scaffolds for cell adhesion, differentiation, the most significant application of NCs is the potential to replace blood vessels [98, 99]. Since NCs are biocompatible materials, relatively less toxic, contains water, and has great mechanical properties, the applications are limitless. It is on the rise, making it a favorite nanomaterial among biomedical engineers and researchers [100]. Secondly, CNCs demonstrated potential in bioimaging, as biosensors and biodetectors, once functionalized using different fluorophores [98, 101]. Furthermore, NCs can also play vital roles in pharmaceutical industries, proving to be of great help for drug production [93].

Applications of NC are limitless. With emerging applications in cosmetics, flexible electronics, and organic light emitting diodes (LEDs), it is evident that NCs and CNCs are huge and broad [66]. Spanning from electronics to paper and textile industries to biomedicine, automotive, and packaging, the functionalization of NCs makes it a stepping stone toward future technologies and a greener environment.

15.9 Nitrogen and Sulfur Co-doped Bio-carbon

Polypeptides reacted with pyrolyzed zinc-based metal–organic coordination polymer (MOCP-Zn) can result in a co-doped bio-carbon. As demonstrated by Fang et al., when zinc salt $Zn(NO_3)_2 \cdot 6H_2O$ and polypeptides (derived from chicken feathers) undergo a hydrothermal reaction, MOCP-Zn is synthesized, which when pyrolyzed results in a co-doped bio-carbon floc [102]. Provided that the synthesis of bio-carbon significantly utilizes biomolecules, the procedure reduces carbon footprinting compared to the complete industrial synthesis of "synthetic-carbon" derived materials. MOCP-Zn is an organic–inorganic hybrid that offers a rich carbon source that holds a porous structure due to coordinate bonding. Finally, once MOCP-Zn is reacted with the polypeptides from keratin rich in amino acids, nitrogen, and sulfur, a considerable waste or biomatter-chicken feather, a Nitrogen and Sulfur Bio-carbon floc is essentially synthesized. Such materials have been experimented with to supervise redox reactions efficiently in both acid and alkaline mediums.

15.9.1 Application Co-doped Bio-carbon

Consisting of an abundant porous structure and a large surface area to volume ratio, giving rise to enhanced oxygen reduction activities, a significant Nitrogen and sulfur co-doped bio-carbon flocs (NSC) application is a proton exchange membrane fuel cell [102]. Such doped bio-carbon materials from waste biomass, with enhanced electrocatalytic activity, can also serve as bio-carbon catalysts for many chemical and treatment-plant applications. Furthermore, utilizing waste biomass and preparing bio-carbons with enhanced electrocatalytic activity is environmentally sustainable. It is also economically deemed as a cheaper route for biocatalysts and membrane fuel cell productions.

15.10 Biochar

The remnant of biomass, usually after pyrolysis, are known as Biochars [103]. Such biochars have amorphous carbon structures, with basal planes of graphene sheets stacked unaligned, providing conductive properties [103, 104]. The biochar conductivity mainly depends on the pyrolysis process of biomass, which can be performed at different high temperatures (Lehmann and Joseph 2012) [103, 105].

15.10.1 Application of Biochar

Biochar consisting of many functional groups enables cross-link formations with various polymer matrices via chemical bonding. One potential application of biochar is to increase thermal stability and tensile properties of other polymers, such as styrene-butadiene rubber, epoxy resin [103, 106, 107], and wood/polypropylene composites [103, 108]. Secondly, the appreciable electrical conductivity and polymer bonding ability enable biochar to be potential microelectronics. Therefore, biochar holds promise to be both cheaper and "greener" replacements of CNTs and graphene fillers in the future [103] upon more research and further study.

15.10.1.1 Application of Bio-carbon Derived from Sisal Leaves

Free-standing bio-carbons can inherit hierarchical texture from their precursor biomass, depending on carbonizing condition, delivering a unique structure to the bio-carbon. Carbonizing free-standing bio-carbon derived from sisal leaves can be observed to have heightened conductivity [109]. It does so by facilitating faster ion exchange pathways. A similar approach to controlling the carbonizing environment can give better-performing electrodes with increased capacitance, longevity, and high performing storage devices [109].

15.11 Biopolymers

Biopolymers are polymeric biomolecules that are biodegradable, produced from living organisms. Unlike synthetic polymers, biopolymers are complex and have definite configurations, making them selective to specific molecules and chemically active, especially *in vivo* ([110, 111], section 20.4). Based on the chemical backbone, primarily polymers can be classified as polysaccharides, polyesters, polycarbonates, vinyl polymers, polyamides. Furthermore, based on repeating units, they can be called proteins, nucleic acids, polysaccharides. Fossil fuel derived polymers do not degrade/decompose naturally and often require to be incinerated, contributing to greenhouse gas production. On the other hand, biopolymers, since they are from renewable sources (starch, gelatin, natural fibers, chitin, or other organic matter), undergo a natural decomposition or degradation process, returning to the environment further aid with greenhouse gas reduction [111]. Synthesis of biopolymers is primarily through (i) microorganisms fermentation, where a special culture environment for the microorganisms is maintained, and (ii) chemical polymerization, where monomeric units are reacted to form polymers. The property of biopolymers consisting of many functional groups such as –COOH, –SH, $-NH_2$ enables the production of bio-nanoparticles and biomolecules from the associated biopolymers, developing many possible applications.

15.11.1 Biopolymers in "Green" Synthesis of Nanoparticles

Historically, silver (alongside gold) was particularly known for its antimicrobial properties, which are important in medicine and biology [112, 113]. Moreover, today, upon the development of nanotechnology, silver nanoparticles weaved their way into optoelectronics, biological probes, display devices, and many more applications [112]. However, the tremendous dominance of silver nanoparticles in various applications, the synthesis of silver-gold nanoparticles, is expensive and hazardous for our health and the environment. However, with the advancement of science, green synthesis of nanoparticles proved to be cheap and eco-friendly at the same time. For example, Dextrose Reduced Silver Nanoparticles, upon complete green synthesis, delivers antimicrobial and sensing properties. Using the biopolymer dextrose (reducing agent), gelatin (stabilizing and capping agent), silver nitrate (precursor of silver), and water as solvent, the method resulted in highly stable silver nanoparticles with better antibacterial and antimicrobial activity [114]. Moreover, since synthetic polymer matrices are energy expensive in general, the use of biopolymers (cellulose of bacterial origin) and CNTs to develop thermoelectric paper is another recent advancement [115]. Such nanomaterials have shown immense thermal resistance that can act as thermoelectric generators, converting waste heat to electricity and exemplifying yet another remarkable application of bio-carbon nanomaterials [115].

15.11.2 Biopolymers in Waste Water Treatments

Starch, cyclodextrin, chitin, and chitosan (CS) are naturally occurring polymers that have unique physicochemical properties, chemical stability, and selectivity toward particular compounds ([29], section 22.6; [84]). Such biopolymers tend to make great adsorbents for wastewater treatment, especially when used together with TiO_2.

15.11.3 Biopolymers as Bioplastics

Plastics are "magical" materials that are durable, malleable, and are almost everything around us. The worst and the greatest problem surrounding plastic is its outstanding durability and non-biodegradable trait. Moreover, from its birth to death, plastic highly contributes to carbon footprint. The unignorable disadvantages of plastics lead to the growing popularity of bioplastics, synthesized from algae. Cheap, high yield, and adaptable to any environment, algae-based plastics can soon be commercialized to have numerous applications once further developed. Contributing to minimal carbon footprint, the era of bioplastics just began and demonstrated the importance of biopolymers and their subsequent demand by nature.

15.11.4 Nanocomposites

Two or more multiphase solid materials (often with different chemical and physical properties) form a nanocomposite, where at least one of its material constituents is in the nanoscale [116]. An emerging material of the twenty-first century, nanocomposites hold a promising future in designing various materials because of their remarkable properties absent in conventional composites [116–121]. Studied to overcome the limitations of microcomposites and monolithic bioinspired nanocomposites are the new frontiers, being relatively eco-friendly ([29], section 21.2; [122–125]).

Nanocomposite consisting of multiple different phase domains can have improved thermal resistive capacities and permeation properties ([29], section 21.3; [50, 51, 126, 127]).

This is achieved by incorporating particulates within the biopolymer matrix. Chitosan, together with CNCs, can be designed to be better films with selective filtrating characteristics. Moreover, another methodology observed delivers bioinspired nanocomposites of silver nanoparticles using tomato fruit extract, which had varied effectiveness based on the nanoparticles' size [112]. Furthermore, with additional research, more innovative synthesis of silver nanoparticles using plant extracts such as *Malpighia emarginata* (West Indian Cherry), *Argemone mexicana* leaf, polymers, and other "green" methodologies, most of "bio" origin were reported [112, 128–130]. For example, the biocidal silver activated charcoal nanocomposites can serve as an effective filtration and can be reused for the water filtration bed columns, as disinfectants, and water treatment purposes for an appreciable period [112, 130, 131]. Cross-linked chitosan/nitrogen doped-graphene quantum dot nanocomposites have promised potential in bone tissue engineering with additional innovative applications [122, 123, 132]. This demonstrates how "green" synthesis and bioinspired nanocomposites are revolutionizing materials for engineers and technologists to develop environment-friendly devices and applications.

15.11.5 Peptide Nanoparticles

Unlike cellulose or graphene that consists of only carbon, peptide nanoparticles have more than varying elements, apart from carbon only, organic origin, and can be categorized under bio carbons. Structural formation or self-assembly of peptides into complex nano-architectures often require specific environmental conditions. Consisting of ionic, hydrogen bonding, hydrophobic, and π–π stacking interactions, together the peptides can be manipulated to self construct into various structures within the nano and micro scale [133, 134]. Peptide nanomaterials can also have their conductivity controlled based on selecting the amino acids used for generating the polymer nanotube fibrils. Photonic devices are primarily functionalized by self-assembling peptide nanotubes [135, 136]. Moreover, these assemblies have the ability to replace conventional switches and wires [135, 136]. The following presents some of the major peptide nanostructures:

15.11.5.1 Dipeptides

They are some of the simplest forms of peptide motifs, which can undergo further assembly to form nanotubes and unlimited applications.

15.11.5.2 Peptide Amphiphiles

Usually, peptide amphiphiles self-assemble into spherical structures, held in conformation by hydrophobic and electrostatic interactions, consisting of a hydrophilic domain [134]. Such peptide architecture has promising applications as nano-vesicle and can be employed for drug delivery. It is composed of native amino acids and completely biodegradable. Moreover, functionalization using several moieties brought on significant functional traits: cholesterol was observed to induce micelles formation and increase stability; glutamic acid to enhance tumor apoptosis *in vivo*, in an acidic environment; rendering cationic peptides to induce antimicrobial properties, serving as antimicrobial nano agents even to treat brain infections, crossing the blood-brain barrier, and other infectious diseases against various microorganisms and many more [137].

15.11.5.3 Dendrimers

Dendrimers are an extension of peptide amphiphiles. They enable surface functionalization. When bound to a linear hydrophobic peptide with a negatively charged C-terminal glutamate residue, dendrimers form nano micelles and encapsulate hydrophobic drug or DNA cargo [138].

In the past decade, gold, magnetic, semiconducting, and mesoporous silicon nanoparticles were common for various biomedical applications. However, such nanoparticles were often toxic, less stable, and mainly non-biodegradable, which could excite biological complications. Therefore, the outbreak of such peptide nanoconjugates offers a cheaper, preferable replacement bypassing past limitations. One such example would be the successful delivery of doxorubicin, an antitumor drug, *in vivo*, hence demonstrating a promising future peptide nanomaterial in drug delivery.

15.11.5.4 Coiled-Coil Peptides

Hydrogen bonds create a right-handed helix composed of β-sheets and α-helices (fundamentals of secondary polypeptide structure), directing the amino acid side chains away from the core [134]. Stabilization can be enhanced upon the addition of particular motifs and peptides that aggregate together, allowing ion exchange between charged amino acids and polar groups, which leads to a helical bundle-like structure [139, 140]. The final assembly of peptide conformation can lead to bioactive materials required in various areas [140–142].

Coiled-coil nanopeptides can behave as immunogens, especially for Severe Acute Respiratory Syndrome (SARS), malaria, HIV [134, 143–146]. It may potentially serve as an immunogen for coronavirus (COVID-19) and lead to an efficient vaccine.

15.11.5.5 Peptide–Nucleic Acid Complexes

The formation of compact nanopeptides, when negatively charged nucleic acids combine with positively charged peptides, results in the peptide–nucleic acid complex [134].

Peptide–nucleic acid nanoparticles were observed to successfully perform gene delivery, transfect isolated cells (*in vivo*). Also, functionalizing the NPs with long cationic peptides have shown increased stability and toxicity. Therefore, although peptide–nucleic nanoparticles are growing attractive in biomedicine, ensuring particle stability and balancing toxicity issues, *in vivo*, further needs to be studied [134, 147].

15.11.5.6 Casein Micelles

Casein (phosphorylated proteins), commonly known as milk protein, has an amphiphilic diblock copolymer structure [148]. The presence of hydrophilic N-terminal and hydrophobic C-terminal domains ensures a globular structure, which can form micelles in an aqueous solution under particular conditions [149].

Casein micelles can carry a range of hydrophobic cargoes. Being non-toxic, non-immunogenic, and cheap, they make excellent nano delivery systems/vehicles. Casein micelles NPs in clinical trials and studies for delivering therapeutic hydroponic drugs for stomach cancer, promising to have improved and widen the scope of its utility.

15.11.5.7 Peptide Nanotubes

Ring-shaped peptide subunits of alternating D and L amino acid residues and intersubunit hydrogen bonding forms a tubular structure, identified as peptide nanotube [150]. Although peptide nanotubes require further characterization with strong mechanical properties and thermal stability, it can be a better alternative for CNTs, in the future.

Peptide nanomaterials are emerging, coming in various architectures, fibers, tubes, crystals, micelles, and vesicles [133, 134, 151–154]. Their biomedical field application is also tailored for imaging, sensing, and monitoring purposes. In chemical industries, peptide nanoparticles can serve as capping ligands for gold nanoparticles and bioproteins for photovoltaic and water splitting photoelectrochemical cells [136]. Recent advancements to sensitize biomolecules for power cells can also be another example with more innovative applications yet to come [155].

15.11.6 Further Application of Bio-carbon Nanomaterials as Devices

Peptides, NC, particular biochars, and biopolymers are all intensely researched and constantly functionalized to innovate devices, primarily to replace CNTs and other synthetic nanomaterials. The most prominent application as a device for biocarbon nanomaterials apart from water treatment plants and drug delivery vehicles is the biosensor.

Biosensors are micro-analytical devices that can be used both quantitatively or qualitatively, depending on the motive. These devices consist of transducers that emit signals to measure specific molecules. The transducers happen to be bio-integrated and with sensitive detection systems. One such example of a biosensor is the gluco-sensor that initially started as a prime monitoring device during cardiovascular surgery. Today, it is readily being used by diabetic patients almost regularly. The gluco-sensor utilizes a platinum electrode, connected to an analytical device able to detect glucose upon oxidase reaction [156–158]. Other notable biosensors include the handheld, smartphone-based biosensors developed based on fluorometric technology, therapeutic enzyme nanodevice biosensors, and bio-labels [158, 159]. Significant characterized properties remain unchanged even after functionalization via polymers enables nanomaterial as efficient biosensors [158].

Supercapacitors or electrochemical capacitors are another promising devices that incorporate and optimize the balance between potassium hydroxide (KOH) system and cellulose, creating a highly efficient cell [160, 161]. Such supercapacitors are an excellent alternative to sustainable energy sources, potentially replacing fossil fuels, saving the depleting environment. A particular nanostructure from KOH and cellulose ensures an excellent porous 3D network with a high surface area, delivering low cell resistivity and high specific capacitance [160].

Applications of biocarbon nanomaterials, especially focused on environmental sustainability and bioremediation, are still developing. Although promising breakthroughs, discoveries, and innovations have been made, more significant research is yet needed before the applications are commercialized at a large scale. However, based on the progress witnessed so far, humanity may optimistically look ahead toward successfully reducing and overcoming the current environmental distresses by maintaining an effective advancement and implementation of nanotechnology.

15.11.7 Characterization Technique

Till functionalization, the complete production process of nanoparticles requires proper characterization to observe, control, and maintain the composition, size, morphology, and structure of the nanoparticles and be on the right course of synthesis ([29], section 17.4). As nanoparticles' formation occurs, a quick qualitative confirmation of nanoparticles' formation is an observation for color change. Secondly, shape, structure, morphology, and size are eventually characterized using the following techniques:

(1) X-ray diffraction (XRD)
(2) Raman spectroscopy
(3) Ultraviolet–visible spectroscopy (UV–vis)
(4) Dynamic light scattering (DLS) spectroscopy
(5) Fourier transform infrared (FTIR) spectroscopy
(6) Transmission electron microscopy (TEM)
(7) Scanning electron microscopy (SEM)
(8) Atomic force microscopy (AFM)

XRD provides data on the crystalline structure, orientation, and phase to elaborate on characterization techniques ([29, 162, 163], section 17.4). Raman spectroscopy provides the vibration modes of molecules to determine surface groups' nanoparticles [164, 165]. UV of 190–380 nm and visible region from 400 to 800 nm interact to incite electron transition, which helps characterize particular nanoparticles. The DLS allows the quantification of charge and size of nanoparticles in a clear medium [165, 166], and FTIR enables to deduce the functional groups present during the synthesis process. As for the TEM, SEM, and AFM, data images can be gathered, providing a direct method of characterizing the structure, size, and morphology of the nanoparticles, unlike the indirect spectroscopic techniques ([29], section 17.4). It should be noted that there are many more characterization and analytical techniques found today, each providing different pieces of the sample puzzle, all to be analyzed together.

The characterization technique of nanomaterials is vital. It directs material scientists and engineers to synthesize the targeted nanomaterial for the various applications, tailoring the process to ultimately generate eco-friendly materials, composites, and substances. Furthermore, characterization techniques enable in monitoring, sensing, and detection of toxins under various condition, exemplified below.

Biologically synthesized nanoparticles can probe when a redox reaction between Hg2+ ions in aqueous media removes a specific biological stabilizer from the AgNP surface, providing a color change of the reaction media. Such color change can be monitored using the UV–vis. A similar application of color change with the aggregation of particular nanoparticles can also be used for sensing the presence of any ions (Cr^{3+} ions, for example) in a solution needed for biological and/or environmental analysis [167–172].

Detection of toxins in food is a significant necessity that utilizes immunochromatographic methods, using gold NP-anti conjugates. A prominent example would be the functionalization of Au-NPs with sugar coast allows the detection and sensing of ricin, a potent toxin that can cause death ([29, 173–176], section 17.5).

15.12 Conclusion

Advancement toward biosynthesis and biomaterials, using bio-carbons, proteins, DNA, viruses, plants, and microorganisms, together with certain fabrication demonstrates the limitless scope, opportunity, and hope to attain a "green" sustainable environment [29, 79]. More study and research is required to bypass the critical parameters and arising challenges in this new frontier of nanotechnology. Functionalizing bio-carbon nanomaterials for "green" techniques and devices based on the never-ending innovative and versatile ideas and routes must be pursued.

Biomolecules and bioinspired synthesis promote and utilize fewer and milder reagents, with little to almost negligible toxicity and exposure to hazardous chemicals, using eco-friendly substrates (and products). Bio-carbons (including other bio-candidates) are often more reusable or recyclable than synthetic substrates, requiring simple separation techniques [29]. Nanotechnology that incorporates bio-carbons curves the road toward a new era and future, as already indicated by the several applications, scopes, and promises of the various emerging nanomaterials in almost every discipline and research. In short, the functionalization of bio-carbon nanomaterials in many applications is humanity's present task to pave tomorrow's technological and environmental possibilities.

Important Websites

https://www.nanowerk.com
https://www.azonano.com
https://www.nature.com
https://www.rsc.org
https://www.nature.com/subjects/nanobiotechnology
https://www.sciencedaily.com/news/matter_energy/nanobiotechnology
https://www.nanotechmag.com
https://www.intechopen.com

References

1 Ray, U. (2018). What are the different types of nanoparticles?. https://www.azonano.com/article.aspx?ArticleID=4938 (accessed 29 December 2020).

2 Cartaxo, A.L. (2015). Nanoparticles types and properties – understanding these promising devices in the biomedical area. Semantic Scholar. Corpus ID: 189955782. https://www.semanticscholar.org/paper/Nanoparticles-types-and-properties-%E2%80%93-understanding-Cartaxo/9570b62e051cf8cff1398915cf14eafcff21fbdd#citing-papers

3 Khan, I., Saeed, K., and Khan, I. (2017). Nanoparticles: properties, applications and toxicities. *Arabian J. Chem.* https://doi.org/10.1016/j.arabjc.2017.05.011.

4 Thomas, S.C., Harshita, Mishra, P.K., and Talegaonkar, S. (2015). Ceramic nanoparticles: fabrication methods and applications in drug delivery. *Curr. Pharm. Des.* 21 (42): 6165–6188. https://doi.org/10.2174/1381612821666151027153246.

5 Bangal, M., Ashtaputre, S., Marathe, S. et al. (2005). Semiconductor nanoparticles. *Hyperfine Interact.* 160: 81–94. https://doi.org/10.1007/s10751-005-9151-y.

6 Dahman, Y. (2017). *Nanotechnology and Functional Materials for Engineers*. Amsterdam: Elsevier.

7 Crucho, C.I.C. and Barros, M.T. (2017). Polymeric nanoparticles: a study on the preparation variables and characterization methods. *Mater. Sci. Eng., C* 80: 771–784. https://doi.org/10.1016/j.msec.2017.06.004.

8 Subbiah, R., Veerapandian, M.S., and Yun, K. (2010). Nanoparticles: functionalization and multifunctional applications in biomedical sciences. *Curr. Med. Chem.* 17 (36): 4559–4577.

9 Gupta, T.K., Budarapu, P.R., Chappidi, S.R. et al. (2019). Advances in carbon based nanomaterials for bio-medical applications. *Curr. Med. Chem.* 26: 6851. https://doi.org/10.2174/0929867326666181126113605.

10 Ong, Y.T., Ahmad, A.L., Zein, S.H.S., and Tan, S.H. (2010). A review on carbon nanotubes in an environmental protection and green engineering perspective. *Braz. J. Chem. Eng.* 27 (2): 227–242.

11 Power, A.C., Gorey, B., Chandra, S., and Chapman, J. (2018). Carbon nanomaterials and their application to electrochemical sensors: a review. *Nanotechnol. Rev.* 7 (1): 19–41.

12 Rahmati, M. and Mozafari, M. (2019). Biological response to carbon-family nanomaterials: interactions at the nano-bio interface. *Front. Bioeng. Biotechnol.* 7: 4.

13 Saikia, N. (2017). Chapter 13: Functionalized carbon nanomaterials in drug delivery: emergent perspectives from application (ed. G. Kyzar, A.C. Mitropoulos), 231–250. In: *Novel Nanomaterials-Synthesis and Applications*. IntechOpen. ISBN: 978-1-83881-460-1.

14 Ahmad, K. and Pan, W. (2015). Microstructure-toughening relation in alumina based multiwall carbon nanotube ceramic composites. *J. Eur. Ceram. Soc.* 35: 663–671.

15 De Volder, M.F., Tawfick, S.H., Baughman, R.H., and Hart, A.J. (2013). Carbon nanotubes: present and future commercial applications. *Science* 339: 535–539.

16 Dasgupta, N., Ranjan, S., and Lichtfouse, E. (2019). *Environmental Nanotechnology*, vol. 2. Cham: Springer International Publishing.

17 Asha, A.B., and Narain, R. (2020). Nanomaterials properties. In: *Polymer Science and Nanotechnology*, 343–359. Elsevier https://doi.org/10.1016/b978-0-12-816806-6.00015-7.

18 Hosseini, S.A., Talebipour, S., Neyestani, M.R. et al. (2018). Graphene oxide MgFe2O4 nanocomposites for Cr(VI) remediation: a comparative modeling study. *Nanotechnol. Environ. Eng.* 3(1), 1–11

19 Maddinedi, S., Mandal, B.K., Patil, S.H. et al. (2017). Diastase induced green synthesis of bilayered reduced graphene oxide and its decoration with gold nanoparticles. *J. Photochem. Photobiol. B Biol.*, 166, 252–258.

20 Amin, M.T., Alazba, A.A., and Manzoor, U. (2014). A review of removal of pollutants from water/wastewater using different types of nanomaterials. *Adv. Mater. Sci. Eng.*, 2014.

21 Zaytseva, O. and Neumann, G. (2016). Carbon nanomaterials: production, impact on plant development, agricultural and environmental applications. *Chem. Biol. Technol. Agric.* 3: 17. https://doi.org/10.1186/s40538-016-0070-8.

22 Krätschmer, W. (2011). The story of making fullerenes. *Nanoscale* 3 (6): 2485–2489. https://doi.org/10.1039/c0nr00925c.

23 Krätschmer, W., Lamb, L.D., Fostiropoulos, K., and Huffman, D.R. (1990). Solid C_{60}: a new form of carbon. *Nature* 347 (6291): 354–358. https://doi.org/10.1038/347354a0.

24 Singh, H. and Srivastava, M. (1995). Fullerenes: synthesis, separation, characterization, reaction chemistry, and applications – a review. *Energy Sources* 17: 615–640.

25 Mallakpour, S. and Khadem, E. (2019). Chapter 7: Carbon nanotubes for heavy metals removal. In: *Composite Nanoadsorbents* (eds. G. Kyzas, A. Mitropouios), 181–210. Elsevier.

26 Mallakpour, S. and Rashidimoghadam, S. (2019). Chapter 9: Carbon nanotubes for dyes removal. In: *Composite Nanoadsorbents* (eds. G. Kyzas and A.C. Mitrpoulos), 211–244. Amsterdam: Elsevier. https://doi.org/10.1016/B978-0-12-814132-8.00010-1. eBook ISBN: 9780128141335.

27 Harris, P.J.F. (1999). *Carbon Nanotubes and Related Structures*. Cambridge: Cambridge University Press.

28 Zhang, Q., Huang, J.Q., Zhao, M.Q. et al. (2011). Carbon nanotube mass production: principles and processes. *ChemSusChem* 4 (7): 864–889. https://doi.org/10.1002/cssc.201100177.

29 Mishra, A.K. and Hussain, C.M. (2018). *Nanotechnology in Environmental Science* (Chapters: 5.2, 8.1, 10.2–10.4, 17.4, 18.4, 18.5, 20.4, 21.2, 21.3, 21.4, 22.6). Wiley-VCH.

30 Brukh, R., Sae-Khow, O., and Mitra, S. (2008). Stabilizing single-walled carbon nanotubes by removal of residual metal catalysts. *Chem. Phys. Lett.* 459: 149–152.

31 Heras, A., Colina, A., López-Palacios, J. et al. (2009). Electrochemical purification of carbon nanotube electrodes. *Electrochem. Commun.* 11: 1535–1538.

32 Gore and Sane, A. (2011). Chapter 7: Flame synthesis of carbon nanotubes. In: *Carbon Nanotubes: Synthesis, Characterization, Applications* (ed. Siva Yellampalli), 16801. IntechOpen. https://doi.org/10.5772/21012.

33 Karwa, M., Iqbal, Z., and Mitra, S. (2006). Scaled-up self-assembly of carbon nanotubes inside long stainless steel tubing. *Carbon* 44: 1235–1242.

34 Jiao, L., Zhang, L., Wang, X. et al. (2009). Narrow graphene nanoribbons from carbon nanotubes. *Nature* 458 (7240): 877–880. https://doi.org/10.1038/nature07919.

35 Novoselov, K.S., Fal'ko, V.I., Colombo, L. et al. (2012). A roadmap for graphene. *Nature* 490 (7419): 192–200. https://doi.org/10.1038/nature11458.

36 Mallakpour, S. and Khadem, E. (2019). Chapter 8: Carbon nanotubes for heavy metals removal. In: *Composite Nanoadsorbents* (eds. G. Kyzas and A.C. Mitrpoulos), 181–210. Amsterdam, Netherlands: Elsevier. https://doi.org/10.1016/B978-0-12-814132-8.00009-5, eBook ISBN: 9780128141335.

37 MacKenzie, K.J., See, C.H., Dunens, O.M., and Harris, A.T. (2008). Do single-walled carbon nanotubes occur naturally? *Nat. Nanotechnol.* 3 (6): 310–310. https://doi.org/10.1038/nnano.2008.139.

38 Velasco-Santos, C., Martínez-Hernández, A.L., Consultchi, A. et al. (2003). Naturally produced carbon nanotubes. *Chem. Phys. Lett.* 373 (3–4): 272–276. https://doi.org/10.1016/s0009-2614(03)00615-8.

39 Mracek, J., Fagan, R.D., Stengelin, R.M., and Hesjedal, T. (2011). Are carbon nanotubes a naturally occurring material? Hints from methane CVD using lava as a catalyst. *Curr. Nanosci.* 7 (3): 294–296. https://doi.org/10.2174/157341311795542543.

40 Su, D.S. and Chen, X.W. (2007). Natural lavas as catalysts for efficient production of carbon nanotubes and nanofibers. *Angew. Chem. Int. Ed.* 46 (11): 1823–1824. https://doi.org/10.1002/anie.200604207.

41 Becker, L., Bada, J., Winans, R., and Bunch, T.E. (1994). Fullerenes in Allende meteorite. *Nature* 372: 507. https://doi.org/10.1038/372507a0.

42 Buseck, P.R., Tsipursky, S.J., and Hettich, R. (1992). Fullerenes from the geological environment. *Science (New York, NY)* 257 (5067): 215–217. https://doi.org/10.1126/science.257.5067.215.

43 Andrade, K., Guerra, S., and Debut, A. (2014). Fullerene-based symmetry in *Hibiscus rosa-sinensis* pollen. *PLoS One* 9 (7): e102123. https://doi.org/10.1371/journal.pone.0102123.

44 Honek, J.F. (2013). Bionanotechnology and bionanomaterials: John Honek explains the good things that can come in very small packages. *BMC Biochem.* 14 (1): 29. https://doi.org/10.1186/1471-2091-14-29.

45 Adewumi, G., Eboka, A., and Inambao, F. (2017). Thermal conductivity and viscosity of bio-based carbon nanotubes: review. https://www.ijrer.org/ijrer/index.php/ijrer/article/view/6253/pdf (accessed 29 December 2020).

46 Singh, G., Chandoha-Lee, C., Zhang, W. et al. (2016). Biodegradation of nanocrystalline cellulose by two environmentally-relevant consortia. *Water Res.* 104: 137–146.

47 Neouze, M.-A. and Schubert, U. (2008). Surface modification and functionalization of metal and metal oxide nanoparticles by organic ligands. *Monatsh. Chem.* 139 (3): 183–195.

48 Wang, J., Zheng, S., Shao, Y. et al. (2010). Amino functionalized $Fe_3O_4@SiO_2$ core–shell magnetic nanomaterial as a novel adsorbent for aqueous heavy metals removal. *J. Colloid Interface Sci.* 349 (1): 293–299.

49 Wang, X., Liu, L.-H., Ramström, O., and Yan, M. (2009). Engineering nanomaterial surfaces for biomedical applications. *Exp. Biol. Med.* 234 (10): 1128–1139.

50 Abdolmaleki, A., Mallakpour, S., and Karshenas, A. (2017). Facile synthesis of glucose-functionalized reduced graphene oxide (GFRGO)/poly(vinyl alcohol) nanocomposites for improving thermal and mechanical properties. *Mater. Sci. Eng., B* 217: 26–35.

51 Abdolmaleki, A., Mallakpour, S., and Karshenas, A. (2017). Synthesis and characterization of new nanocomposites films using alanine-Cu-functionalized graphene oxide as nanofiller and PVA as polymeric matrix for improving their properties. *J. Solid State Chem.* 253: 398–405.

52 Abdolmaleki, A., Mallakpour, S., and Mahmoudian, M. (2017). Preparation and evaluation of edge selective sulfonated graphene by chlorosulfuric acid as an active metal-free electrocatalyst for oxygen reduction reaction in lkaline media. *ChemistrySelect* 2 (34): 11211–11217.

53 Gautam, R.K. and Chattopadhyaya, M.C. (2016). *Nanomaterials for Wastewater Remediation*. Oxford: Butterworth-Heinemann.

54 Chowdhury, A.K.M.R.H., Tavangar, A., Tan, B., and Venkatakrishnan, K. (2017). Biofunctionalized 3-D carbon nano-network platform for enhanced fibroblast cell adhesion. *Sci. Rep.* 7 (1) https://doi.org/10.1038/srep44250.

55 Datta, P. and Genzer, J. (2016). "Grafting through" polymerization involving surface-bound monomers. *J. Polym. Sci., Part A: Polym. Chem.* 54 (2): 263–274.

56 Sejoubsari, R.M., Martinez, A.P., Kutes, Y. et al. (2016). "Grafting-through": growing polymer brushes by supplying monomers through the surface. *Macromolecules* 49 (7): 2477–2483.

57 Georgakilas, V., Otyepka, M., Bourlinos, A.B. et al. (2012). Functionalization of graphene: covalent and non-covalent approaches, derivatives and applications. *Chem. Rev.* 112 (11): 6156–6214.

58 Zhao, J., Lu, J.P., Han, J., and Yang, C.-K. (2003). Noncovalent functionalization of carbon nanotubes by aromatic organic molecules. *Appl. Phys. Lett.* 82 (21): 3746–3748.

59 Grubbs, R.B. (2007). Roles of polymer ligands in nanoparticle stabilization. *Polym. Rev.* 47 (2): 197–215.

60 Hussain, C.M. and Kecili, R. (2019). *Modern Environmental Analysis Techniques for Pollutants*. Elsevier.

61 Chen, K., Lu, G., Chang, J. et al. (2012). Hg(II) ion detection using thermally reduced graphene oxide decorated with functionalized gold nanoparticles. *Anal. Chem.* 84 (9): 4057–4062.

62 Lu, Y., Goldsmith, B.R., Kybert, N.J., and Johnson, A.T.C. (2010). DNA decorated graphene chemical sensors. *Appl. Phys. Lett.* 97 (8): 083107.

63 Palit, S. and Hussain, C.M. (2020). Functionalization of nanomaterials for industrial applications: recent and future perspectives. In: *Handbook of Functionalized Nanomaterials for Industrial Applications* (ed. C.M. Hussain), 3–14. Elsevier.

64 Jasmani, L. and Thielemans, W. (2018). Preparation of nanocellulose and its potential application. *Forest Res.* 7: 222.

65 Jin, S.F., Chen, Y., and Liu, M. (2013). Nanocellulose applications in environmental protection. *Adv. Mater. Res.* 662: 198–201. https://doi.org/10.4028/www.scientific.net/amr.662.198.

66 Soni, B., Hassan, E.B., Schilling, M.W., and Mahmoud, B. (2016). Transparent bionanocomposite films based on chitosan and TEMPO-oxidized cellulose nanofibers with enhanced mechanical and barrier properties. *Carbohydr. Polym.* 151: 779–789. https://doi.org/10.1016/j.carbpol.2016.06.022.

67 Pruden, A., Vikesland, P., and Renneckar, S. (2012). Environmental implications of nanocellulose: biodegradation and toxicity potential. http://grantome.com/ grant/NSF/CBET-1236005 (accessed 29 December 2020).

68 Ruiz-Palomero, C., Soriano, M.L., and Valcárcel, M. (2016). Nanocellulose as analyte and analytical tool: opportunities and challenges. *Trends Anal. Chem.* 87: 1–18.

69 Filson, P.B., Dawson-Andoh, B.E., and Schwegler-Berry, D. (2009). Enzymaticmediated production of cellulose nanocrystals from recycled pulp. *Green Chem.* 11: 1808–1814.

70 Hayashi, N., Kondo, T., and Ishihara, M. (2005). Enzymatically produced nanoordered short elements containing cellulose Iβ crystalline domains. *Carbohydr. Polym.* 61: 191–197.

71 Håkansson, H. and Ahlgren, P. (2005). Acid hydrolysis of some industrial pulps: effect of hydrolysis conditions and raw material. *Cellulose* 12: 177–183.

72 Turbak, A.F., Snyder, F.W., and Sandberg, K.R. (1983). Microfibrillated cellulose, a new cellulose product: properties, uses, and commercial potential. *J. Appl. Polym. Sci.* 37: 325–343.

73 Dufresne, A. and Belgacem, M.N. (2013). Cellulose-reinforced composites from micro-to nanoscale. *Polímeros* 23 (3): 227–286.

74 Atalla, R.H. (2000). The individual structures of native celluloses. *Lenzinger Ber.* 79: 5–14.

75 Sheikhi, A., Kakkar, A., and Van de Ven, T.G.M. (2016). A leaf out of nature's book: hairy nanocelluloses for bioinspired mineralization. *Cryst. Growth Des.* 16 (8): 4627–4634.

76 Van de Ven, T.G.M. and Sheikhi, A. (2016). Hairy nanocellulose nano crystalloids: a novel class of nanocellulose. *Nanoscale* 8: 15101–15114.

77 Jozala, A.F., de Lencastre-Novaes, L.C., Lopes, A.M. et al. (2016). Bacterial nanocellulose production and application: a 10-year overview. *Appl. Microbiol. Biotechnol.* 100: 2063–2072.

78 Zheng, X., Pan, J., Zhang, F. et al. (2016). Fabrication of free-standing bio-template mesoporous hybrid film for high and selective phosphate removal. *J. Chem. Eng. Data* 284: 879–887.

79 Rajasekhar, C. and Kanchi, S. (2018). Green nanomaterials for clean environment. In: *Handbook of Ecomaterials*, (eds. L.M. Torres Martínez, O.V. Kharissova, and B.I. Kharisov). 1–18. Springer. https://doi.org/10.1007/978-3-319-48281-1_73-1.

80 Jin, L., Li, W., Xu, Q., and Sun, Q. (2015). Amino-functionalized nanocrystalline cellulose as an adsorbent for anionic dyes. *Cellulose* 22: 2443–2456.

81 Karim, Z., Mathew, A.P., Grahn, M. et al. (2014). Nanoporous membranes with cellulose nanocrystals as functional entity in chitosan: removal of dyes from water. *Carbohydr. Polym.* 112: 668–676.

82 Hussain, C.M. and Kharisov, B. (eds.) (2016). *Advanced Environmental Analysis: Applications of Nanomaterials*, vols. 1 and 9. Royal Society of Chemistry.

83 Hussain, C.M. and Hussain, C.G. (2018). Future of industrial development and nanomaterials (concluding notes). In: *Handbook of Nanomaterials for Industrial Applications* (ed. C.M. Hussain), 1073–1076. Elsevier.

84 Hussain, C.M. and Mishra, A.K. (2018). *Nanotechnology in Environmental Science*, vols. 1 and 2. Wiley.

85 Liu, P., Sehaqui, H., Tingaut, P. et al. (2013). Cellulose and chitin nanomaterials for capturing silver ions (Ag^+) from water via surface adsorption. *Cellulose* 21: 449–461.

86 Suopajärvi, T., Liimatainen, H., Karjalainen, M. et al. (2015). Lead adsorption with sulfonated wheat pulp nanocelluloses. *J. Water Process Eng.* 5: 136–142.

87 Nomngongo, P.N., Ngila, J.C., Musyoka, S.M. et al. (2013). A solid phase extraction procedure based on electrospun cellulose-goxolane-2,5-dione nanofibers for trace determination of Cd, Cu, Fe, Pb and Zn in gasoline samples by ICP-OES. *Anal. Methods* 5: 3000–3008.

88 Sirvio, J.A., Hasa, T., Leiviska, T. et al. (2016). Bisphosphonate nanocellulose in the removal of vanadium(V) from water. *Cellulose* 23: 689–697.

89 Stephen, M., Catherine, N., Brenda, M. et al. (2011). Oxolane-2,5-dione modified electrospun cellulose nanofibers for heavy metals adsorption. *J. Hazard. Mater.* 192: 922–927.

90 Taha, A.A., Wu, Y., Wang, H., and Li, F. (2012). Preparation and application of functionalized cellulose acetate/silica composite nanofibrous membrane via electrospinning for Cr(VI) ion removal from aqueous solution. *J. Environ. Manage.* 112: 10–16.

91 Sato, A., Wang, R., Ma, H.Y. et al. (2011). Novel nanofibrous scaffolds for water filtration with bacteria and virus removal capability. *J. Electron Microsc.* 60 (3): 201–209.

92 Zadeh, M.H.B. and Shahdadi, H. (2015). Nanocellulose coated with various free fatty acids can adsorb fumonisin B1. *Colloids Surf.* 134: 26–30.

93 Du, X., Zhang, Z., Liu, W., and Deng, Y. (2017). Nanocellulose-based conductive materials and their emerging applications in energy devices – a review. *Nano Energy* 35: 299–320. https://doi.org/10.1016/j.nanoen.2017.04.001.

94 Zhou, Y., Fuentes-Hernandez, C., Khan, T.M. et al. (2013). Recyclable organic solar cells on cellulose nanocrystal substrates. *Sci. Rep.* 3: 1536. https://doi.org/10.1038/srep01536.

95 Zhou, Y., Khan, T.M., Liu, J.-C. et al. (2014). Efficient recyclable organic solar cells on cellulose nanocrystal substrates with a conducting polymer top electrode deposited by film-transfer lamination. *Org. Electron.* 15 (3): 661–666. https://doi.org/10.1016/j.orgel.2013.12.018.

96 Völkel, L., Ahn, K., Hähner, U. et al. (2017). Nano meets the sheet: adhesive-free application of nanocellulosic suspensions in paper conservation. *Herit. Sci.* 5 (1) https://doi.org/10.1186/s40494-017-0134-5.

97 Manikkam, V. (2018). What are the applications of nanocellulose in packaging? – PreScouter – custom intelligence from a Global Network of Experts. PreScouter. https://www.prescouter.com/2018/02/nanocellulose-applications-packaging/ (accessed 29 December 2020).

98 Abitbol, T., Rivkin, A., Cao, Y. et al. (2016). Nanocellulose, a tiny fiber with huge applications. *Curr. Opin. Biotechnol.* 39: 76–88. https://doi.org/10.1016/j.copbio.2016.01.002.

99 Klemm, D., Schumann, D., Udhardt, U., and Marsch, S. (2001). Bacterial synthesized cellulose – artificial blood vessels for microsurgery. *Prog. Polym. Sci.* 26 (9): 1561–1603. https://doi.org/10.1016/s0079-6700(01)00021-1.

100 Klemm, D., Schumann, D., Kramer, F. et al. (2006). Nanocelluloses as innovative polymers in research and application. *Polysaccharides II Adva. Polym. Sci.* 205: 49–96. https://doi.org/10.1007/12_097.

101 Dong, S. and Roman, M. (2007). Fluorescently labeled cellulose nanocrystals for bioimaging applications. *J. Am. Chem. Soc.* 129 (45): 13810–13811. https://doi.org/10.1021/ja076196l.

102 Fang, Y., Wang, H., Yu, H., and Peng, F. (2016). From chicken feather to nitrogen and sulfur co-doped large surface bio-carbon flocs: an efficient electrocatalyst for oxygen reduction reaction. *Electrochim. Acta* 213: 273–282. https://doi.org/10.1016/j.electacta.2016.07.121.

103 Nan, N., DeVallance, D.B., Xie, X., and Wang, J. (2015). The effect of bio-carbon addition on the electrical, mechanical, and thermal properties of polyvinyl alcohol/biochar composites. *J. Compos. Mater.* 50 (9): 1161–1168. https://doi.org/10.1177/0021998315589770.

104 Xie, X., Goodell, B., Qian, Y. et al. (2008). Significance of the heating rate on the physical properties of carbonized maple wood. *Holzforschung* 62 (5) https://doi.org/10.1515/hf.2008.090.

105 Lehmann, J., and Joseph, S. (eds.). (2012). Biochar for Environmental Management: Science and Technology. Routledge.

106 Ahmetli, G., Kocaman, S., Ozaytekin, I., and Bozkurt, P. (2013). Epoxy composites based on inexpensive char filler obtained from plastic waste and natural resources. *Polym. Compos.* 34 (4): 500–509. https://doi.org/10.1002/pc.22452.

107 Peterson, S.C. (2012). Utilization of low-ash biosschar to partially replace carbon black in styrene–butadiene rubber composites. *J. Elastomers Plast.* 45 (5): 487–497. https://doi.org/10.1177/0095244312459181.

108 Das, O., Sarmah, A.K., and Bhattacharyya, D. (2015). A novel approach in organic waste utilization through biochar addition in wood/polypropylene composites. *Waste Manage. (New York, NY)* 38 https://doi.org/10.1016/j.wasman.2015.01.015.

109 Li, Y., Zhang, Q., Zhang, J. et al. (2015). A top-down approach for fabricating free-standing bio-carbon supercapacitor electrodes with a hierarchical structure. *Sci. Rep.* 5 (1) https://doi.org/10.1038/srep14155.

110 Mishra, R.K., Ha, S.K., Verma, K., and Tiwari, S.K. (2018). Recent progress in selected bio-nanomaterials and their engineering applications: an overview. *J. Sci. Adv. Mater. Devices* https://doi.org/10.1016/j.jsamd.2018.05.003.

111 Mohan, S., Oluwafemi, O.S., Kalarikkal, N. et al. (2016). Biopolymers – application in nanoscience and nanotechnology. In: *Recent Advances in Biopolymers* (ed. F.K. Perveen), 47–66. https://doi.org/10.5772/62225.

112 Arputha Kiruba, V.S., Dakshinamurthy, A., and Selvakumar, P.M. (2013). Eco-friendly biocidal silver-activated charcoal nanocomposite: antimicrobial application in water purification. *Synth. React. Inorg. Met.-Org. Nano-Met. Chem.* 43 (8): 1068–1072. https://doi.org/10.1080/15533174.2012.749906.

113 Atiyeh, B.S., Costagliola, M., Hayek, S.N., and Dibo, S.A. (2007). Effect of silver on burn wound infection control and healing: review of the literature. *Burns* 33 (2): 139–148. https://doi.org/10.1016/j.burns.2006.06.010.

114 Mohan, S., Oluwafemi, O.S., George, S.C. et al. (2014). Completely green synthesis of dextrose reduced silver nanoparticles, its antimicrobial and sensing properties. *Carbohydr. Polym.* 106: 469–474. https://doi.org/10.1016/j.carbpol.2014.01.008.

115 Abol-Fotouh, D., Dörling, B., Zapata-Arteaga, O., et al. (2019). Farming thermoelectric paper. *Energy Environ.* Sci. 12 (2), 716–726.

116 Camargo, P.H.C., Satyanarayana, K.G., and Wypych, F. (2009). Nanocomposites: synthesis, structure, properties and new application opportunities. *Mater. Res.* 12: 1–39.

117 Braun, T., Schubert, A., and Sindelys, Z. (1997). Nanoscience and nanotechnology on the balance. *Scientometrics* 38: 321–325.

118 Khare, P., Patel, R.K., and Shankar, R. (2019). Recent developments in nanocarbon-polymer composites for environmental and energy applications. In: *Green Chemistry Applications* (ed. M. Eyvaz and E. Yüksel). IntechOpen.

119 Khare, R. (2005). Carbon nanotube based composites: a review. *J. Miner. Mater. Charact. Eng.* 4 (01): 31.

120 Sanchez, C., Julian, B., Belleville, P., and Popal, M. (2005). Applications of hybrid organic–inorganic nanocomposites. *J. Mater. Chem.* 15: 3559–3592.

121 Yao, H.B., Fang, H.Y., Wang, X.H., and Yu, S.H. (2011). Hierarchical assembly of micro-/nanobuilding blocks: bio-inspired rigid structural functional materials. *Chem. Soc. Rev.* 40: 3764.

122 Mallakpour, S. and Behranvand, V. (2018). Synthesis of mesoporous recycled poly(ethylene terephthalate)/MWNT/carbon quantum dot nanocomposite from sustainable materials using ultrasonic waves: application for methylene blue removal. *J. Cleaner Prod.* 190: 525–537.

123 Mallakpour, S. and Khadem, E. (2018). Construction of crosslinked chitosan/nitrogen-doped graphene quantum dot nanocomposite for hydroxyapatite biomimetic mineralization. *Int. J. Biol. Macromol.* 120: 1451–1460.

124 Mallakpour, S. and Khadem, E. (2016). Carbon nanotube–metal oxide nanocomposites: fabrication, properties and applications. *Chem. Eng. J.* 302: 344–367.

125 Mallakpour, S. and khodadadzadeh L. (2018). Chapter 7: Biocompatible and biodegradable chitosan nanocomposites loaded with carbon nanotubes. In: *Biodegradable and Biocompatible Polymer Composites Processing, Properties and Applications* (ed. N.G. Shimpi), 187–221. Sawston, Cambridge: Elsevier, Woodhead Publishing. https://doi.org/10.1016/B978-0-08-100970-3.00007-9.

126 Abdolmaleki, A., Mallakpour, S., and Borandeh, S. (2016). Improving interfacial interaction of l-phenylalanine-functionalized graphene nanofiller and poly(vinyl alcohol) nanocomposites for obtaining significant membrane properties: morphology, thermal, and mechanical studies. *Polym. Compos.* 37 (6): 1924–1935.

127 Benshitrit, R.C., Levi, C.S., Tal, S.L. et al. (2012). Development of oral food-grade delivery systems: current knowledge and future challenges. *Food Funct.* 3 (1): 10–21. https://doi.org/10.1039/c1fo10068h.

128 Singh, A., Jain, D., Upadhyay, M. et al. (2010). Green synthesis of silver nanoparticles using *Argemone Mexicana* leaf extract and evaluation of their antimicrobial activities. *Dig. J. Nanomater. Biostruct.* 5: 483–489.

129 Sivaraman, S., Elango, I., Kumar, S., and Santhanam, V. (2009). A green protocol for room temperature synthesis of silver nanoparticles in seconds. *Curr. Sci.* 97. (7): 00113891

130 Sherly Arputha Kiruba, V., Dakshinamurthy, A., Subramanian, P.S., and Mosae Selvakumar, P. (2015). Green synthesis of biocidal silver-activated charcoal nanocomposite for disinfecting water. *J. Exp. Nanosci.* 10 (7): 532–544. https://doi.org/10.1080/17458080.2013.848295.

131 Kiruba, V.S.A., Selvakumar, P.M., and Dakshinamurthy, A. (2014). Biocidal nano-silver reinforced activated charcoal in water treatment. *Synth. React. Inorg. Met.-Org. Nano-Met. Chem.* 45 (10): 1570–1575. https://doi.org/10.1080/15533174.2013.865221.

132 Mallakpour, S. and Khadem, E. (2019). Linear and nonlinear behavior of crosslinked chitosan/N-doped graphene quantum dot nanocomposite films in cadmium cation uptake. *Sci. Total Environ.* 690: 1245–1253.

133 Rad-Malekshahi, M., Lempsink, L., Amidi, M. et al. (2016). Biomedical applications of self-assembling peptides. *Bioconjugate Chem.* 27 (1): 3–18. https://doi.org/10.1021/acs.bioconjchem.5b00487.

134 Spicer, C.D., Jumeaux, C., Gupta, B., and Stevens, M.M. (2018). Peptide and protein nanoparticle conjugates: versatile platforms for biomedical applications. *Chem. Soc. Rev.* 47 (10): 3574–3620. https://doi.org/10.1039/c7cs00877e.

135 Mcgimpsey, W. G. (2005). Cyclic peptide structures for molecular scale electronic and photonic devices. U.S. Patent No. 6 902 720. Washington, DC: U.S. Patent and Trademark Office. https://patents.google.com/patent/US6902720B2/en?q=Cyclic+peptide+nanotube+structures+molecular+scale+electronic+and+photonic+devices&language=ENGLISH#patentCitations (accessed 29 December 2020).

136 Lévy, R., Thanh, N.T.K., Doty, R.C. et al. (2004). Rational and combinatorial design of peptide capping ligands for gold nanoparticles. *J. Am. Chem. Soc.* 126 (32): 10076–10084. https://doi.org/10.1021/ja0487269.

137 Liu, L., Xu, K., Wang, H. et al. (2009). Self-assembled cationic peptide nanoparticles as an efficient antimicrobial agent. *Nat. Nanotechnol.* 4 (7): 457–463. https://doi.org/10.1038/nnano.2009.153.

138 Xu, X., Yuan, H., Chang, J. et al. (2012). Cooperative hierarchical self-assembly of peptide dendrimers and linear polypeptides into nanoarchitectures mimicking viral capsids. *Angew. Chem. Int. Ed.* 51 (13): 3130–3133. https://doi.org/10.1002/anie.201106080.

139 Burkhard, P., Meier, M., and Lustig, A. (2000). Design of a minimal protein oligomerization domain by a structural approach. *Protein Sci.* 9 (12): 2294–2301. https://doi.org/10.1110/ps.9.12.2294.

140 Wu, Y. and Collier, J.H. (2017). α-Helical coiled-coil peptide materials for biomedical applications. *Wiley Interdiscip. Rev. Nanomed. Nanobiotechnol.* 9 (2) https://doi.org/10.1002/wnan.1424.

141 Apostolovic, B., Danial, M., and Klok, H.A. (2010). Coiled coils: attractive protein folding motifs for the fabrication of self-assembled, responsive and bioactive materials. *Chem. Soc. Rev.* 39 (9): 3541–3575. https://doi.org/10.1039/b914339b.

142 Stevens, M.M., Flynn, N.T., Wang, C. et al. (2004). Coiled-coil peptide-based assembly of gold nanoparticles. *Adv. Mater.* 16 (11): 915–918. https://doi.org/10.1002/adma.200306430.

143 Boato, F., Thomas, R.M., Ghasparian, A. et al. (2007). Synthetic virus-like particles from self-assembling coiled-coil lipopeptides and their use in antigen display to the immune system. *Angew. Chem. Int. Ed.* 46 (47): 9015–9018. https://doi.org/10.1002/anie.200702805.

144 Kaba, S.A., Brando, C., Guo, Q. et al. (2009). A nonadjuvanted polypeptide nanoparticle vaccine confers long-lasting protection against rodent malaria. *J. Immunol. (Baltimore, MD: 1950)* 183 (11): 7268–7277. https://doi.org/10.4049/jimmunol.0901957.

145 Pimentel, T.A., Yan, Z., Jeffers, S.A. et al. (2009). Peptide nanoparticles as novel immunogens: design and analysis of a prototypic severe acute respiratory syndrome vaccine. *Chem. Biol. Drug Design* 73 (1): 53–61. https://doi.org/10.1111/j.1747-0285.2008.00746.x.

146 Wahome, N., Pfeiffer, T., Ambiel, I. et al. (2012). Conformation-specific display of 4E10 and 2F5 epitopes on self-assembling protein nanoparticles as a potential HIV vaccine. *Chem. Biol. Drug Des.* 80 (3): 349–357. https://doi.org/10.1111/j.1747-0285.2012.01423.x.

147 Adami, R.C. and Rice, K.G. (1999). Metabolic stability of glutaraldehyde cross-linked peptide DNA condensates. *J. Pharm. Sci.* 88 (8): 739–746. https://doi.org/10.1021/js990042p.

148 Horne, D.S. (2002). Casein structure, self-assembly and gelation. *Curr. Opin. Colloid Interface Sci.* 7 (5–6): 456–461. https://doi.org/10.1016/s1359-0294(02)00082-1.

149 Bachar, M., Mandelbaum, A., Portnaya, I. et al. (2012). Development and characterization of a novel drug nanocarrier for oral delivery, based on self-assembled β-casein micelles. *J. Controlled Release* 160 (2): 164–171. https://doi.org/10.1016/j.jconrel.2012.01.004.

150 Hartgerink, J.D., Granja, J.R., Milligan, R.A., and Ghadiri, M.R. (1996). Self-assembling peptide nanotubes. *J. Am. Chem. Soc.* 118 (1): 43–50. https://doi.org/10.1021/ja953070s.

151 Fan, T., Yu, X., Shen, B., and Sun, L. (2017). Peptide self-assembled nanostructures for drug delivery applications. *J. Nanomater.* 2017: 1–16. https://doi.org/10.1155/2017/4562474.

152 Habibi, N., Kamaly, N., Memic, A., and Shafiee, H. (2016). Self-assembled peptide-based nanostructures: smart nanomaterials toward targeted drug delivery. *Nano Today* 11 (1): 41–60. https://doi.org/10.1016/j.nantod.2016.02.004.

153 Hamley, I.W. (2011). Self-assembly of amphiphilic peptides. *Soft Matter* 7 (9): 4122. https://doi.org/10.1039/c0sm01218a.

154 Rubert Pérez, C.M., Stephanopoulos, N., Sur, S. et al. (2015). The powerful functions of peptide-based bioactive matrices for regenerative medicine. *Ann. Biomed. Eng.* 43 (3): 501–514. https://doi.org/10.1007/s10439-014-1166-6.

155 Gao, X. and Matsui, H. (2005). Peptide-based nanotubes and their applications in bionanotechnology. *Adv. Mater.* 17 (17): 2037–2050. https://doi.org/10.1002/adma.200401849.

156 Goode, J.A., Rushworth, J.V., and Millner, P.A. (2015). Biosensor regeneration: a review of common techniques and outcomes. *Langmuir* 31 (23): 6267–6276. https://doi.org/10.1021/la503533g.

157 Turner, A.P.F. (2013). Biosensors: sense and sensibility. *Chem. Soc. Rev.* 42 (8): 3184. https://doi.org/10.1039/c3cs35528d.

158 Vigneshvar, S. and Senthilkumaran, B. (2018). Current technological trends in biosensors, nanoparticle devices and biolabels: Hi-tech network sensing applications. *Med. Devices Sens.*: e10011. https://doi.org/10.1002/mds3.10011.

159 Carlson, M.A., Bargeron, C.B., Benson, R.C. et al. (2000). An automated, handheld biosensor for aflatoxin. *Biosens. Bioelectron.* 14 (10–11): 841–848. https://doi.org/10.1016/s0956-5663(99)00057-3.

160 Abouelamaiem, D.I., He, G., Parkin, I. et al. (2018). Synergistic relationship between the three-dimensional nanostructure and electrochemical performance in biocarbon supercapacitor electrode materials. *Sustain. Energy Fuels* 2 (4): 772–785. https://doi.org/10.1039/c7se00519a.

161 Lin, X.-Q., Lü, Q.-F., Li, Q. et al. (2018). Fabrication of low-cost and ecofriendly porous biocarbon using konjaku flour as the raw material for high-performance supercapacitor application. *ACS Omega* 3 (10): 13283–13289. https://doi.org/10.1021/acsomega.8b01718.

162 Barrett, C.S., Cohen, J.B., Faber, J. et al. (1986). *Advances in X-Ray Analysis*, vol. 29. New York: Plenum Press.

163 Klug, H.P. and Alexander, L.E. (1974). *X-Ray Diffraction Procedures for Polycrystalline Amorphous Materials*, 2e. New York: Wiley.

164 Kneipp, K., Wang, Y., Kneipp, H. et al. (1997). Single molecule detection using surface-enhanced Raman scattering (SERS). *Phys. Rev. Lett.* 78: 1667–1670.

165 Poinern, G.E.J. (2014). *A Laboratory Course in Nanoscience and Nanotechnology*, 1e. Boca Raton, FL: CRC Press Taylor & Francis.

166 Jiang, J., Oberdörster, G., and Biswas, P. (2009). Characterization of size, surface charge, and agglomeration state of nanoparticle dispersions for toxicological studies. *J. Nanopart. Res.* 11: 77–89.

167 Balavigneswaran, C.K., Kumar, T.S.J., Packiaraj, R.M., and Prakash, S. (2014). Rapid detection of Cr(VI) by AgNPs probe produced by *Anacardium occidentale* fresh leaf extracts. *Appl. Nanosci.* 4: 367–378.

168 Elavarasi, M., Rajeshwari, A., Chandrasekaran, N., and Mukherjee, A. (2013). Simple colorimetric detection of Cr^{3+} in aqueous solution by as synthesized citrate capped gold nanoparticles and development of a paper based assay. *Anal. Methods* 5: 6211–6218.

169 Fargasova, A. (2008). Phytotoxicity of chromium and nickel. *Ecol. Chem. Eng. S* 15: 335–348.

170 Farhadi, K., Forougha, M., Molaeia, R. et al. (2012). Highly selective Hg^{2+} colorimetric sensor using green synthesized and unmodified silver nanoparticles. *Sens. Actuators, B* 161: 880–885.

171 Ha, W., Yu, J., Wang, R. et al. (2014). "Green" colorimetric assay for the selective detection of trivalent chromium based on *Xanthoceras sorbifolia* tannin attached to gold nanoparticles. *Anal. Methods* 6: 5720–5726.

172 Zhang, J., Zhang, L., Wei, Y. et al. (2013). A selective carbazole based fluorescent probe for Cr^{3+}. *Anal. Methods* 5: 5549–5554.

173 Liu, B.H., Tsao, Z.J., Wang, J.J., and Yu, F.Y. (2008). Development of a monoclonal antibody against ochratoxin A and its application in enzyme-linked immunosorbent assay and gold nanoparticle immunochromatographic strip. *Anal. Chem.* 80: 7029–7035.

174 Shim, W.B.O., Kim, K.Y., and Chung, D.H. (2009). Development and validation of a gold nanoparticle immunochromatographic assay (ICG) for the detection of zearalenone. *J. Agric. Food Chem.* 57: 4035–4041.

175 Wang, X.H., Liu, T., Xu, N. et al. (2007). Enzyme-linked immunosorbent assay and colloidal gold immunoassay for ochratoxin A: investigation of analytical conditions and sample matrix on assay performance. *Anal. Bioanal.Chem.* 289: 903–911.

176 Uzawa, H., Ohga, K., Shinozaki, Y. et al. (2008). A novel sugar-probe biosensor for the deadly plant proteinous toxin, ricin. *Biosens. Bioelectron.* 24: 923–927.

16

Carbon-Based Nanomaterials in Drug Delivery Systems

Subhendu Chakroborty[1] *and Suban K. Sahoo*[2]

[1] *Auro University, School of Sciences, Surat, Gujarat, 394510, India*
[2] *Sardar Vallabhbhai National Institute of Technology (SVNIT), Department of Applied Chemistry, Surat, Gujarat, 395007, India*

16.1 Introduction

With the advent of nanotechnology, various allotropes and forms of carbon were developed (Figure 16.1). The carbon-based nanomaterials [1–3] have gained high popularity in the arena of drug delivery [4], bioimaging [5], diagnostic [6], cancer therapy [7], and biosensing [8], owing to their unusual outstanding chemical versatility [9] and mechanical properties [10]. Besides, the ease of tunable surface modification [11] makes them attractive drug carriers for intracellular transport. Carbon nanomaterials (CNMs) with unique nanostructure are the highly popular carrier in the arena of drug delivery [12–15]. These provide a huge surface area [16] and unique magnetic [17], optical [18], and electrical properties [19]. In the past decades, the applications of nanostructured carbon materials have boosted rapidly in the various domains. The most valuable application of CNMs is in the development of nanocarriers to deliver specific drug molecules to target cells.

The most famous explored CNMs as a modern drug delivery system (DDS) are carbon nanotubes (CNTs), carbon dots (CDs), graphene oxide (GO), and nanodiamond (ND). Although these types of CNMs vary in their morphology, they all are nanostructured materials with various compelling qualities that promote them as an outstanding drug delivery candidates. Some of the notable features include dispersal of the CNMs in aqueous solutions [20] and cellular uptake of CNMs via endocytosis pathways [21] along with its ability to directly translocate through the cell membranes.

16.2 Carbon Nanomaterials in Drug Delivery

16.2.1 Carbon Nanotubes (CNTs) in Drug Delivery

CNTs carry great prospective to this burgeoning area of DDSs and other potential applications in a wide range of fields [22–24], owing to their exceptional properties like high

Environmental Applications of Carbon Nanomaterials-Based Devices, First Edition.
Edited by Shadpour Mallakpour and Chaudhery Mustansar Hussain.

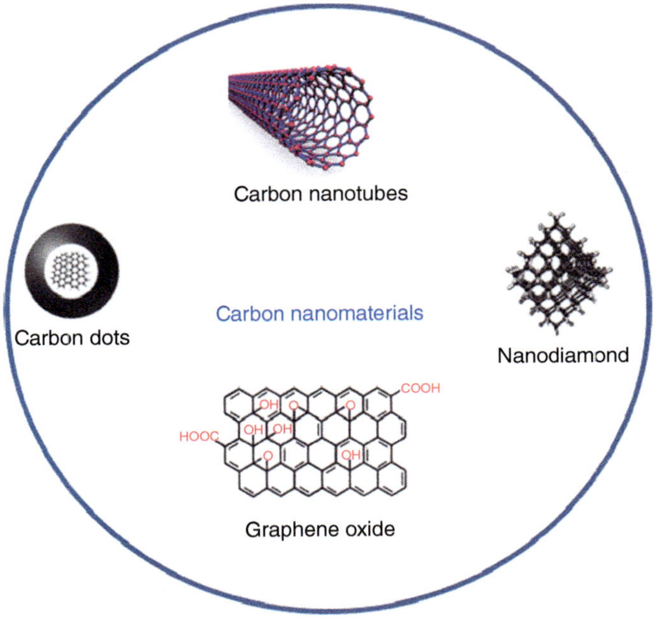

Figure 16.1 Carbon-based nanomaterials as a drug delivery carrier.

surface area, needlelike structure, photothermal ablation, and high loading capacity that strengthen them as a carriers in biological systems [25, 26]. Further, CNTs have the capability to cross many biologic barriers, passing via the plasma membrane and penetrating the cytoplasm through a "cellular needle" process, which enables the transport and delivery of the drug molecules into the target tissue and can be removed from the body through renal or fecal excretion. CNTs, the nanometer-sized tubes made of carbon, can be single-walled carbon nanotubes (SWCNTs), double-walled carbon nanotubes (DWCNTs), and multi-walled carbon nanotubes (MWCNTs). CNTs have the ability to attach with specific drugs by chemical bonding or adsorption. Apart from this, the low dispersion and toxicity of CNTs in biocompatible media can be overcome through a variety of surface functionalization or bioconjugation of the nanotubes [27, 28].

Lim and coworkers [29] reported a unique approach for the surface modification of SWCNTs for improving the DDS. The detailed preparation process of polyampholytic alternating polymers (PMT) and PMT/SWCNTs is illustrated in Figure 16.2. The doxorubicin (DOX) loading capacity was 150%. The resulting PMT/SWCNT hybrid system provides an initial burst release at pH 5.5, followed by controlled drug delivery at pH 7.4. The cytotoxicity effect was also demonstrated by comparison of PMT1/SWCNTs, DOX/PMT1/SWCNTs, and DOX toward HeLa cells by varying the concentration. In another work, Mandal and coworker [30] prepared a hybrid system containing silk protein and DOX-loaded folic acid (FA)-functionalized SWCNT-FA/DOX. This system exhibited pH-responsive and temperature-depending release of DOX and, thus, could represent a promising tool for the treatment for cancer. They also present excellent on-demand DOX release using intermittent near-infrared (NIR) light to the hybrid gel system. In addition, the covalently modified

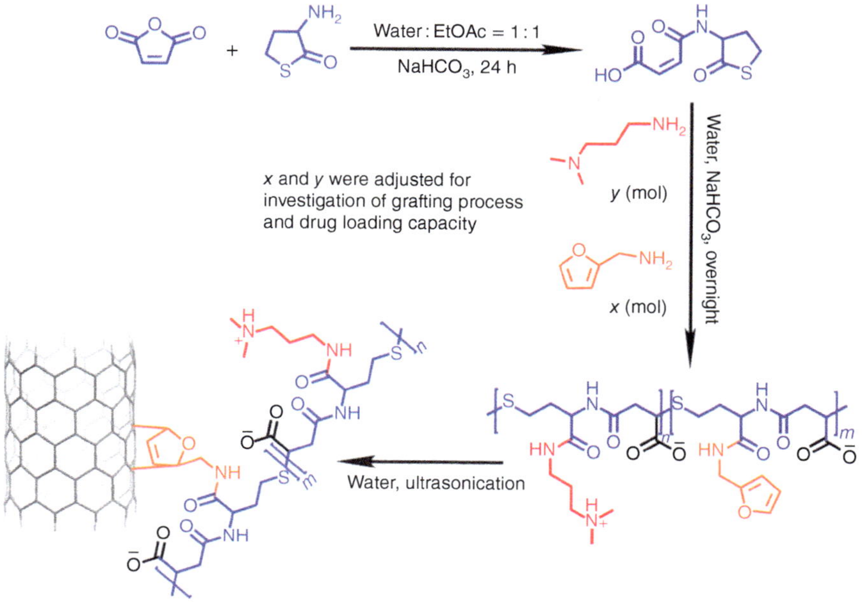

Figure 16.2 Preparation of polyampholytic PMT polymers and PMT/SWCNTs.

SWCNTs with poly(ethyleneglycol) (PEG) and polyethylenimine (PEI) were applied as a carrier for DOX delivery and the treatment for breast cancer cells [31]. The modified SWNTs possess an extremely high DOX loading capacity. The release of DOX was pH stimulated and independent of dilution. Thus, this system shows great potential for the treatment of cancer.

The chitosan (CS)-functionalized SWCNT was synthesized for enhancing nanocarriers in water [32]. To prepare the DDS, docetaxel (DTX) was loaded on SWCNT through π–π stacking interaction followed by attaching with arginine–glycine–aspartic (RGD)-coated CS to make RGD-CS-SWCNT-DTX and then evaluated as an effective DDS. The maximum release of DTX was observed at pH 5.0 at a release rate of 68%. *In vitro* studies show that RGD-CSSWCNT-DTX has better growth inhibition than CS-SWCNT-DTX against A549 cells. Overall results showed significant antitumor efficacies of the prepared nanocarrier systems in the animal models. Subsequently, Chen and coworkers reported the peptide-containing asparagine-glycine-arginine (NGR) motif linked with SWCNTs via a simple non-covalent technique [33]. In this system, DOX and Gd-DTPA were attached through π–π and *n*–π interactions, respectively. The *in vivo* experiment was done on S180-bearing mice model, and the biodistribution of DOX/NGR-SWCNTs/Gd-DTPA was observed by treating the model mice via tail-vein intravenous route injection. The *in vivo* antitumor efficiency results showed a higher apoptosis level for DOX/NGR-SWCNTs/Gd-DTPA group at 61.1%. Also, the signal intensity is increased during *in vivo* MRI, resulting in the change in MR imaging at the tumor site. In another work, Peighambardoust and coworkers [34] developed SWCNTs doped with iron oxide nanoparticles (Fe_3O_4-iSWCNT). In this system, iron oxide nanoparticles are decorated with poly(lactic acid)-*co*-methoxy polyethylene glycol copolymeric (PLA-co-mPEG) micelles and loaded with the puissant chemotherapy agent DTX for controlled release of drug delivery. To design this system, the polymeric micelle is prepared first and then sonicated with Fe_3O_4-SWNTs to form Fe_3O_4-SWNTs/PLA-mPEG nanocomposites. The *in vitro* experiment shows that the DTX release was 19.98% in the first 6.5 hours. The *in vitro* testing of the DTX-loaded nanocomposites showed significant anticancer effects toward MCF-7 cells. The cytotoxicity study on cancer cells in culture was also studied on the nanocomposite.

The levofloxacin (LVF) delivery system was developed using MWCNTs decorated with a liquid crystalline molecularly imprinted polymer (LC-MIP) as an oral floating drug delivery system (FDDS) [35]. MWCNTs were initially functionalized with 9-vinylanthracene (VA), and then polymerization of LC-MIPs was achieved on the surface of MWCNTs (MWCNT@LC-MIP) (Figure 16.3). The *in vitro* LVF release behavior showed release rate accelerated to 89.8% in 21 hours. The *in vivo* pharmacokinetic experiment was carried out on male Wistar rats, and the drug was administrated intragastrically. Based on the research data, this nanocarrier could be used as suitable candidates for gastroretentive DDS.

Wang et al. [36] developed an FA conjugated with oxidized large-inner-diameter multi-walled carbon nanotubes (O-LID-MWCNTs) for loading/release of antitumor agent cisplatin (CDDP). The drug entries were further blocked by the ultrashort oxidized LID-MWCNT (UST) nanoparticles. The resultant DDS was found to have excellent loading efficiency of 70.97%. The cytotoxicity experiment was estimated against CAL-27 cells and MCF-7 cells using CCK-8 assay, and the obtained data showed the inhibition of tumor cell proliferation with an IC_{50} value of 4.21 and 3.76 μM, respectively. In another work, Farahani group proposed a new DDS using PEG-functionalized MWNTs attached to DOX [37]. The *in vitro*

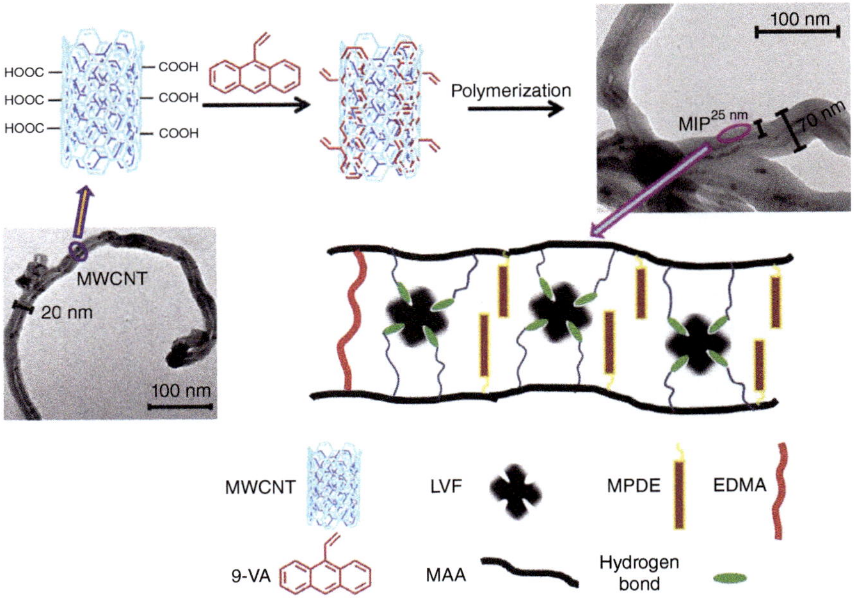

Figure 16.3 Schematic representation of MWCNT@LC-MIP preparation. Source: Adapted with permission from Zhang et al. [35]. Elsevier.

release experiment revealed that this DDS achieved a pH-dependent release profile, showing more release of DOX at pH 5.5.

16.2.2 Graphene Oxide (GO) in Drug Delivery

The GO and its derivatives turn out to be a "rising star" in the arena of DDS and other potential applications due to their excellent physiochemical properties and unique planar structure with sp^2 hybridization [38–41]. Moreover, GO has a water contact angle of 30.7°, which is efficient for constructing hydrogen bonds in the presence of carboxylate moiety on the edge site and hydroxide moiety on the basal plane. In reduced graphene oxide (rGO), some functional oxygen moieties are still maintained, rendering it less basal reactivity than GO [42]. Compared with CNTs, GO exhibits some crucial features such as easy fabrication and modification, absence of toxic metal particles, low cost, and high surface area with two external surfaces. In biomedical applications, both the oxidized and reduced forms of graphene are found to be viable for the DDSs. The conjugation of GO with various drug molecules can be tuned by physical adsorption via π–π stacking interaction or covalent cross-linking. GO demonstrates pH-dependent fluorescence emission in visible or NIR light, which can also be beneficial for pH-sensing and molecular imaging [43].

Recently, Namazi and coworkers [44] prepared magnetic graphene oxide (MGO) via chemical coprecipitation of Fe^{3+} and Fe^{2+} ions in the presence of GO, followed by fabrication of mesoporous silica ($mSiO_2$) to improve the loading capacity. The resulted $MGOSiO_2$ had a high specific surface area (375.62 m^2/g), and it exhibited a loading capacity of 15.7% for ibuprofen (IBU). The drug release profile showed that the burst release of IBU from $MGOmSiO_2$-IBU was 86% within the first two hours. The burst release problem was avoided by coating carboxymethyl cellulose (CMC) on $MGOmSiO_2$-IBU. Thus, the burst release of the drug enhanced to eight hours following the first-order kinetics through a non-Fickian mechanism. The vibrating sample magnetometric (VSM) results suggest that bio-nanocomposite hydrogel beads could be displaced employing an external magnetic field and can be guided to the target site.

Mauro et al. modified GO with PEG moiety and FA to form GO-PEG-Fol and to enhance drug delivery for cancer cells [45]. The obtained GO-PEG-Fol showed efficiently loading of the drug doxorubicin hydrochloride (DOX) at a rate of more than 33% and further demonstrated NIR light-stimulated heaters capable of activating local hyperthermia and doxo release. The results from the photothermal cytotoxic study showed that breast carcinoma cells, i.e. MCF7 and MDA-MB-231, were 3–12 times higher than healthy cells. In another work, Luan and coworkers [46] reported modified GO as a new nanovehicle composed of redox-sensitive prodrug molecules such as PEG, disulfide bond (—SS—), and polycaprolactone (PCL) to form graphene oxide/prodrug nanohybrids (GOPNs) using π–π stacking interactions and hydrophobic effect. The cellular uptake and cytotoxicity were investigated *in vitro* on different cancer cell lines A549 and B16. The results revealed that the GOPN system has higher cytotoxicity (A549, 24 hours; B16, 48 hours) and more effective cellular uptake, as shown from fluorescence intensity (red) on A549 cell. In the *in vivo* study with B16 cells and Kunming mice, GOPN was injected intravenously, resulting in the efficient delivery of DOX to tumor-bearing mice through an enhanced permeability and retention (EPR) effect.

Javanbakht et al. [47] prepared a novel DOX delivery system by modification of GO with CMC and zinc-based metal–organic framework (MOF-5) using a one-pot solvothermal method. The results of this experiment showed that the DOX release rate from the DDS in a tumor cell microenvironment (i.e. pH 5) was higher than that in normal conditions (i.e. pH 7.4). The cytotoxicity study of DOX@CMC/MOF-5/GO was performed on K562 cells using MTT assay showing significant cytotoxicity. To study the controlled release of DOX, Xie et al. [48] developed layer-by-layer functionalized MGO nanocomposites by coating with chitosan/sodium alginate to form MGO-CS/SA (Figure 16.4). The nanocomposites loaded with DOX through electrostatic attraction and π–π stacking interaction exhibit a high loading capacity (137%, w/w). The releasing rate of DOX from the MGO-CS/SA system was observed to be faster in acidic than normal physiological conditions showing characteristic pH-dependent releasing property. The resulting nanocomposites indicated magnetically enhanced cellular uptake, concentration-dependent cytotoxicity, and excellent photothermal properties. In another work, Li and Xie modified GO nanosheets by natural peptide protamine sulfate (PRM) and sodium alginate (SA) using a layer-by-layer self-assembly method to develop a cancer DDS [49]. The resulting system indicated suppression of protein adhesion and pH-sensitive release of DOX. The nanocomposites show a significant cellular uptake on MCF-7 breast cancer cells.

Palai et al. [50] established a green process to conjugate the PEGylated nanoceria with reduced graphene using aqueous leaf extract of *Azadirachta indica* (neem) and following syn-graphenization strategy to obtain rGO-CeNP-PEG nanocomposite as a nanocarrier for the delivery of DOX to the cancer cell. *In vitro* investigations indicated that the rate of DOX release was higher and controlled with low pH. The authors carried out *in vitro* cytotoxicity

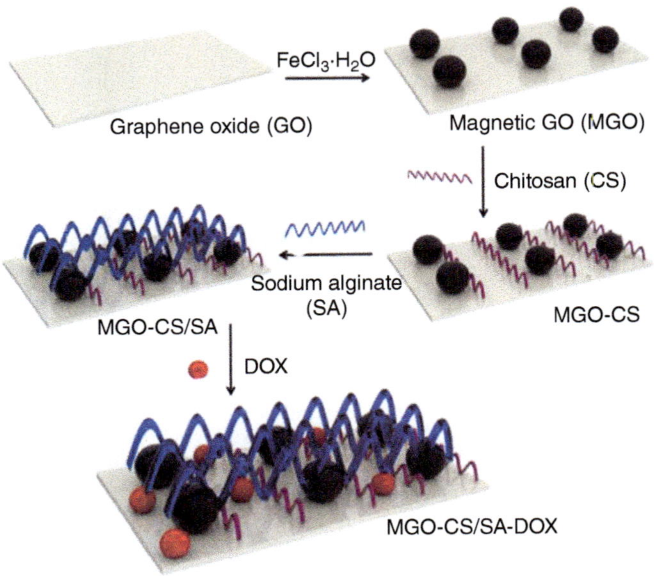

Figure 16.4 Depiction of the synthesis of MGO-CS/SA and DOX loading. Source: Adapted with permission from Xie et al. [48]. Elsevier.

and cellular uptake studies using HeLa cells to evaluate that this nanoplatform had enhanced antitumor activity. Li and Meng modified GO nanomaterial with RGD acid, aminated polyethylene glycol (6ARM-PEG-NH_2, PEG-NH_2), disulfide linkages (3,3-dithiodipropionic acid [DTPA]), and DOX [51]. The experimental results showed that the release of DOX was triggered at pH 5.5 upon exposure to 10 mm glutathione (GSH). The *in vitro* experiment proved that the tumor inhibition rate of the RGD-GO-PEG-S-S-DOX system was 78% on human hepatocyte cancer (Hep-G2) cells at 1.56 μg/ml concentration.

Kim and coworkers have fabricated rGO with pH-responsive polymers like poly(ethylene imine) (PEI)-poly(L-lysine)-poly(L-glutamic acid) (PKE polymers) for tumor-targeted DOX delivery systems [52]. Zeta potential analysis confirmed the pH-responsive charge-conversional property of PK_5E_7(PEI-rGO). The release of DOX from PK_5E_7(PEI-rGO) system was strongly pH responsive. At a mildly acidic pH, the release rate of DOX was 36.3% in 24 hours. The anticancer activity of DOX-loaded PK_5E_7(PEI-rGO) increased against HeLa and A549 cell lines and was tested using the MTT assay at pH 6.0. The efficient nanoparticle-based DDS was developed using galactosylated chitosan/graphene oxide/doxorubicin (GC–GO–DOX) for tumor therapy (Figure 16.5) [53]. The study revealed that the system possesses a high drug loading capacity (1.08 mg/mg drug per polymer). The drug-loaded nanoparticles were stable at standard conditions and released the DOX at a rate of 33.1% in the acidic pH environment (pH 5.5).

Vimala and coworker [54] functionalized GO with PEG and FA to develop a potential DDS by attaching the camptothecin (CPT) drug to the surface area of GO through π–π stacking. The pH-responsive drug release behavior was observed with the help of UV–Vis spectral analysis. The cytotoxicity study of GO-PEGFA-CPT showed that cell viability assay of about 95% at a concentration of 100 μg/ml in breast cancer cell lines (MCF-7) cells. A new DDS was reported in which mPEG-PLGA (PP), DOX, and mPEG-FA were integrated with GO (Figure 16.6) [55]. The *in vitro* drug release behavior of the GO-based system was

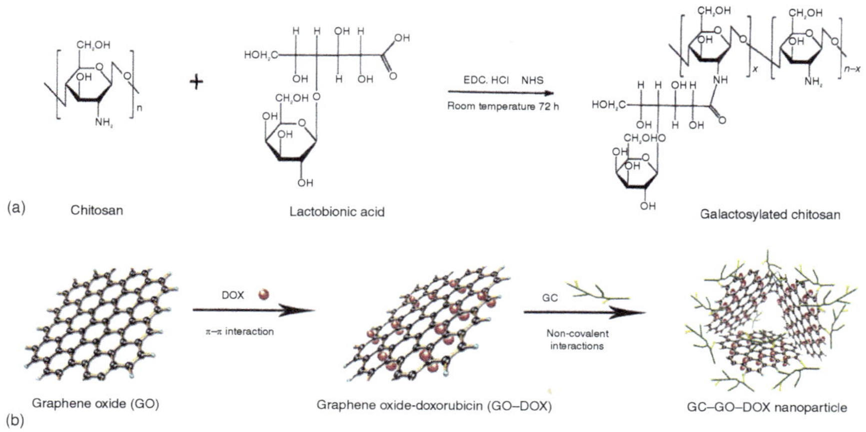

Figure 16.5 Schematic representation of the synthesis of galactosylated chitosan and the fabrication of GC–GO–DOX nanoparticles. (a) Synthesis of galactosylated chitosan (GC). (b) Fabrication of GC–GO–DOX nanoparticle. Source: Adapted with permission from Wang et al. [53]. Elsevier.

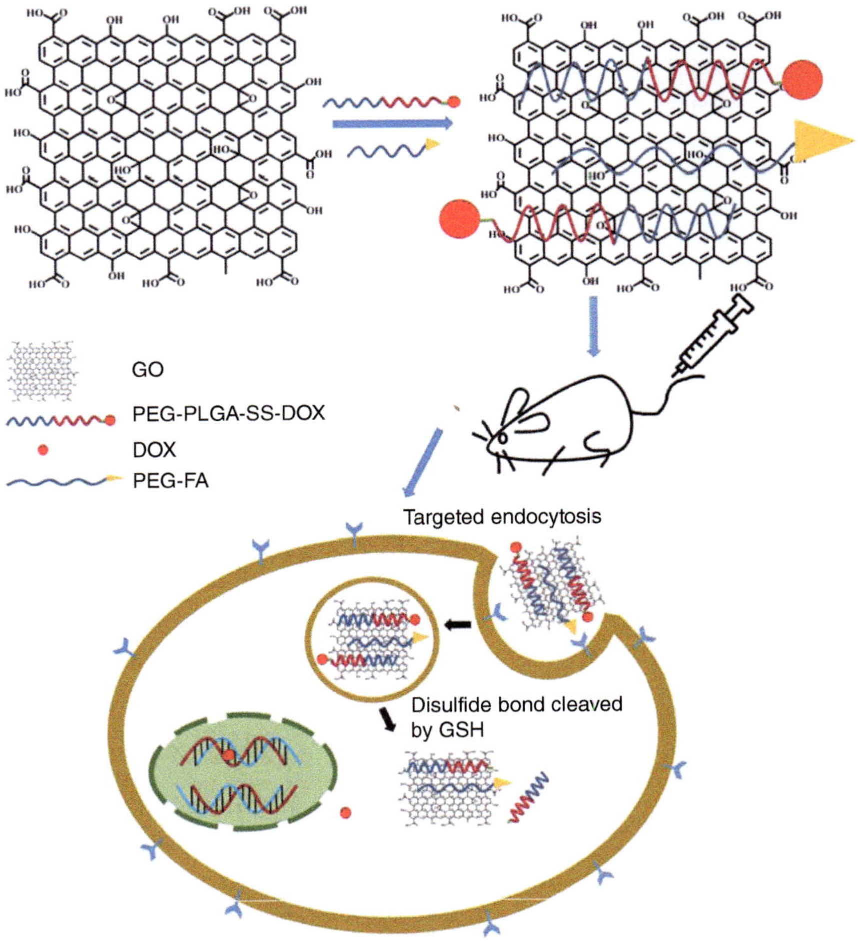

Figure 16.6 Illustration of active tumor-targeting and sufficient redox-sensitive release of GO/PP-SS-DOX/PEG-FA nanohybrids. Source: Adapted with permission from Huang et al. [55]. Elsevier.

observed at a rate of 90% within 72 hours. The *in vivo* study with B16 cells shows that this system enhanced the anticancer effect and reduced the severe side effects in tumor tissues through FR-mediated accumulation.

Pooresmaeil and Namazi [56] reported a facile method for coating MGO with β-cyclodextrin (β-CD) to obtain β-CD-MG as a nanocarrier for the transfer of hydrophilic drug DOX and hydrophobic drug methotrexate (MTX). The *in vitro* release profiles suggest that both drugs are pH responsive and were simulated in both physiological and cancerous tissue conditions. Further, the drug loading efficiency in this process for DOX and MTX was found to be ~37.4% and ~23.4%, respectively. The screening of toxicity from MTT assay revealed that the nanocomposite, where cell viability found to be more than 80%, indicated nontoxicity for K562 cells. Chen and coworkers [57] prepared dual targeting DDS for targeted delivery of the anticancer drug. The MGO was first modified by coprecipitation of Fe_3O_4, followed by coating with PEG and cetuximab (CET) to form MGO-PEG-CET. The antitumor efficacy of the system was explored using *in vivo* xenograft tumor-bearing model mice. Further, NIR laser light was used as photothermal therapy to study the effect on cancer cells. The combining effect of magnetic targeting and laser treatment efficaciously reduces the tumor cells to 42% of its initial volume within 14 days.

Zhai and coworkers [58] reported the modified GO using heparin (Hep) and adipic dihydrazide (ADH) attached through amido bonds to obtain GO-ADH-Hep. The resulting system indicated the pH-sensitive release of DOX. The cytotoxicity study of fabricated nanocomplexes exhibited cytotoxic activity against MCF-7 and HepG2 cells. Further, the biodistribution of the nanomaterial shows less cardiotoxicity and pulmonary toxicity compared with free DOX and unmodified GO. In another work, Ma and coworkers [59] efficaciously used GO for co-delivery of DOX and MMP-9 shRNA plasmid to tumor cells. The attachment of dendrimer to GO takes place through Michael's addition of tris(2-aminoethyl)amine and methyl acrylate (MA) followed by amidation with ethylenediamine (Figure 16.7). Furthermore, the system displayed low cytotoxicity and excellent biocompatibility.

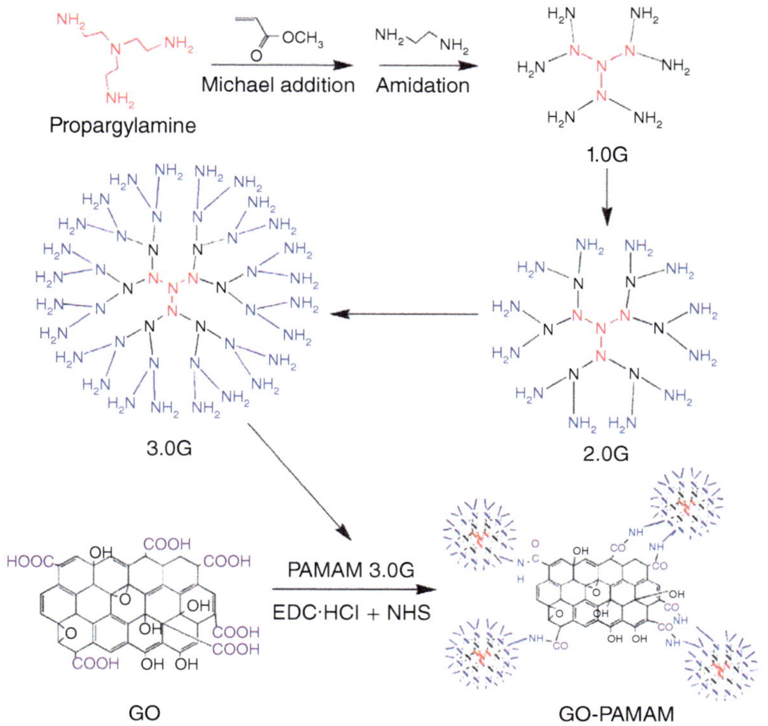

Figure 16.7 Synthetic route of PAMAM 3.0G and GO-PAMAM. Source: Adapted with permission from Gu et al. [59]. Elsevier.

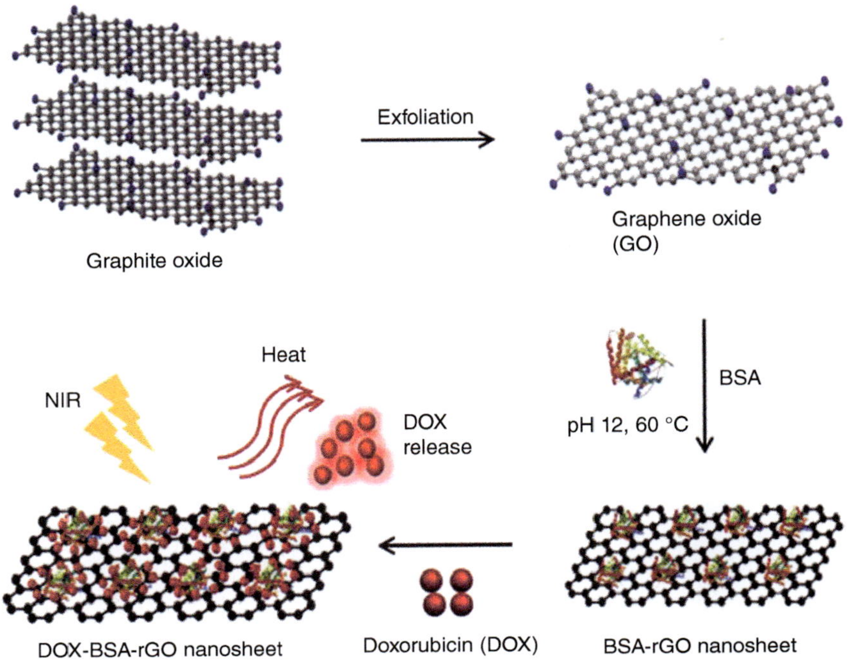

Figure 16.8 Schematic illustration of the synthesis of DOX-BSA-rGO nanosheets. Source: Adapted with permission from Cheon et al. [61]. American Chemical Society.

Lin and coworkers utilized the atom transfer radical polymerization (ATRP) technique to formulate a novel grafted GO with 2-methacryloyloxy ethyl phosphorylcholine (MPC) and 2-aminoethyl methacrylate (AMA) to obtain GO-MPC-AMA [60]. The GO-MPC-AMA was reacted further with FA to conjugate via amino group and the carboxy groups to acquire (GO-PCn-FA). DOX was loaded on to the DOX@GO-PCn-FA, which could precisely target the cancer cells and induced significant antitumor effect.

Chung and coworkers [61] developed a straightforward technique wherein bovine serum albumin (BSA) at pH 12 was used to obtain BSA-rGO nanosheets (Figure 16.8). They showed that the use of BSA not only reduced GO to rGO but also acts as a stabilizer. Due to the strong optical NIR absorbance of DOX-BSA-rGO nanosheets, it was used as a powerful tool for a photothermal agent and induced chemo-photothermal therapy.

16.2.3 Carbon Dots (CDs) in Drug Delivery

CDs are zero-dimensional spherical carbon-based material with a size smaller than 10 nm. In addition, the CDs has unique tunable surface properties, high photostability, low toxicity, tunable emission spectra, small size, good biocompatibility, and high selectivity for the target in drug delivery [62]. These possesses pH-sensitive properties and can quickly disperse in water without further surface modification. The *in vitro* and *in vivo* fluorescence-based imaging experiments further demonstrated that CDs could be

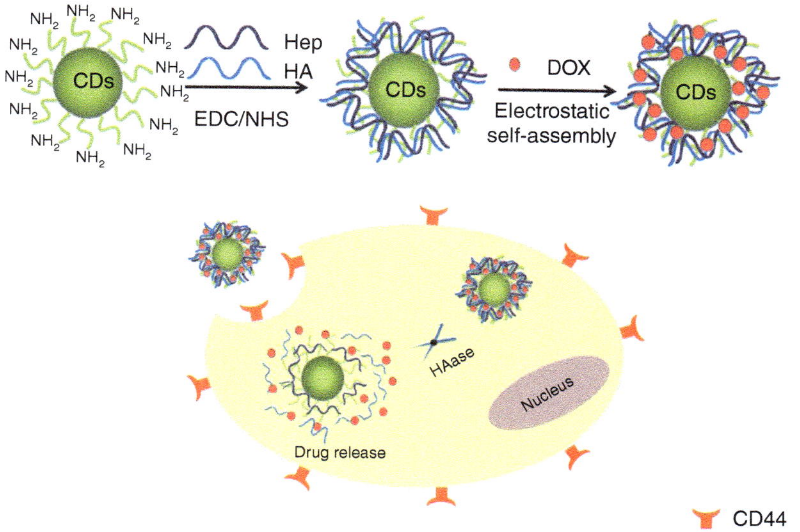

Figure 16.9 Synthesis, functionalization, and release of CDs-HA-Hep/DOX drug delivery system. Source: Adapted with permission from Duan et al. [65]. Elsevier.

employed as a nanocarrier for image-guided drug delivery [63]. For the DDS, target releasing can promote the internalization of the carriers by responding to acidic tumor microenvironments. CDs have a large surface-area-to-volume ratio, which makes them the best applicants for cross-linking and drug loading. Amide bond attaching is a famous cross-linking approach that depends on interactions between primary amine and carboxyl moiety [64].

Li and coworkers [65] fabricated a CD-based DDS by covalent functionalization of hyaluronic acid (HA) and Hep followed by loading model drug DOX on the surface through electrostatic interactions (Figure 16.9). The *in vitro* release study revealed that the CD-HAHep/DOX system achieved a dual-responsive release profile, showing more release of DOX at pH 5.6. MTT and scratch assay experiment revealed that the resultant CD system inhibits cancer cell growth and cellular migration. He and coworkers employed HA via a one-step hydrothermal treatment to synthesize HA-CD for an anticancer drug carrier [66]. The DOX-loaded HA-CD conjugated via 4-carboxybenzaladehyde (*p*-CBA) revealed pH-triggered release behavior and displayed high cytotoxicity on 4T1 cells, good hemocompatibility, and excellent serum stability. DOX was encapsulated in the (HA-CD@*p*-CBA-DOX) system with a loading rate of 18.12%, and the release of DOX was observed in the presence of acid medium. Cong et al. [67] prepared CD with red emission from hydrothermal reaction and the resultant system covered by PEG polymer (mPEG-CHO) to obtain R-CD@mPEG-CHO. This system showed controlled DOX release profiles and pH-responsive property in tumor acid microenvironment.

Na and coworkers [68] have introduced the pH-sensitive carbon dots (pSCDs) prepared by reacting citric acid with 1-(3-aminopropyl)imidazole (API) (Figure 16.10). The imidazole groups coating the CDs enhanced the loading efficiency of the drug up to 78.55%. The results showed the release of drugs was seven times higher than the control

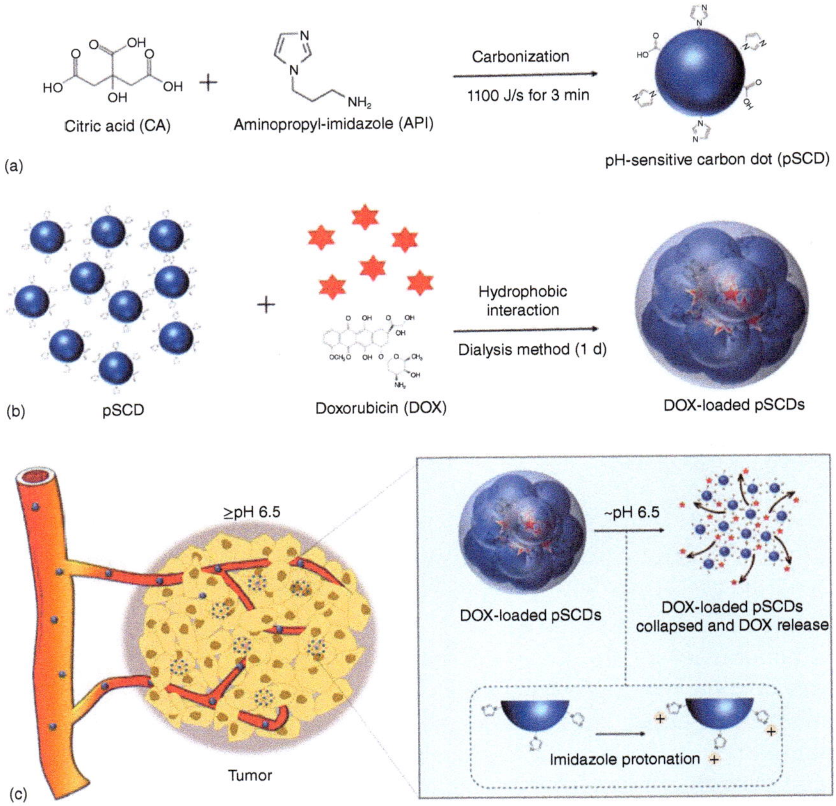

Figure 16.10 Schematic synthesis of the pSCDs (a) and preparation of DOX-loaded pSCDs (b) and (c) schematic illustration for the drug delivery process of DOX-loaded pSCDs. Source: Adapted with permission from Seo et al. [68]. American Chemical Society.

group under acidic pH conditions. The *in vivo* study showed that the growth of tumor size in an HCT-116-bearing mouse model was only retarded in the presence of DOX-loaded pSCDs.

Shu and coworkers [69] reported a facile hydrothermal procedure for preparing FA-CDots using citric acid as the carbon source. The attachment of boronic acid ligands on FA-CDots takes place by reacting with 3-aminophenyl boronic acid (3-APBA), 1-ethyl-3-(3-dimethylaminopropyl) carbodiimide (EDC) and *N*-hydroxysuccinimide (NHS) followed by coupling with β-CD to obtain β-CD/CDots as nanocarrier. DOX molecule was infused in the β-CD cavity with a drug loading ratio of 27.3% at pH 7.4. The intracellular uptake of the attached DOX against HEK 293T and HeLa cells and sustained release of free DOX were visualized by confocal microscopy and flow cytometry. Similarly, Zhu and coworkers [70] used a facile hydrothermal method to synthesize CDs using citric acid and ethylenediamine. The *in vitro* studies show that the system possessed potential proliferative function, high drug payload capacity, better drug release profile, increased cell uptake, and good biocompatibility. The cytotoxicity was investigated using inhibitory concentration values (IC_{50}) of CD-DOX group, and free DOX against MCF-7 cells was 0.652 and 0.356 μg/ml and 0.983 and 0.939 μg/ml for 24 and 48 hours, respectively. These results show great promise for cancer treatment.

Gu and coworkers [71] have reported green and simple technique for the preparation of CDs via hydrothermal route using milk as a carbon source. Then, the DOX was attached to CDs by electrostatic interactions and multiple surface CD functional groups. The *in vitro* DOX release experiment revealed that CD-DOX complexes have a pH-dependent profile. The system also showed higher cytotoxicity and apoptosis against adenoid cystic carcinoma cell line (ACC-2) than that of bare DOX. Recently, Leblanc and coworkers [72] developed a simple strategy for C-dots to conjugate with transferrin, epirubicin, and temozolomide to achieve excellent anticancer activity. The *in vitro* experiments showed that the transferrin-attached samples decrease the cell viability compared with the non-transferrin. In the cytotoxicity study against glioblastoma brain tumor cells, the C-DT exhibited an upsurge to 86% in SJGBM2 at a concentration of 0.01 μM, while the C-ET and C-TT only reduced to 33% and 8% near the same concentration.

16.2.4 Nanodiamonds (NDs) in Drug Delivery

NDs consist of a diamond core crystalline structure and possess a unique surface property. They have a high capacity for adsorption, large specific surface area, and chemically inert [73]. NDs have been recently introduced as a template for drug delivery application, owing to its small primary particle size (~4–5 nm), mechanical and thermal properties, easy surface modification, rapid conjugations with the drug, fast transmembrane transport, feasibility to scale up, and excellent biocompatibility both *in vitro* and *in vivo* [74]. NDs possess pretty good characteristics of fluorescence that imparts diagnostic feature and thus can be utilized as image probes injected into a tumor. Furthermore, it has been demonstrated that NDs at a concentration of 25 μg/ml neither stimulate apoptosis nor reduce proliferation at the level of the transcriptional response. Previous investigations have revealed that NDs are not cytotoxic against neuroblastoma and macrophage cell lines. Furthermore, the stability and safety of NDs were established by biodistribution studies in mice and *Caenorhabditis elegans* [75, 76].

Feng and Wei have introduced the pH-dependent ND-based composites via thiol-ene click reaction [77]. The hydrazone bonds were prepared by reacting between ester groups and hydrazine. Incorporation of CHO-PEG and DOX was integrated with ND–SO–NH to afford ND–P–D composites (Figure 16.11). The DOX release patterns of the ND–P–D composites are pH dependent, and acidic tumor microenvironments triggered the DOX release. The cytotoxicity experiment of ND–P–D composites displayed that cell viability decreased significantly in the highest concentration in HepG2 cells after 24 hours' incubation. This study indicates that the ND-based composites show excellent biocompatibility and therapeutic effect.

Liu and coworkers [78] have reported the ND-based composites, which is first modified with ethylenediamine to tether the amino moiety on the surface of ND followed by Michael addition reaction of poly(ethylene glycol) methyl ether methacrylate (PEGMA) to acquire ND@PEGMA. The modified ND exhibited excellent biocompatibility in the HepG2 cells. The assay data of cell viability indicate that the DOX-loaded composite was effective in

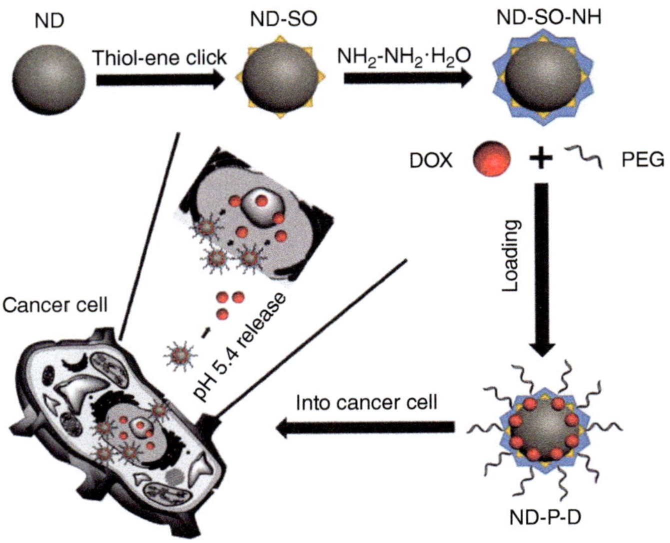

Figure 16.11 The schematic procedure for preparation of ND–SO–NH. Source: Adapted with permission from Long et al. [77]. Elsevier.

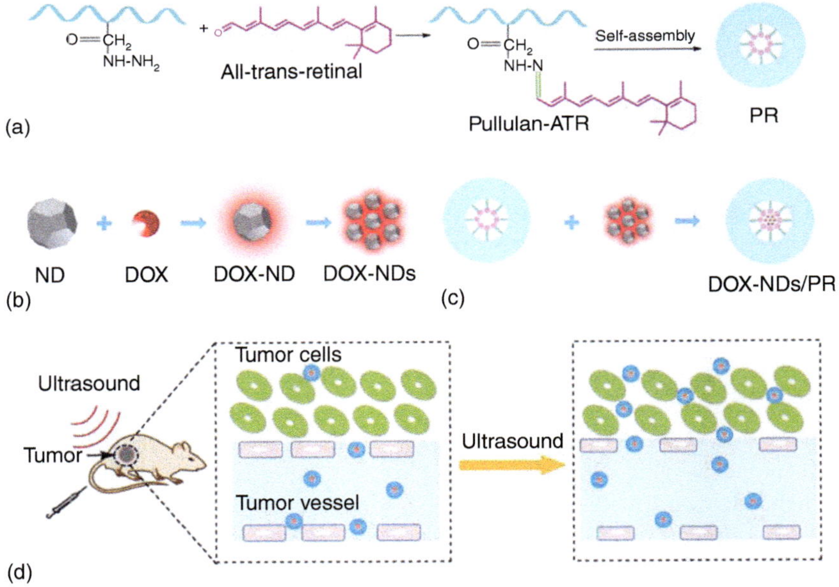

Figure 16.12 Step-wise (a)–(d) illustrations of ultrasound-enhanced delivery of DOX-NDs/PR nanoparticles into the tumors. Source: Adapted with permission from Li et al. [79]. American Chemical Society.

killing cancer cells. In another work, the pH-responsive pullulan-ATR-DOX conjugate NDs were synthesized and applied as a drug delivery carrier (Figure 16.12). The results demonstrated improved intracellular retention, and *in vivo* studies showed DOX in the cancer cell reached 17.3% within 24 hours of injection [79].

16.3 Conclusions

In this chapter, we concise the latest advance of modified carbon-based nanomaterials as an attractive material for drug delivery, and they include CNTs, GO, CDs, and NDs. Surprisingly, the EPR effect of carbon-based nanomaterials caught the attention of drug researchers to use it as a carrier in the DDS. We thoroughly reviewed the drug loading and releasing efficiency, cytotoxicity, and biocompatibility of modified CNTs, GO, CDs, and NDs from acidic to normal physiological conditions. The efficiency of attached and unattached drugs *in vitro* and *in vivo* with the system was thoroughly discussed. The modified DDS loaded with a particular drug showed good biocompatibility and therapeutic effect. Although exciting progression was made in the DDS, the toxicity and biocompatibility issue have always been a significant obstacle to the development of this system. From the *in vivo* studies, it has been found that some nanocarriers interact with the various organs, and a certain level of toxicity was also observed. For an ideal DDS, it should be nontoxic and selective to the target cells.

Important Websites

https://www.nanowerk.com/spotlight/spotid=51090.php.
https://onlinelibrary.wiley.com/doi/abs/10.1002/jcp.26899.
https://pubs.acs.org/doi/10.1021/ie402360f.
https://link.springer.com/article/10.1007/s40097-018-0265-6.
https://pubs.rsc.org/en/content/articlelanding/2016/tb/c6tb01259k#!divAbstract.
https://www.thno.org/v02p0302.htm.

References

1 Hussain, C.M. (2018). *Handbook of Nanomaterials for Industrial Applications*. Elsevier.
2 Hussain, C.M. (2020). *Handbook of Functionalized Nanomaterials for Industrial Applications*. Elsevier.
3 Hussain, C.M. (2020). *Handbook of Manufacturing Applications of Nanomaterials*. Elsevier.
4 Hosnedlova, B., Kepinska, M., Fernandez, C. et al. (2019). Carbon nanomaterials for targeted cancer therapy drugs: a critical review. *Chem. Rec.* 19: 502–522.
5 Kailasa, S.K., Ha, S., Baek, S.H. et al. (2019). Tuning of carbon dots emission color for sensing of Fe^{3+} ion and bioimaging applications. *Mater. Sci. Eng., C* 98: 834–842.
6 Loh, K.P., Ho, D., Chiu, G.N.C. et al. (2018). Clinical applications of carbon nanomaterials in diagnostics and therapy. *Adv. Mater.* 30: 1802368.
7 Saleem, J., Wang, L., and Chen, C. (2018). Carbon-based nanomaterials for cancer therapy via targeting tumor microenvironment. *Adv. Healthcare Mater.* 7: 1800525.
8 Wang, Z. and Dai, Z. (2015). Carbon nanomaterial-based electrochemical biosensors: an overview. *Nanoscale* 7: 6420–6431.
9 Barg, S., Perez, F.M., Ni, N. et al. (2014). Mesoscale assembly of chemically modified graphene into complex cellular networks. *Nat. Commun.* 5: 4328.

10 Hayashi, T., Kim, Y.A., Natsuki, T., and Endo, M. (2007). Mechanical properties of carbon nanomaterials. *ChemPhysChem* 8: 999–1004.

11 Mei, Q., Zhang, K., Guan, G. et al. (2010). Highly efficient photoluminescent graphene oxide with tunable surface properties. *Chem. Commun.* 46: 7319–7321.

12 Shen, Z., Wu, J., Yu, Y. et al. (2019). Comparison of cytotoxicity and membrane efflux pump inhibition in HepG2 cells induced by single-walled carbon nanotubes with different length and functional groups. *Sci. Rep.* 9: 1–9.

13 Amenta, V. and Aschberger, K. (2015). Carbon nanotubes: potential medical applications and safety concerns. *Wiley Interdiscip. Rev. Nanomed. Nanobiotechnol.* 7: 371–386.

14 Seabra, A.B., Paula, A.J., de Lima, R. et al. (2014). Nanotoxicity of graphene and graphene oxide. *Chem. Res. Toxicol.* 27: 159–168.

15 Chng, E.L.K. and Pumera, M. (2013). The toxicity of graphene oxides: dependence on the oxidative methods used. *Chemistry* 19: 8227–8235.

16 Wen, J., Xu, Y., Li, H. et al. (2015). Recent applications of carbon nanomaterials in fluorescence biosensing and bioimaging. *Chem. Commun.* 51: 11346–11358.

17 Qi, X., Qin, C., Zhong, W. et al. (2010). Large-scale synthesis of carbon nanomaterials by catalytic chemical vapor deposition: a review of the effects of synthesis parameters and magnetic properties. *Materials* 3: 4142–4174.

18 Yang, F., Ren, X., LeCroy, G.E. et al. (2018). Zero-dimensional carbon allotropes—carbon nanoparticles versus fullerenes in functionalization by electronic polymers for different optical and redox properties. *ACS Omega* 3: 5685–5691.

19 Jintoku, H., Matsuzawa, Y., and Yoshida, M. (2018). Switching the optical and electrical properties of carbon nanotube hybrid films using a photoresponsive dispersant as a dopant. *RSC Adv.* 8: 11186–11190.

20 Hou, J., Du, W., Meng, F. et al. (2018). Effective dispersion of multi-walled carbon nanotubes in aqueous solution using an ionic-gemini dispersant. *J. Colloid Interface Sci.* 512: 750–757.

21 Shuang, E., Mao, Q.X., Yuan, X.L. et al. (2018). Targeted imaging of the lysosome and endoplasmic reticulum and their pH monitoring with surface regulated carbon dots. *Nanoscale* 10: 12788–12796.

22 Mallakpour, S. and khodadadzadeh, L. (2018). Biocompatible and biodegradable chitosan nanocomposites loaded with carbon nanotubes. In: *Biodegradable and Biocompatible Polymer Composites Processing, Properties and Applications*, Chapter 7 (ed. N.G. Shimpi), 187–221. Sawston: Woodhead Publishing. https://doi.org/10.1016/B978-0-08-100970-3.00007-9.

23 Mallakpour, S. and Khadem, E. (2019). Carbon nanotubes for heavy metals removal. In: *Composite Nanoadsorbents*, Chapter 8 (eds. G. Kyzas and A.C. Mitrpoulos), 181–210. Amsterdam: Elsevier. https://doi.org/10.1016/B978-0-12-814132-8.00009-5; eBook ISBN: 9780128141335.

24 Mallakpour, S. and Khadem, E. (2016). Carbon nanotube–metal oxide nanocomposites: fabrication, properties and applications. *Chem. Eng. J.* 302: 344–367.

25 Rosen, Y. and Elman, N.M. (2009). Carbon nanotubes in drug delivery: focus on infectious diseases. *Expert Opin. Drug Delivery* 6: 517–530.

26 Yang, D., Yang, F., Hu, J. et al. (2009). Hydrophilic multi-walled carbon nanotubes decorated with magnetite nanoparticles as lymphatic targeted drug delivery vehicles. *Chem. Commun.* (29): 4447–4449.

27 Palanisamy, S., Velusamy, V., Balu, S. et al. (2020). Sonochemical synthesis and anchoring of zinc oxide on hemin-mediated multi-walled carbon nanotubes-cellulose nanocomposite for ultra-sensitive biosensing of H_2O_2. *Ultrason. Sonochem.* 63: 104917.

28 Vardharajula, S., Ali, S.Z., Tiwari, P.M. et al. (2012). Functionalized carbon nanotubes: biomedical applications. *Int. J. Nanomed.* 7: 5361.

29 Phan, Q.T., Patil, M.P., Tu, T.T. et al. (2020). Polyampholyte-grafted single walled carbon nanotubes prepared via a green process for anticancer drug delivery application. *Polymer* 193: 122340.

30 Gangrade, A. and Mandal, B.B. (2019). Injectable carbon nanotube impregnated silk based multifunctional hydrogel for localized targeted and on-demand anticancer drug delivery. *ACS Biomater. Sci. Eng.* 5: 2365–2381.

31 Farvadi, F., Tamaddon, A., Sobhani, Z., and Abolmaali, S.S. (2017). Polyionic complex of single-walled carbon nanotubes and PEG-grafted-hyperbranched polyethyleneimine (PEG-PEI-SWNT) for an improved doxorubicin loading and delivery: development and in vitro characterization. *Artif. Cells Nanomed. Biotechnol.* 45: 855–863.

32 Li, B., Zhang, X.X., Huang, H.Y. et al. (2018). Effective deactivation of A549 tumor cells in vitro and in vivo by RGD-decorated chitosan-functionalized single-walled carbon nanotube loading docetaxel. *Int. J. Pharm.* 543: 8–20.

33 Yan, C., Chen, C., Hou, L. et al. (2017). Single-walled carbon nanotube-loaded doxorubicin and Gd-DTPA for targeted drug delivery and magnetic resonance imaging. *J. Drug Targeting* 25: 163–171.

34 Panahi, F.H., Peighambardoust, S.J., Davaran, S., and Salehi, R. (2017). Development and characterization of PLA-mPEG copolymer containing iron nanoparticle-coated carbon nanotubes for controlled delivery of docetaxel. *Polymer* 117: 117–131.

35 Zhang, L.P., Tan, X.X., Huang, Y.P., and Liu, Z.S. (2018). Floating liquid crystalline molecularly imprinted polymer coated carbon nanotubes for levofloxacin delivery. *Eur. J. Pharm. Biopharm.* 127: 150–158.

36 Wang, P., Meng, A., Yang, T. et al. (2016). Ultrashort nanotube blocked large-inner-diameter multi-walled carbon nanotubes based drug delivery system for tumor targeted therapy. *J. Nanosci. Nanotechnol.* 16: 6755–6765.

37 Farahani, B.V., Behbahani, G.R., and Javadi, N. (2016). Functionalized multi walled carbon nanotubes as a carrier for doxorubicin: drug adsorption study and statistical optimization of drug loading by factorial design methodology. *J. Braz. Chem. Soc.* 27: 694–705.

38 Abdolmaleki, A., Mallakpour, S., and Karshenas, A. (2017). Facile synthesis of glucose-functionalized reduced graphene oxide (GFRGO)/poly(vinyl alcohol) nanocomposites for improving thermal and mechanical properties. *Mater. Sci. Eng., B* 217: 26–35.

39 Mallakpour, S., Abdolmaleki, A., and Karshenas, A. (2017). Graphene oxide supported copper coordinated amino acids as novel heterogeneous catalysts for epoxidation of norbornene. *Catal. Commun.* 92: 109–113.

40 Abdolmaleki, A., Mallakpour, S., and Karshenas, A. (2017). Synthesis and characterization of new nanocomposites films using alanine-Cu-functionalized graphene oxide as nanofiller and PVA as polymeric matrix for improving of their properties. *J. Solid State Chem.* 253: 398–405.

41 Yang, Y., Zhang, Y.M., Chen, Y. et al. (2012). Construction of a graphene oxide based non-covalent multiple nanosupramolecular assembly as a scaffold for drug delivery. *Chemistry* 18: 4208–4215.

42 Shams, M., Guiney, L.M., Huang, L. et al. (2019). Influence of functional groups on the degradation of graphene oxide nanomaterials. *Environ. Sci.: Nano* 6: 2203–2214.

43 Campbell, E., Hasan, M.T., Pho, C. et al. (2019). Graphene oxide as a multifunctional platform for intracellular delivery, imaging, and cancer sensing. *Sci. Rep.* 9: 1–9.

44 Pooresmaeil, M., Javanbakht, S., Nia, S.B., and Namazi, H. (2020). Carboxymethyl cellulose/mesoporous magnetic graphene oxide as a safe and sustained ibuprofen delivery bio-system: synthesis, characterization, and study of drug release kinetic. *Colloids Surf., A* 594: 124662.

45 Mauro, N., Scialabba, C., Agnello, S. et al. (2020). Folic acid-functionalized graphene oxide nanosheets via plasma etching as a platform to combine NIR anticancer phototherapy and targeted drug delivery. *Mater. Sci. Eng., C* 107: 110201.

46 Ma, N., Song, A., Li, Z., and Luan, Y. (2019). Redox-sensitive prodrug molecules meet graphene oxide: an efficient graphene oxide-based nanovehicle toward cancer therapy. *ACS Biomater. Sci. Eng.* 5: 1384–1391.

47 Javanbakht, S., Pooresmaeil, M., and Namazi, H. (2019). Green one-pot synthesis of carboxymethylcellulose/Zn-based metal-organic framework/graphene oxide bio-nanocomposite as a nanocarrier for drug delivery system. *Carbohydr. Polym.* 208: 294–301.

48 Xie, M., Zhang, F., Peng, H. et al. (2019). Layer-by-layer modification of magnetic graphene oxide by chitosan and sodium alginate with enhanced dispersibility for targeted drug delivery and photothermal therapy. *Colloids Surf., B* 176: 462–470.

49 Xie, M., Zhang, F., Liu, L. et al. (2018). Surface modification of graphene oxide nanosheets by protamine sulfate/sodium alginate for anticancer drug delivery application. *Appl. Surf. Sci.* 440: 853–860.

50 Palai, P.K., Mondal, A., Chakraborti, C.K. et al. (2019). Doxorubicin loaded green synthesized nanoceria decorated functionalized graphene nanocomposite for cancer-specific drug release. *J. Cluster Sci.* 30: 1565–1582.

51 Wei, M., Lu, T., Nong, Z. et al. (2019). Reductive response and RGD targeting nano-graphene oxide drug delivery system. *J. Drug Delivery Sci. Technol.* 53: 101202.

52 Ryu, K., Park, J., and Kim, T.I. (2019). Effect of pH-responsive charge-conversional polymer coating to cationic reduced graphene oxide nanostructures for tumor microenvironment-targeted drug delivery systems. *Nanomaterials* 9: 1289.

53 Wang, C., Zhang, Z., Chen, B. et al. (2018). Design and evaluation of galactosylated chitosan/graphene oxide nanoparticles as a drug delivery system. *J. Colloid Interface Sci.* 516: 332–341.

54 Deb, A. and Vimala, R. (2018). Camptothecin loaded graphene oxide nanoparticle functionalized with polyethylene glycol and folic acid for anticancer drug delivery *J. Drug Delivery Sci. Technol.* 43: 333–342.

55 Huang, C., Wu, J., Jiang, W. et al. (2018). Amphiphilic prodrug-decorated graphene oxide as a multi-functional drug delivery system for efficient cancer therapy. *Mater. Sci. Eng., C* 89: 15–24.

56 Pooresmaeil, M. and Namazi, H. (2018). β-Cyclodextrin grafted magnetic graphene oxide applicable as cancer drug delivery agent: synthesis and characterization. *Mater. Chem. Phys.* 218: 62–69.

57 Lu, Y.J., Lin, P.Y., Huang, P.H. et al. (2018). Magnetic graphene oxide for dual targeted delivery of doxorubicin and photothermal therapy. *Nanomaterials* 8: 193.

58 Zhang, B., Yang, X., Wang, Y., and Zhai, G. (2017). Heparin modified graphene oxide for pH-sensitive sustained release of doxorubicin hydrochloride. *Mater. Sci. Eng., C* 75: 198–206.

59 Gu, Y., Guo, Y., Wang, C. et al. (2017). A polyamidoamne dendrimer functionalized graphene oxide for DOX and MMP-9 shRNA plasmid co-delivery. *Mater. Sci. Eng., C* 70: 572–585.

60 Qin, Y., Wang, C., Jiang, Y. et al. (2017). Phosphorylcholine oligomer-grafted graphene oxide for tumor-targeting doxorubicin delivery. *RSC Adv.* 7: 41675–41685.

61 Cheon, Y.A., Bae, J.H., and Chung, B.G. (2016). Reduced graphene oxide nanosheet for chemo-photothermal therapy. *Langmuir* 32: 2731–2736.

62 Zeng, Q., Shao, D., He, X. et al. (2016). Carbon dots as a trackable drug delivery carrier for localized cancer therapy in vivo. *J. Mater. Chem. B* 4: 5119–5126.

63 Feng, T., Ai, X., An, G. et al. (2016). Charge-convertible carbon dots for imaging-guided drug delivery with enhanced in vivo cancer therapeutic efficiency. *ACS Nano* 10: 4410–4420.

64 Beack, S., Kong, W.H., Jung, H.S. et al. (2015). Photodynamic therapy of melanoma skin cancer using carbon dot–chlorin e6–hyaluronate conjugate. *Acta Biomater.* 26: 295–305.

65 Duan, Q., Ma, L., Zhang, B. et al. (2020). Construction and application of targeted drug delivery system based on hyaluronic acid and heparin functionalised carbon dots. *Colloids Surf., B* 188: 110768.

66 Li, J., Li, M., Tian, L. et al. (2020). Facile strategy by hyaluronic acid functional carbon dot-doxorubicin nanoparticles for CD44 targeted drug delivery and enhanced breast cancer therapy. *Int. J. Pharm.* 578: 119122.

67 Cong, H.L., Wang, M.Z., Xu, Y.H. et al. (2020). Multifunctional carbon dots based nanoparticles coupling optical and pH-dependent drug release properties as drug delivery platforms. *Integr. Ferroelectr.* 206: 151–159.

68 Seo, J., Lee, J., Lee, C.B. et al. (2019). Nonpolymeric pH-sensitive carbon dots for treatment of tumor. *Bioconjugate Chem.* 30: 621–632.

69 Yang, T., Huang, J.L., Wang, Y.T. et al. (2019). β-Cyclodextrin-decorated carbon dots serve as nanocarriers for targeted drug delivery and controlled release. *ChemNanoMat* 5: 479–487.

70 Kong, T., Hao, L., Wei, Y. et al. (2018). Doxorubicin conjugated carbon dots as a drug delivery system for human breast cancer therapy. *Cell Prolif.* 51: e12488.

71 Yuan, Y., Guo, B., Hao, L. et al. (2017). Doxorubicin-loaded environmentally friendly carbon dots as a novel drug delivery system for nucleus targeted cancer therapy. *Colloids Surf., B* 159: 349–359.

72 Hettiarachchi, S.D., Graham, R.M., Mintz, K.J. et al. (2019). Triple conjugated carbon dots as a nano-drug delivery model for glioblastoma brain tumors. *Nanoscale* 11: 6192–6205.

73 Kaur, R. and Badea, I. (2013). Nanodiamonds as novel nanomaterials for biomedical applications: drug delivery and imaging systems. *Int. J. Nanomed.* 8: 203.

74 Gao, G., Guo, Q., and Zhi, J. (2019). Nanodiamond-based theranostic platform for drug delivery and bioimaging. *Small* 15: 1902238.

75 Rojas, S., Gispert, J.D., Martín, R. et al. (2011). Biodistribution of amino-functionalized diamond nanoparticles. In vivo studies based on 18F radionuclide emission. *ACS Nano* 5: 5552–5559.

76 Mohan, N., Chen, C.S., Hsieh, H.H. et al. (2010). In vivo imaging and toxicity assessments of fluorescent nanodiamonds in *Caenorhabditis elegans*. *Nano Lett.* 10: 3692–3699.

77 Long, W., Ouyang, H., Wan, W. et al. (2020). "Two in one": simultaneous functionalization and DOX loading for fabrication of nanodiamond-based pH responsive drug delivery system. *Mater. Sci. Eng., C* 108: 110413.

78 Yang, G., Long, W., Yan, W. et al. (2020). Surface PEGylation of nanodiamond through a facile Michael addition reaction for intracellular drug delivery. *J. Drug Delivery Sci. Technol.* 57: 101644.

79 Li, H., Ma, M., Zhang, J. et al. (2019). Ultrasound-enhanced delivery of doxorubicin-loaded nanodiamonds from pullulan-all-trans-retinal nanoparticles for effective cancer therapy. *ACS Appl. Mater. Interfaces* 11: 20341–20349.

17

Functionalized Carbon Nanomaterials (FCNMs): A Green and Sustainable Vision

Upasana Issar[1] *and Richa Arora*[2]

[1] *University of Delhi, Kalindi College, Department of Chemistry, Delhi, 110008, India*
[2] *University of Delhi, Shivaji College, Department of Chemistry, Delhi, 110027, India*

17.1 Introduction

Carbon-based nanomaterials (CNMs) have gained immense popularity since the inception of nanotechnology owing to their unique electronic, optical, thermal, and mechanical properties. The carbon in these CNMs is either sp^2 or sp^3 hybridized. These materials can be classified based on their dimensionalities as zero-dimensional fullerene, carbon nanodots (CNDs), nanodiamonds (NDs), one-dimensional carbon nanotubes (CNTs), and two-dimensional graphene. There is an entire dedicated family of graphene-based nanomaterials, including pristine graphene (monolayer graphene), graphene oxide (GO), reduced graphene oxide (rGO), and graphene quantum dots (GQDs) [1].

Changes of any form in the original structure of CNMs are called functionalization and the obtained nanomaterials are called functionalized carbon nanomaterials (FCNMs). The functionalization can be covalent or non-covalent attachment of functional groups/small organic molecules/biomolecules/polymers, etc. on the surface/side or end walls of CNMs or it can be the defects created within the CNMs. The other type of functionalization includes doping of CNMs with atoms such as nitrogen or oxygen. The covalent functionalization involves the formation of strong covalent bond between different chemical groups and carbon of CNMs by utilizing common organic reactions such as radical addition, cycloaddition, oxidation reaction, esterification, or amidation reactions [2]. The CNTs prepared experimentally do not possess perfect hexagonal structures, resulting in heptagon and pentagon rings due to side/end defects that increase the reactivity and mechanical properties of CNTs. Functionalization at these places are feasible. GO is a highly oxidized form of graphene where the edges and plane of graphene are occupied by carboxyl and epoxy/hydroxyl groups, respectively [3]. Compared to graphene, GO displays less electrical conductivity due to the presence of many oxygen-containing functional groups on its surface. Therefore, to improve this property, GO is subjected to partial reduction. The obtained rGO somewhat restores the intrinsic electrical properties of graphene. Likewise,

Environmental Applications of Carbon Nanomaterials-Based Devices, First Edition.
Edited by Shadpour Mallakpour and Chaudhery Mustansar Hussain.

functionalization can be carried out on fullerene, NDs, and nanodots [1]. In non-covalent functionalization, π–π stacking interactions dominate and the hybridization of carbon remains intact keeping the electronic structure of CNMs completely preserved. Bio-functionalization is the loading of biomolecules such as DNA and proteins non-covalently on the surface of CNMs. Polymer wrapping and surfactant functionalization also come under non-covalent functionalization [4]. Another type is metal/metal oxide nanoparticle functionalization on CNMs that could either be covalent or non-covalent [5].

CNMs display strong van der Waals interaction within themselves, and due to their hydrophobic nature, these materials tend to form aggregates especially in hydrophilic environment [6]. This aggregation changes their size and geometry affecting its inherent properties. Since most of the reactions/processes are favored in aqueous medium, it is required to modify CNMs with groups that are hydrophilic in nature thereby increasing the dispersion ability of CNMs in hydrophilic/lipophobic environment [7]. Functionalization is also required:

- To make efficient probes for analytical purposes especially as electrochemical sensors and bio-sensors.
- To design novel nanomaterials for specific applications such as fuel cells, batteries, supercapacitors, and energy storage devices.
- To increase the inherent electronic, optical, thermal, and mechanical properties of these nanomaterials.
- To surpass the toxic hazards imposed by pristine carbon nanomaterials.

17.2 Environment-Friendly and Greener Ways to Synthesize FCNMs

It is the need of the hour to use FCNMs for generation and storage of renewable energy and also in the purification and remediation processes required for a greener and sustainable environment. The greener route for synthesis of nanomaterials can be chosen based on the selection of nontoxic capping/reducing agents, the choice of safe solvents, and the development of energy-efficient synthetic techniques. The use of benign solvents such as water, ionic liquids, or supercritical fluids is employed alternatively in the synthesis of FCNMs. Some of the green and energy-efficient techniques to produce FCNMs involve use of microwave, ultrasonic, and solar radiations. "One-pot synthesis" is a synthetic process in which all the reactants are subjected to sequential chemical reactions in a single pot thereby increasing the reaction efficiency and yield of the product. In this way, the waste formed during the synthesis is minimized considerably. Nowadays, scientists are combining all the tools/techniques in order to develop with the best protocol for synthesizing FCNMs in a most eco-friendly manner.

In search for solvent-less synthetic route for functionalizing CNMs, many attempts have been made till date. However, in true sense, a totally solvent-free synthesis is hard to achieve as auxiliary solvents are still required in the separation and purification steps. Various solvent-free covalent functionalization of CNMs reactions such as nucleophilic additions, 1,3-dipolar cycloaddition reactions, Diels–Alder reactions, and free radical reactions and oxidations have been reported so far. These additions can be initiated by microwave, plasma, thermal, or mechano-chemical treatments [8]. A solvent-free microwave-assisted synthesis is an efficient greener way of synthesizing functionalized carbon nanotubes (FCNTs). Here, in the absence of solvent, the CNTs are directly exposed to microwave radiations. Being microwave active materials, they rapidly absorb the radiations resulting in achieving a very high temperature in few seconds. For example, the conventional way of 1,3-dipolar cycloaddition of azomethine ylide on CNTs takes five days [9], whereas in using solvent-free microwave-assisted methodology, it takes only an hour to obtain functionalized single-walled carbon nanotubes (FSWCNTs) [10]. Similar strategies can also be employed to get functionalized carbon nanohorns (FCNHs) [11].

The use of ionic liquids combined with microwave irradiation imposes a controlled functionalization of aryl-diazonium salt on single-walled carbon nanotube (SWCNT) as uncontrolled excessive functionalization sometimes leads to the disruption of conjugated geometry of SWCNT, thereby affecting its properties [12]. The oxidized CNMs are most widely used in catalysis [13]. The oxidizing conditions are ambient to create defects onto the CNMs that help in increasing the reactivity making FCNMs a good catalyst. A green strategy of using ozone in solvent vapor is carried out. The use of solvent vapor (mixture of water, hydrogen peroxide, and ethanol) along with ozone effectively oxygenates the sidewalls of CNT with high selectivity [14].

A one-pot synthesis of copper functionalized CNDs is achieved using microwave radiations. The prepared Cu-CNDs prove to be an efficient bio-marker for histidine. The Cu atoms (loaded on CND) interact with histidine and shows quenching in the fluorescence spectra [15]. Such functionalized carbon nanodots (FCNDs) hold immense use in biomedicines to detect histidine or other amino acids present in biological fluid samples. A procedure becomes all the more facile and eco-friendly when the precursors used are renewable in nature. Xylan (third most abundant biopolymer on Earth) and branched polyethyleneimine (BPEI) are renewable materials that can be utilized as precursors. A microwave-assisted preparation of xylan-CND and BPEI-CND has been performed. These FCNDs act as promising fluorescent probes for sensitively and selectively detect tannic acid [16]. A microwave hydrothermal route is adopted to functionalize CNT with gallotannin, a class of tannin compounds that are nontoxic, biodegradable, and abundantly available in nature. The study shows that prepared FCNT allows the removal of gallium ion (Ga^{3+}), a pollutant, in a green and efficient manner [17]. Covalent and non-covalent functionalization of graphene is also reported widely using microwave-assisted techniques. Some functional moieties are metal form like Ni foam, while others are biodegradable materials like chitosan and ethylene glycol [18, 19]. These functionalized graphene are diversely used in the biomedical research and as supercapacitors.

Ultrasound irradiation-assisted synthesis of FCNMs is also a more viable technique that involves the growth of nanoparticle in the heat generated through ultrasonic energy. This technique is also called sonochemical synthetic process and the branch is called sono-nano chemistry. Circulating the aqueous solution of CNT and graphene in high-power ultrasonic cell and simultaneously injecting ozone results in production of oxidized CNMs via environmentally friendly ozonolysis route. The ultrasonication restricts the agglomeration of the nanoparticles that remain dispersed in the medium. Hence, the oxidized CNMs can be readily used for electrophoretic deposition [20]. Fullerenes have truncated icosahedron geometry, normally very stable for functionalization. But to fully explore the applications of fullerene-based materials, functionalization is required. With the advances in the ultrasonic techniques, fullerenes (C_{60}) have shown to undergo fast functionalization in presence of ultrasonic radiations. Fullerenes have been functionalized with hydroxy- and oxygen-containing functional groups using the potentials of sonochemistry. These prepared polyhydroxylated fullerenes are now widely used as drug carrier in biomedical applications [21].

17.3 Applications of FCNMs for a Green and Sustainable Environment

The applications shown by FCNMs ranging from removing pollutants to acting as catalysts are largely based on the combination of properties shown by pristine CNMs and due to the modifications/functional groups present [22–28].

17.3.1 FCNMs in Wastewater Remediation

CNMs have high surface to volume ratio, a key requirement for them to act as adsorbents. However, addition of more functional groups enhances the chemical adsorption activity shown by CNMs. One of the most promising applications of FCNMs is its usage in wastewater remediation. The contaminants such as heavy metal ions, organic molecules, and dyes can be adsorbed on the surface of FCNMs and helps in cleaning water. The contaminants are usually chemisorbed based on electrostatic or coordinate interactions with –COOH, $-NH_2$, and –OH groups present on the surface of FCNMs. Sometimes functional groups on CNMs also act as ion exchangers and effectively exchange heavy metal ions from water. As far as wastewater remediation process is concerned, it has been found that most of the literature is dedicated to FSWCNTs, functionalized multiwalled carbon nanotubes (FMWCNTs), and graphene-based materials. Nonetheless, composites, hybrids, and other FCNMs are also examined currently in research [29–37].

Metal ions enter into water channels through effluents drawn from industries manufacturing leather, cement, dyes, metal alloys, steel, etc. These are non-degradable, highly toxic, and carcinogenic in nature. The removal of ions can be via adsorption, sorption, or photodegradation of pollutants on FCNMs. The role and mechanism of FCNMs in photodegradation of pollutants is discussed in Section 17.3.4. Lead, chromium, cadmium, and mercury are the most toxic heavy metals even in trace amounts as these are non-biodegradable and bio-accumulative. Magnetic functionalized CNMs ($Fe(NO_3)_3$ or Fe_3O_4–CNMs) have received a lot of attention in removing these heavy metal ions and to recover the FCNMs after the adsorption process. The heavy metal ions are adsorbed on the surface to these FCNMs and then the solution is kept under the influence of high magnetic field. The desorption of metal ions take place freeing FCNMs. This is the best way to retrieve/recycle the nanomaterials and avoid environmental risk. Among metalloids, As(III) and As(V), are most carcinogenic ions found in contaminated ground water. The removal of these ions has also been done through magnetic functionalized CNMs viz CNTs-based iron oxides and GO-$MnFe_2O_4$ [38]. Many metal ions such as Cd(II), Cr(III), Cr(IV), Cu(II), Ni(II), Pb(II), and Zn(II) are found to show adsorption potency toward functionalized CNMs, namely triphenylphosphine-linked multiwalled carbon nanotubes (MWCNTs), polyacrylic acid-modified MWCNTs, NaClO-SWCNTs/MWCNTs, MWCNTs-iron oxide, ethylene diamine tetra-acetic acid-modified fullerenes/graphene/MWCNTs, GO, GO-$MnFe_2O_4$, and magnetic-$CoFe_2O_4$-rGO [39, 40]. Arsenic ion, As(V) and manganese ion, Mn(VII) also display enhanced adsorption on MWCNTs-OCH_2COOH (functionalized MWCNTs) [41].

Recently, a hydroxamic acid-functionalized GO is used to adsorb Cd(II) ions from aqueous solution with high adsorption capacity of 502 mg/g. The adsorbent can be successfully recovered using magnetic surfactant. This magneto-responsive surfactant interacts with functionalized GO to form non-covalent magnetic functionalized graphene. Further the change in pH results in desorption of the surfactant freeing the adsorbent [42]. rGO decorated with 4-sulfophenylazo groups is found to adsorb Pb(II), Cu(II), Ni(II), Cd(II), and Cr(III) from aqueous solution. The adsorption involves ion exchange and coordination of metal ion with adsorbents [43]. Recently, a bis(2-pyridylmethyl)amino groups substituted GO proves to be a remarkable solid phase adsorbent for the removal of heavy metal ions. The removal of Cu(II), Ni(II), and Co(II) from wastewater is successfully performed. The desorption/regeneration of adsorbent is sufficiently substantial, suggesting a new path of using solid phase adsorbent toward the adsorption of toxic metal ions from wastewater [44].

There are few poisonous rare earth metal ions present in wastewater, whose removal needs to be addressed. Gd(III), La(III), Y(III), and Nd(III) ions display high sorption capacity on GO surface, thereby allowing their removal from bodies of water [45]. In another study, Nd(III) and Fe(III) ions coexisting in aqueous solution are adsorbed by acid-functionalized MWCNTs. Desorption studies revealed reusability of FMWCNT adsorbent. The selectivity of adsorbent toward Nd(III) increases in the presence of ascorbic acid indicating selective recovery of Nd(III). The whole process involved no use of toxic solvents [46].

Industries such as food, cosmetic, pharmaceutical, textile, and plastic extensively use organic dyes that then enter the water system through the industrial drain off and cause severe damage to the ecosystem. These are highly carcinogenic and toxic even at low concentration. Removal of dyes from wastewater can be achieved through CNMs [47]. Several studies are available on the action of FCNMs on various types of dyes such as acid, natural, basic, synthetic, disperse, and azo dyes. Organic dye pollutants being amphipilic in nature interact with both functional groups and the surface of FCNMs, creating more adsorption sites resulting in strong adsorption. Dyes such as basic red 46, bromo thymol blue, congo red, direct red 23, malachite green, methyl orange, and methylene blue are shown to adsorb on functionalized CNT such as NH_2-SWCNT, COOH-SWCNT, oxidized-SWCNT, oxidized MWCNT, magnetic Fe_3C-MWCNT, TiO_2-CNT, thiol-MWCNT, *p*-toluene sulfonic acid, and polyaniline GO-CNT in an aqueous medium [48, 49]. The sorption of the dyes on GO is strongly governed by π–π stacking and electrostatic interactions. The pH and ionic strength of the aqueous solution play significant role in the removal process [50]. Graphene-magnetite nano-hybrid (G/Fe_3O_4) is another material used for the removal of dyes. A study shows 99% sorption capacity of fuchsine dye within 30 minutes of exposure [51]. The theoretical studies using conductor-like screening model for real solvents (COSMO–RS) on the interaction of crystal violet with MWCNTs–OH and MWCNTs–COOH suggest that the hydrogen bonding interactions dominate during the adsorption process [52].

Apart from dyes, there are many harmful organic chemicals that leach from the industries such as automobile, oil-refining, and pharmaceutical. The organic compounds include phenols, amines, polycyclic aromatic compounds, polychlorinated biphenyls, antibiotics, and pesticides. On the same sorption principle, FCNTs and graphene-based materials have been exploited for removal of these organic pollutants. Sulfonated graphene nanosheets are shown to soak α-naphthol effectively via π–π interactions. GO also remove naphthalene, β-naphthol, 1,2,4-trichlorobenzene, and 2,4,6-trichlorophenol via hydrogen bonding [50]. Antibiotics enter the water channel as the waste from pharmaceutical industries, hospitals, etc. The removal of antibiotics have been reported through adsorption or catalytic oxidation processes. Since the antibiotics are large organic molecules consisting of variety of substituent, the FCNMs can provide effective ways of interactions. The adsorption phenomenon involves π–π stacking, hydrogen bonding, and hydrophobic and electrostatic interactions. The functionalization, pH, and ionic strength of the medium play key role in adsorption studies. Doxycycline, tetracycline, oxytetracycline, norfloxacin, and ciprofloxacin can be eradicated from wastewater using GO and rGO-magnetite hybrid [50, 53].

Besides sorption and filtration of pollutants, FCNMs have also been found to possess strong antimicrobial properties. These are shown to disinfect water by removing not only chemical contaminants but also microbial pathogens such as virus and bacteria, making water fit for drinking purposes. Surface membrane functionalized SWCNT attack and inhibit the action of *Escherichia coli*. Oxidized-SWCNT with oxidants such as H_2O_2 and NaOCl exhibit high tendency to kill *Bacillus antharacis* spores. SWCNT-OH and SWCNT-COOH display strong antimicrobial activity toward *Gram-positive* and *Gram-negative* bacteria [54].

17.3.2 FCNMs in Air Pollution Remediation

There are mainly three roles that any potential nanomaterial should satisfy in order to become useful in remediation process especially toward air pollutants. FCNMs are expected to detect the pollutants present in air (with as low as few ppb concentration of pollutant), to adsorb/degrade the pollutants and to filter the pollutants and pass the clean air. As per World Health Organization, the air is said to be polluted if it contains particulate matter, nitrogen-based oxides (NOx), ground-level ozone, sulfur-based oxides (SOx), volatile organic compounds (VOCs), carbon monoxide, and chlorofluorocarbons beyond the permissible limit. In fact there is also a concern regarding the increase in the amount of greenhouse gases present in the atmosphere. Apart from these gases, there are harmful VOCs mainly benzene, toluene, ethylbenzene, and xylene, present in atmosphere.

Some of the FCNMs have shown potential in sensing these pollutants. The defects and carbon vacancy in functionalized graphene and functionalized GO make them better sensors. The sensing ability of the adsorbents (FCNMs) is based on the change in the conductance value it experiences on adsorbing/absorbing gaseous molecules. Graphene-based materials and doped graphene sheets are shown to have a potential to be gas sensors owing to their large surface area and high electrical conductivity [55]. Selectivity and sensitivity during the sensing is the key factor. The functionalization of graphene not only allows selectivity but also induces difficulty in desorption of gas.

Removal of VOCs has been achieved by macrocyclic-functionalized CNT and graphene/GO/rGO. The macrocyclic groups are mainly cyclodextrins and calixarene derivatives [56]. The mechanism of adsorption of non-polar VOCs on FCNMs is mainly physisorption and less chemisorption. The polar functional groups on FCNMs reduce the adsorptive power of VOCs. For instance, comparison of GO and rGO as adsorbents for VOCs in water vapor environment reveal that VOCs shows a strong interaction with rGO that contains less oxygenated groups on the surface than GO [57]. On similar lines, the zwitterionic CNT showed less sorption toward gases (CO_2, CH_4, and N_2) compared to carboxylic acid CNT [58].

Greenhouse gas, CO_2 can be captured from flue gas by 3-aminopropyl-triethoxysilane-MWCNT. The amine rich surface of this FMWCNT provides more sites for CO_2 adsorption even at low temperatures [59]. A study on adsorptive capacity of nanoparticles toward CO_2 and CH_4 suggest that Fe_3O_4-graphene display greatest sorption capacity toward CO_2 due to its high thermal stability and porosity [60]. Metal nanoparticle-functionalized CNMs are also the promising candidates in environment gas monitoring. Pd nanocluster-MWCNT effectively monitors NO_2, H_2S, NH_3, and CO gases, even if they are present in amounts as low as sub-ppm level [61]. The carbon nanofibers (CNFs) that are also known as vapor grown nanofibers are rolled graphene sheets arranged in shapes as stacked cups, cones, or plates. Functionalization of CNFs with carboxylic acid and ester exhibit greater sensing ability toward NH_3 compared to pristine CNF at room temperature [62]. These types of sensors can be easily employed in a gas-monitoring device for detecting harmful gases in the environment.

17.3.3 FCNMs in Nuclear Waste Management

Nuclear wastes containing f-block metals enter the environment through illegal disposing of nuclear waste, electrical waste, and nuclear fallouts. The exposure of these radionuclides imposes heavy threat on living organism. The toxicity may induce bone, lung, or liver cancer in humans. Nuclear waste management deals with the separation, processing, and recovery of valuable radioactive isotopes of actinides such as thorium, uranium, plutonium, and neptunium present in nuclear wastes. FCNM-based sorbents are not only efficient but are also selective in adsorbing/absorbing toxic ions such as Pu^{4+}, $PuO_2{}^{2+}$, $UO_2{}^{2+}$, Th^{4+}, $NpO_2{}^{+}$, Am^{3+}, $NpO_2{}^{2+}$, and Eu^{3+}. Magnetic Fe_2O_3-MWCNTs and $CoFe_2O_4$-MWCNTs have been widely used for adsorption of radionuclides. Diglycloamide-MWCNT also adsorbs Th^{4+} and $UO_2{}^{2+}$ effectively. This FMWCNT shows brilliant sorptivity toward $UO_2{}^{2+}$ compared to other amide-functionalized MWCNTs. Another FMWCNT for adsorption of $UO_2{}^{2+}$ are carboxymethyl cellulose-functionalized MWCNT. Am^{3+} can be extracted by dendrimer functionalized MWCNT. Addition of functionalized fullerenes into the solution of FMWCNT has been found to affect the sorption performance of Th^{4+} on FMWCNT. It has been reported that addition of hydroxylated fullerene ($C_{60}(OH)_n$) and carboxylated fullerene ($C_{60}(C(COOH)_2)_n$) increases the sorption of Th^{4+} on oxidized MWCNT. The mechanism of action could be due to the enhanced π–π electron donor–acceptor interactions between functionalized fullerenes and FMWCNT giving a competitive path for the sorption of Th^{4+} on FMWCNT. Pu^{4+} and $PuO_2{}^{2+}$ can be adsorbed on oxidized MWCNT [63].

Similar results are obtained with graphene-based nanomaterials. GO, GO-supported ethyleneimine polymer composite, cetyltrimethyl ammonium bromide, and phytic acid-modified GO and phenanthroline diamide GO are some of the adsorbents for uranyl ions. Other radionuclides are also adsorbed efficiently on functionalized GO and rGO [64]. Though FCNMs have gained popularity in removing these radionuclides, several current research efforts aim to magnify the results at larger scale. More sustainable approach is still required for the large scale production of these FCNMs in order to address waste management issues.

17.3.4 FCNMs as Electrocatalysts and Photocatalysts

Here, few of the environment-friendly processes involving FCNMs would be described, namely water splitting, carbon dioxide (CO_2) conversion process, and nitrogen reduction process. FCNMs can act as both electrocatalyst as well as photocatalyst for these processes [65]. The above said processes provide alternative energy resources by simultaneously helping in reducing pollutants from the environment followed by degradation process of pollutants, for a clean and green environment.

In water splitting, water is broken down into hydrogen (used as fuel, discussed in Section 17.3.5.4) and oxygen via hydrogen evolution reaction (HER) and oxygen evolution reaction (OER) taking place at the cathode and the anode, respectively, under the influence of electricity. Once the hydrogen has been created electrolytically, another hindrance in its use as energy source is its storage, which can also be dealt with FCNMs (discussed in Section 17.3.5.3). Electrocatalytic reduction of nitrogen to ammonia is another important process as ammonia is essential for agriculture and industries. Nitrogen reduction reaction (NRR) takes place at electrocatalyst. It is interesting to note that for both electrolytic processes, water splitting and ammonia reduction, water is used as an electrolyte. For OER, HER, and NRR, multiple heteroatom-doped graphene sheets have proved to be a very good electrocatalyst [66]. Other modifications of graphene like GO, rGO and even their hybrids with metal oxides and other nanoparticles can be used efficiently for OER and HER [67, 68]. Computational studies have shown that first-row transition metal (from Sc to Zn) single atoms attached to graphyne nanosheet (having acetylene linkage) prove to be an efficient electrocatalyst in OER and HER for water splitting [69]. Another unique FCNM, C_{60} adsorbed on SWCNT also provide potential surface for both OER and HER [70].

Water splitting process can also take place using photocatalyst, by using solar energy. One such photocatalyst is N-doped CNDs (prepared by hydrothermal method). Doping introduces n- and p-type domains in CNDs leading to the ease in transfer of electrons between such domains and increased charge separation on irradiation of light on it. CNDs can be further functionalized using different moieties like methanol, oligomeric poly(ethylene glycol) diamine, and 1,1′-bi(2-naphthalene)-modified CNDs to enhance their activity. Such FCNDs have intramolecular Z-scheme structure, wherein photoreduction reaction occurs on p-type domain and photooxidation reaction on n-type domain [71]. Another FCNM for this process is graphane (hydrogenated graphene or graphene hydride), obtained from covalent functionalization of graphene involving covalent bond formation between hydrogen and carbon atoms of graphene [72].

In CO_2 conversion process, CO_2 reduction reaction (CO_2RR) is an important reaction which takes place at cathode of the electrolysis cell. The products of this electrolysis include the reduction of CO_2 to CO, formic acid, oxalic acid, formaldehyde, methanol, methane, ethylene, ethane, and ethanol, to name a few. This electrolytic process can also be catalyzed by FCNMs. Doped CNMs like CNT, graphene, GQDs dots with heteroatoms like nitrogen, boron, oxygen, phosphorus, sulfur, fluorine, chlorine, etc. prove to be efficient electrocatalysts for CO_2RR [68, 73, 74].

For converting multi-step CO_2 conversion process into a simpler, single step process, photocatalysts come into picture. This process is called artificial photosynthesis. Here, FCNDs play a vital role as photocatalyst owing to oxygen and nitrogen containing functionalities on their surface which acts as Lewis base. CNDs obtained from GO are used to produce methane from CO_2 [71]. Various functionalized graphene, GO, rGO, GQDs, and their hybrids also enhance the photocatalytic CO_2RR [67].

One major concern today is the increasing pollution levels in air and water, caused by waste chemicals coming out of factories as well as households. A simple way out would be to treat such waste chemicals and convert them to harmless products. One such environment-friendly technique is photocatalysis that involves the degradation of pollutants like harmful organic dyes, heavy metal ions, etc. over green photocatalyst FCNMs. Some important FCNMs in this regard include GO and rGO that exhibit photocatalytic activity in UV and visible range. They have large surface area, layered structure, and oxygen functionalities on the surface, which is responsible for their covalent and non-covalent interactions with pollutants. Their activity can be further improved by introducing various structural modifications like doping. One such example involves Cu-doped rGO [75]. Hybrid structures involving GO are also potent photocatalysts [67]. Another class of CNM-based photocatalyst is CNDs that are nontoxic, low-cost, and photostable materials. Various functionalities like carboxyl, hydroxyl, and amine can be introduced on the surface of CNDs to enhance its photocatalytic activity. Structural modifications further improve their activity owing to improved photogenerated charge separation. This makes transfer of electrons much easier. N-doped CNDs can be prepared from glucose by heating it in microwave that can be used for photocatalytic degradation of toxic organic dye like Rose Bengal and methyl orange. S,N-codoped CNDs can be used as photosensitizers for the conversion of hazardous pollutant like hydrogen sulfide to hydrogen (which may be further used as an energy source). Phenylhydrazine-functionalized CNDs, are highly efficient for the degradation of organic dye like methylene blue [71, 76–78].

17.3.5 FCNMs for Energy Storage

In the current century, the major source of power generation is still fossil fuels, which are bad for our environment. Therefore, major research goals in today's world, is to look for better, effective, and greener options. Various environment-friendly alternate sources include wind, solar, and geothermal energy, to name a few. The major problem with them apart from the high cost is the storage of energy generated by these sources. The most common medium of storing the generated energy are batteries. However, the conventional batteries use mainly environment-unfriendly metals like lithium and lead and also have a short life span. So, the researchers engaged in look-out for more efficient and greener sources for energy and its storage stumbled upon FCNMs. They not only have a role in improving already existing renewable source of energy like solar cell but also energy storage materials like batteries, fuel cell, and superconductor.

17.3.5.1 FCNMs and Solar Cells

Solar energy technology is one of the most important technologies for the development of clean renewable energy for future. As a result, solar conversion technologies currently attract intense interest, research, and development, mostly due to their environmental advantages, commercial scalability, and high energy conversion efficiency. Different new-age solar cells explained in this chapter include dye-sensitized solar cells (DSSCs) [79], organic solar cells (OSCs) [80], and perovskite solar cells (PSCs) [81]. These solar cells comprise of cathode, photoactive layer, and anode. Two more components, i.e. electron transport layer (ETL) and hole transport layer (HTL), are present in PSCs and OSCs. In DSSCs, electrolytes are included for charge transport and redox reaction [82].

DSSCs have been rigorously studied as a promising low-cost solar cell technology. A typical DSSC comprise of conductive glass substrates, a dye-sensitized metal oxide semiconductor electrode, a catalyst counter electrode, and an electrolyte solution inserted between the two electrodes. Usually, very efficient Pt counter electrodes are used in DSSCs. However, now, the use of scarce, costly, and environment-unfriendly Pt may be replaced by low-cost, sustainable, and green FCNMs. Also, these can be used to make photoanode electrode. The role of FCNMs as both cathode and anode in DSSC can be attributed to their high surface area, stability, and good electron conductivity [83]. Studies have shown that DSSCs containing TiO_2 nanoparticles and FCNMs as photoanode electrode have high potential [84]. Some of the examples include TiO_2 nanoparticles incorporated plasma-functionalized carbon nanocapsules (CNCs), CNTs and graphene, acid-treated carboxyl group-SWCNTs, and carboxyl group-MWCNTs. These hybrids show an increased conversion efficiency and photocurrent density because incorporation of FCNMs increases the surface area of TiO_2 film [85]. Other advantages of such arrangement include few steps involved in their manufacturing process, therefore reducing the time, cost, and number of chemicals involved in the synthesis [83, 86, 87].

FCNMs, especially FCNTs have been synthesized using various simple, environment-friendly methods resulting in the formation of efficient, sustainable, green metal-free FCNTs to be used as counter electrode for DSSCs. Water-soluble, polyelectrolyte-grafted-MWCNTs were synthesized via a low-cost, facile deposition technique using an environmentally friendly solution [88]. Apart from this, noble metals (Ag, Au, Pt) functionalized single wall CNFs are promising as counter electrode [89]. FCNMs can be used as electrolyte to increase the efficiency DSSCs. Carboxy-functionalized MWCNTs containing poly(vinyl alcohol)/poly(ethylene glycol) gel electrolyte reduces the resistance to charge transfer and hence prove to be efficient electrolyte [90]. Ionic liquids, like 1-propyl-3-methylimidazolium iodide, are another environment-friendly electrolyte to be used in DSSCs, but the main problem is their high viscosity. This can be overcome by the use of another potential candidate for DSSCs, which is carboxyl group-functionalized-MWCNTs. This reduces the viscosity of ionic liquid without compromising on its efficiency or stability [91].

To improve efficiency and to obtain more sustainable solar cells, graphene has been modified over the years by introducing various functionalities or by making some composites or even by doping. Some of these modifications include environment-friendly edge-functionalized graphene nanoplatelets, which has –OH, –COOH functionalities on the edge, thereby maintaining the basic graphitic structure of graphene [92], functionalized graphene nanosheets (in the form of composite with CoS nanoparticles) [93], and functionalization of graphene with poly(3,4-ethylenedioxythiophene) and polystyrenesulfonate [94]. All these could be used as a counter electrode in DSSC resulting in the formation of low-cost and platinum-free electrode having a high-power conversion efficiency.

OSCs are another class of potential renewable energy sources. This unique class of solar cell is made from inexpensive organic materials which can be deposited on a variety of substrates. Their efficiency can be improved by replacing harmful organic materials with an advanced and efficient bulk heterojunction (BHJ) of semiconducting polymers and fullerene derivatives [95]. But this arrangement is inherently flawed, owing to slow transport properties of semiconductors. Incorporation of FCNTs as active photolayer in BHJ can act as a charge transporter and hence increases the interaction between the electrodes because of ease of charge transfer. One such example includes blending of acid-treated SWCNTs into the composites of poly(3-hexylthiophene) (P3HT) and [6,6]-phenyl-C_{61}-butyric acid methyl ester (PCMB) resulting in improved maximum power efficiency [96]. 3,5-Dinitrobenzoyl-functionalized graphene nanoflakes (GNFs) can also be used as charge carrier in organic ternary BHJ solar cells. This kind of functionalization helps in fine tuning energy levels of graphene in concordance with that of polymer and fullerene used for OSC [97]. Another potential candidate is functionalized NDs. Carboxylated-ND can be used to transport charge in BHJ based solar cells by forming a composite with C_{60} via covalent bond. This unique arrangement takes advantage of C_{60} as electron acceptor and ND as transporter of electron [98]. A remarkable device performance enhancement has been observed in BHJ solar cells by employing N- or B-doped MWCNTs in photoactive layer composed of P3HT (polymer):PCMB (fullerene). Its efficiency and performance are increased because of ease of charge transfer. B-doped MWCNT is p-type semiconductor having highest occupied molecular orbital close to polymer (electron donor), enhancing easy hole transfer. On the other hand, N-doped MWCNT is n-type semiconductor having lowest unoccupied molecular orbital close to fullerene (electron acceptor), enhancing easy electron transfer [99, 100]. GO can also be used as hole transport material by incorporating in P3HT:PCMB photolayer [101].

FCNTs, graphene, and derivatives of fullerene can be used to frame anode, cathode, photoactive layer, ETL or HTL layer to enhance the activity of PSCs. Only CNM-based PSC has also been fabricated in which graphene acts as anode and CNT acts as cathode [82].

17.3.5.2 FCNMs and Supercapacitors

Recently, it has been shown that FCNTs and graphene-based materials have the potential to improve the electrochemical performance of the supercapacitors because of their mesoporous network with a large surface area, responsible for efficient charge storage [102]. CNTs can be functionalized in simple, cost-effective, and greener ways to be used as electrode in supercapacitors. Functionalization of CNTs can also be done to improve their solubility and efficiency as an electrode. A very simple approach to functionalize CNTs (oxidize CNTs) involves the use of UV-generated ozone. Oxidized CNTs are hydrophilic and hence, can use low-cost, hydrophilic, and highly conducting electrolytes. This results in increase in the value of specific capacitance and power density, which can be attributed to the reversible redox reaction of oxygen-containing functional groups [103]. Another simple and green hydrothermal technique to manufacture FCNTs has been used to integrate 2D nanostructured MoS_2 on to the surface of alkali-activated FCNTs. This results in an increase in surface area of FCNTs, thereby reducing the interfacial resistance leading to ease of electron transfer through it. Also, the power density and specific conductance increases [104]. Functionalized CNTs (having aldehydic, ketonic, and carboxyl group) can be prepared by electrochemical method to make portable and wearable solid-state supercapacitor [105].

Another interesting method involves the use of functionalized candle flame soot to make supercapacitor electrodes. Candle flame soot contains carbon nanoparticles. The electrochemical studies showed that functionalized candle flame soot (especially oxygen-containing functional groups) have high specific capacitance, high current density, high specific energy density, and low internal resistance properties, making it suitable as supercapacitor electrodes [106]. Another green and sustainable candidate for supercapacitor electrode is phosphate-functionalized CNT-containing carbon composites, prepared by one-pot soft-template hydrothermal reaction and fructose as source of carbon [107]. Other such modifications include water-soluble arylsulfonic acid-functionalized single-walled nanotubes (SWNTs) [108], carboxylic and ester functionalized SWNT and multiwalled nanotube (MWNT) electrodes using low-cost spray deposition technique [109]. The addition of transition metal oxides to the surface of CNTs has been studied extensively during the past decade. Iron oxides, being most abundant leading to its low cost, have emerged as one of the most environment-friendly addition to CNTs [110, 111]. Ionic liquids (like having imidazolium cation), owing to their thermal and chemical stability provide a safer option to enhance the charge storage capacity of supercapacitor. Its function is further enhanced if a supercapacitor use electrodes made of functionalized double-walled CNT and GO [112, 113].

17.3.5.3 FCNMs and Hydrogen Storage

Hydrogen, over the decade, has evolved as one the most clean, green, and safe fuel. It is not a primary fuel but has to be generated from other sources. The main problem related to the use of hydrogen as a fuel is the daunting task of storing and transporting it. The most appropriate way of transporting and storing it involves the use of solid-state materials. Over the years, researchers have been focusing on designing new and efficient materials that could be used as a carrier for hydrogen. Hydrogen carrier should interact with hydrogen reversibly or there should be optimum interaction between them so that it can store hydrogen and release it when necessary. Functionalization of CNMs enhances their interaction with hydrogen by acting as anchor between hydrogen and CNMs, thereby increasing their hydrogen storing capacity. Some of the modifications include BH_3-functionalized SWCNTs, acid-functionalized CNTs, polyvinylpyrrolidone functionalization of MWCNTs, and TiO_2-functionalized SWCNT. These functionalized, nanoengineered three-dimensional CNT fibers have at least two times more hydrogen storing capacity than the other simple carbon materials. This can be attributed to their high density, high surface area, and excellent thermal conductivity of FCNMs [114–117]. Introduction of various metals like lithium, scandium, titanium, and palladium in CNMs including CNTs, graphene, and fullerene can also enhance their hydrogen storage capacity. One thing needs to be kept in mind is that, increasing the amount of metal in CNMs does not necessarily increase its hydrogen storing capacity but it is the change in morphology of the carrier, both structurally (creating of more space for hydrogen in CNMs) and chemically (increasing the interaction between CNMs and hydrogen), which is responsible for enhanced hydrogen storage capacity of CNMs [118].

Functionalized graphene including doped graphene, acid-functionalized graphene, GO, and rGO are an important carrier of hydrogen [119]. The hydrogen storing capacity of rGO can be modified by various metals like Ni, Pd, and Ag [120]. rGO can also be modified by doping with nitrogen. The hydrogen storage capacity of such doped graphene increases many fold because H—H bond in H_2 becomes weak and C—H bond between graphene and hydrogen increases because of the increase in polarity in graphene due to the substitution of carbon by nitrogen in graphene sheets [121]. Certain acids like chlorosulfonic acid or oleum super acids have dual role in enhancing the hydrogen storage power of graphene sheets. Firstly, because of the introduction of aryl groups in graphene, space between subsequent graphene sheets increases, making more room for hydrogen. Secondly, presence of sulfonic groups leads to partial charge transfer between graphene sheets, further enhancing the capacity to take hydrogen [122, 123]. Graphane and functionalized graphane are also potential candidates for hydrogen storage [72, 124].

17.3.5.4 FCNMs and Fuel Cell

The polymer electrolyte membrane fuel cell or proton exchange membrane fuel cell (PEMFC) has over the years emerged as a clean and renewable energy source. The three main components of PEMFC are anode (at which hydrogen oxidation reaction [HOR] takes place), cathode (at which oxygen reduction reaction [ORR] takes place), and proton exchange membrane. But, like any other renewable source, it also works using Pt as both cathode and anode, which is available in limited quantities in nature, leading to its high cost. The main problem is ORR, as reduction of oxygen is not easy, O=O being a strong bond. ORR takes place following two different pathways: one is four-electron pathway leading to the formation of water as by-product and other is two-electron pathway leading to the formation of hydrogen peroxide as a by-product. Both these pathways are useful as one gives clean water and other gives an important oxidizing agent used in industries as bleach and for removing pollutants from wastewater. Although the main focus has been to design more potent electrocatalyst for four-electron pathway, there are some materials like N/O-codoped graphitic carbon catalysts which selectively follows two-electron pathway leading to the formation of hydrogen peroxide [125].

In order to search for designing alternative for Pt-based electrocatalyst, FCNMs showed high potential, which can either act as metal-free electrocatalyst or can be used in conjugation with Pt, thereby reducing the usage of Pt. Some of the modifications include carboxyl functionalized SWCNT, non-covalent functionalization by using solvents, surfactants, aromatic compounds, polymers, or polyelectrolytes [126, 127]. Development of metal-free electrocatalyst involve the use of poly(diallyldimethylammonium chloride) functionalized CNTs [128]. Another modification of CNM is FCNDs which can be used as electrocatalyst for ORR. It possesses –OH, –COOH, and $-NH_2$ groups on the surface. They can be further modified by replacing carbon with nitrogen, sulfur, phosphorous, etc. to enhance their activity. Graphene-based FCNMs like doped graphene and GO are also potential FCNMs in this regard [129–131]. Apart from use of FCNMs as electrocatalyst in PEMFC, graphene-based CNMs can also be employed as very durable and stable proton exchange membrane with high proton conductivity [129].

Microbial fuel cell (MFC) is another important fuel cell. It has two-pronged function, one in wastewater treatment (as the source of chemical energy in this cell is organic waste and biomass) and another in cost-effective and sustainable energy generating process because it uses freely available microbes as catalyst. Here also, the main problem is ORR. In this, the electrocatalyst used should be compatible with microbes as well apart from providing surface for ORR. In this respect, FCNMs including FCNTs and graphene-based composites are much in use [132–134]. A very useful, economical, and environment-friendly CNM has been designed, namely N-doped hierarchical porous carbon nanosheets using water lettuces as source of carbon. The major advantage of such FCNM is that it can be used to replace Pt as electrocatalyst for ORR not only in MFC but other fuel cells like Zn-air (alkaline) and direct methanol fuel cell [135]. Another such green electrocatalyst include novel hierarchical nitrogen-doped porous carbon derived from *Borassus flabellifer* male inflorescences [136].

FCNMs can be used as electrocatalyst in other fuel cells also. Platinum nanoparticle-functionalized carbon composites supported at the carbon–ceramic electrode as an electrocatalyst and amino-rich, cationic poly(ethyleneimine) polyelectrolyte FMWCNTs supported electrocatalyst in direct ethanol fuel cell [137, 138], FCNTs supported electrocatalyst for glucose fuel cell [139] and acid-functionalized CNT composites for formic acid fuel cell [140].

17.3.6 FCNMs and Biofuels

Biodiesel is a renewable, biodegradable, and eco-friendly fuel made by transesterification and esterification of vegetable oils, animal fats, or free fatty acids. The reaction is a catalysis-driven process. Selection of catalyst plays an important role in achieving maximum yield of biodiesel under mild conditions. Acidic/basic FCNMs are potential candidates as catalysts in transesterification process owing to their excellent surface properties [141]. Acid-functionalized CNMs have been explored in detail. Sulfonic-functionalized MWCNT is one of potential catalyst in the production of biodiesel [142, 143]. Sulfonic acid grafted on rGO (rGO-SO_3H) promises to be an efficient, recyclable heterogeneous catalyst for a conversion reaction of soybean oil to biodiesel [144]. Sulfonic or phosphate groups-functionalized rGO catalyst is found to be far superior to commercially available catalyst, Amberlyst-15, for transesterification of rapeseed oil to biodiesel [145].

Biofuels are the fuels derived from natural/biodegradable oils. Biodiesel is a class of biofuel that contains more oxygenated groups as compared to diesel. This feature helps in better combustion of the fuel and thereby reduces the particulate matter, CO, and other unburned hydrocarbons content. At present, the use of biodiesel is needed to control the excessive air pollution in the world. However, there is one disadvantage of using biodiesel over diesel and that is its high viscosity and density. On burning viscous and denser fuel, brake-specific fuel consumption increases along with the increase in the NOx and smoke emission [146]. Since in the near future, biofuels are recommended at a larger scale, it is imperative to look for ways to overcome this disadvantage. Additives such as ethanol and specific catalysts help in reducing the viscosity and density of biodiesel. A lot of research is going on in selecting the catalyst as additive since the addition of catalyst not only reduces NOx and smoke emission but also leads to the speeding up of fuel combustion reactions, thereby, improving engine performance [147]. FCNMs with polar (oxygenated) functionalization have exceptional properties like shape, size, aspect ratio, high thermal and electrical conductivity, high mechanical strength, and better solubility and therefore become suitable as catalysts. Addition of hydroxylated-MWCNT into a biofuel blend (diesel, biodiesel, and ethanol), significantly improves the engine performance and emissions. As far as emission of harmful gases are concerned, the study reports reduction in CO and unburned hydrocarbon emission but increase in NOx level [148]. This is a matter of concern. A study where, GO is used as an additive in catalyzing the combustion of biodiesel (made from jatropha oil) showed satisfactory results with ultra-low NOx, CO, and unburned hydrocarbon emissions. This blend of biodiesel, GO, and *n*-butanol maintains a high thermal efficiency of the engine as well [149]. Nonetheless, advanced studies are still required to come up with more effective FCNMs that cater to all the drawbacks of using biodiesel as a fuel.

17.3.7 FCNMs as Nanofertilizers

With the vision of clean and green environment, FCNMs can be proposed to serve as fertilizers creating a revolution in agricultural industry. Due to the large surface to volume ratio and high pore size, FCNMs have the ability to carry nutrients effectively. The main advantage of nanofertilizer over conventional fertilizer is the slow and controlled release of nutrients. This slow nutrient delivery system ensures proper and consistent crop growth and guarantees no excess nitrogen leach from fertilizers [150]. FCNMs act on plants differently based on the nature and type of functionalization. The acidic/basic/neutral functionalization would affect the uptake and release of nutrients, thereby affecting the rate of growth of plants. Though transportation of FCNMs within the cell is somewhat difficult, owing to the cellular membrane and structural constraints, some FMWCNTs have been found to successfully penetrate the plant cell walls of tomato and gram crop [151, 152]. In another hydroponic study, the uptake and translocation of two FMWCNTs in maize and soybean plants is investigated. The roots, stems, and leaves of maize and soybean plants take up and translocate the positively charged NH_2-MWCNT and negatively charged COOH-MWCNT, successfully. The results showed increased growth in maize while suppressed growth in soybeans after 18 days exposure [153]. On the other hand, urea functionalized MWCNT proves to be effective nutrient delivery systems in rice plant [154]. The germination of seeds is also affected by FCNMs. GO is found to stimulate the germination of chive and spinach in soil. The effects are governed by the uptake of water by functional groups on GO and its transportation by hydrophobic graphene layer to the seeds [155]. It is important to mention that all functionalized fullerenes are unable to promote cell growth in plants. The phytotoxicity of these nanomaterials is high. An impact of carboxyfullerene on *Nicotiana tabacum in vitro* cell cultures showed high phytotoxicity with cell wall spoiling [156]. An extensive research is still required to understand the mechanism of action of these FCNMs as nutrient carriers. This is needed to propose and generate novel nanofertilizers for future agricultural sector.

17.3.8 Miscellaneous Applications

FCNMs being biocompatible and eco-friendly hold enormous applications in diverse fields. FCNMs can attack and adsorb oil from water based on the functionalization. This feature can be utilized in removing dangerous oily layer above water surface which could result from oil spill during petroleum extraction. Magnetic FCNMs like ferric oxide nanoparticle-doped CNTs, super-paramagnetic iron oxide nanoparticle-decorated MWCNTs, and magnetic porous graphene/MWCNTs beads have been used to adsorb gasoline oil, diesel-in-water emulsions, and oil emulsions [157]. Many more FCNMs can be designed to check their feasibility toward oil spill remediation.

FCNMs play variety of roles ranging from drug carrier, gene therapy to biosensor and bioimaging. Functionalized NDs such as ND-Si, ND-OH, ND-COOH, ND-NH_2, ND-poly(ethylene glycol) (PEG), ND-poly-L-lysine, and many more have been shown to carry drugs like 10-hydroxy camptothecin, paclitaxel, doxorubicin, α-bungarotoxin, oligonucleotides [158]. FCNTs such as Pt conjugated-SWCNTs, folate-decorated magnetic MWCNTs, phospholipid–polyethylene glycol–SWCNTs, and Au–Ag/MWCNTs have been exploited in real-time biomedical applications. These FCNMs are potential materials toward developing sustainable medical devices [159].

17.4 Summary

In this chapter, we have summarized recent progress involved in development and usage of FCNMs toward green and sustainable environment. Functionalization of many inorganic and organic groups can be easily grafted on CNMs through environmentally friendly techniques. These fabricated FCNMs exhibit enhanced reactivity and selectivity by displaying features of both functional groups as well as CNMs synergistically. This widens their scope and application toward addressing existing environmental issues and also in developing novel nanomaterials to be used in energy storage devices. FCNTs and graphene-based materials display extraordinary affinity toward inorganic and organic pollutants and hence are much superior to conventional adsorbents. Nonetheless research is still needed to understand the mechanism of action of these FCNMs on pollutants and also to develop nanomaterials that can show reaction efficiency as close to 100% without loss of nanomaterials. FCNMs are useful not only as an energy source but for energy storage as well. Although applicability of FCNMs for commercial energy storage devices is still in nascent stage. Overall, it can be concluded that FCNMs are the unprecedented materials of the green and sustainable future.

Important Websites

http://www.pcrg.unsw.edu.au/functionalized-carbon-based-materials-for-electrochemical-energy-applications

https://www.sciencedaily.com/releases/2020/08/200803092131.htm

References

1 Karfa, P., De, S., Majhi, K.C. et al. (2019). Chapter 2.07: Functionalization of carbon nanostructures. In: *Comprehensive Nanoscience and Nanotechnology*, 2e (eds. D. Andrews, T. Nann and R. Lipson), 123–144. Academic Press. https://doi.org/10.1016/B978-0-12-803581-8.11225-1.

2 Kawamoto, M., He, P., and Ito, Y. (2016). Green processing of carbon nanomaterials. *Adv. Mater.* 28: 1–29. https://doi.org/10.1002/adma.201602423.

3 Dreyer, D.R., Park, S., Bielawski, C.W. et al. (2010). The chemistry of graphene oxide. *Chem. Soc. Rev.* 39: 228–240. https://doi.org/10.1039/B917103G.

4 Fujigaya, T. and Nakashima, N. (2015). Non-covalent polymer wrapping of carbon nanotubes and the role of wrapped polymers as functional dispersants. *Sci. Technol. Adv. Mater.* 16 (2): 024802. https://doi.org/10.1088/1468-6996/16/2/024802.

5 Modugno, G., Ménard-Moyon, C., Prato, M. et al. (2015). Carbon nanomaterials combined with metal nanoparticles for theranostic applications. *Br. J. Pharmacol.* 172 (4): 975–991. https://doi.org/10.1111/bph.12984.

6 Al-Hamadani, Y.A.J., Chu, K.H., Son, A. et al. (2015). Stabilization and dispersion of carbon nanomaterials in aqueous solutions: a review. *Sep. Purif. Technol.* 156 (2): 861–874. https://doi.org/10.1016/j.seppur.2015.11.002.

7 Naqvi, S.T.R., Rasheed, T., Hussain, D. et al. (2020). Modification strategies for improving the solubility/dispersion of carbon nanotubes. *J. Mol. Liq.* 297: 111919. https://doi.org/10.1016/j.molliq.2019.111919.

8 Basiuk, E.V. and Basiuk, V.A. (2015). Solvent-free functionalization of carbon nanomaterials. In: *Green Processes for Nanotechnology* (eds. V. Basiuk and E. Basiuk), 163–205. Cham: Springer. https://doi.org/10.1007/978-3-319-15461-9_6.

9 Georgakilas, V., Kordatos, K., Prato, M. et al. (2002). Organic functionalization of carbon nanotubes. *J. Am. Chem. Soc.* 124: 760–761. https://doi.org/10.1021/ja016954m.

10 Brunetti, F.G., Herrero, M.A., Muñoz, J.deM. et al. (2007). Reversible microwave-assisted cycloaddition of aziridines to carbon nanotubes. *J. Am. Chem. Soc.* 129 (47): 14580–14581. https://doi.org/10.1021/ja077927k.

11 Rubio, N., Herrero, M.A., Meneguetti, M. et al. (2009). Efficient functionalization of carbon nanohorns via microwave irradiation. *J. Mater. Chem.* 19: 4407–4413. https://doi.org/10.1039/B900776H.

12 Barrejón, M., Syrgiannis, Z., and Prato, M. (2018). Ionic liquids plus microwave irradiation: a general methodology for the retro-functionalization of single-walled carbon nanotubes. *Nanoscale* 10 (33): 15782–15787. https://doi.org/10.1039/C8NR04590A.

13 Mallakpour, S., Abdolmaleki, A., and Karshenas, A. (2017). Graphene oxide supported copper coordinated amino acids as novel heterogeneous catalysts for epoxidation of norbornene. *Catal. Commun.* 92: 109–113. https://doi.org/10.1016/j.catcom.2017.01.017.

14 Luo, J., Liu, Y., Wei, H. et al. (2017). A green and economical vapor-assisted ozone treatment process for surface functionalization of carbon nanotubes. *Green Chem.* 19 (4): 1052–1062. https://doi.org/10.1039/C6GC02806C.

15 Che, Y., Pang, H., Li, H. et al. (2019). Microwave-assisted fabrication of copper-functionalized carbon quantum dots for sensitive detection of histidine. *Talanta* 196: 442–448. https://doi.org/10.1016/j.talanta.2018.12.098.

16 Yang, P., Zhu, Z., Chen, M. et al. (2019). Microwave-assisted synthesis of polyamine-functionalized carbon dots from xylan and their use for the detection of tannic acid. *Spectrochim. Acta, Part A* 213: 301–308. https://doi.org/10.1016/j.saa.2019.01.043.

17 Xiong, Y., Cui, X., Zhang, M. et al. (2020). Microwave hydrothermal synthesis of gallotannin/carbon nanotube composites for the recovery of gallium ion. *Appl. Surf. Sci.* 510: 145414. https://doi.org/10.1016/j.apsusc.2020.145414.

18 Hu, H., Wang, X., Wang, J. et al. (2011). Microwave-assisted covalent modification of graphene nanosheets with chitosan and its electrorheological characteristics. *Appl. Surf. Sci.* 257 (7): 2637–2642. https://doi.org/10.1016/j.apsusc.2010.10.035.

19 Huang, Z., Zhang, H., Chen, Y. et al. (2013). Microwave-assisted synthesis of functionalized graphene on Ni foam as electrodes for supercapacitor application. *Electrochim. Acta* 108: 421–428. https://doi.org/10.1016/j.electacta.2013.06.080.

20 Yeo, E.S.Y., Mathys, G.I., Brack, N. et al. (2017). Functionalization and dispersion of carbon nanomaterials using an environmentally friendly ultrasonicated ozonolysis process. *J. Vis. Exp.* 123: e55614. https://doi.org/10.3791/55614.

21 Afreen, S., Muthoosamy, K., and Manickam, S. (2019). Sono-nano chemistry: a new era of synthesising polyhydroxylated carbon nanomaterials with hydroxyl groups and their industrial aspects. *Ultrason. Sonochem.* 51: 451–461. https://doi.org/10.1016/j.ultsonch.2018.07.015.

22 Hussain, C.M. (2018). *Handbook of Nanomaterials for Industrial Applications*. Elsevier.

23 Hussain, C.M. (2019). *Handbook of Environmental Materials Management*, 1e. Springer International Publishing.

24 Hussain, C.M. (2020). *Handbook of Functionalized Nanomaterials for Industrial Applications*. Elsevier.

25 Hussain, C.M. (2020). *Handbook of Manufacturing Applications of Nanomaterials*. Elsevier.

26 Hussain, C.M. and Keçili, R. (2019). *Modern Environmental Analysis Techniques for Pollutants*, 1e. Elsevier.

27 Hussain, C.M. and Kharisov, B. (2017). *Advanced Environmental Analysis: Application of Nanomaterials*. The Royal Society of Chemistry.

28 Hussain, C.M. and Mishra, A.K. (2019). *Nanotechnology in Environmental Science*, 2 Volumes. Wiley.

29 Abdolmaleki, A., Mallakpour, S., and Borandeh, S. (2016). Improving interfacial interaction of l-phenylalanine-functionalized graphene nanofiller and poly(vinyl alcohol) nanocomposites for obtaining significant membrane properties: morphology, thermal, and mechanical studies. *Polym. Compos.* 37: 1924–1935. https://doi.org/10.1002/pc.23371.

30 Abdolmaleki, A., Mallakpour, S., and Karshenas, A. (2017). Synthesis and characterization of new nanocomposites films using alanine-Cu-functionalized graphene oxide as nanofiller and PVA as polymeric matrix for improving of their properties. *J. Solid State Chem.* 253: 398–405. https://doi.org/10.1016/j.jssc.2017.06.014.

31 Abdolmaleki, A., Mallakpour, S., and Karshenas, A. (2017). Facile synthesis of glucose-functionalized reduced graphene oxide (GFRGO)/poly(vinyl alcohol) nanocomposites for improving thermal and mechanical properties. *Mater. Sci. Eng., B* 217: 26–35. https://doi.org/10.1016/j.mseb.2017.01.002.

32 Mallakpour, S., Abdolmaleki, A., and Borandeh, S. (2017). Fabrication of amino acid-based graphene-zinc oxide (ZnO) hybrid and its application for poly(ester–amide)/graphene-ZnO nanocomposite synthesis. *J. Thermoplast. Compos. Mater.* 30: 358–380. https://doi.org/10.1177/0892705715598365.

33 Mallakpour, S. and Khadem, E. (2016). Carbon nanotube–metal oxide nanocomposites: fabrication, properties and applications. *Chem. Eng. J.* 302: 344–367. https://doi.org/10.1016/j.cej.2016.05.038.

34 Mallakpour, S. and Khadem, E. (2018). Construction of crosslinked chitosan/nitrogen-doped graphene quantum dot nanocomposite for hydroxyapatite biomimetic mineralization. *Int. J. Biol. Macromol.* 120: 1451–1460. https://doi.org/10.1016/j.ijbiomac.2018.09.127.

35 Mallakpour, S. and Khadem, E. (2019). Chapter 8: Carbon nanotubes for heavy metals removal. In: *Composite Nanoadsorbents* (eds. G. Kyzas and A.C. Mitrpoulos), 181–210. Amsterdam, Netherlands: Elsevier. https://doi.org/10.1016/B978-0-12-814132-8.00009-5.

36 Mallakpour, S. and Khadem, E. (2019). Linear and nonlinear behavior of crosslinked chitosan/N-doped graphene quantum dot nanocomposite films in cadmium cation uptake. *Sci. Total Environ.* 690: 1245–1253. https://doi.org/10.1016/j.scitotenv.2019.06.431.

37 Mallakpour, S. and Khodadadzadeh, L. (2018). Chapter 7: Biocompatible and biodegradable chitosan nanocomposites loaded with carbon nanotubes. In: *Biodegradable and Biocompatible Polymer Composites Processing, Properties and Applications* (ed. N.G. Shimpi), 187–221. Sawston, Cambridge: Elsevier, Woodhead Publishing. https://doi.org/10.1016/B978-0-08-100970-3.00007-9.

38 Xu, J., Cao, Z., Zhang, Y. et al. (2018). A review of functionalized carbon nanotubes and graphene for heavy metal adsorption from water: preparation, application, and mechanism. *Chemosphere* 195: 351–364. https://doi.org/10.1016/j.chemosphere.2017.12.061.

39 Ahmad, S.Z.N., Salleh, W.N.W., Ismail, A.F. et al. (2020). Adsorptive removal of heavy metal ions using graphene-based nanomaterials: toxicity, roles of functional groups and mechanisms. *Chemosphere* 248: 126008. https://doi.org/10.1016/j.chemosphere.2020.126008.

40 Lau, Y.J., Khan, F.S.A., Mubarak, N.M. et al. (2019). Chapter 10: Functionalized carbon nanomaterials for wastewater treatment. In: *Micro and Nano Technologies, Industrial Applications of Nanomaterials* (eds. S. Thomas, Y. Grohens and Y.B. Pottathara), 283–311. Elsevier. https://doi.org/10.1016/B978-0-12-815749-7.00010-4.

41 Egbosiuba, T.C., Abdulkareem, A.S., Kovo, A.S. et al. (2020). Enhanced adsorption of As(V) and Mn(VII) from industrial wastewater using multi-walled carbon nanotubes and carboxylated multi-walled carbon nanotubes. *Chemosphere* 254: 126780. https://doi.org/10.1016/j.chemosphere.2020.126780.

42 Jafaryan, A., Sadjadi, S., Gharib, A. et al. (2019). Optimization of cadmium adsorption from aqueous solutions by functionalized graphene and the reversible magnetic recovery of the adsorbent using response surface methodology. *Appl. Organomet. Chem.* 33: e5085. https://doi.org/10.1002/aoc.5085.

43 Zhang, C.-Z., Chen, B., Bai, Y. et al. (2018). A new functionalized reduced graphene oxide adsorbent for removing heavy metal ions in water via coordination and ion exchange. *Sep. Sci. Technol.* 53 (18): 2896–2905. https://doi.org/10.1080/01496395.2018.1497655.

44 Chaabane, L., Beyou, E., Ghali, A.E. et al. (2020). Comparative studies on the adsorption of metal ions from aqueous solutions using various functionalized graphene oxide sheets as supported adsorbents. *J. Hazard. Mater.* 389: 121839. https://doi.org/10.1016/j.jhazmat.2019.121839.

45 Ashour, R.W., Abdelhamid, H.N., Abdel-Magied, A.F. et al. (2017). Rare earth ions adsorption onto graphene oxide nanosheets. *Solvent Extr. Ion Exch.* 35 (2): 91–103. https://doi.org/10.1080/07366299.2017.1287509.

46 Dasgupta, K., Vijayalakshmi, R., and Anitha, M. (2016). Recovery of Nd(III) in coexistence with Fe(III) ions from aqueous phase using functionalized multiwalled carbon nanotubes: an environmental benign approach. *J. Environ. Chem. Eng.* 4: 2103–2113. https://doi.org/10.1016/j.jece.2016.03.037.

47 Mallakpour, S. and Rashidimoghadam, S. (2019). Chapter 9: Carbon nanotubes for dyes removal. In: *Composite Nanoadsorbents* (eds. G. Kyzas and A.C. Mitrpoulos), 211–244. Amsterdam, Netherlands: Elsevier. https://doi.org/10.1016/B978-0-12-814132-8.00010-1.

48 Jun, L.Y., Mubarak, N.M., Yee, M.J. et al. (2018). An overview of functionalised carbon nanomaterial for organic pollutant removal. *J. Ind. Eng. Chem.* 67: 175–186. https://doi.org/10.1016/j.jiec.2018.06.028.

49 Mallakpour, S. and Behranvand, V. (2018). Synthesis of mesoporous recycled poly(ethylene terephthalate)/MWNT/carbon quantum dot nanocomposite from sustainable materials using ultrasonic waves: application for methylene blue removal. *J. Cleaner Prod.* 190: 525–537. https://doi.org/10.1016/j.jclepro.2018.04.120.

50 Ali, I., Basheer, A.A., Mbianda, X.Y. et al. (2019). Graphene based adsorbents for remediation of noxious pollutants from wastewater. *Environ. Int.* 127: 160–180. https://doi.org/10.1016/j.envint.2019.03.029.

51 Wang, C., Feng, C., Gao, Y. et al. (2011). Preparation of a graphene based magnetic nanocomposite for the removal of an organic dye from aqueous solution. *Chem. Eng. J.* 173 (1): 92–97. https://doi.org/10.1016/j.cej.2011.07.041.

52 Sellaoui, L., Dotto, G.L., Peres, E.C. et al. (2017). New insights into the adsorption of crystal violet dye on functionalized multi-walled carbon nanotubes: experiments, statistical physics and COSMO–RS models application. *J. Mol. Liq.* 248: 890–897. https://doi.org/10.1016/j.molliq.2017.10.124.

53 Wang, X., Yin, R., Zeng, L. et al. (2019). A review of graphene-based nanomaterials for removal of antibiotics from aqueous environments. *Environ. Pollut.* 253: 100–110. https://doi.org/10.1016/j.envpol.2019.06.067.

54 Thines, R.K., Mubarak, N.M., Nizamuddin, S. et al. (2017). Application potential of carbon nanomaterials in water and wastewater treatment: a review. *J. Taiwan Inst. Chem. Eng.* 72: 116–133. https://doi.org/10.1016/j.jtice.2017.01.018.

55 Yuan, W. and Shi, G. (2013). Graphene-based gas sensors. *J. Mater. Chem. A* 1 (35): 10078–10091. https://doi.org/10.1039/C3TA11774J.

56 Clément, P. and Llobet, E. (2020). Chapter 7: Carbon nanomaterials functionalized with macrocyclic compounds for sensing vapors of aromatic VOCs. In: *Woodhead Publishing Series in Electronic and Optical Materials, Semiconductor Gas Sensors*, 2e (eds. R. Jaaniso and O.K. Tan), 223–237. Woodhead Publishing. https://doi.org/10.1016/B978-0-08-102559-8.00007-0.

57 Bai, Y., Huang, Z.-H., and Kang, F. (2013). Synthesis of reduced graphene oxide/phenolic resin-based carbon composite ultrafine fibers and their adsorption performance for volatile organic compounds and water. *J. Mater. Chem. A* 1 (33): 9536–9543. https://doi.org/10.1039/C3TA10545H.

58 Surapathi, A., Chen, H.-Y., Marand, E. et al. (2013). Gas sorption properties of zwitter ion-functionalized carbon nano tubes. *J. Membr. Sci.* 429: 88–94. https://doi.org/10.1016/j.memsci.2012.11.021.

59 Su, F., Lu, C., Cnen, W. et al. (2009). Capture of CO_2 from flue gas via multiwalled carbon nanotubes. *Sci. Total Environ.* 407 (8): 3017–3023. https://doi.org/10.1016/j.scitotenv.2009.01.007.

60 Alonso, A., Moral-Vico, J., Markeb, A.A. et al. (2017). Critical review of existing nanomaterial adsorbents to capture carbon dioxide and methane. *Sci. Total Environ.* 595: 51–62. https://doi.org/10.1016/j.scitotenv.2017.03.229.

61 Penza, M., Rossi, R., Alvisi, M. et al. (2008). Pt- and Pd-nanoclusters functionalized carbon nanotubes networked films for sub-ppm gas sensors. *Sens. Actuators, B* 135: 289–297. https://doi.org/10.1016/j.snb.2008.08.024.

62 Kasim, N.A.M., Abdullah, N., Yasin, F.M. et al. (2019). Functionalized carbon nanofibers (CNFs) for ammonia gas detection at room temperature. *Mater. Today: Proc.* 19 (4): 1459–1466. https://doi.org/10.1016/j.matpr.2019.11.169.

63 Sengupta, A. and Gupta, N.K. (2017). MWCNTs based sorbents for nuclear waste management: a review. *J. Environ. Chem. Eng.* 5 (5): 5099–5114. https://doi.org/10.1016/j.jece.2017.09.054.

64 Gupta, N.K., Choudhary, B.C., Gupta, A. et al. (2019). Graphene-based adsorbents for the separation of f-metals from waste solutions: a review. *J. Mol. Liq.* 289: 111121. https://doi.org/10.1016/j.molliq.2019.111121.

65 Abdolmaleki, A., Mallakpour, S., and Mahmoudian, M. (2017). Preparation and evaluation of edge selective sulfonated graphene by chlorosulfuric acid as an active metal-free electrocatalyst for oxygen reduction reaction in alkaline media. *ChemistrySelect* 34: 11211–11217. https://doi.org/10.1002/slct.201702409.

66 Chen, D., Zou, Y., and Wang, S. (2019). Surface chemical-functionalization of ultrathin two-dimensional nanomaterials for electrocatalysis. *Mater. Today Energy* 12: 250–268. https://doi.org/10.1016/j.mtener.2019.01.006.

67 Stergiou, A., Cantón-Vitoria, R., Psarrou, M.N. et al. (2020). Functionalized graphene and targeted applications – highlighting the road from chemistry to applications. *Prog. Mater Sci.* 114: 100683. https://doi.org/10.1016/j.pmatsci.2020.100683.

68 Tao, H., Fan, Q., Ma, T. et al. (2020). Two-dimensional materials for energy conversion and storage. *Prog. Mater Sci.* 111: 100637. https://doi.org/10.1016/j.pmatsci.2020.100637.

69 Gao, X., Zhou, Y., Tan, Y. et al. (2019). Graphyne doped with transition-metal single atoms as effective bifunctional electrocatalysts for water splitting. *Appl. Surf. Sci.* 492: 8–15. https://doi.org/10.1016/j.apsusc.2019.06.050.

70 Xu, T., Shen, W., Huang, W. et al. (2020). Fullerene micro/nanostructures: controlled synthesis and energy applications. *Mater. Today Nano* 11: 100081. https://doi.org/10.1016/j.mtnano.2020.100081.

71 Yang, P. and Yang, Z. (2020). Modulating charge separation and transfer kinetics in carbon nanodots for photoredox catalysis. *J. Energy Chem.* 50: 365–377. https://doi.org/10.1016/j.jechem.2020.03.068.

72 Fei, Y., Fang, S., and Hu, Y.H. (2020). Synthesis, properties and potential applications of hydrogenated graphene. *Chem. Eng. J.* 397: 125408. https://doi.org/10.1016/j.cej.2020.125408.

73 Liu, S., Yang, H., Su, X. et al. (2019). Rational design of carbon-based metal-free catalysts for electrochemical carbon dioxide reduction: a review. *J. Energy Chem.* 36: 95–105. https://doi.org/10.1016/j.jechem.2019.06.013.

74 Wang, Y., He, D., Chen, H. et al. (2019). Catalysts in electro-, photo- and photoelectrocatalytic CO_2 reduction reactions. *J. Photochem. Photobiol., C* 40: 117–149. https://doi.org/10.1016/j.jphotochemrev.2019.02.002.

75 Ikram, M., Ali, S., Aqeel, M. et al. (2020). Reduced graphene oxide nanosheets doped by Cu with highly efficient visible light photocatalytic behavior. *J. Alloys Compd.* 837: 155588. https://doi.org/10.1016/j.jallcom.2020.155588.

76 Prekodravac, J., Vasiljević, B., Marković, Z. et al. (2019). Green and facile microwave assisted synthesis of (metal-free) N-doped carbon quantum dots for catalytic applications. *Ceram. Int.* 45: 17006–17013. https://doi.org/10.1016/j.ceramint.2019.05.250.

77 Xie, Z., Yu, S., Fan, X.-B. et al. (2021). Wavelength-sensitive photocatalytic H_2 evolution from H_2S splitting over g-C_3N_4 with S,N-codoped carbon dots as the photosensitizer. *J. Energy Chem.* 52: 234–242. https://doi.org/10.1016/j.jechem.2020.04.051.

78 Han, W., Li, D., Zhang, M. et al. (2020). Photocatalytic activation of peroxymonosulfate by surface-tailored carbon quantum dots. *J. Hazard. Mater.* 395: 122695. https://doi.org/10.1016/j.jhazmat.2020.122695.

79 Bonaccorso, F., Colombo, L., Yu, G. et al. (2015). 2D materials. Graphene, related two-dimensional crystals, and hybrid systems for energy conversion and storage. *Science* 347 (6217): 1246501. https://doi.org/10.1126/science.1246501.

80 Cheng, P. and Zhan, X. (2016). Stability of organic solar cells: challenges and strategies. *Chem. Soc. Rev.* 45 (9): 2544–2582. https://doi.org/10.1039/c5cs00593k.

81 Abate, A., Correa-Baena, J.P., Saliba, M. et al. (2018). Solar cells: from the laboratory to the assembly line. *Chem. Eur. J.* 24 (13): 3083–3100. https://doi.org/10.1002/chem.201704507.

82 Fu, X., Xu, L., Li, J. et al. (2018). Flexible solar cells based on carbon nanomaterials. *Carbon* 139: 1063–1073. https://doi.org/10.1016/j.carbon.2018.08.017.

83 Karim, N.A., Mehmood, U., Zahid, H.F. et al. (2019). Nanostructured photoanode and counter electrode materials for efficient dye-sensitized solar cells (DSSCs). *Sol. Energy* 185: 165–188. https://doi.org/10.1016/j.solener.2019.04.057.

84 Falla, P.H., Dias, M.O.G., Vieira, H. et al. (2014). Optimization the efficiency photovoltaic solar cells using synthesized TiO_2 semiconductor nanomaterials and functionalized carbon nanotubes. *Energy Procedia* 57: 1860–1867. https://doi.org/10.1016/j.egypro.2014.10.050.

85 Mallakpour, S., Behranvand, V., and Mallakpour, F. (2019). Synthesis of alginate/carbon nanotube/carbon dot/fluoroapatite/TiO_2 beads for dye photocatalytic degradation under ultraviolet light. *Carbohydr. Polym.* 224: 115138. https://doi.org/10.1016/j.carbpol.2019.115138.

86 Chan, Y.-F., Wang, C.-C., Chen, B.-H. et al. (2011). Incorporation of plasma-functionalized carbon nanocapsules into a nanocrystalline TiO_2 photoanode for use in dye-sensitized solar cells. *Carbon* 49: 4898–4910. https://doi.org/10.1016/j.carbon.2011.07.012.

87 Sreekala, C.K., Indiramma, J., and Subramanya, K.B. (2013). Functionalized multi-walled carbon nanotubes for enhanced photocurrent in dye-sensitized solar cells. *J. Nanostruct. Chem.* 3: 19. https://doi.org/10.1186/2193-8865-3-19.

88 Han, J., Kim, H., Kim, D.Y. et al. (2010). Water-soluble polyelectrolyte-grafted multiwalled carbon nanotube thin films for efficient counter electrode of dye-sensitized solar cells. *ACS Nano* 4 (6): 3503–3509. https://doi.org/10.1021/nn100574g.

89 Susmitha, K., Kumar, M.N., Giribabu, L. et al. (2016). Noble metals (Ag, Au, Pt) functionalized carbon nanohorns as alternate counter electrodes for dye sensitized solar cells. *J. Mater. Sci. - Mater. Electron.* 27 (6): 5802–5809. https://doi.org/10.1007/s10854-016-4495-4.

90 Khannam, M., Nath, B.C., Mohan, K.J. et al. (2017). Development of quasi-solid-state dye-sensitized solar cells based on a poly(vinyl alcohol)/poly(ethylene glycol)/functionalized multi-walled carbon nanotubes gel electrolyte. *ChemistrySelect* 2: 673–679. https://doi.org/10.1002/slct.201601766.

91 Neo, C.Y. and Ouyang, J. (2012). Functionalized carbon nanotube-induced viscosity reduction of an ionic liquid and performance improvement of dye-sensitized solar cells. *Electrochim. Acta* 85: 1–8. https://doi.org/10.1016/j.electacta.2012.08.041.

92 Kweon, D.H. and Baek, J.-B. (2019). Edge-functionalized graphene nanoplatelets as metal-free electrocatalysts for dye-sensitized solar cells. *Adv. Mater.* 31: 1804440. https://doi.org/10.1002/adma.201804440.

93 Miao, X., Pan, K., Wang, G. et al. (2014). Well-dispersed CoS nanoparticles on a functionalized graphene nanosheet surface: a counter electrode of dye-sensitized solar cells. *Chem. Eur. J.* 20: 474–482. https://doi.org/10.1002/chem.201303558.

94 Yue, G., Wu, J., Xiao, Y. et al. (2013). Functionalized graphene/poly(3,4-ethylenedioxythiophene): polystyrenesulfonate as counter electrode catalyst for dye-sensitized solar cells. *Energy* 54: 315–321. https://doi.org/10.1016/j.energy.2013.01.037.

95 Yu, G., Gao, J., Hummelen, J.C. et al. (1995). Polymer photovoltaic cells: enhanced efficiencies via a network of internal donor–acceptor heterojunctions. *Science* 270 (5243): 1789–1791. https://doi.org/10.1126/science.270.5243.1789.

96 Yan, J., Ni, T., Zou, F. et al. (2014). Towards optimization of functionalized single-walled carbon nanotubes adhering with poly(3-hexylthiophene) for highly efficient polymer solar cells. *Diamond Relat. Mater.* 41: 79–83. https://doi.org/10.1126/science.270.5243.1789.

97 Bonaccorso, F., Balis, N., Stylianakis, M.M. et al. (2015). Functionalized graphene as an electron-cascade acceptor for air-processed organic ternary solar cells. *Adv. Funct. Mater.* 25 (25): 3870–3880. https://doi.org/10.1002/adfm.201501052.

98 Lau, X.C., Desai, C., and Mitra, S. (2013). Functionalized nanodiamond as a charge transporter in organic solar cells. *Sol. Energy* 91: 204–211. https://doi.org/10.1016/j.solener.2013.01.024.

99 Keru, G., Ndungu, P.G., Mola, G.T. et al. (2016). Organic solar cells with boron- or nitrogen-doped carbon nanotubes in the P3HT: PCBM photoactive layer. *J. Nanomater.* https://doi.org/10.1155/2016/5923402.

100 Lee, J.M., Park, J.S., Lee, S.H. et al. (2011). Selective electron- or hole-transport enhancement in bulk-heterojunction organic solar cells with N- or B-doped carbon nanotubes. *Adv. Mater.* 23: 629–633. https://doi.org/10.1002/adma.201003296.

101 Li, S.-S., Tu, K.-H., Lin, C.-C. et al. (2010). Solution-processable graphene oxide as an efficient hole transport layer in polymer solar cells. *ACS Nano* 4 (6): 3169–3174. https://doi.org/10.1021/nn100551j.

102 Wang, B., Hu, C., and Dai, L. (2016). Functionalized carbon nanotubes and graphene-based materials for energy storage. *Chem. Commun.* 52 (100): 14350. https://doi.org/10.1039/C6CC05581H.

103 Iglesias, D., Senokos, E., Alemán, B. et al. (2018). Gas-phase functionalization of macroscopic carbon nanotube fiber fabrics: reaction control, electrochemical properties and use for flexible supercapacitors. *ACS Appl. Mater. Interfaces* 10 (6): 5760–5770. https://doi.org/10.1021/acsami.7b15973.

104 Gupta, H., Mothkuri, S., McGlynn, R. et al. (2020). Activated functionalized carbon nanotube and 2D nanostructured MoS_2 hybrid electrode material for high performance supercapacitor application. *Phys. Status Solidi A* 217: 1900855. https://doi.org/10.1002/pssa.201900855.

105 Xiao, X., Li, T., Peng, Z. et al. (2014). Freestanding functionalized carbon nanotube-based electrode for solid-state asymmetric supercapacitors. *Nano Energy* 6: 1–9. https://doi.org/10.1016/j.nanoen.2014.02.014.

106 Raj, C.J., Kim, B.C., Cho, B.-B. et al. (2016). Electrochemical supercapacitor behaviour of functionalized candle flame carbon soot. *Bull. Mater. Sci.* 39 (1): 241–248. https://doi.org/10.1007/s12034-015-1113-7.

107 Fan, X., Yu, C., Ling, Z. et al. (2013). Hydrothermal synthesis of phosphate-functionalized carbon nanotube-containing carbon composites for supercapacitors with highly stable performance. *ACS Appl. Mater. Interfaces* 5: 2104–2110. https://doi.org/10.1021/am303052n.

108 Zhou, C., Kumar, S., Doyle, C.D. et al. (2005). Functionalized single wall carbon nanotubes treated with pyrrole for electrochemical supercapacitor membranes. *Chem. Mater.* 17 (8): 1997–2002. https://doi.org/10.1021/cm047882b.

109. Zhao, X., Chu, B.T.T., Ballesteros, B. et al. (2009). Spray deposition of steam treated and functionalized single-walled and multi-walled carbon nanotube films for supercapacitors. *Nanotechnology* 20: 065605. https://doi.org/10.1088/0957-4484/20/6/065605.

110 Fisher, R.A., Watt, M.R., and Ready, W.J. (2013). Functionalized carbon nanotube supercapacitor electrodes: a review on pseudocapacitive materials. *ECS J. Solid State Sci. Technol.* 2 (10): M3170–M3177.

111 Reddy, M.V., Yu, T., Sow, C.H. et al. (2007). α-Fe_2O_3 nanoflakes as an anode material for Li-ion batteries. *Adv. Funct. Mater.* 17 (15): 2792–2799. https://doi.org/10.1002/adfm.200601186.

112 Borges, R.S., Ribeiro, H., Lavall, R.L. et al. (2012). Temperature stable supercapacitors based on ionic liquid and mixed functionalized carbon nanomaterials. *J. Solid State Electrochem.* 16 (11): 3573–3580. https://doi.org/10.1007/s10008-012-1785-5.

113 Mallakpour, S., Abdolmaleki, A., Mahmoudian, M. et al. (2017). Synergetic effect of synthesized sulfonated polyaniline/quaternized graphene and its application as a high-performance supercapacitor electrode. *J. Mater. Sci.* 52: 9683–9695. https://doi.org/10.1007/s10853-017-1118-2.

114 Kanmani, M., Lavanya, R., Silambarasan, D. et al. (2014). First principles studies on hydrogen storage in single-walled carbon nanotube functionalized with TiO_2. *Solid State Commun.* 183: 1–7. https://doi.org/10.1016/j.ssc.2013.12.017.

115 Rajaura, R.S., Srivastava, S., Sharma, P.K. et al. (2018). Structural and surface modification of carbon nanotubes for enhanced hydrogen storage density. *Nano-Struct. Nano-Objects* 14: 57–65. https://doi.org/10.1016/j.nanoso.2018.01.005s.

116 Silambarasan, D., Vasu, V., Iyakutti, K. et al. (2014). Reversible hydrogen storage in functionalized single-walled carbon nanotubes. *Physica E* 60: 75–79. https://doi.org/10.1016/j.physe.2014.02.006.

117 Singh, P., Kulkarni, M.V., Gokhale, S.P. et al. (2012). Enhancing the hydrogen storage capacity of Pd-functionalized multi-walled carbon nanotubes. *Appl. Surf. Sci.* 258: 3405–3409. https://doi.org/10.1016/j.apsusc.2011.11.075.

118 Mohan, M., Sharma, V.K., Kumar, E.A. et al. (2019). Hydrogen storage in carbon materials – a review. *Energy Storage* 1: e35. https://doi.org/10.1002/est2.35.

119 Jain, V. and Kandasubramanian, B. (2020). Functionalized graphene materials for hydrogen storage. *J. Mater. Sci.* 55 (1): 1865–1903. https://doi.org/10.1007/s10853-019-04150-y.

120 Wei, L. and Mao, Y. (2016). Enhanced hydrogen storage performance of reduced graphene oxide hybrids with nickel or its metallic mixtures based on spillover mechanism. *Int. J. Hydrogen Energy* 41 (27): 11692–11699. https://doi.org/10.1016/j.ijhydene.2016.04.030.

121 Gohari-Bajestani, Z., Akhlaghi, O., Yürüm, Y. et al. (2017). Synthesis of anatase TiO_2 with exposed (001) facets grown on N-doped reduced graphene oxide for enhanced hydrogen storage. *Int. J. Hydrogen Energy* 42: 6096–6103. https://doi.org/10.1016/j.ijhydene.2016.11.069.

122 Gangu, K.K., Maddila, S., Mukkamala, S.B. et al. (2019). Characteristics of MOF, MWCNT and graphene containing materials for hydrogen storage: a review. *J. Energy Chem.* 30: 132–144. https://doi.org/10.1016/j.jechem.2018.04.012.

123 Jin, Z., Lu, W., O'Neill, K.J. et al. (2011). Nano-engineered spacing in graphene sheets for hydrogen storage. *Chem. Mater.* 23: 923–925. https://doi.org/10.1021/cm1025188.

124 Hussain, T., Sarkar, A.D., and Ahuja, R. (2012). Strain induced lithium functionalized graphane as a high capacity hydrogen storage material. *Appl. Phys. Lett.* 101: 103907. https://doi.org/10.1063/1.4751249.

125 Chen, E., Bevilacqua, M., Tavagnacco, C. et al. (2019). High surface area N/O co-doped carbon materials: selective electrocatalysts for O_2 reduction to H_2O_2. *Catal. Today* https://doi.org/10.1016/j.cattod.2019.06.034.

126 Jha, N., Ramesh, P., Bekyarova, E. et al. (2013). Functionalized single-walled carbon nanotube-based fuel cell benchmarked against US DOE 2017 technical targets. *Sci. Rep.* 3: 2257. https://doi.org/10.1038/srep02257.

127 Yuan, W., Lu, S., Xiang, Y. et al. (2014). Pt-based nanoparticles on non-covalent functionalized carbon nanotubes as effective electrocatalysts for proton exchange membrane fuel cells. *RSC Adv.* 4 (86): 46265–46284. https://doi.org/10.1039/C4RA05120C.

128 Wang, S., Yu, D., and Dai, L. (2011). Polyelectrolyte functionalized carbon nanotubes as efficient metal-free electrocatalysts for oxygen reduction. *J. Am. Chem. Soc.* 133: 5182–5185. https://doi.org/10.1021/ja1112904.

129 Hossain, S., Abdalla, A.M., Suhaili, S.B.H. et al. (2020). Nanostructured graphene materials utilization in fuel cells and batteries: a review. *J. Energy Storage* 29: 101386. https://doi.org/10.1016/j.est.2020.101386.

130 Kiani, M., Zhang, J., Luo, Y. et al. (2018). Recent developments in electrocatalysts and future prospects for oxygen reduction reaction in polymer electrolyte membrane fuel cells. *J. Energy Chem.* 27 (4): 1124–1139. https://doi.org/10.1016/j.jechem.2018.01.019.

131 Mohideen, M.M., Liu, Y., and Ramakrishna, S. (2020). Recent progress of carbon dots and carbon nanotubes applied in oxygen reduction reaction of fuel cell for transportation. *Appl. Energy* 257: 114027. https://doi.org/10.1016/j.apenergy.2019.114027.

132 Iannaci, A., Mecheri, B., D'Epifanio, A. et al. (2016). Iron nitrogen-functionalized carbon as efficient oxygen reduction reaction electrocatalyst in microbial fuel cells. *Int. J. Hydrogen Energy* 41 (43): 19637–19644. https://doi.org/10.1016/j.ijhydene.2016.04.154.

133 Kaur, R., Marwah, A., Chhabra, V.A. et al. (2019). Recent developments on functional nanomaterial-based electrodes for microbial fuel cells. *Renewable Sustainable Energy Rev.* 119: 109551. https://doi.org/10.1016/j.rser.2019.109551.

134 Zhu, N., Lu, Y., Liu, B. et al. (2017). $CoMn_2O_4$-supported functionalized carbon nanotube: efficient catalyst for oxygen reduction in microbial fuel cells. *J. Nanopart. Res.* 19 (10): 331. https://doi.org/10.1007/s11051-017-4023-3.

135 Liu, L., Zeng, G., Chen, J. et al. (2018). N-doped porous carbon nanosheets as pH-universal ORR electrocatalyst in various fuel cell devices. *Nano Energy* 49: 393–402. https://doi.org/10.1016/j.nanoen.2018.04.061.

136 Kumaresan, T.K., Gunasekaran, S.S., Elumalai, S.K. et al. (2019). Promising nature-based nitrogen-doped porous carbon nanomaterial derived from *Borassus flabellifer* male inflorescence as superior metal-free electrocatalyst for oxygen reduction reaction. *Int. J. Hydrogen Energy* 44 (47): 25918–25929. https://doi.org/10.1016/j.ijhydene.2019.08.044.

137 Cheng, Y. and Jiang, S.P. (2013). Highly effective and CO-tolerant PtRu electrocatalysts supported on poly(ethyleneimine) functionalized carbon nanotubes for direct methanol fuel cells. *Electrochim. Acta* 99: 124–132. https://doi.org/10.1016/j.electacta.2013.03.081.

138 Habibi, B. and Emanzadeh, H. (2018). Platinum nanoparticles/functionalized carbon nanoparticles composites supported on the carbon-ceramic electrode and their electroactivity for ethanol oxidation. *Iran. J. Hydrog. Fuel Cell* 1: 35–47. https://doi.org/10.22104/ijhfc.2018.2920.1175.

139 Hussein, L., Feng, Y.J., Alonso-Vante, N. et al. (2011). Functionalized-carbon nanotube supported electrocatalysts and buckypaper-based biocathodes for glucose fuel cell applications. *Electrochim. Acta* 56: 7659–7665. https://doi.org/10.1016/j.electacta.2011.06.067.

140 Themsirimongkon, S., Sarakonsri, T., Lapanantnoppakhun, S. et al. (2015). Synthesis of Pt_xPd_y nanoparticles decorated functionalized carbon nanotubes as highly anodic catalysts for formic acid fuel cells. *Int. J. Electrochem. Sci.* 10: 5737–5746.

141 Konwar, L.K., Boro, J., and Deka, D. (2014). Review on latest developments in biodiesel production using carbon-based catalysts. *Renewable Sustainable Energy Rev.* 29: 546–564. https://doi.org/10.1016/j.rser.2013.09.003.

142 Shu, Q., Zou, W., He, J. et al. (2019). Preparation of the F^--SO42−/MWCNTs catalyst and kinetic studies of the biodiesel production via esterification reaction of oleic acid and methanol. *Renewable Energy* 135: 836–845. https://doi.org/10.1016/j.renene.2018.12.067.

143 Shuit, S.H., Yee, K.F., Lee, K.T. et al. (2013). Evolution towards the utilisation of functionalised carbon nanotubes as a new generation catalyst support in biodiesel production: an overview. *RSC Adv.* 3 (24): 9070–9094. https://doi.org/10.1039/C3RA22945A.

144 Santos, T.C.D., Santos, E.C.S., Dias, J.P. et al. (2019). Reduced graphene oxide as an excellent platform to produce a stable Bronsted acid catalyst for biodiesel production. *Fuel* 257: 116015. https://doi.org/10.1016/j.fuel.2019.115793.

145 Gaidukevič, J., Barkauskas, J., Malaika, A. et al. (2018). Modified graphene-based materials as effective catalysts for transesterification of rapeseed oil to biodiesel fuel. *Chin. J. Catal.* 39 (10): 1633–1645. https://doi.org/10.1016/S1872-2067(18)63087-6.

146 Seraç, M.R., Aydın, S., Yılmaz, A. et al. (2020). Evaluation of comparative combustion, performance, and emission of soybean-based alternative biodiesel fuel blends in a CI engine. *Renewable Energy* 148: 1065–1073. https://doi.org/10.1016/j.renene.2019.10.090.

147 Hosseini, S.H., Taghizadeh-Alisaraei, A., Ghobadian, B. et al. (2017). Effect of added alumina as nano-catalyst to diesel-biodiesel blends on performance and emission characteristics of CI engine. *Energy* 124: 543–552. https://doi.org/10.1016/j.energy.2017.02.109.

148 Shekofteh, M., Gundoshmian, T.M., Jahanbakhshi, A. et al. (2020). Performance and emission characteristics of a diesel engine fueled with functionalized multi-wall carbon nanotubes (MWCNTs-OH) and diesel–biodiesel–bioethanol blends. *Energy Rep.* 6: 1438–1447. https://doi.org/10.1016/j.egyr.2020.05.025.

149 EL-Seesy, A.I. and Hassan, H. (2019). Investigation of the effect of adding graphene oxide, graphene nanoplatelet, and multiwalled carbon nanotube additives with *n*-butanol-Jatropha methyl ester on a diesel engine performance. *Renewable Energy* 132: 558–574. https://doi.org/10.1016/j.renene.2018.08.026.

150 Solanki, P., Bhargava, A., Chhipa, H. et al. (2015). Nano-fertilizers and their smart delivery system. In: *Nanotechnologies in Food and Agriculture* (eds. M. Rai, C. Ribeiro, L. Mattoso, et al.), 81–101. Cham: Springer, Springer International Publishing. https://doi.org/10.1007/978-3-319-14024-7_4.

151 Tripathi, S., Sonkar, S.K., and Sarkar, S. (2011). Growth stimulation of gram (*Cicer arietinum*) plant by water soluble carbon nanotubes. *Nanoscale* 3: 1176–1181. https://doi.org/10.1039/C0NR00722F.

152 Villagarcia, H., Dervishi, E., Silva, K.D. et al. (2012). Surface chemistry of carbon nanotubes impacts the growth and expression of water channel protein in tomato plants. *Small* 8: 2328–2334. https://doi.org/10.1002/smll.201102661.

153 Zhai, G., Gutowski, S.M., Walters, K.S. et al. (2015). Charge, size, and cellular selectivity for multiwall carbon nanotubes by maize and soybean. *Environ. Sci. Technol.* 49 (12): 7380–7390. https://doi.org/10.1021/acs.est.5b01145.

154 Yatim, N.M., Shaaban, A., Dimin, M.F. et al. (2019). Urea functionalized multiwalled carbon nanotubes as efficient nitrogen delivery system for rice. *Adv. Nat. Sci.: Nanosci. Nanotechnol.* 10 (1): 015011. https://doi.org/10.1088/2043-6254/ab0881.

155 Hu, C., Liu, L., Li, X. et al. (2018). Effect of graphene oxide on copper stress in *Lemna minor* L.: evaluating growth, biochemical responses, and nutrient uptake. *J. Hazard. Mater.* 341: 168–176. https://doi.org/10.1016/j.jhazmat.2017.07.061.

156 Liu, Q., Zhang, X., Zhao, Y. et al. (2013). Fullerene-induced increase of glycosyl residue on living plant cell wall. *Environ. Sci. Technol.* 47: 7490–7498. https://doi.org/10.1021/es4010224.

157 Singh, H., Bhardwaj, N., Arya, S.K. et al. (2020). Environmental impacts of oil spills and their remediation by magnetic nanomaterials. *Environ. Nanotechnol. Monit. Manage.* 14: 100305. https://doi.org/10.1016/j.enmm.2020.100305.

158 Sireesha, M., Jagadeesh, V., and Ramakrishna, B.S. (2017). Functionalized carbon nanotubes in bio-world: applications, limitations and future directions. *Mater. Sci. Eng., B* 223: 43–63. https://doi.org/10.1016/j.mseb.2017.06.002.

159 Jariwala, D.H., Patel, D., and Wairkar, S. (2020). Surface functionalization of nanodiamonds for biomedical applications. *Mater. Sci. Eng., C* 113: 110996. https://doi.org/10.1016/j.msec.2020.110996.

18

Functionalized Carbon Nanomaterials for Impending Pharmaceutical Applications: A Green and Sustainable Vision

Vaneet Kumar[1], *Saruchi*[2], *and Harsh Kumar*[3]

[1] *CT Group of Institutions, Department of Applied Sciences, (CTIEMT), Shahpur, Jalandhar 144020, Punjab, India*
[2] *CT Group of Institutions, Department of Biotechnology (CTIPS), Shahpur, Jalandhar 144020, Punjab, India*
[3] *Dr. B R Ambedkar National Institute of Technology, Department of Chemistry, Jalandhar 144011, Punjab, India*

18.1 Introduction

Pharmaceutical research and its applications on functionalized carbon nanomaterials have been popular in the last few years. Many studies focus on treating the toxicity, carcinogenicity of cell, and tissue function by the use of carbon nanotube (CNT) and functionalized CNT. The influences of particle size, morphology, surface charge on their interfaces with tissues, immune system cells uptake, and correlations with toxicity or carcinogenicity lack consensus. Sometimes, several serious problems transform inert nano-sized particles from instructive learning to manufacturing processes that follow with superiority trade, governmental standards, and some authoritarian perspectives for nature and all living being. In broad, mass and nature of inorganic nanoparticles, the temperature of water, pH level, DO, oxygen diffusion, conductivity, HNO_3, orto-phosphate, SO_4, Cl^{-1}, interfacial and surface properties that result in an array of protein corona on the outer surfaces of material [1].

Many researchers show interest in using these functionalized carbon nanomaterials in different pharmaceutical applications such as medicine in recent studies. They are also applied in different phototherapy and imaging types in biofield and biosensors [2, 3]. In the available literature on nanoscience, many researchers have studied nanoparticles of Fe_2O_3. Its chemical and biological properties, substance stability, non-toxicity, biocompatibility, elevated saturation magnetization, and high magnetic vulnerability are better than other nanoparticles [4, 5].

Because of these unique properties, CNTs have numerous pharmaceutical applications such as bioimaging [6–12], hyperthermia [13–15], drug delivery [16–20], cell labeling [21], and gene delivery [22–26]. In the recent review of literature, it has been detected that other nanomaterials (magnetic) such as Iron–Cobalt, Copper–Nickel, Iron–Nickel, Cobalt–Iron oxide, and Manganese–Iron oxide, are being studied for bioimaging, hyperthermia, and drug delivery as probable choices to Fe_2O_4 nanoparticles.

Environmental Applications of Carbon Nanomaterials-Based Devices, First Edition.
Edited by Shadpour Mallakpour and Chaudhery Mustansar Hussain.

In 1991, a very famous scientist, Iijima, gave a brief description of the cylindrical structure of CNTs. He explained CNTs in a very unique and excellent way [27]. According to him, CNTs are of two types: single-walled carbon nanotubes (SWCNTs), which made up of a solitary sheet of graphite flawlessly enfold into a follicle cylinder with a width amid 0.4 and 2.5 nm, and multi-wall carbon nanotubes (MWCNTs), which include an additional sheet of graphite material with divergent widths up to 100 nm. The unique structure of CNTs and exceptional physical and chemical properties [28] permit its wide industrial applications. The different types of structural arrangement, unique mechanical and electronic properties of CNTs are because of their nano-size and mass and very high electrical and thermal capability.

These techniques are beneficial for alleviating tolerant risk and ailment progression and recognizing modified drugs. Restorative carriers to nano are two types of families (liposomes and albumin) that have before and are now used in global research practice. There are many other restorative agents on a trial basis.

CNTs have a large capability of loading to take different bioactive agents, such as drugs, to penetrate the plasma membrane [29]. Their essential spectroscopic properties, and Raman and photoluminescence spectroscopy technique could be precious means of analyzing imaging diseases. CNTs can also monitor the status of *in vivo* therapy, pharmaceutical behavior, and drug delivery effectiveness. Its unique optical and thermodynamic properties could be used directly in diagnostics of pharmaceutical and therapy. The greatest obstacle of CNTs used in nanomedicine was poor dispersibility. Many functionalization routes have been completed to solubilize CNTs and better their biocompatibility [30–34].

Nanotechnology is a fast-growing field in various fields for the last many years. The different properties of CNTs in the pharmaceutical industry have been presented in numerous research papers (literature) with a spotlight on cancer treatment. Therefore, in an imperfect time, CNTs have become a topic of awareness by many scientists in various disciplines worldwide. Scientists have found an excellent result with CNTs as effective antioxidants for healthiness treatment in the outlook [35]. The other major use of CNTs is that they can be an effective tool for separating enantiomer in chiral drugs and chemicals and removing many drugs and other harmful materials by solid-phase extraction (SPE) technique before analysis [36, 37].

This chapter summarizes several pharmaceutical applications of functionalized carbon nanomaterials in drug delivery and disease treatment.

This chapter will focus on the synthesis of functionalized CNTs, having many applications in the pharmaceutical field. Its main applications are for the treatment of medicine by different modification with CNTs and functionalized CNTs. Its high surface and capacity of adsorbing many therapeutic and diagnostic agents such as medicines, genetic materials, vaccination, antibodies, and biosensors Figure 18.1.

18.2 Carbon Nanotubes: Functionalization for Biomedical Applications

Smooth surfaces with no lynching bonds compose pristine CNTs chemically inert and contrary to almost all types of organic and inorganic liquids. A liquid-type CNT procedure is not easy to achieve, and this is a chief disadvantage for CNTs' applications in the pharmaceutical

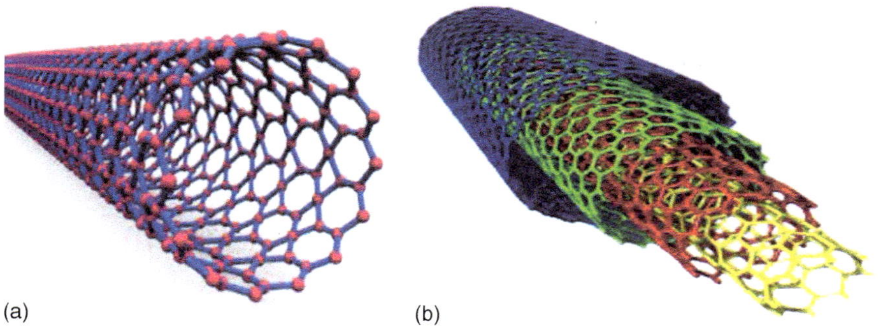

Figure 18.1 Theoretical presentation of (a) (SWCNT) and (b) (MWCNT).

industry. In the latest study, many researchers have observed that carbon atoms in both single-wall CNTs and multi-wall CNTs can show chemical reactivity in the direction of numerous reagents. Both CNTs can be believed as the narrative macromolecular appearance of carbon. To enhance knowledge in the field of CNT, booklovers can study Niyogi's and Tasis's reviews [28, 34]. When certain modifications are done in the CNTs, it will illustrate exceptional characteristics and solubility behavior [38].

Functionalized CNTs could increase miscibility in water and get better biocompatibility behavior. Covalent and noncovalent functionalization of CNTs are two main techniques to enhance miscibility of water. Covalent functionalization can also be called "imperfect functionalization" because strong oxidants can tarnish atoms of carbon (defective) on the wall side or at the end of CNTs to produce –COOH or carboxylated fractions. They can be chemically modified using amidation or esterification [39] . The surface of carboxylated CNTs can be grafted by various polymers [39], metals [40], and biological molecules [41]. Other additional reactions could be done by modifications (covalent and noncovalent) of CNTs obtained from fullerenes or graphite surface [42]. It is easy to control covalent functionalization of CNTs. Multiple functionalities probability can be introduced. This process is useful for functionalized CNTs, which is used as a novel drug delivery system. Since covalent functionalized CNTs' have unique surface properties, this type has a wide range of pharmacy, agriculture, textile, and other industrial applications.

Covalent functionalization of CNTs could be obtained by a simple method of oxidation with nitric acid. By this method, acetic acid groups can be obtained at the unlock sides and at the defects on the sidewalls of SWCNT or MWCNT, later, extra covalent conjugation with an amino acid. The formation of –COOH group on the sidewalls of CNTs, many researchers have also explained the cycloaddition of nitrene, arylation using R–N+ 2X or 1,3-dipolar cycloadditions in different ways [43–46]. A general method of synthesis of covalent functionalized CNTs has been depicted in Figure 18.2.

The noncovalent functionalization of CNTs is also drawing attention in pharmaceuticals and other industries. For example, properties of CNTs, such as near-IR, fluorescence and Raman scattering, can be restrained because the electronic structure of the conjugated CNT will not be weakened [47]. By covering some polymers around CNT sidewall [31] and the other is in the course of π–π interface among aromatic rings of the loaded materials and π-electrons of sheets (graphite) on the outer surface of CNTs [48].

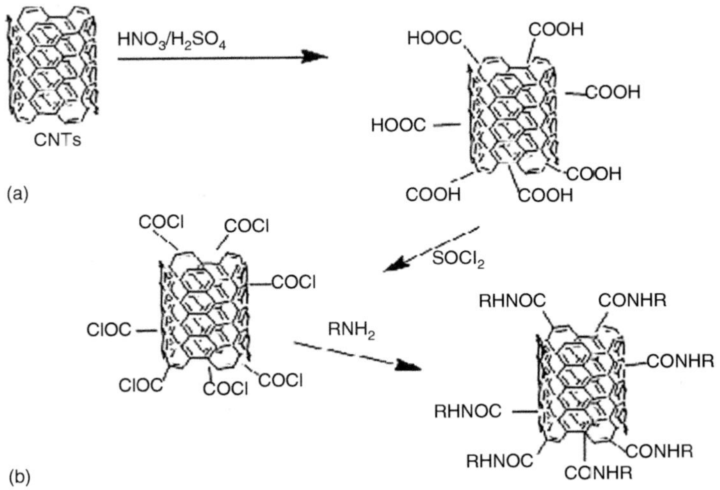

Figure 18.2 (a) Functionalized carbon nano tube. (b) Formation of covalent functionalization of carbon nanotubes.

Functionalized CNTs (non-covalent) could be processed by fleecing CNTs with surfactant molecules (amphiphilic) or polyethylene glycol. The large aromatic (π-electrons) CNTs (hydrophobic surface) them ideal partners made for noncovalent interactions with appropriate corresponding molecules and macrobiomolecules (DNA). This type of interaction will take place in CNTs from both sides. However, macromolecules could not be bonded on their inside [46, 49]. Figure 18.3 indicated the noncovalent functionalization of CNTs. After functionalization, CNTs become hydrophilic and are prepared to be connected with a molecule of drugs or biomolecular (RNA, carbohydrates, catalyst, and biosensors) discharge into the body organ. After functionalization, CNTs can show signs of increased water miscibility and reduced toxicity [50].

18.2.1 Applications of Functionalization Carbon Nanotubes in the Pharmaceutical Field

The major applications are in the pharmaceutical field consist of medicine, biomolecule, gene therapy, cells or organs delivery, regeneration of tissue, diagnostics of biosensors, and analysis. They are also shown briefly in Figure 18.4. Figure 18.4 indicated the application of functionalized CNT in various fields. The common procedure using CNTs can be used for drug delivery. The drug may be attached on the outside or inside of functionalized CNTs. The obtained conjugate initiates into the body of mammal by common ways like oral/injection or straightforwardly to the target site from side to side using a conjugate magnet, i.e. the target organ as lymphatic nodes will be guided by an external magnet. The mammal cell consumes the drug CNT tablet, and lastly, the nanotubes tablet its materials into the cell, and therefore the drug will be delivered [51, 52]. Usually, functionalized CNTs can cart interest of molecules athwart the membrane of cytoplasm and membrane of nuclear without generating a toxic effect; consequently, the drug with CNT conjugate confirms to be

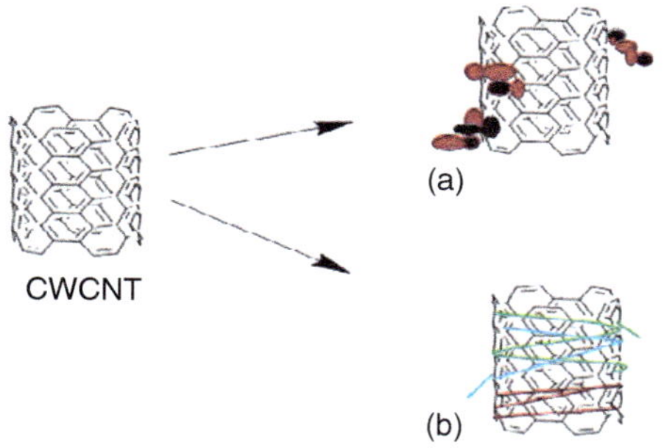

Figure 18.3 Formation of noncovalent functionalization of carbon nanotubes.

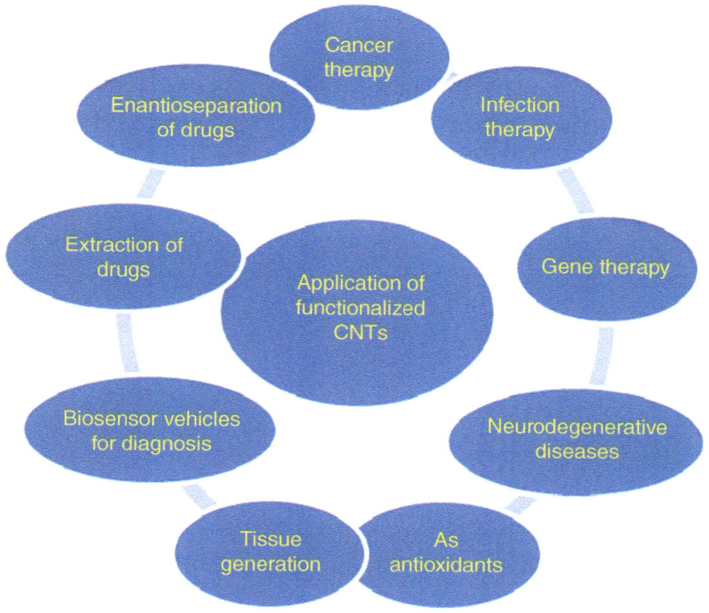

Figure 18.4 Different applications of functionalized CNTs in pharmaceutical field.

secure and additional valuable than a drug used only by old preparation [43, 44]. The drug can be delivered with two possibilities after arriving at the targeted cell. Possibilities are: the drug goes into the cells exclusive of internalization of the CNT carrier, or both the drug and the CNT carrier go into the cells. The second method is important compared to the first method because, in the second method, drug carriers proceed to release medicine molecule *in situ* within cells. Whereas, the non-internalization method of the environment of extracellular facilitates consuming medicine carrier conjugates and after that medicine irritable itself the membrane of lipid to go into the cells. There is a probability of drug consummation through this infiltration by itself. It has been observed that functionalized CNT could treat different diseases in pharmaceutical sciences. Functionalized CNT is a recent area of research for many researchers worldwide.

18.2.2 Treatments of Tumors by Functionalized CNT

One of the dangerous diseases in the world is cancer. For cancer therapy, efficiency, and specificity are two necessities. Therapy of tumors by passive and active targeting mechanisms as drug carriers can improve by applying nanomaterials.

At present essential method for the treatment of tumors is Chemotherapy, nearby surgical discovery, and radiation. Treatment of tumors by traditional chemotherapy technique had some poor result because of 02 main drawbacks: chemotherapeutic drugs having low down specificity, that means toxicity of systemic, and resistance of drug directs to little efficacy. CNTs have been found as narrative medicine with low effect of toxicity and immunogenicity. In the previous research, various curative molecules have been detected treated with CNTs, and the result was very excellent compared to the previous result (without CNTs). There are many research papers available on the above statement [53–55]. Using CNT to treat tumors by chemotherapy, the elevated outside area of CNTs permits proficient loading of medicines. Because the improved retention effect and permeability are common in hard tumors, CNTs encumbered with drugs could reduce tumor tissues in time; the absorption in the tumors will attain many folds elevated compared to plasma. Dai's group [56] reported that a 13% accumulation of injected PEGylated SWCNTs in tumors, and the result was excellent. An additional broadly used anticancer drug like cisplatin has been injected with SWCNT as a "delivery system of longboat" and attained enhanced uptake in cancer cells.

When functionalized CNTs with antibodies of antigens were introduced on the cancer cell surface, then the treatment results were more effective than CNT. It can be concluded from the above discussion that the antitumor activity of CNTs may be because of warm transduction, leading to a reduction of cellular cytotoxicity. The same has been given in Figure 18.5.

18.2.3 Treatment of Infectious Diseases by Functionalized CNT

Recent diseases in the whole world like Covid-19, SARS, Spanish flu, and Swine flu have shown that communicable diseases are very harmful from a public health perspective. Several other infectious diseases, like AIDS, have become deadly, and to date, no treatment has been discovered. The pharmaceutical application of functionalized CNT has spread brightness on the rapid treatment and useful therapy.

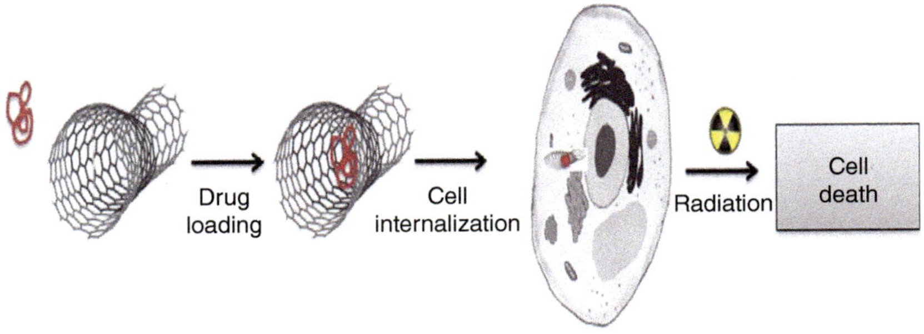

Figure 18.5 Cancer cell treatment by hyperthermia as an adjuvant therapy.

The antimicrobial effect and proposed CNT mechanism have been found by Kang et al. during 2007, 2008 [57, 58]. They performed the study with single-wall CNT and multi-wall CNT and concluded that single-wall CNT is very toxic to bacteria compared to multi-wall CNT. When there is some organic amendment on the outside of CNTs that can produce sites for the connection of bioactive molecules, the alternative structure can be conserved and, therefore, obtain explicit antibodies of anti-epitope [59].

CNTs are playing a crucial part in these viral diseases by indicating elevated-sensitive devices of detection. This can be easily understood with the help of a coordinated bio-sensor [60] built of silver nanoparticles, and single-wall CNTs studied for observing the nanoscale of HIV-1 PR, accountable for virion assemblage and maturation. The growth of effectual HIV-1 PR inhibitors of this protease was promising to expedite with the realization of high-sensitive detection. Hepatitis C virus RNA is the second example of a diagnosis of the viral disease by electrical detection method. CNTs of large surface and volume ratio with unique electronic properties through a reception component for constructing elevated-sensitive biodetectors, which were critically required to identify viral type diseases and discover new antiviral drugs. It was expected. As a result, functionalized CNTs may add considerably to the healing of infectious diseases like Covid-19, SARS, Spanish flu, and Swine flu in the future.

18.2.4 Functionalized CNT as Antioxidants

The impending role of functionalized CNTs as antioxidants is nowadays a promising area of recent research. When CNTs, with carboxylated SWCNTs, have been taken, it results in antioxidants and may have constructive pharmaceuticals applications in putting off chronic ailments, aging, and foodstuff conservation well. Therefore, probable calls for more examinations of dissimilar types of CNTs to develop their valuable result as a free radical forager.

18.2.5 Functionalized CNTs as Diagnostics

Diagnostics means biosensors. They are playing a vital role in the field of public health and science. Biosensors always analyze chemical, physical, or biological magnitude and convert it into the digital analog called a signal. The high speed of electron chemistry with the elastic outside, the high surface area of CNTs, could give better results in the biosensor. CNT can also control the blood sugar of human beings when it combines with glucose-oxidase biosensors. A variety of mechanisms and CNTs properties have been used in the building of these molecular machineries. Figure 18.6 indicated the sensing mechanisms of CNT-made biosensors.

Recently, research has shown that CNTs could also produce QDs or might act like QDs. Mostly, these are nanocrystal-made semiconductors with a range of nano size. They may emit light of various colors because of diode substances. Because of their light-generating characteristics, they have potential application for different body parts [61]. CNTs can mimic these radiance-generating nanoparticles. Functionalized CNT will be more effective as compared to CNT. In the field of diagnostics biosensors for scientific research and healthcare.

Figure 18.6 Indicated dissimilar sensing mechanisms of CNT-made biosensors.

18.2.6 Solid Phase Extraction of Drugs and Biochemical's with CNTs

As we know that CNT has strong interaction through of other molecules, mainly containing C_6H_6 rings, surfaces of the CNTs always exhibit exceptional adsorption ability. Non functionalized or functionalized CNT appearance would be examined as adsorbents of SPE used for conjugation with SPE sorbents of the classical form (C18 silica, XAD-2 copolymer) for the diagnostic removal of drugs, pesticides, or natural compounds in a different outline such as organic fluids, drug synthesis, surroundings, places, and organs of mammal [37]. During numerous proportional CNTs, studies show the alike or elevated capacity of adsorption as compared to sorbents made of silica or macroporous resins. CNTs having vital SPE applications could be observed in dissimilar latest review papers dealing with common features [62]. Many drugs worldwide can be extracted in different physicochemical methods by SPE using either SWCNTs or MWCNTs as adsorbents.

Functionalized CNT could be prospects in the field of agriculture. Furthermore, CNTs can also be used to eliminate many inorganic ions and organometallic compounds for the synthesis of stationary phases of gas chromatograph or high-performance liquid chromatography columns. In spectroscopy, if we used functionalized CNT with more modification, it can be used to remove many harmful and toxic drugs from the environment. Functionalized CNT may be the future material for the column filling and removing many other organic and inorganic waste materials from the environment.

Functionalized carbon nanomaterials have made a great evolution in sciences to remove, reduce, and eliminate the micro toxic and carcinogenic materials from our environment.

Functionalized carbon nanomaterials have been used in the biological field for imaging, chemical side sensing, directing, distribution, therapy, reaction catalysis, and energy harvesting. The above-mentioned applications always require tailored surface functionalization nanoparticles [60, 63–82].

18.2.7 Toxicity Contemplation of CNTs

In contrast to bulky objects, CNT has an exceptional feature that includes an exponentially elevated aspect ratio that escorts to toxicity. Accretion of nanoscale particles might increase tissue uptake. Therefore the biological function of cells may be influenced critically. Sometimes, if metal traces is existing in CNT will an additional contributory cause for toxicity [83].

CNTs exposure may also cause many faults in different organs of the body and organ systems. There may be temporary organ damage, predominantly to the lungs and heart, attributable to their delayed clearance from the body due to sturdy agglomeration if premature systemic exposure of MWCNTs. Inflammatory events will happen when constant accretion of agglomerated MWCNTs in the lungs initiates [84]. Likewise, single-wall CNTs have been observed to influence the CNS due to an elevated degree of accretion in the spinal cord and dorsal root ganglia. Additionally, single-wall CNTs considerably reduce DNA content while the cells were exposed to scatter single-wall CNT packages due to their accretion in the material of nuclear.

Furthermore, exposure to single-wall CNTs has an outcome in ultrastructural and morphological changes in the cultured form [85]. Multi-wall CNTs are additional toxic than single-wall CNTs and persuade enormous loss of cell viability during programmed cell death while comparing the evaluation of toxicity studies. The toxicity of multi-wall CNTs is because of many factors like oxidative stress, cellular toxicity, formation of free radicals, accretion of peroxidative products, as indicated in Figure 18.7. CNTs also show primary genetic toxicity due to the straight contact of atoms with cells or secondary genetic toxicity due to creating a surplus of ROS. Multi-wall CNTs have genetic toxicity effects due to ROS damage, leading to enhance mutation frequencies [86].

Sometimes corporal interference of CNT using cellular and extracellular ingredients is one of the appropriate mechanisms of damage. According to past studies, toxicity in CNT is not enough, and many other additional toxicity studies are required on mammal models, including human beings. Functionalization of single-wall CNTs and multi-wall CNTs with new modifications and their effects on genotoxicity may be a recent requirement of nature.

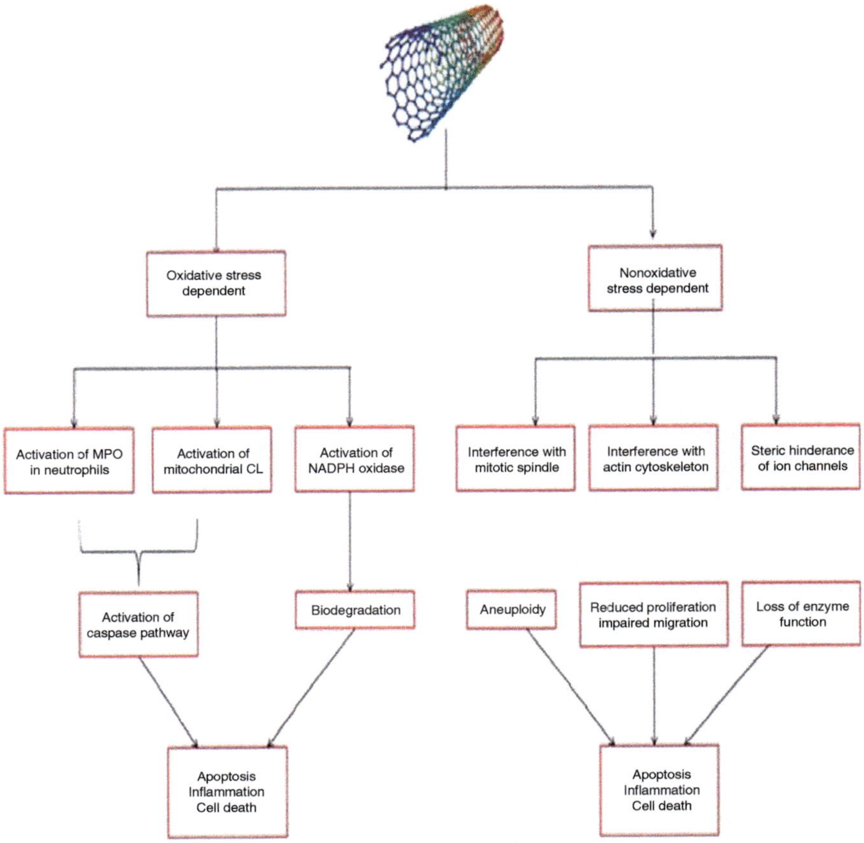

Figure 18.7 Pathways of cellular toxicity induced by CNTs.

18.3 Conclusion and Future Perspectives

This chapter concludes that chemically modification of CNTs as functionalized CNTs and its unique properties like surface area, tunable length, different physical properties, formulate CNTs advance material in agriculture, textile, transport, and pharmacy industry. Functionalized CNTs' as medicine carriers in marked delivery has made them prime aspirants for the analysis and healing of many diseases like cancer, CNS disorders, infectious diseases (Covid-19, SARS, Spanish flu, and Swine flu), Extraction of drugs and biochemical's in Solid Phase, antioxidants, and HIV. Functionalized CNT is also playing a significant role in tissue engineering. After modification to functionalization, CNTs are becoming hydrophilic and linked with medicine or many biomolecules.

As a result, in a span of very little time, CNTs have pinched the awareness of nonscientists from the manufacturing unit and academic world. In the previous three decades, an extraordinary effort has been made to utilize CNTs for applications in the field of biomedical. CNTs in the form of functionalization have again unlocked new viewpoints for applying CNTs in various pharmaceutical fields.

Although, despite their major role in nanosciences, CNTs still necessitate more research examinations to assure their security outline in delivering the drug. Studies on toxicity are serious about launching the occupied *in vivo* potential of CNTs to deliver drugs earlier than their actual application and marketing. Additionally, before the commercialization of CNTs, firm directives are compulsory, taking into their realm the environmental, physical condition, and safety issues. On the other hand, as frighteningly elevated numbers of results are piling up, it could be sensibly expected that functionalized CNTs have a diamond opportunity in the pharmaceutical field.

Important Websites

https://www.beilstein-journals.org/bjoc/series/42.
https://pubs.acs.org/doi/10.1021/ar010160v.
https://www.ncbi.nlm.nih.gov/pmc/articles/PMC3471599.
https://www.mdpi.com/2311-5629/5/4/72/htm.
https://www.intechopen.com/books/novel-nanomaterials-synthesis-and-applications/functionalized-carbon-nanomaterials-in-drug-delivery-emergent-perspectives-from-application.
https://www.sciencedirect.com/science/article/pii/B9780128157497000104.

References

1 Nobile, S. and Nobile, L. (2017). Nanotechnology for biomedical applications: recent advances in neurosciences and bone tissue engineering. 57: 644–650. https://doi.org/10.1002/pen.24595.

2 McNamara, K. and Tofail, S.A.M. (2013). Biomedical applications of nanoalloys. In: *Nanoalloys: From Fundamentals to Emergent Applications* (ed. C. Florent), 345. Oxford: Elsevier.

3 McNamara, K. and Tofail, S.A.M. (2015). Nanosystems: the use of nanoalloys, metallic, bimetallic, and magnetic nanoparticles in biomedical applications. *Phys. Chem. Chem. Phys.* 17 (42): 27981–27995.

4 Sun, C., Lee, J.S., and Zhang, M. (2008). Magnetic nanoparticles in MR imaging and drug delivery. *Adv. Drug Delivery Rev.* 60 (11): 1252–1265.

5 Chen, Z.P., Zhang, Y., Zhang, S. et al. (2008). Preparation and characterization of water-soluble monodisperse magnetic iron oxide nanoparticles via surface double-exchange with DMSA. *Colloid Surf., A* 316: 210–216.

6 Hong, R.Y., Feng, B., Chen, L.L. et al. (2008). Synthesis, characterization and MRI application of dextran-coated Fe_3O_4 magnetic nanoparticles. *Biochem. Eng. J.* 42: 290–300.

7 Zhang, Y., Liu, J.Y., Ma, S. et al. (2010). Synthesis of PVP-coated ultra-small Fe_3O_4 nanoparticles as a MRI contrast agent. *J. Mater. Sci. - Mater. Med.* 21 (4): 1205–1210.

8 Arsalani, N., Fattahi, H., and Nazarpoor, M. (2010). Synthesis and characterization of PVP-functionalized superparamagnetic Fe_3O_4 nanoparticles as an MRI contrast agent. *Express Polym. Lett.* 4 (6): 329–338.

9 Maeng, J.H., Lee, D.H., Jung, K.H. et al. (2010). Multifunctional doxorubicin loaded superparamagnetic iron oxide nanoparticles for chemotherapy and magnetic resonance imaging in liver cancer. *Biomaterials* 31 (18): 4995–5006.

10 Chen, W., Yi, P., Zhang, Y. et al. (2011). Composites of aminodextran-coated Fe_3O_4 nanoparticles and graphene oxide for cellular magnetic resonance imaging. *ACS Appl. Mater. Interfaces* 3 (10): 4085–4091.

11 Arsalani, N., Fattahi, H., Laurent, S. et al. (2012). Polyglycerol-grafted superparamagnetic iron oxide nanoparticles: highly efficient MRI contrast agent for liver and kidney imaging and potential scaffold for cellular and molecular imaging. *Contrast Media Mol. Imaging* 7: 185–194.

12 Ahmad, T., Bae, H., Rhee, I. et al. (2012). Gold-coated iron oxide nanoparticles as a T_2 contrast agent in magnetic resonance imaging. *J. Nanosci. Nanotechnol.* 12 (7): 5132–5137.

13 Jiang, Q.L., Zheng, S.W., Hong, R.Y. et al. (2014). Folic acid-conjugated Fe_3O_4 magnetic nanoparticles for hyperthermia and MRI in vitro and in vivo. *Appl. Surf. Sci.* 307: 224–233.

14 Nordmeyer, D., Stumpf, P., Groger, D. et al. (2014). Iron oxide nanoparticles stabilized with dendritic polyglycerols as selective MRI contrast agents. *Nanoscale* 6 (16): 9646.

15 Suresh, G., Ramanujan, R.V., Chong, T.W. et al. (2015). Poly(*N*-isopropyl acrylamide) coated magnetite nanoparticles as contrast agents for magnetic resonance imaging. *Nanosci. Nanotechnol. Lett.* 7: 15–19.

16 Jia, Y., Yuan, M., Yuan, H. et al. (2012). Co-encapsulation of magnetic Fe_3O_4 nanoparticles and doxorubicin into biodegradable PLGA nanocarriers for intratumoral drug delivery. *Int. J. Nanomed.* 7: 1697–1708.

17 Akbarzadeh, A., Samiei, M., Joo, S.W. et al. (2012). Synthesis, characterization and in vitro studies of doxorubicin-loaded magnetic nanoparticles grafted to smart copolymers on A_{549} lung cancer cell line. *J. Nanobiotechnol.* 18: 10–46.

18 Mohammadi-Samani, S., Miri, R., Salmanpour, M. et al. (2013). Preparation and assessment of chitosan-coated superparamagnetic Fe_3O_4 nanoparticles for controlled delivery of methotrexate. *Res. Pharm. Sci.* 8 (1): 25–33.

19 Ding, W. and Guo, L. (2013). Immobilized transferrin $Fe_3O_4@SiO_2$ nanoparticle with high doxorubicin loading for dual-targeted tumor drug delivery. *Int. J. Nanomed.* 8: 4631–4639.

20 Sato, A., Itcho, N., Ishiguro, H. et al. (2013). Magnetic nanoparticles of Fe_3O_4 enhance docetaxel-induced prostate cancer cell death. *Int. J. Nanomed.* 8: 3151–3160.

21 Barrow, M., Taylor, A., Murray, P. et al. (2015). SPIONs for cell labelling and tracking using MRI: magnetite or maghemite? *Chem. Soc. Rev.* 6 (1): 101–106.

22 Peng, X., Chen, J., Cheng, T. et al. (2012). PLGA modified Fe_3O_4 nanoclusters for siRNA delivery. *Mater. Lett.* 81: 102.

23 Zheng, S.W., Liu, G., Hong, R.Y. et al. (2012). An advanced high-energy sodium ion full battery based on nanostructured $Na_2Ti_3O_7/VOPO_4$ layered materials. *Appl. Surf. Sci.* 259: 201–207.

24 Li, D., Zhang, Y., Jin, S. et al. (2014). Development of a redox/pH dual stimuli-responsive MSP@P(MAA-Cy) drug delivery system for programmed release of anticancer drugs in tumour cells. *Mater. Chem. B* 2: 5187–5192.

25 Yang, H., Li, Y., Li, T. et al. (2014). Polyaniline-carbon nanotubes@zeolite imidazolate framework67-carbon cloth hierarchical nanostructures for supercapacitor. *Sci. Rep.* 4: 1–11.

26 Duan, J., Dong, J., Zhang, T. et al. (2014). Polyethyleneimine-functionalized iron oxide nanoparticles for systemic siRNA delivery in experimental arthritis. *Nanomedicine (London)* 9 (6): 789–801.

27 Iijima, S. (1991). Helical microtubules of graphitic carbon. *Nature* 354: 56–58.

28 Niyogi, S., Hamon, M.A., Hu, H. et al. (2002). Chemistry of single-walled carbon nanotubes. *Acc. Chem. Res.* 35: 1105–1113.

29 Pantarotto, D., Singh, R., McCarthy, D. et al. (2004). Functionalized carbon nanotubes for plasmid DNA gene delivery. *Angew. Chem. Int. Ed.* 43 (39): 5242–5246.

30 Liu, Y.T., Zhao, W., Yu Huang, Z. et al. (2006). Noncovalent surface modification of carbon nanotubes for solubility in organic solvents. *Carbon* 44 (8): 1613–1616.

31 Star, A., Stoddart, J.F., Steuerman, D. et al. (2001). Preparation and properties of polymer-wrapped single-walled carbon nanotubes. *Angew. Chem. Int. Ed.* 40 (9): 1721–1725.

32 Tasis, D., Tagmatarchis, N., Georgakilas, V., and Prato, M. (2003). Soluble carbon nanotubes. *Chem. Eur. J.* 9: 4000–4008.

33 Lee, G.W., Kim, J., Yoon, J. et al. (2008). Structural characterization of carboxylated multi-walled carbon nanotubes. *Thin Solid Films* 516: 5781–5784.

34 Tasis, D., Tagmatarchis, N., Bianco, A., and Prato, M. (2006). Chemistry of carbon nanotubes. *Chem. Rev.* 106 (3): 1105–1136.

35 Galano, A. (2010). Carbon nanotubes: promising agents against free radicals. *Nanoscale* 2 (3): 373–380.

36 Yu, J., Huang, D., Huang, K., and Hong, Y. (2011). Preparation of hydroxypropyl-β-cyclodextrin cross-linked multi-walled carbon nanotubes and their application in enantioseparation of clenbuterol. *Chin. J. Chem.* 29 (5): 893–897.

37 El-Sheikh, A.H. and Sweileh, J.A. (2011). Recent applications of carbon nanotubes in solid phase extraction and preconcentration: a review. *Jordan J. Chem.* 6 (1): 1–16.

38 Chen, J., Hamon, M.A., Hu, H. et al. (1998). Solution properties of single-walled carbon nanotubes. *Science* 282: 95–98.

39 Sun, Y.P., Fu, K., Lin, Y., and Huang, W. (2002). Functionalized carbon nanotubes: properties and applications. *Acc. Chem. Res.* 35 (12): 1096–1104.

40 Lordi, V., Yao, N., and Wei, J. (2001). Method for supporting platinum on single-walled carbon nanotubes for a selective hydrogenation catalyst. *Chem. Mater.* 13: 733–737.

41 Huang, W., Taylor, S., Fu, K. et al. (2002). Attaching proteins to carbon nanotubes via diimide-activated amidation. *Nano Lett.* 2: 311–314.

42 Dyke, C.A. and Tour, J.M. (2004). Overcoming the insolubility of carbon nanotubes through high degrees of sidewall functionalization. *Chem. Eur. J.* 10: 812–817.

43 Zhang, Y., Bai, Y., and Yan, B. (2010). Functionalized carbon nanotubes for potential medicinal applications. *Drug Discovery Today* (11–12): 15, 428–435.

44 Kateb, B., Yamamoto, V., Alizadeh, D. et al. (2010). Multi-walled carbon nanotube (MWCNT) synthesis, preparation, labeling, and functionalization. *Methods Mol. Biol.* 651: 307–317.

45 Liu, Z., Sun, X., Nakayama-Ratchford, N., and Dai, H. (2007). Supramolecular chemistry on water-soluble carbon nanotubes for drug loading and delivery. *ACS Nano* 1 (1): 50–56.

46 Yang, W., Thordarson, P., Gooding, J.J. et al. (2007). Carbon nanotubes for biological and biomedical applications. *Nanotechnology* 18: 12, 412001.

47 Choi, J.H., Nguyen, F.T., Barone, P.W. et al. (2007). Multimodal biomedical imaging with asymmetric single-walled carbon nanotube/iron oxide nanoparticle complexes. *Nano Lett.* 7 (4): 861–867.

48 Petrov, P., Stassin, F., Pagnoulle, C., and Jérôme, R. (2003). Noncovalent functionalization of multi-walled carbon nanotubes by pyrene containing polymers. *Chem. Commun.* 17: 2094–2095.

49 Digge, M.S., Moon, R.S., and Gattani, S.G. (2012). Applications of carbon nanotubes in drug delivery: a review. *Int. J. Pharm. Tech. Res.* 4 (2): 839–847.

50 Nguyen, L.A., He, H., and Pham-Huy, C. (2006). Chiral drugs. An overview. *Int. J. Biomed. Sci.* 2 (2): 85–100.

51 Singh, B.G.P., Baburao, C., and Pispati, V. (2012). Carbon nanotubes. A novel drug delivery system. *Int. J. Res. Pharm. Chem.* 2 (2): 523–532.

52 Usui, Y., Haniu, H., Tsuruoka, S., and Saito, N. (2012). Carbon nanotubes innovate on medical technology. *Med. Chem.* 2 (1): 1–6.

53 Klumpp, C., Kostarelos, K., Prato, M., and Bianco, A. (2006). Functionalized carbon nanotubes as emerging nanovectors for the delivery of therapeutics. *Biochim. Biophys. Acta, Biomembr.* 1758: 404–412.

54 Bianco, A., Kostarelos, K., and Prato, M. (2005). Applications of carbon nanotubes in drug delivery. *Curr. Opin. Chem. Biol.* 9: 674–679.

55 Pastorin, G. (2009). Crucial functionalizations of carbon nanotubes for improved drug delivery: a valuable option? *Pharm. Res.* 26: 746–749.

56 Liu, Z., Cai, W., He, L. et al. (2007). In vivo biodistribution and highly efficient tumour targeting of carbon nanotubes in mice. *Nat. Nanotechnol.* 2: 47–52.

57 Kang, S., Herzberg, M., Rodrigues, D.F., and Elimelech, M. (2008). Antibacterial effects of carbon nanotubes: size does matter. *Langmuir* 24: 6409–6413.

58 Kang, S., Pinault, M., Pfefferle, L.D., and Elimelech, M. (2007). Single-walled carbon nanotubes exhibit strong antimicrobial activity. *Langmuir* 23: 8670–8673.

59 Yandar, N., Pastorin, G., Prato, M. et al. (2008). Immunological profile of a plasmodium vivax AMA-1 N-terminus peptide–carbon nanotube conjugate in an infected plasmodium berghei mouse model. *Vaccine* 26 (46): 5864–5873.

60 Mallakpour, S. and Khodadadzadeh, L. (2018). Biocompatible and biodegradable chitosan nanocomposites loaded with carbon nanotubes. In: *Biodegradable and Biocompatible Polymer Composites Processing, Properties and Applications*, Chapter 7 (ed. N.G. Shimpi), 187–221. Sawston, Cambridge: Elsevier, Woodhead Publishing. https://doi.org/10.1016/B978-0-08-100970-3.00007-9.

61 Lim, Y.T., Kim, S., Nakayama, A. et al. (2003). Selection of quantum dot wavelengths for biomedical assays and imaging. *Mol. Imaging* 2: 50–64.

62 Ravelo-Perez, L.M., Herrera-Herrera, A.V., Hernandez-Borges, J., and Rodrıguez-Delgado, M.A. (2010). Carbon nanotubes:solid-phase extraction. Review. *J. Chromatogr. A* 1217: 2618–2641.

63 Mallakpour, S. and Khadem, E. (2019). Carbon nanotubes for heavy metals removal. In: *Composite Nanoadsorbents*, Chapter 8 (eds. G. Kyzas and A.C. Mitrpoulos), 181–210. Amsterdam, Netherlands: Elsevier. https://doi.org/10.1016/B978-0-12-814132-8.00009-5; eBook ISBN: 9780128141335.

64 Mallakpour, S. and Rashidimoghadam, S. (2019). Carbon nanotubes for dyes removal. In: *Composite Nanoadsorbents*, Chapter 9 (eds. G. Kyzas and A.C. Mitrpoulos), 211–244. Amsterdam, Netherlands: Elsevier. https://doi.org/10.1016/B978-0-12-814132-8.00010-1; eBook ISBN: 9780128141335.

65 Mallakpour, S. and Khadem, E. (2016). Carbon nanotube–metal oxide nanocomposites: fabrication, properties and applications. *Chem. Eng. J.* 302: 344–367.

66 Mallakpour, S., Abdolmaleki, A., and Borandeh, S. (2017). Fabrication of amino acid-based graphene-zinc oxide (ZnO) hybrid and its application for poly(ester–amide)/graphene-ZnO nanocomposite synthesis. *J. Thermoplast. Compos. Mater.* 30: 358–380.

67 Mallakpour, S. and Behranvand, V. (2018). Synthesis of mesoporous recycled poly(ethylene terephthalate)/MWNT/carbon quantum dot nanocomposite from sustainable materials using ultrasonic waves: application for methylene blue removal. *J. Cleaner Prod.* 190: 525–537.

68 Mallakpour, S., Behranvand, V., and Mallakpour, F. (2019). Synthesis of alginate/carbon nanotube/carbon dot/fluoroapatite/TiO_2 beads for dye photocatalytic degradation under ultraviolet light. *Carbohydr. Polym.* 224: 115–138.

69 Abdolmaleki, A., Mallakpour, S., and Karshenas, A. (2017). Facile synthesis of glucose-functionalized reduced graphene oxide (GFRGO)/poly(vinyl alcohol) nanocomposites for improving thermal and mechanical properties. *Mater. Sci. Eng., B* 217: 26–35.

70 Mallakpour, S., Abdolmaleki, A., and Karshenas, A. (2017). Graphene oxide supported copper coordinated amino acids as novel heterogeneous catalysts for epoxidation of norbornene. *Catal. Commun.* 92: 109–113.

71 Abdolmaleki, A., Mallakpour, S., and Karshenas, A. (2017). Synthesis and characterization of new nanocomposites films using alanine-Cu-functionalized graphene oxide as nanofiller and PVA as polymeric matrix for improving of their properties. *J. Solid State Chem.* 253: 398–405.

72 Abdolmaleki, A., Mallakpour, S., and Borandeh, S. (2016). Improving interfacial interaction of l-phenylalanine-functionalized graphene nanofiller and poly(vinyl alcohol) nanocomposites for obtaining significant membrane properties: morphology, thermal, and mechanical studies. *Polym. Compos.* 37: 1924–1935.

73 Mallakpour, S., Abdolmaleki, A., Mahmoudian, M. et al. (2017). Synergetic effect of synthesized sulfonated polyaniline/quaternized graphene and its application as a high-performance supercapacitor electrode. *J. Mater. Sci.* 52: 9683–9695.

74 Abdolmaleki, A., Mallakpour, S., and Mahmoudian, M. (2017). Preparation and evaluation of edge selective sulfonated graphene by chlorosulfuric acid as an active metal-free electrocatalyst for oxygen reduction reaction in alkaline media. *ChemistrySelect* 34: 11211–11217.

75 Mallakpour, S. and Khadem, E. (2018). Construction of crosslinked chitosan/nitrogen-doped graphene quantum dot nanocomposite for hydroxyapatite biomimetic mineralization. *Int. J. Biol. Macromol.* 120: 1451–1460.

76 Mallakpour, S. and Khadem, E. (2019). Linear and nonlinear behavior of crosslinked chitosan/N-doped graphene quantum dot nanocomposite films in cadmium cation uptake. *Sci. Total Environ.* 690: 1245–1253.

77 Hussain, C.M. and Kharisov, B. (2017). *Advanced Environmental Analysis-Application of Nanomaterials*. The Royal Society of Chemistry.

78 Hussain, C.M. (2018). *Handbook of Nanomaterials for Industrial Applications*. Elsevier.

79 Hussain, C.M. and Mishra, A.K. (2019). *Nanotechnology in Environmental Science*, vol. 2. Wiley.

80 Hussain, C.M. (2019). *Handbook of Environmental Materials Management*, 1e. Springer International Publishing.

81 Hussain, C.M. and Keçili, R. (2019). *Modern Environmental Analysis Techniques for Pollutants*, 1e. Elsevier.

82 Hussain, C.M. (2020). *Handbook of Functionalized Nanomaterials for Industrial Applications*. Elsevier.

83 Beg, S., Rizwan, M., Sheikh, A.M. et al. (2011). Advancement in carbon nanotubes: basics, biomedical applications and toxicity. *J. Pharm. Pharmacol.* 63: 141–163.

84 Takagi, A., Hirose, A., Nishimura, T. et al. (2008). Induction of mesothelioma in p53 1/2 mouse by intraperitoneal application of multi-wall carbon nanotube. *J. Toxicol. Sci.* 33: 105–116.

85 Yacobi, N.R., Phuleria, H.C., Demaio, L. et al. (2007). Nanoparticle effects on rat alveolar epithelial cell monolayer barrier properties. *Toxicol. in Vitro* 21: 1373–1381.

86 Sargent, L.M., Reynolds, S.H., and Castranova, V. (2010). Potential pulmonary effects of engi-neered carbon nanotubes: *in vitro* genotoxic effects. *Nanotoxicology* 4: 396–408.

Index

Environmental Applications of Carbon Nanomaterials-Based Devices, First Edition.
Edited by Shadpour Mallakpour and Chaudhery Mustansar Hussain.

d